T0227736

Bacterial Disease Resistance in Plants
Molecular Biology and Biotechnological Applications

FOOD PRODUCTS PRESS®

Crop Science
Amarjit S. Basra, PhD
Senior Editor

New, Recent, and Forthcoming Titles of Related Interest:

Mineral Nutrition of Crops: Fundamental Mechanisms and Implications by Zdenko Rengel

Conservation Tillage in U.S. Agriculture: Environmental, Economic, and Policy Issues by Noel D. Uri

Cotton Fibers: Developmental Biology, Quality Improvement, and Textile Processing edited by Amarjit S. Basra

Heterosis and Hybrid Seed Production in Agronomic Crops edited by Amarjit S. Basra

Intensive Cropping: Efficient Use of Water, Nutrients, and Tillage by S. S. Prihar, P. R. Gajri, D. K. Benbi, and V. K. Arora

Physiological Bases for Maize Improvement edited by María E. Otegui and Gustavo A. Slafer

Plant Growth Regulators in Agriculture and Horticulture: Their Role and Commercial Uses edited by Amarjit S. Basra

Crop Responses and Adaptations to Temperature Stress edited by Amarjit S. Basra

Plant Viruses As Molecular Pathogens by Jawaid A. Khan and Jeanne Dijkstra

In Vitro Plant Breeding by Acram Taji, Prakash P. Kumar, and Prakash Lakshmanan

Crop Improvement: Challenges in the Twenty-First Century edited by Manjit S. Kang

Barley Science: Recent Advances from Molecular Biology to Agronomy of Yield and Quality edited by Gustavo A. Slafer, José Luis Molina-Cano, Roxana Savin, José Luis Araus, and Ignacio Romagosa

Tillage for Sustainable Cropping by P. R. Gajri, V. K. Arora, and S. S. Prihar

Bacterial Disease Resistance in Plants: Molecular Biology and Biotechnological Applications by P. Vidhyasekaran

Handbook of Formulas and Software for Plant Geneticists and Breeders edited by Manjit S. Kang

Postharvest Oxidative Stress in Horticultural Crops edited by D. M. Hodges

Encyclopedic Dictionary of Plant Breeding and Related Subjects by Rolf H. G. Schlegel

Handbook of Processes and Modeling in the Soil-Plant System edited by D. K. Benbi and R. Nieder

The Lowland Maya Area: Three Millennia at the Human-Wildland Interface edited by A. Gómez-Pompa, M. F. Allen, S. Fedick, and J. J. Jiménez-Osornio

Bacterial Disease Resistance in Plants

Molecular Biology and Biotechnological Applications

P. Vidhyasekaran, PhD

Food Products Press®
An Imprint of The Haworth Press, Inc.
New York • London • Oxford

Published by

Food Products Press®, an imprint of The Haworth Press, Inc., 10 Alice Street, Binghamton, NY 13904-1580.

Cover design by Jennifer M. Gaska.

Library of Congress Cataloging-in-Publication Data

Vidhyasekaran, P.
Bacterial disease resistance in plants: molecular biology and biotechnological applications / P. Vidhyasekaran.
p. cm.
Includes bibliographical references (p.) and index.
ISBN 1-56022-924-1 (alk. paper)—ISBN 1-56022-925-X (pbk. : alk. paper)
1. Bacterial diseases of plants. 2. Plants—Disease and pest resistance. I. Title.

SB734 .V54 2002
632.3—dc21

2001054346

CONTENTS

Chapter 3. Active Oxygen Species 167

Chapter 4. Inducible Plant Proteins 185

ABOUT THE AUTHOR

Dr. P. Vidhyasekaran, PhD, FNA, is the former Director of the Center for Plant Protection Studies, Tamil Nadu Agricultural University, Coimbatore, India. He is the author of seven books from leading publishers in the United States and India. He has published more than 400 articles in 32 international journals as well as several book chapters. Dr. Vidhyasekaran has served on the editorial boards of several journals and has been elected President of the Indian Society of Plant Pathologists. He is a Fellow of the National Academy of Agricultural Sciences and several other scientific societies, and a member of the New York Academy of Sciences. Dr. Vidhyasekaran has won several national awards as well as the Tamil Nadu Best Scientist Award. He has served as a visiting scientist in 14 countries, including the United States, Philippines, and Denmark.

Foreword

The book *Bacterial Disease Resistance in Plants: Molecular Biology and Biotechnological Applications* by Dr. P. Vidhyasekaran is a very timely contribution. Spectacular progress has been made in recent years in breeding for resistance to bacterial diseases in plants using recombinant DNA technologies. Genome maps are also being prepared in several microorganisms.

This book comprehensively covers recent developments in the field of biotechnology and resistance to biotic stresses with particular reference to diseases caused by bacteria. Both the theoretical and practical aspects of the problem have been dealt with in detail.

The last chapter on molecular manipulation of bacterial disease resistance will stimulate young scholars to undertake research in this important field. We are indebted to both Dr. Vidhyasekaran and The Haworth Press for the excellence of both the content and the production of this important publication.

M. S. Swaminathan, PhD
Former Director General
International Rice Research Institute, Philippines
World Food Prize and Magsaysay
Award Laureate

Preface

Bacterial diseases cause huge losses in various crops, and the economies of several countries have been shattered due to the outbreak of some of these diseases. In the United States, citrus canker has occurred very severely, and eradication of entire orchards had to be undertaken to prevent its spread. Brown rot and blackleg of potato, bacterial wilt of tomato, black rot of cabbage, cauliflower, radish, turnip, broccoli, mustard, and other cruciferous plants, and soft rot of various vegetables are widespread in North and South America, Europe, Australia, and South Asia. Bacterial blight of rice has affected the economies of several Asian countries. Chemical control of bacterial diseases is almost impossible, although several antibiotics have been tried. Antibiotics, which are used for treatment of human and livestock diseases, are not recommended for crop disease management due to the possibility of development of antibiotic-resistant mammalian bacterial strains. Attempts to develop bacterial disease-resistant varieties have resulted in only limited success, as no resistance genes have been identified against most of the bacterial pathogens. Even in cases of those diseases against which resistance genes have been identified, the disease is highly race specific. Several races of bacterial pathogens exist, and hence resistance breaks down quickly in the field.

During the 1990s, molecular biology of bacterial pathogenesis was given importance, and several universities and research institutes around the world have undertaken intensive research in this field with the aim of developing new strategies for bacterial disease management. In recent years, several disease-resistance genes have been cloned; interestingly, it has been found that the resistance genes are not functional genes. They do not inhibit bacterial growth and suppress disease symptom development. This job is done by defense genes, which encode for different inducible proteins and secondary metabolites. Defense genes are universal; they are present in both resistant and susceptible varieties throughout the entire plant kingdom. But they are sleeping genes and quiescent in healthy plants. Specific signals are needed to activate these genes, and a complex signal transduction system exists. By manipulating this signal system, defense genes can be activated, and synthesis of bactericidal proteins and secondary metabolites can be induced in any plant, making the plants resistant to diseases. All the cloned and characterized resistance genes (except one) have been shown to be involved in the signal transduction system. Thus the resistance genes are not functional genes; they are only regulatory genes. Homologues of resistance genes also have been detected in susceptible plants. An intensive knowledge of the signal transduction system in plant-bacterial pathogen interactions will help in development of effective control measures against bacterial diseases. In fact, several Nobel prizes for medicine have been awarded recently for the discovery of the signaling system in human beings. This book describes, in depth, this signaling system, discussing the most recent experimental results and hypotheses.

Several biotic and abiotic inducers of the signal transduction system have been identified, and they can be exploited to induce systemic resistance in the susceptible but high-yielding varieties. Transgenic plants which overexpress the required signals for induction of disease resistance have also been developed. Several endophytic bacteria have been shown to induce systemic resistance, and they have been developed as biocontrol agents. This book critically discusses these novel strategies for disease management.

The interaction of plants and bacterial pathogens appears to be a carefully regulated complex of biological relationships. During the interactions, plant-derived molecules function as signals that induce expression of specific bacterial genes; the products of these bacterial genes in turn induce changes in plant gene expression. When elicitors of both bacteria and plants are carefully regulated through a suitable signal transduction system, the plants can be made resistant to bacterial diseases. This book describes these plant-bacterial interactions in depth and suggests future lines of research needed to exploit them for disease management.

Another approach is to directly make the quiescent defense genes express constitutively without any need for a signal transduction system. Transgenic plants that constitutively express the defense genes have been developed, and these plants show enhanced resistance to bacterial diseases. This book describes various defense-gene products and their usefulness in disease management.

This book is a valuable source for researchers in plant pathology, plant molecular biology, plant biotechnology, plant biochemistry, plant physiology, applied biology, and other branches of plant science. This book is also highly useful to students of botany and plant pathology who take courses in host plant resistance, plant pathogenic bacteria, molecular approaches in plant pathology, plant disease physiology, principles of plant pathology, and molecular plant pathology.

Chapter 1

Molecular Recognition Processes Between Plant and Bacterial Pathogens

INTRODUCTION

All plants are endowed with defense mechanisms, and disease resistance is not an exception, but a rule (Vidhyasekaran, 1988a, 1997, 1998). Still, bacterial pathogens cause diseases in some plants under certain conditions. Modern techniques of molecular biology have contributed to the unraveling of the mechanisms that regulate plant-bacterial pathogen interactions leading to disease or disease resistance. Many plant-bacteria interactions are highly specific. This specificity depends on cell-cell recognition in which complementary molecules on the cell surface of both organisms interact in specific ways. There is increasing evidence that the outcome of a plant-bacteria interaction involves a series of molecular recognition processes between the plant and the bacterial pathogen. The continuous exchange of information between both partners and the appropriate combination in space and time of the many events induced by the pathogen attack determine, finally, whether an interaction will lead to disease or disease resistance (Benhamou, 1991).

Plants recognize pathogens and synthesize various defense chemicals, while pathogens recognize host plants and evolve mechanisms to overcome defenses of the host by evading host recognition and response. The interaction of bacterial pathogens with plants requires active metabolism of both pathogens and plants. During the interaction, plant-derived molecules function as signals that induce expression of bacterial genes; the products of these bacterial genes in turn induce changes in plant gene expression, resulting in metabolic changes in plants (Ream, 1989). This type of interaction has been seen in almost all bacterial pathogen and plant interactions, irrespective of whether they are compatible or incompatible interactions. Similar interaction is not observed when saprophytic bacteria are inoculated on plants, and this interaction appears to be characteristically induced by bacterial pathogens. These complex interactions are discussed in this chapter.

PHYSICAL CONTACT OF PLANT CELLS IS NECESSARY FOR BACTERIAL RECOGNITION

Contact between bacterial pathogens and plants appears to be necessary either for pathogenesis or for induction of disease resistance. Adhesion of bacteria to the leaf surface seems to be important in pathogenesis of *Pseudomonas savastanoi* pv. *phaseolicola* as the mutants of the bacteria with greatly reduced adherence to the leaf surface of bean showed lower incidence of halo blight disease (Romantschuk and Bamford, 1986; Romantschuk et al., 1991).

For *Agrobacterium tumefaciens,* an early and essential step for successful infection is attachment to the host plants (Lippincott and Lippincott, 1969). Nonattaching mutants that have been isolated are known to be avirulent (Douglas et al., 1982; Thomashow et al., 1987; Matthysse et al., 1996). Transposon mutants of *A. tumefaciens* and *A. rhizogenes* that fail to bind to suspension-cultured carrot cells were found to be avirulent (Robertson Crews et al., 1990). *Erwinia rhapontici,* which infects rhubarb *(Rheum rhaponticum)* and causes pink wheat *(Triticum aestivum)* grains, adsorbs to the rhubarb leaf surface and to wheat grains (Louhelainen et al., 1990). *Xanthomonas hyacinthi,* the pathogen of hyacinth, adheres to stomata (Van Doorn et al., 1994).

Pathogens attach to nonhost cell surfaces also. Attachment of the incompatible bacterium *Pseudomonas syringae* pv. *pisi* in intercellular spaces of tobacco has been reported (Goodman et al., 1976; Politis and Goodman, 1978). Physical contact of plant cells by incompatible bacterial pathogens seems to be necessary to induce hypersensitive reaction (HR) (rapid cell necrosis, the phenotypic expression of host defense mechanism) in plants (Hutcheson et al., 1989). When incompatible bacteria were prevented from contacting plant cells by immobilization in water-agar, normal HR failed to develop (Stall and Cook, 1979).

Thus, both compatible and incompatible bacterial pathogens adhere to plants for their action. Five different pathovars of *P. syringae* adsorbed equally well to both the susceptible host and to nonhost plants (Romantschuk et al., 1991). *A. tumefaciens* attached similarly to both resistant and susceptible grapevine cells (Pu and Goodman, 1993). When the bacterial pathogens were infiltrated into pear and bean leaves, no appreciable electrolyte leakage was observed. But when the infiltrated leaves were dried for one hour, the maximum induction of electrolyte leakage was observed in both compatible and incompatible interactions. The drying phase would have allowed close contact between bacteria and plant cells (Brisset and Paulin, 1991), and this kind of contact between plant and bacterial cells seems to be necessary for pathogenesis or disease resistance.

MOLECULES RESPONSIBLE FOR PHYSICAL CONTACT

The outer membrane of the cell wall of gram-negative plant pathogenic bacteria is a complex lipid bilayer containing glycoproteins, lipopolysaccharides (LPS), and proteins. The bacterial pathogens have flagella and project threads of pili, and in some cases are covered by a loose extracellular exopolysaccharide slime (Wingate et al., 1990). Fimbriae and pili are the filamentous nonflagellar appendages of many bacterial pathogens, and they may enable the bacteria to attach to plant cells (Romantschuk, 1992). Fimbriae have been detected in *Xanthomonas campestris* pv. *campestris, X. hyacinthi, X. translucens* pv. *graminis, X. axonopodis* pv. *begoniae, X. axonopodis* pv. *manihotis, X. vesicatoria, X. vasicola* pv. *holcicola, X. arboricola* pv. *pruni, X. axonopodis* pv. *citri, X. axonopodis* pv. *phaseolicola, X. fragariae, X. maltophila, X. albilineans, X. arboricola* pv. *populi, Xylophilus ampelina* (Van Doorn et al., 1994), *Ralstonia solanacearum* (Romantschuk, 1992), and *Erwinia rhapontici* (Korhonen et al., 1988).

The bulk of the filament of fimbriae is made up of helically arranged subunits of a major protein ranging in size from 9 to 22 kilodaltons (kDa) in different fimbrial types (Ottow, 1975; Jann and Jann, 1990). These proteins may be responsible for adhesion of bacterial cells to plant cells. The fimbrial protein subunit with a molecular mass of 17 kDa has been detected in *X. hyacinthi* (Van Doorn et al., 1994). Amino acid composition of the fimbrial subunit revealed that half of the total number of amino acids are hydrophobic in nature. As viewed by fluorescence microscopy, purified native fimbriae of *X. hyacinthi* gave a strong signal at the stomata. Antifimbrial antiserum inhibited the attachment (Van Doorn et al., 1994). These results suggest a function for this type of fimbriae in mediating the adherence of bacteria to the plant surface.

The antiserum raised against purified fimbriae from *X. hyacinthi* reacted with a 17 kDa protein band from various *X. campestris* pathovars and many other *Xanthomonas* spp. (Van Doorn et al., 1994). The 17 kDa protein seems to be conserved in all *Xanthomanas* spp. tested (Van Doorn et al., 1994). The fimbrial protein of *Ralstonia solanacearum* has a molecular mass of 9.5 kDa (Romantschuk, 1992). *Erwinia rhaponctici* expresses fimbriae of a novel type (Korhonen et al., 1988). Fimbriated bacteria and isolated fimbria efficiently adsorb to the rhubarb leaf surface and wheat grains, the hosts of *E. rhapontici* (Louhelainen et al., 1990).

Although fimbriae have been shown to be involved in adhesion of many bacterial pathogens (Haahtela et al., 1985; Korhonen et al., 1988; Young and Sequeira, 1986; Romantschuk and Bamford, 1986; Vesper, 1987; Stemmer and Sequeira, 1987), there is a report that they may not be essential for adhesion of some bacterial pathogens (Stemmer and Sequeira, 1987). Both highly fimbriated and nonfimbriated strains of *Ralstonia solanacearum* are found. The virulent strain K60 expressed a low number of fimbriae per cell, whereas the highly fimbriated strain B1 is avirulent. However, Stemmer and Sequeira (1987) have not characterized the fimbriae of virulent and avirulent strains. The presence of some specific adhesive proteins in virulent isolate alone cannot be ruled out.

Pili are also involved in adhesion of bacteria. Various strains of *P. syringae* and *P. savastanoi* adsorb to leaves of the host plants using their pili (Romantschuk et al., 1991). The importance of pilus in adhesion and pathogenicity has been demonstrated by using pilus-negative spontaneous and transposon mutants. These pilus-negative mutants of various pathovars of *P. syringae* and *P. savastanoi* showed greatly reduced adherence to both the leaf surface of the susceptible host plant and to nonsusceptible plants (Romantschuk and Bamford, 1986; Romantschuk et al., 1991; Nurmiaho-Lassila et al., 1991). Lack of pili was correlated to lower incidence of halo blight disease of beans incited by *P. savastanoi* pv. *phaseolicola* (Romantschuk and Bamford, 1986). Nonpiliated mutants of *P. savastanoi* pv. *phaseolicola* adhered to the surface of bean leaves with an approximately 70 percent reduced efficiency compared to the wild type (Romantschuk et al., 1991). The mutants were complemented with cosmids from a wild-type *phaseolicola* library. The complemented strains express pili at approximately three times the wild-type level and had increased ability to adsorb to bean leaf surface (Romantschuk et al., 1991).

Formation of the pilus was dependent on at least two hypersensitive response pathogenicity *(hrp)* genes, *hrpS* and *hrpH (hrcC)* in *Pseudomonas syringae* pv. *tomato*. (Roine et al., 1997). HrpA was identified as a major structural protein of the pilus. A nonpolar *hrpA* mutant of *P. syringae* pv. *tomato* was unable to form the pilus or cause either a hypersensitive response or disease in plants (Roine et al., 1997). It suggests that surface appendages may be involved in bacterial invasion.

Some extracellular polysaccharides of bacteria have been implicated in adhesion. Two linked chromosomal loci, *ChvA* and *ChvB,* have been shown to be involved in the attachment of *Agrobacterium tumefaciens* to plants (Douglas et al., 1982; Swart et al., 1994). *ChvB* mutants defective in attachment to plant cells are avirulent on a variety of hosts (Douglas et al., 1982). In addition, *ChvB* mutants fail to synthesize a neutral polysaccharide, cyclic β-1,2-linked D-glucan (Puvanesarajah et al., 1985). *ChvA* mutants contain intracellular neutral glucan in amounts equivalent to cells of virulent strains but lack extracellular neutral glucan (O'Connell and Handelsman, 1989). When *ChvB* is directly involved in glucan synthesis (Zorreguieta and Ugalde, 1986; Zorreguieta et al., 1988), *ChvA* appears to be required for export of the glucan from the cell (O'Connell and Handelsman, 1988). All of the mutants that map in virulence loci that affect neutral glucan production or export *(chvA, chvB, pscA,* and *exoC)* are also defective in attachment to plant cells, suggesting that extracellular neutral glucan is required for attachment and pathogenesis (Thomashow et al., 1987). Lipopolysaccharides have been implicated in the initial attachment of *A. tumefaciens* to plant cell surface (Whatley et al., 1976). Rhicadhesin produced by *A. tumefaciens* is also involved in adhesion. Calcium chloride appears to be required for attachment, virulence, and activity of rhicadhesin (Swart et al., 1994). Presence of a Ti plasmid favorably influences the attachment of agrobacteria (Pu and Goodman, 1993). Transposon mutants of *A. tumefaciens* that were unable to attach to plant cells were avirulent. A clone from a library of *A. tumefaciens* DNA, which was able to complement these chromosomal *att* mutants, was identified (Matthysse et al., 1996). A 10 kb region of this clone was sequenced, and a putative operon containing nine open reading frames *(attA1A2BCDEFGH)* was found. The second and third open reading frames *(attA2* and *attB)* showed homology to genes encoding the membrane-spanning proteins of periplasmic binding protein-dependent (ABC) transport systems from different gram-negative bacteria (Matthysse et al., 1996).

Plant molecules are also required for attachment of bacterial cells to plant cells. Proteolytic treatments of the plant cells reduce adherence of *A. tumefaciens* (Gurlitz et al., 1987). *A. tumefaciens* biotype 2 strains are bound by ionic interaction between the negatively charged bacterial surface and positively charged glycoproteins on the plant surface (Sykes and Matthysse, 1988). The carbohydrate pectin has been suggested to function as a receptor for attachment of *Agrobacterium* (Neff et al., 1987). Adhesion to wheat grains by *Erwinia rhapontici* was reduced in the presence of β-galactosides (Louhelainen et al., 1990). It suggests that β-galactosyl residues of an unidentified plant cell wall glyco-conjugate function as the receptor for the bacterial cell.

Vitronectin-like proteins have been reported in plants. Binding of *A. tumefaciens* by vitronectin has been reported (Wagner and Matthysse, 1992). Mutant strains of *A. tumefaciens* with reduced ability to bind vitronectin were much less able to bind to carrot cells and were avirulent. Binding by the wild-type bacteria was inhibited by antivitronectin antibodies and added vitronectin. Cell wall proteins extractable with dilute detergent mediate bacterial binding and cross-react with human vitronectin. These results suggest that vitronectin-like binding sites are functional in plant cells (Wagner and Matthysse, 1992).

In general, attachment of bacterial cells to plant cells seems to be important in pathogenesis. The function of attachment is not known. However, in plant-fungus interactions, attachment seems to be involved in the signal transduction process (Vidhyasekaran, 1997).

MANY BACTERIAL PATHOGENS INDUCE NECROSIS ON HOSTS AND NONHOSTS

Plant pathogenic bacteria, particularly those belonging to *Xanthomonas, Pseudomonas, Erwinia,* and *Ralstonia* genera, generally induce two kinds of reactions in plants. In susceptible host species, disease reactions typically indicated by water-soaked lesion appear; but soon the lesions become necrotic (Ercolani and Crosse, 1966). In nonhost plants or resistant cultivars of susceptible host plants, a rapid necrosis of the infected tissue known as hypersensitive reaction occurs (Klement, 1982; Sequeira, 1983; Willis et al., 1991). Vascular pathogens also induce a rapid and localized necrotic resistant response in the vascular system of resistant cultivars. This response is referred to as vascular HR or VHR (Bretschneider et al., 1989; Reimers and Leach, 1991).

The mechanisms that control disease susceptibility and HR appear to be related in these plant-pathogen systems. In both susceptible and resistant interactions, necrosis is observed. The only difference in the two types of plant reactions is in its timing (Klement, 1982). Bacterial mutants that do not cause disease symptoms in susceptible hosts fail to cause HR in nonhosts and resistant cultivars of the susceptible host (Lindgren et al., 1984, 1986; Boucher et al., 1986; Huang et al., 1988, 1991). Hence, a single factor may be necessary for bacterial elicitation of both the HR and disease symptoms (Huang et al., 1991). This factor may be controlled by bacterial genes, which would have received the plant signals for activation.

BACTERIAL PATHOGENS GROW IN BOTH HOST AND NONHOST PLANTS

Another basic character of bacterial pathogens seems to be their ability to grow in intercellular spaces of both hosts and nonhosts/incompatible hosts. Bacterial populations increase dramatically within 48 h of inoculation, and final cell numbers can increase 10^{-5} fold over initial inoculum levels in many compatible interactions. But the bacterial populations in incompatible interactions increase only 10- to 100-fold within the first 48 h (Klement, 1982; Sequeira, 1983; Hodson et al., 1995a). *Xanthomonas vesicatoria* strain 87-7 is virulent in pepper *(Capsicum annuum)* but incompatible in tomato *(Lycopersicon esculentum).* When plants were inoculated with 10^4

colony forming unit (cfu)/ml, the population of the bacteria reached 10^8 cfu/cm^2 of leaf tissue by six days in pepper. In contrast, the population of the same strain increased to 10^6 cfu/cm^2 of leaf tissue by four days in tomato, and then the population decreased after six days (Canteros et al., 1991).

Transposon *Tn 4431* was used to introduce *lux CDABE* operon into *X. campestris* pv. *campestris,* the causal agent of black rot of crucifers. When the bacteria were wound-inoculated, high bioluminescent bacterial population levels were observed within three days, and maximum population was observed within five days in the susceptible variety. In resistant hosts, the population increased slowly, and at 12 days after inoculation the population of the bacteria was almost equal in both susceptible and resistant varieties (Figure 1.1; Dane and Shaw, 1993).

Both in susceptible and resistant reactions, *Pseudomonas savastanoi* pv. *phaseolicola* multiplied almost similarly in bean pods (Table 1.1; Mansfield et al., 1994). Bacterial numbers in compatible and incompatible interactions of *X. oryzae* pv. *oryzae* and rice *(Oryza sativa)* increased equally until levels reached 10^7-10^8 cfu/leaf (Barton-Willis et al., 1989). A mutant of *X. oryzae* pv. *oryzae* obtained by insertion of transposon showed less virulence, but multiplied almost as well as the wild isolate in rice leaves (Nakayachi, 1995).

Initially, both compatible and incompatible bacterial pathogens may grow almost equally in plants. But subsequently the compatible pathogen may multiply severalfold while incompatible pathogens may remain almost static. *Pantoea stewartii* subsp. *stewartii* multiplied well in maize leaves and

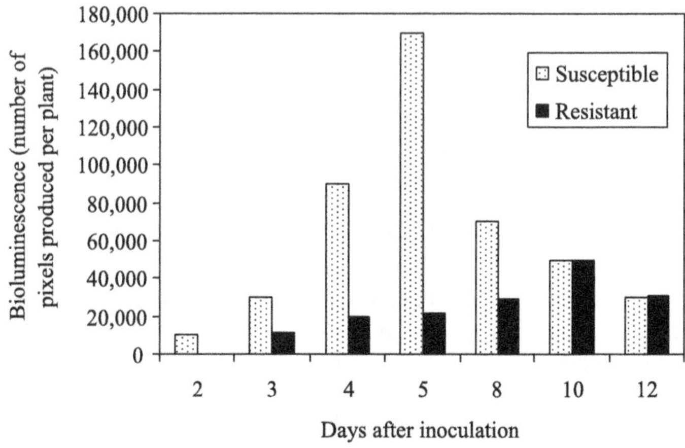

FIGURE 1.1. Bioluminescence in black-rot-susceptible or resistant cabbage seedlings at different times after inoculation with the bioluminescent strain of *Xanthomonas campestris* pv. *campestris* (*Source:* Dane and Shaw, 1993).

TABLE 1.1. Bacteria Recovered from Infection Sites in Bean Pods Four Days After Inoculation

Pseudomonas savastanoi **pv. *phaseolicola* strain**	**Plant reaction**	**Log_{10}cfu recovered**
Race 6	S	8.9
Race 6 with *AvrPphB*	HR	7.5
Race 6 with *AvrPphE*	HR	7.6

Source: Mansfield et al., 1994.

S = susceptible large water-soaked lesion; HR = hypersensitive reaction

caused disease symptoms while its mutant did not produce any disease symptoms. The parent strain multiplied well in maize leaves during the first 48 h and the population reached 10^9 cfu/plant. The mutant also grew well during the first 48 h and after 48 h, the mutant populations remained fairly constant at 1.3×10^8 cfu/plant (Coplin et al., 1992). Both compatible (race 2) and incompatible (race 1) races of *X. vesicatoria* grew almost equally in tomato cultivar Hawaii 7998 up to two days after inoculation. Subsequently, the compatible race multiplied severalfold while the population of the incompatible race remained almost static (Figure 1.2; Whalen et al., 1993).

Multiplication of incompatible bacteria may not only become static after initial increase, but also may decline at later stages of infection. *P. syringae* pv. *pisi* strain PF 247, which is virulent on both pea cultivars Kelvedon Wonder and Martus and *P. syringae* pv. *pisi* strain PF 247 carrying an avirulence gene from *P. savastanoi* pv. *phaseolicola* race 4 strain 1302A, which is avirulent on 'Kelvedon Wonder' but virulent on 'Martus,' multiplied equally well in stems of both the cultivars three days after inoculation. But the population of the strain PF 247 carrying avirulence gene *pp PY40* declined at eight days after inoculation in the resistant cultivar 'Kelvedon Wonder' but not in the susceptible cultivar 'Martus' (Table 1.2; Wood et al., 1994).

When *X. axonopodis* pv. *glycines* mutant NP1 (which is nonpathogenic) was inoculated into a susceptible soybean cultivar, the number of cells remained constant for about seven days and then declined to approximately 10 percent of the cells that were present at the time of inoculation. In contrast, the number of cells of the parent strain increased about tenfold during the first six days after inoculation and then remained nearly constant (Hwang et al., 1992). A nonpathogenic isolate of *Erwinia amylovora* had only a brief period of multiplication in plants (Walters et al., 1990).

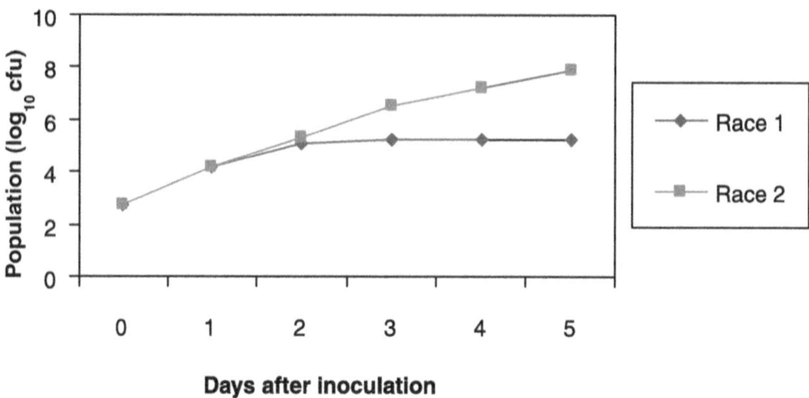

FIGURE 1.2. Relative growth of incompatible (race 1) and compatible (race 2) races of *Xanthomonas vesicatoria* in tomato cultivar Hawaii 7998 (*Source:* Whalen et al., 1993).

TABLE 1.2. Bacteria Recovered from Inoculation Sites in Stems of the Pea Cultivars Kelvedon Wonder and Martus, Three and Eight Days After Inoculation with *Pseudomonas syringae* pv. *pisi*

***Pseudomonas syringae* pv. *pisi* strain**	**log_{10} cfu recovered per site**			
	Kelvedon Wonder		**Martus**	
	3 days	**8 days**	**3 days**	**8 days**
PF 247	8.2	8.1	8.4	8.5
PF 247 carrying avirulence gene	7.1	6.6	8.2	8.4

Source: Wood et al., 1994.

Basically, ability to grow in planta is not always linked to pathogenicity of bacterial pathogens. A group of nonpathogenic mutants of *X. campestris* pv. *campestris* multiplied well in turnip seedlings (Daniels et al., 1984). A nonpathogenic mutant of *X. campestris* pv. *campestris,* generated by Tn 4431 insertional mutagenesis, was able to grow in excised leaves of cauliflower (Shaw et al., 1987). When a virulent strain of *Ralstonia solanacearum* was inoculated into soil and tomato plants were grown in the inoculated soil, the bacteria could be isolated from the stems within 23 h after inoculation (Table 1.3; Trigalet and Demery, 1986). The bacterial concentration estimated at the cotyledon level increased progressively up to about 10^8 cfu/stem piece. When inoculated under same conditions, Tn5 transposon-induced avirulent mutant of the bacteria colonized the stem at the cotyledon level within 15 days after inoculation. But population of the avirulent bacteria never increased above 2×10^4 even at 15 days after inoculation (Table 1.3; Trigalet and Demery, 1986).

Thus, it appears that bacterial pathogens can multiply in planta irrespective of the plant, a host or nonhost. Although the bacteria grow in nonhosts, they do not produce disease symptoms in the nonhosts. The lack of correlation between disease symptoms and growth in planta has been observed with many plant pathogens (Smidt and Kosuge, 1978; Willis et al., 1990; Hrabak and Willis, 1992; Young et al., 1996).

For the growth of bacterial pathogens in intercellular spaces of plants, host nutrients should be made available to the bacteria in the intercellular spaces. Leakage of nutrients from plant cells may make nutrients available to bacteria. Leakage of electrolytes is associated with the leakage of nutrients (Atkinson and Baker, 1987b).

BACTERIAL PATHOGENS INDUCE LEAKAGE OF NUTRIENTS IN BOTH HOST AND NONHOST PLANTS

Leakage of nutrients appears to be a prerequisite for bacterial multiplication in the plant (Youle and Cooper, 1987). *Pseudomonas savastanoi* pv. *phaseolicola* is virulent on bean and avirulent on pear and *Erwinia amylovora* is virulent on pear but avirulent on bean (Brisset and Paulin, 1991). In both compati-

TABLE 1.3. Relative Invasiveness of Virulent and Tn5-Induced Avirulent Mutant of *Ralstonia solanacearum* in Tomato Plants

	Bacterial concentration (cfu per stem piece)	
Time after inoculation (h)	**Virulent strain**	**Avirulent strain**
18	0	0
23	1×10^6	0
71	4×10^6	0
96	4×10^8	1×10^3
168	—*	2×10^4

Source: Trigalet and Demery, 1986.

*No analyses were made as the plants wilted.

ble (*P. savastanoi* pv. *phaseolicola* on bean and *E. amylovora* on pear) and incompatible (*P. savastanoi* pv. *phaseolicola* on pear and *E. amylovora* on bean) interactions, electrolyte leakage was observed. In an incompatible interaction (*E. amylovora* on bean), increase in electrolyte leakage was greater during the first 10 to 30 h; but subsequently increase in electrolyte leakage was almost similar to a compatible interaction (*P. savastanoi* pv. *phaseolicola*) (Brisset and Paulin, 1991). Similar results were obtained in pear leaves inoculated with compatible and incompatible bacteria (Brisset and Paulin, 1991). Increased electrolyte leakage has been observed in apple leaves inoculated with both virulent and avirulent strains of *E. amylovora* (Burkowicz and Goodman, 1969; Addy, 1976). A dramatic increase in electrolyte leakage in leaves of tomato cultivar Hawaii 7998 occurred after 12 h postinfiltration with *X. vesicatoria* carrying the avirulence gene *avrhxv* (Whalen et al., 1993). In leaves infiltrated with *X. vesicatoria* race with insertional mutation in the avirulence gene, *avrRxv,* the increase in electrolyte leakage was observed but it was slower than that induced by the avirulent strain (Whalen et al., 1993).

Thus both compatible and incompatible pathogens induced leakage of nutrients in plants. But when these pathogens lose their pathogenicity and ability to induce resistance, the avirulent mutants lose their ability to provoke electrolyte leakage. When an avirulent strain of *E. amylovora* was infiltrated into pear leaf discs, no significant increase in electrolyte leakage was observed (Brisset and Paulin, 1991).

All these studies conducted so far suggest that many bacterial pathogens have two common properties: (1) they are able to grow in plants irrespective of whether they are going to cause disease or not, and (2) they are able to obtain nutrients either from their host or nonhosts. The major differences between susceptible and resistant reactions are seen only after their growth in planta and release of host nutrients. At that time host defense mechanism seems to be activated. Host defense mechanisms seem to be induced in both compatible and incompatible interactions. The exchange of signals between *X. vesicatoria* and pepper plant cells as well as between *P. savastanoi* pv. *phaseolicola* and bean cells during both compatible and incompatible interactions was indicated by nonspecific localized papilla deposition (Brown and Mansfield, 1988, 1991; Brown et al., 1993). All defense genes are activated in both compatible and incompatible interactions (Pontier et al., 1994). Compatible pathogens avoid defense mechanisms and hence their growth in planta is not suppressed at later stages of infection. Incompatible pathogens are unable to suppress defense mechanisms of the host and hence their growth is retarded at later stages. Compatible pathogens may sustain a regulated release of nutrients from host cells for their growth. Incompatible pathogens may accelerate the sudden release of nutrients from cells, resulting in cell membrane damage and cell death and hence at later stages enough nutrients will not be available for growth. Compatible pathogens may release different virulence factors to produce typical disease symptoms. Incompatible pathogens may release elicitors to induce defense mechanisms. How do all these reactions take place? All these interactions appear to be controlled by various bacterial and plant genes. These are described in this chapter.

BACTERIAL GENES INVOLVED IN RECOGNITION OF HOSTS AND NONHOSTS

hrp Genes

Dual Function of hrp *Genes*

The first group of bacterial genes involved in recognition are strictly required for growth in planta, initiation of disease symptoms on host plants, and induction of defense mechanisms as expressed by hypersensitive reactions on nonhost plants. The denomination *hrp* (*h*ypersensitive *r*esponse and *p*athogenicity) has been proposed to designate these genes (Lindgren et al., 1986). The group of genes that are required for the elicitation of hypersensitive necrosis by *Pseudomonas savastanoi* pv. *phaseolicola* on nonhost plants and on a resistant cultivar of the susceptible host, bean, are also required for pathogenicity on a susceptible cultivar of bean (Lindgren et al., 1986). Loss of the ability of *P. savastanoi* pv. *phaseolicola* to elicit a hypersensitive response on tobacco and other nonhost plants was associated with loss of pathogenicity on the susceptible host bean (Lindgren et al., 1986). Some strains of *Ralstonia solanacearum* caused wilting in tomato but elicited an HR on tobacco. Some mutants of the tomato pathogenic strain differed from the wild type in loss of pathogenicity in tomato, and these mutants lost ability to induce the HR on tobacco (Boucher et al., 1986). These results suggest that HR and pathogenicity may share one or more steps or components controlled by the bacterial genome. The apparent dual role of *hrp* genes suggests that plants have developed resistance mechanisms based on recognition of bacterial determinants that control pathogenicity. It is also possible that some HR "suppressors" produced by plant cells may block HR elicitation until removal by parasitic activities of the pathogen (Hutcheson et al., 1989).

Several hrp Genes Are Required for a Bacterial Pathogen to Induce HR and Pathogenesis

Many genes responsible for pathogenesis and HR (*hrp* genes) have been detected in a single bacterial pathogen. *hrp* genes are typically clustered in the bacterial genome or on a large plasmid (Anderson and Mills, 1985; Panopoulos et al., 1985; Niepold et al., 1985; Lindgren et al., 1986; Cuppels, 1986; Boucher et al., 1987; Huang et al., 1988; Bonas et al., 1991). The *hrp* genes of *Pseudomonas syringae* pv. *syringae* strain 61, the pathogen of tomato and bean, appear to be clustered in a 25 kb region of the genome (Huang et al., 1991). A cosmid, pHIR11, was constructed containing the 25 kb region of the bacterial genome (Huang et al., 1988). The cosmid was inserted into two saprophytic bacteria, *Pseudomonas fluorescens* and *Escherichia coli,* which are not able to cause HR or induce disease symptoms in plants. Both the bacteria carrying the cosmid elicited HR in tobacco (a nonhost of *P. syringae* pv. *syringae*) (Huang et al., 1991).

To locate functional loci, the cosmid pHIR11 was mutagenized with a transposon *TnphoA* in *E. coli,* and transposon insertion derivatives were immobilized into *P. fluorescens* and screened for their HR phenotype (Huang et al., 1991). Forty-one insertions that affected HR elicitation were identified. HR$^-$pHIR11:*Tnpho*A derivatives were transferred into the *P. syringae* pv. *syringae* strain 61 genome by marker exchange to generate a collection of mutants. Merodiploids were constructed by mobilizing various *Tnpho*A-mutated derivatives of pHIR11 into the mutants. Based on the HR phenotypes on tobacco leaves (incompatible host) caused by these merodiploids, 13 complementation groups were defined. Any insertion of *Tnpho*A in complementation groups II-XIII inactivated all *hrp*-associated phenotypes in *P. syringae* pv. *syringae* 61, including the HR on tobacco leaves and disease on bean leaves (Huang et al., 1991). The complementation group I, *(hrmA)* (*h*ypersensitive *r*esponse *m*odulators locus) is required for the HR phenotype in *P. fluorescens* but not in *P. syringae* pv. *syringae* 61 (Huang et al., 1991). *E. coli* carrying the *P. syringae* pv. *syringae* 61 *hrp* gene cluster requires the *hrm*A locus to initiate the HR in tobacco, whereas the parental *P. syringae* pv. *syringae* does not (Heu and Hutcheson, 1991b). The *hrm*A locus appears to modulate the phenotypic expression of the *hrp* genes during incompatible interactions (Huang et al., 1991). The nucleotide sequence of the *hrm*A locus revealed an open reading frame that encodes for a protein of 41400 Da (Heu and Hutcheson, 1991a). Elicitation of the HR by bacteria carrying the gene cluster appears to occur via the concerted action of the entire *hrp/hrm* cluster (Heu and Hutcheson, 1991a,b; Huang et al., 1991).

A set of chromosomal *hrp-uid* (encodes β-glucuronidase:GUS) fusions was constructed to investigate the transcriptional organization and expression of this cluster (Xiao and Hutcheson, 1991b). The *hrp/hrmA* cluster has been shown to be organized into eight transcriptional units and potentially encode 26 polypeptides (including HrmA) (Xiao and Hutcheson, 1991b; Huang et al., 1992, 1993, 1995; Xiao et al., 1992; He et al., 1993; Lidell and Hutcheson, 1994).

DNA sequences have been reported for many *hrp* genes of *P. syringae* pv. *syringae* strain 61 such as *hrpK, hrpL* (Xiao et al., 1994), *hrpJ* (Huang et al., 1993; Lidell and Hutcheson, 1994), *hrpU* (the first two open reading frames, ORFs) (Lidell and Hutcheson, 1994), *hrp*H (last ORF) (Huang et al., 1992), *hrpZ* (second ORF) (He et al., 1993), and *hrpRS* (Xiao et al., 1994). The sequence of a 3.7 kb DNA fragment covering complementation groups VIII and IX revealed five ORFs in the same transcript, designated as *hrpC, hrpW, hrpO, hrpX,* and *hrpY* and predicted to encode proteins of 14795, 23211, 9381, 28489, and 39957 Da, respectively. The sequence of a 2.9 kb DNA fragment containing mainly group XI revealed five ORFs, designated *hrpC, hrpD, hrpE, hrpF,* and *hrpG,* predicted to encode proteins of 29096, 15184, 21525, 7959, and 13919 Da, respectively (Huang et al., 1995). The complete inventory of *hrp* genes of *P. syringae* pv. *syringae* strain 61 has been obtained, and the twenty-five Hrp proteins encoded by the *hrp* genes are HrpA, HrpB, HrpC, HrpD, HrpE, HrpF, HrpG, HrpH, HrpI, HrpJ, HrpJ-3, HrpJ-4, HrpJ-5, HrpK, HrpL, HrpO, HrpR, HrpS, HrpU, HrpU1, HrpU2, HrpW, HrpX, HrpY, and HrpZ (Huang et al., 1995).

Tn5 mutants of another strain of *Pseudomonas syringae* pv. *syringae,* strain PS 3020 were obtained (Anderson and Mills, 1985; Bertoni and Mills, 1987). The prototrophic mutant PS 9021 was unable to induce disease symptoms on bean and elicit HR on tobacco. An 8.5 kb region of the cosmid clone pOSU 3102 complemented the mutant phenotype (Niepold et al., 1985). Transposon mutagenesis delimited the *hrp* region that restored mutant PS9021 to a length of 3.9 kb (Mills and Niepold, 1987). The sequence analysis revealed two open reading frames, ORF1 and ORF2 (Mukhopadhyay et al., 1988). The predicted sizes of the two polypeptides were 40 and 83 kDa, respectively (Mills et al., 1985). An *E. coli* consensus promoter sequence was present upstream of and overlapping with ORF1, while a sequence containing the features of a transcriptional terminator was found downstream of ORF2 (Mills et al., 1985).

The *hrp* cluster of *Pseudomonas savastanoi* pv. *phaseolicola,* the bean pathogen, spans a chromosomal region of approximately 22 kb and consists of nine complementation groups *(hrpL, hrp AB, hrpC, hrpD, hrpE, hrpF,* and *hrpSR)* (Lindgren et al., 1986, 1989). Mutations in all *hrp* loci, except *hrpC,* greatly reduced the ability of the bacteria to multiply in bean (Rahme et al., 1991). An *hrpC* mutant did not elicit the HR on nonhost but did grow, although reduced about 10^2-fold and produced fewer lesions on bean leaves (Rahme et al., 1991).

The nucleotide sequence of an *hrp* locus *(hrpS)* located near the right end of the *hrp* cluster has been determined (Grimm and Panopoulos, 1989). The largest open reading frame (ORF302) in *hrpS* has a coding capacity for a protein of 302 aminoacids. The ORF 302 may be part of a transcriptional unit extending further upstream (Grimm and Panopoulos, 1989). Three *hrp* genes unlinked to the 22 kb cluster, namely *hrpM, hrpT,* and *hrpQ* have also been reported (Grimm and Panopoulos, 1989; Mills and Mukopadhyay, 1990; Rahme et al., 1991, 1992).

Several *hrp* genes have been detected in *Ralstonia solanacearum,* the pathogen affecting several plants belonging to more than 30 families including crops such as potato, tobacco, and banana (Boucher et al., 1988; Xu et al., 1988). By the use of Tn5 random mutagenesis, 13 mutants of *R. solanacearum,* which had lost pathogenicity toward their homologous host (tomato), were isolated (Boucher et al., 1985). Nine of them were simultaneously affected in their ability to induce HR on a nonhost plant tobacco and therefore carried a mutation in a mutation gene (Boucher et al., 1985, 1986; Trigalet and Demery, 1986). Most of the mutated pathogenicity genes were located in a large region of a megaplasmid which was deleted in block in Acr^r mutants selected as being resistant to acridine orange. In addition, the mutations which affect *hrp* genes all map in this region, suggesting that the genes conferring this property could be clustered (Boucher et al., 1986).

A pLAFR3 cosmid clone designated pVir2 containing a 25 kb DNA insert was isolated from a wild-type *R. solanacearum* GMI 1000 genomic library. This cosmid was shown to complement all but one of nine Tn5-induced mutants which have been isolated after random mutagenesis and which have lost both pathogenicity toward tomato and ability to induce HR on tobacco (*hrp* mutants) (Boucher et al., 1987). These mutants carry a Tn5 insertion, which maps within a 25 kb region of DNA which is carried on the pVir2 insert. Localized mutagenesis of this region led to the identification of a set of genes that confer pathogenicity to *R. solanacearum* (Boucher et al., 1987). These genes map within a 20 kb region of DNA (Boucher et al., 1987). Mutagenesis of pVir2 with Tn5-*lac* further localized the *hrp* gene cluster to a 17.5 kb region at the left end of the insert in pVir2 and extending out of the cloned region (Boucher et al., 1987). Isolation of plasmid pAE8, which carries an insert partly overlapping with the left-hand end of pVir2 followed by transposon mutagenesis of this plasmid, showed that the left border of the *hrp* gene cluster is located about 2 kb to the left of the region carried on plasmid pVir2. Isolation of additional mutants in the right-hand end of the pVir2 insert demonstrated that the 3 kb region located to the right-hand end of the pVir2 insert is also required for the normal development of the HR. Insertions occurring in these 3 kb regions led to leaky mutations affecting both pathogenicity on tomato and ability to induce the HR on tobacco. Therefore, the size of the entire *hrp* gene may be about 22 kb (Arlat et al., 1990). The use of transposon Tn5-B20 indicated that the *hrp* gene cluster is organized in a minimum of six transcriptional units (Arlat et al., 1992).

Huang et al. (1990b) provided evidence for a new cluster of *hrp* genes that is clearly different from that previously described by Boucher et al. (1987) in *R. solanacearum* strain K60. The DNA region, which encodes these functions, showed no homology with pVir2. In this new *hrp* region, there were at least two transcriptional units located in a 7.0 kb fragment of the *R. solanacearum* genome. In the wild-type strain, K60, transposon insertion in either transcriptional unit simultaneously affected the ability to induce the HR in potato and to cause wilting of tobacco (Huang et al., 1990). Thus at least two separate *hrp* clusters may be present in *R. solanacearum,* and these clusters may involve several genes which control both pathogenicity and HR.

A cluster of genes involved in pathogenesis and elicitation of the HR has been identified in *Erwinia amylovora* (Bauer and Beer, 1991; Beer et al., 1991). Most of the genes are required for both pathogenicity on the host plants, pear and apple, and for elicitation of the HR on a nonhost plant, tobacco. Based on mapping and complementation results, the cluster of *hrp* genes spans at least 30 kb of *E. amylovora* chromosomal DNA (Bauer and Beer, 1987, 1991; Steinberger and Beer, 1988; Laby et al., 1989; Barny et al., 1990). Bauer and Beer (1987) identified a strain of *E. amylovora,* P66, as unable to elicit the HR and reported its complementation by DNA from the pathogenic strain, Ea 322. The lesion in P66 that results in the Hrp phenotype maps in the *hrp* cluster (Bauer and Beer, 1987). All the *hrp* mutants of *E. amylovora* were restored by a single cosmid, designated as pCPP430, that contains a 45 kb chromosomal insert (Laby et al., 1989; Laby and Beer, 1990). The pCPP430 imparted the ability to elicit HR on tobacco and other plants to *E. coli* and all other members of the Enterobacteriaceae tested (Wei and Beer, 1990). Walters et al. (1990) has cloned a determinant of pathogenicity and ability to cause the HR from *E. amylovora.* The *hrp* gene lies within the 2.1 kb fragment subcloned in pEAY1310. At least six complementation regions in the *hrp* gene cluster may be involved in pathogenicity of host plants and elicitation of the HR in a nonhost plant by *E. amylovora* (Bauer and Beer, 1991).

Several *hrp* genes, *hrCJ, hrcV, hrcN, hrpO, hrpP,* $hrcQ_A$, $hrcQ_B$, *hrcR, hrcS, hrcT,* and *hrcU* have been identified in *Erwinia herbicola* pv. *gypsophilae,* the gall-forming pathogen of gypsophila *(Gypsophila paniculata)* and table beet *(Beta vulgaris)* (Nizan et al., 1997). Sequences related to HrpH and HrpS of *Pseudomonas syringae* have also been detected in *E. herbicola* pv. *gypsophilae* (Nizan et al., 1997). Both *hrpL* and *hrpS* have been detected in *Erwinia* (=*Pantoea*) *stewartii* (Nizan et al., 1997).

The natural isolates of *Xanthomonas vesicatoria* can be classified into three groups based on their host specificity for either pepper or tomato or both plant species (Bonas et al., 1991). Mutants of *X. vesicatoria* defective for growth in susceptible plants could be restored to wild-type growth by complementation with plasmids harboring a genomic region of about 30 kb. The genes localized in the isolated DNA region were designated *hrp* genes since the mutants not only affected in their interaction with the susceptible host plant but also in induction of an HR on resistant host cultivars or on nonhost plants (Bonas et al., 1991). The *hrp* genes of *X. vesicatoria* are located on the chromosome (Bonas et al., 1991). The *hrp* region appears to be highly conserved among different strains of *X. vesicatoria* irrespective of their host specificity (pepper or tomato or both) (Bonas et al., 1991).

The *hrp* region of *X. vesicatoria* contained at least six complementation groups which were designated *hrpA, hrpB, hrpC, hrpD, hrpE,* and *hrp F.* They varied in length from less than 1 kb to 5 kb (Bonas et al., 1991). The length and number of complementation groups suggest the presence of several operons (Bonas et al., 1991). The *hrp* cluster spans to 23 kb, and all six loci in the cluster are required for full pathogenicity (Bonas et al., 1991). The *hrp* gene cluster of *X. vesicatoria* is predicted to contain 21 genes, based on sequence analysis (Fenselau and Bonas, 1995). The *hrp* regulatory genes *hrpX* and *hrpG* are situated outside of the cluster (Wengelnik and Bonas, 1996; Wengelnik et al., 1996b).

In *Xanthomonas campestris* pv. *campestris,* the causal agent of black rot of crucifers, the *hrp* genes form a cluster covering approximately 25 kb, which is organized into three regions controlling HR-inducing ability and pathogenicity, separated by small domains (Arlat et al., 1991). The *hrp* cluster may be located on the chromosome (Arlat et al., 1991).

hrp Genes Are Conserved Among Several Bacterial Pathogens

The *hrp* genes appear to be highly conserved in several bacterial pathogens. The *Ralstonia solanacearum* strain GMI 1000 *hrp* gene cluster from the cosmid clone pVir2 containing a 25 kb *hrp* region showed structural homology by Southern hybridization analysis with all 53 *R. solanacearum* isolates and with all *Xanthomonas isolates* tested, viz., *X. campestris* pv. *campestris, X. vesicatoria, X. arboricola* pv. *juglandis, X. axonopodis* pv. *begoniae,* and *X. translucens* pv. *graminis* (Boucher et al., 1987, 1988). The core of *hrp* cluster from *P. syringae* pv. *syringae* 61 is homologous with a cluster of *hrp* genes from *P. savastanoi* pv. *phaseolicola* NPS 3121 (Huang et al., 1991). It was shown to be conserved in other *P. syringae* pathogens also (Lindgren et al., 1988). An approximately 18 kb *hrp* region is broadly conserved in many *P. syringae* pathovars (Huang et al., 1991). The existence of homology between the *hrp* clusters of *Erwinia amylovora* and *P. syringae* has been reported (Beer et al., 1991). A DNA segment from the *P. savastanoi* pv. *phaseolicola hrp* cluster hybridized to genomic DNA isolated from *P. savastanoi* pv. *glycinea, P. syringae* pv. *tabaci, P. syringae* pv. *angulata,* and *P. syringae* pv. *tomato* (Lindgren et al., 1988). Sequences homologous to the *hrp* region in *P. syringae* pv. *syringae* PS 9021 are present in *P. savastanoi* pv. *phaseolicola* (Niepold et al., 1985). Homologs to Hrp H and Hrp I of *P. syringae* pv. *syringae* have been identified in the *hrp* clusters of *X. vesicatoria* (Fenselau et al., 1992), and *R. solanacearum* (Gough et al., 1992).

The *hrp* region of *X. vesicatoria* was highly conserved among different strains of *X. vesicatoria* (Bonas et al., 1991). DNA from *X. axonopodis* pv. *alfalfae, X. axonopodis* pv. *citri, X. axonopodis* pv. *begoniae, X. axonopodis* pv. *diffenbachiae, X. axonopodis* pv. *glycines, X. axonopodis* pv. *malvacearum, X. oryzae* pv. *oryzae, X. axonopodis* pv. *phaseoli,* and *X. axonopodis* pv. *vignicola* gave strong hybridization signals when probed with sequences from the *X. vesicatoria hrp* region (Bonas et al., 1991). Out of 21 proteins predicted to be encoded in the *hrp* cluster of *X. vesicatoria,* 10 proteins show significant homology to Hrp proteins in *Pseudomonas syringae, Erwinia amylovora,* and *Ralstonia solanacearum* (Fenselau et al., 1992; Fenselau and Bonas, 1995; Wengelnik et al., 1996a).

The *hrp* gene cluster from *Erwinia amylovora* showed homology with the *hrp* gene cluster of *Pantoea stewartii* (Beer et al., 1990). The predicted protein product of one of the *E. amylovora hrp* genes was found to be related to the *hrpS* gene product of *P. savastanoi* pv. *phaseolicola* (Sneath et al., 1990). An *hrp* gene, *hrp-2,* has been cloned from *E. chrysanthemi* (Bauer et al., 1994). The *hrp-2* shows high similarity to the *P. syringae* pv. *syringae* 61 *hrpX* gene. The gene is also a homolog of ORF4 in complementation group VIII of the *E. amylovora hrp* cluster (Bauer et al., 1994).

Even the size of *hrp* gene clusters of different bacterial pathogens appears to be similar. The size of *hrp* gene clusters of *P. syringae* and *P. savastanoi* pathovars (Rahme et al., 1991), *X. axonopodis* pathovars (Bonas et al., 1991; Arlat et al., 1991), and *E. amylovora* (Barny et al., 1990; Beer et al., 1991) is also about 22 kb.

The *hrp* regions of one bacterial pathogen can be introduced into another bacteria and *hrp* expression can be demonstrated. Clones carrying gene clusters have been isolated from genomic libraries of *X. campestris* pv. *campestris* and *X. axonopodis* pv. *vitians,* the causal agents of black rot of crucifers and leaf spot of lettuce, respectively (Arlat et al., 1991). Mutagenesis of the corresponding genomic regions of both pathovars gave strains defective in pathogenicity and HR induction (Arlat et al., 1991). The cosmid PIJ3225 carrying *X. campestris* pv. *campestris* DNA was introduced into the *X. axonopodis* pv. *vitians* mutant 9010. The resulting strain produced symptoms on lettuce and induced a partial and delayed HR on turnip (Arlat et al., 1991). Similarly, pIJ 3221 carrying *X. axonopodis* pv. *vitians* DNA was introduced into Hrp mutants of *X. campestris* pv. *campestris,* and the cosmid restored the ability to induce HR on pepper to the mutants (Arlat et al., 1991). The exchange of an *hrp* AB:: Tn5 mutation via homologous recombination from *P. savastanoi* pv. *phaseolicola* into either *P. syringae* pv. *tabaci* or *P. savastanoi* pv. *glycinea* resulted in the loss of pathogenicity on tobacco or soybean, the respective hosts of these pathovars, and the loss of elicitation of the HR on nonhost plants (Lindgren et al., 1988). Thus, *hrp* genes appear to be interchangeable (Willis et al., 1991; Bonas, 1994).

Although *hrp* genes are conserved in several bacterial pathogens, some differences between *hrp* genes have also been reported. The *hrp* gene cluster from *R. solanacearum* did not show homology with the *hrp* gene cluster in *E. carotovora* subsp. *carotovora, E. carotovora* subsp. *atroseptica, P. syringae* pv. *syringae,* and *P. savastanoi* pv. *phaseolicola* (Boucher et al., 1987, 1988). DNA of strains of *P. savastanoi* pv. *glycinea, P. savastanoi* pv. *phaseolicola, P. syringae* pv. *tomato,* and *Agrobacterium tumefaciens* did not hybridize to the *hrp* region of *X. vesicatoria* (Bonas et al., 1991). DNA of *Ralstonia solanacearum* showed only very weak homology to the *X. vesicatoria hrp* sequences (Bonas et al., 1991). The *hrp* genes from *P. savastanoi* pv. *phaseolicola* failed to hybridize to DNA from *Xanthomonas* spp. (Lindgren et al., 1988). In some cases the *hrp* genes are not interchangeable. The *hrp* region of *P. savastanoi* pv. *phaseolicola* failed to complement any of the nonpathogenic mutants of *X. vesicatoria* (Bonas et al., 1991).

Avirulence (avr) Genes

What Are Avirulence Genes?

Pathogen genes that are critical in determining whether a bacterial strain will be virulent or avirulent on a specific host have been identified. These genes are designated avirulence *(avr)* genes (Wanner et al., 1993). Avirulence genes restrict pathogen host range by specifying the hypersensitive response on plant hosts carrying complementary disease resistance genes (Dangl, 1994; Staskawicz et al., 1995). Some of the *avr* genes, called nonhost *avr* genes (Whalen et al., 1993), induce resistance in a cultivar-specific manner on nonhost species (Whalen et al., 1988, 1991; Kobayashi et al., 1989). Thus an avirulence gene, when carried by a normally virulent strain, may induce cultivar-specific resistance in natural host and nonhost species (Fillingham et al., 1992).

The cloned *avr* gene converts a virulent strain of a bacterial pathogen to avirulence in a cultivar-specific manner (Staskawicz et al., 1984). Experimental removal of *avr* genes increases host ranges to include otherwise resistant cultivars and, unlike *hrp* genes, such strains lacking the *avr* genes are usually not impaired in pathogenicity on susceptible hosts (Gabriel and Rolfe, 1990; Keen, 1990). Hence, it seems that *hrp* genes function in bacterial parasitism while *avr* genes are determinants that control host range. However, some *avr* genes pleiotropically contribute to virulence on susceptible hosts (Kearney and Staskawicz, 1990; Swarup et al., 1992; Lorang et al., 1992, 1994; Yang et al., 1994; Ritter and Dangl, 1995).

Avirulence Genes Induce Resistance in a Cultivar-Specific Manner on Both Host and Nonhost Species

Induction of resistance in cultivar-specific manner in host species. Avirulence genes that induce resistance in a cultivar-specific manner on host species are called host avirulence genes, while some avirulence genes that induce resistance in a genotype-specific manner also on nonhost species are termed nonhost avirulence genes (Whalen et al., 1993). The first host avirulence *(avr)* gene was cloned by Staskawicz et al. (1984) from *Pseudomonas savastanoi* pv. *glycinea.* A plethora of *avr* genes have been isolated subsequently using the same experimental regions: conversion of virulence of a recipient strain to avirulence in a manner dependent on the presence of a corresponding resistance *(R)* gene in the test plant (Keen and Staskawicz, 1988; Long and Staskawicz, 1993; Dangl, 1994). De Feyter and Gabriel (1991) isolated six avirulence genes from *Xanthomonas axonopodis* pv. *malvacearum* strain H, the cotton pathogen. These cloned genes have been designated *avrB4, avrb6, avrBIn, avrb7, avrB101,* and *avrB102*. Each of the cloned genes converted virulent strains of *X. axonopodis* pv. *malvacearum* to avirulence in a cultivar-specific manner. *X. axonopodis* pv. *malvacearum* strain H contains a single plasmid of 90.4 kb. All six *avr* genes are clustered on this plasmid (De Feyter and Gabriel, 1991). Ten *avr* genes—*avrB4, avrB5, avrb6, avrb7, avrBIn3, avrBn, avrB101, avrB102, avrB103,* and *avrB104*—have been cloned from *X. axonopodis* pv. *malvacearum* (Yang et al., 1996).

From *X. oryzae* pv. *oryzae* PX086, Kelemu and Leach (1990) have cloned a 2.5 kb DNA fragment containing a functional gene, *avr10,* which controls bacterial elicitation of resistance in rice cultivars carrying the *Xa10* gene. The presence of cosmid pSK8-4, which contains *avr10,* in a strain of race 1 (PX061) resulted in both shorter lesions and lower final bacterial numbers specifically in cultivar Cas 209 carrying the *Xa10* gene for resistance. The response in other cultivars was not altered. Race 1 merodiploids carrying inactivated *avr10* (pSK8-4::Tn4431) were pathogenic on Cas209 (Kelemu and Leach, 1990). Two more *avr* genes, *avrXa5* and *avrXa7,* have been isolated and characterized from *X. oryzae* pv. *oryzae* (Hopkins et al., 1992).

Six avirulence genes have been cloned from *X. vesicatoria,* a pathogen of tomato and pepper (Swanson et al., 1988; Whalen et al., 1988, 1993; Minsavage et al., 1990; Canteros et al., 1991). They are designated *avr Bs3, avrBsp, avrBs1, avrBxv, avrBs2,* and *avrBsT.* Three avirulence genes *(avrA, avrB,* and *avrC)* have been isolated from *Pseudomonas savastanoi* pv. *glycinea* races (Staskawicz et al., 1984, 1987; Napoli and Staskawicz, 1987; Tamaki et al., 1988). They interact with distinct resistance genes in soybean cultivars (Staskawicz et al., 1987; Keen and Buzzell, 1991). These avirulence genes, when transferred to previously compatible *P. savastonoi* pv. *glycinea* races, confer the ability to elicit HR on soybean cultivars containing the corresponding resistance genes. Another *avr* gene, *avrD,* has been cloned from *P. savastanoi* pv. *glycinea* (Huynh et al., 1989; Yucel et al., 1994a).

P. savastanoi pv. *phaseolicola,* the bean pathogen, contains avirulence genes, *avrD* and *avrcPphC,* which occur on an approximately 120 kb indigenous plasmid (Yucel et al., 1994c). Several other *avr* genes (*avrA1, avrA2, avrA3, avrA4, avrA5, avrPphD, avrPphE,* and *avrPphB,* previously named *avrPph3*) have been isolated from *P. savastanoi* pv. *phaseolicola* (Hitchin et al., 1989; Shintaku et al., 1989; Jenner et al., 1991; Mansfield et al., 1994; Yucel et al., 1994a,b; Wood et al., 1994; Pirhonen et al., 1996).

Four *avr* genes, *avrPpiA1* (previously named *avrAspi1*), *avrPpiB, avrPpi3,* and *avrRps4* have been isolated and characterized from the pea pathogen *P. syringae* pv. *pisi* (Vivian et al., 1989; Bavage et al., 1991; Dangl et al., 1992; Cournoyer et al., 1995; Hinch and Staskawicz, 1996). The *avrRpm1* gene has been cloned from *P. syringae* pv. *maculicola,* the *Arabidopsis* pathogen (Debener et al., 1991; Ritter and Dangl, 1995). Another *avr* gene, *avrRpt2,* was cloned from *P. syringae* pv. *maculicola* strain MM1065 (Dong et al., 1991; Whalen et al., 1991). Six *avr* genes *(avrA, avrB, avrD, avrE, avrPto,* and *avrRpt2)* have been isolated from the tomato pathogen *P. syringae* pv. *tomato* (Ronald et al., 1992, 1993; Innes et al.,

1993; Lorang et al., 1994; Ritter and Dangl, 1995), *avrD* has been isolated from *P. syringae* pv. *lachrymans,* the cucumber pathogen (Yucel et al., 1994a). Thus several *avr* genes have detected in various *P. syringae* and *P. savastanoi* pathovars (Davis et al., 1991; Kunkel et al., 1993; Parker et al., 1993; Wanner et al., 1993; Yu et al., 1993).

Induction of resistance in a cultivar-specific manner in nonhost species

The transfer of individual clones from a genomic library of *Xanthomonas vesicatoria,* a tomato pathogen to *X. axonopodis* pv. *phaseoli,* led to the isolation of the avirulence gene *avrRxv.* Cultivar specificity was found for the interaction of *avrRxv* with bean cultivars. This was the first demonstration that the gene-for-gene concept traditionally applied to race-cultivar interactions could be extended to interactions between pathogens and nonhost plants (Whalen et al., 1988). Avirulence gene *avrD* cloned from *Pseudomonas syringae* pv. *tomato* causes *P. savastanoi* pv. *glycinea* race 4 to elicit the HR on soybean cultivars containing the resistance gene *Rpg4* (Keen and Buzzell, 1991).

The *avrPpiA* gene from *P. syringae* pv. *pisi* was found to alter the virulence of *P. savastanoi* pv. *phaseolicola* to bean in a cultivar-specific manner (Dangl et al., 1992). When the *avrPpiA* gene from *P. syringae* pv. *pisi* was transferred to the *Arabidopsis* pathogen *P. syringae* pv. *maculicola,* the transformant induced HR in an ecotype-specific manner (Dangl et al., 1992). The *avrPphB* gene from the bean pathogen *P. savastanoi* pv. *phaseolicola* races 3 and 4 was found to confer avirulence on the pea pathogen *P. syringae* pv. *pisi* in all cultivars of pea examined (Fillingham et al., 1992; Vivian and Mansfield, 1993).

P. syringae pv. *pisi* strain 151 induces HR when inoculated on the *Arabidopsis* accession Po1. A genomic cosmid library was constructed from DNA from *P. syringae* pv. *pisi* strain 151, and a cosmid was identified that causes the normally virulent *P. syringae* pv. *tomato* strain DC 3000 to induce an HR on Po1 (Hinch and Staskawicz, 1996). The 1.2 kb fragment was designated *avrRp2s* (Hinch and Staskawicz, 1996). When the *avrRps4* from *P. syringae* pv. *pisi* was transferred to *P. savastanoi* pv. *glycinea* race 4, a cultivar-specific reaction to the pathogen was observed in soybean (Hinch and Staskawicz, 1996). Soybean cultivars Harosoy and Hardee are both susceptible to *P. savastanoi* pv. *glycinea.* When *P. savastanoi* pv. *glycinea* carrying *avrRps4* was inoculated into 'Harosoy,' a brown, necrotic HR appeared in two to three days. *P. savastanoi* pv. *glycinea* race 4 carrying *avrRps4* inoculated into 'Hardee' produced water soaking typical of the susceptible interaction (Hinch and Staskawicz, 1996). Five avirulence genes from *P. syringae* pv. *tomato, avrA, avrD, avrE,* and *avrPto* (from strain PT23), and *avrRpt2* (from strain JL1065), have been shown to be recognized by soybean (Kobayashi et al., 1989; Keen et al., 1990; Ronald et al., 1992). Collectively, these avirulence genes elicit the HR in all tested soybean cultivars (Lorang et al., 1994).

The avirulence gene *pthA* has been cloned from *X. axonopodis* pv. *citri,* the citrus pathogen (Swarup et al., 1991, 1992). The gene *pthA* is not known to function in *X. axonopodis* pv. *citri* for avirulence on citrus. No races of *X. axonopodis* pv. *citri* are known, and no genes governing resistance in citrus have been documented. Hence no avirulence gene can be documented in *X. axonopodis* pv. *citri*. But, when introduced into *X. axonopodis* pv. *citri (= X. campestris* pv.*citrumelo)* strain 3048Sp, *pthA* conferred the ability to elicit a strong HR on bean leaves 48 h after inoculation (Swarup et al., 1992). When introduced into *X. axonopodis* pv. *phaseoli* strain G27Sp, *pthA* conferred the ability to elicit stronger HR than the HR observed with *X. axonopodis* pv. *citri (= X. campestris pv. citrumello)* and *pthA*. The same strains without *pthA* are compatible on bean with obvious water soaking (Swarup et al., 1992). When introduced into *X. axonopodis* pv. *malvacearum* strain Xcm 1003, *pthA* conferred cultivar-specific avirulence on a set of nine different congenic resistant lines of cotton (Table 1.4; Swarup et al., 1992). The cultivar specificity of *pthA* was different from that of all previously cloned *X. axonopodis* pv. *malvacearum avr* genes tested in *X. axonopodis* pv. *malvacearum* N strain (Swarup et al., 1992).

The avirulence gene *avrPph3* from *P. savastanoi* pv. *phaseolicola* was tested for its ability to convert virulent *P. syringae* pv. *tomato* strain D3000 to avirulence on *Arabidopsis*. *Arabidopsis* ecotypes Col-0 and Bla-2 were resistant to *P. syringae* pv. *tomato* carrying *avrPph3*. However, ecotype Ler

TABLE 1.4. Specificity of *pthA* on Cotton cv. Acala 44 and Congenic Resistant Lines

Xanthomonas axonopodis	**Cotton cv. Acala 44 congenic resistant lines**				
pv. *malvacearum* strain	**Ac44**	**B1**	**B2**	**B4**	**Bln3**
Xcm 1003	+	+	+	+	+
Xcm 1003/*pthA*	+	–	–	+	–

Source: Swarup et al., 1992.

+ indicates a compatible interaction; – indicates an incompatible interaction was highly susceptible (Simonich and Innes, 1995). Thus recognition of *avr* genes by nonhost plant species appears to be common (Whalen et al., 1988, 1991; Dangl et al., 1992; Fillingham et al., 1992; Ronald et al., 1992; Innes et al., 1993a,b; Wood et al., 1994).

But several other *avr* genes do not recognize nonhost plant species. Clones harboring the *avr* gene *(avrPphE)* from *P. savastanoi* pv. *phaseolicola* were conjugated into *P. syringae* pv. *maculicola* and *P. syringae* pv. *pisi* and transconjugants tested for ability to cause the HR in a range of genotypes of their hosts *Arabidopsis thaliana* and pea, respectively, but no significant alterations in virulence were observed (Mansfield et al., 1994).

An *avr*-like gene, *hrmA,* has been detected in *Pseudomonas syringae* pv. *syringae* strain 61. *hrmA* is located within the 25 kb DNA region adjacent to a conserved cluster of *hrp* genes. *hrmA* is essential for nonpathogens to elicit the HR (Alfano et al., 1997). DNA sequences hybridizing with *hrmA* are not present in *P. syringae* pv. *tabaci, P. savastanoi* pv. *glycinea,* and *P. syringae* pv. *syringae* B728a, but are present in *P. syringae* pv. *tomato* (Alfano et al., 1997). *P. syringae* pv. *tabaci* 11528 causes wildfire disease of tobacco, and *P. syringae* pv. *syringae* B728a is a highly virulent agent of brown spot of bean. The *hrmA* converted the interaction of *P. syringae* pv. *tabaci* and tobacco to incompatibility, but it had no apparent effect on the interaction of *P. syringae* pv. *syringae* B728a with a bean cv. Eagle (Alfano et al., 1997).

avr Gene Locus May Be Conserved Among Many Plant Pathogens

The *avr* genes appear to be conserved among several bacterial pathogens. *Pseudomonas syringae* pv. *tomato,* the tomato pathogen, contained an avirulence gene homologous to *avrA,* the avirulence gene first cloned from *P. savastanoi* pv. *glycinea,* the soybean pathogen (Kobayashi et al., 1989). Homologs of *avrPphE,* the *avr* gene from *P. savastanoi* pv. *phaseolicola,* have been detected in *P. syringae* pv. *tabaci,* the tobacco pathogen (Mansfield et al., 1994). Homologs of *avrPpiB,* the *avr* gene from *P. syringae* pv. *pisi,* were detected in *P. savastanoi phaseolicola, P. syringae* pv. *maculicola,* and *P. syringae* pv. *tomato* (Cournoyer et al., 1995). Homologs of *avrPpiA,* the other *P. syringae* pv. *pisi* avirulence gene characterized, also were shown to be conserved between different *P. syringae* and *P. savastanoi* pathovars (Dangl et al., 1992). A probe consisting of the open reading frame (ORF) of the avirulence gene *avrRps4* from the pea pathogen *P. syringae* pv. *pisi* hybridizes to *P. savastanoi* pv. *phaseolicola,* the bean pathogen (Hinsch and Staskawicz, 1996). The *avr* gene has been detected among different races of

P. syringae pv. *pisi* and *P. savastanoi* pv. *phaseolicola* (Hinsch and Staskawicz, 1996). *avrPphC* was detected in *P. savastanoi* pv. *phaseolicola* and *P. savastanoi* pv. *glycinea* (Yucel et al., 1994b).

Two avirulence genes, *avrB* and *avrC,* from race O of *P. savastanoi* pv. *glycinea,* were sequenced. The coding regions of the two genes share considerable amino acid identity. While *avrB* is located on race O chromosome, *avrC* appears to occur on a plasmid (Tamaki et al., 1988). The *avrD* region of *P. syringae* pv. *tomato* seems highly conserved in various *P. syringae* and *P. savastanoi* isolates (Yucel et al., 1994a). *P. savastanoi* pv. *glycinea* harbors DNA sequences with considerable similarity to *avrD* (Kobayashi et al., 1990b). The presence of *avrD* in a collection of *P. syringae* and *P. savastanoi* pathogens was tested by performing Southern blots using an *avrD*-specific probe (Yucel et al., 1994a). The *avrD* could be detected in *P. savastanoi* pv. *phaseolicola* and *P. syringae* pv. *lachrymans* (Yucel and Keen, 1994).

P. syringae pv. *lachrymans* carried two *avrD* alleles on plasmids of different sizes. Allele 1 from *P. syringae* pv. *lachrymans* was 95 percent identical to *avrD* from *P. syringae* pv. *tomato* at the amino acid level. The allele 2 was 98 percent identical with the *P. savastanoi* pv. *glycinea* allele, while the *avrD* gene from *P. savastanoi* pv. *phaseolicola* was 97 percent identical with the *P. savastanoi* pv. *glycinea* allele at the amino acid level (Yucel et al., 1994a). The *avrD* allele from *P. savastanoi* pv. *glycinea* encodes a protein that has 86 percent amino acid identity with the *avrD* protein from *P. syringae* pv. *tomato* (Kobayashi et al., 1990a). Four difference races of *P. savastanoi* pv. *glycinea* contained DNA sequences showing moderate hybridization to the *avrD* gene of *P. syringae* pv. *tomato* (Kobayashi et al., 1990b).

Hendson et al. (1992) used clones from the *P. syringae* pv. *tomato avrE* region to study the relatedness of *P. syringae* pv. *tomato, P. syringae* pv. *maculicola,* and *P. syringae* pv. *antirrhini*. All these pathovars showed no polymorphisms in the 2.3 and 5.7 kb EcoRI fragments from the *avrE* locus. Lorang and Keen (1995) also observed DNA homologous to *avrE* in nine other *P. syringae* pv. *savastanoi* pathovars.

Many *avr* genes in *Xanthomonas* spp. are closely related. The *Xanthomonas* avirulence gene family is a large one (De Feyter et al., 1993; Yang and Gabriel, 1995). This gene family comprises nearly all published *Xanthomonas avr* genes. Members of the gene family include *avrB4, avrb7, avrBIn, avrB101,* and *avrB102* of *X. axonopodis* pv. *malvacearum* (De Feyter and Gabriel, 1991; De Feyter et al., 1993), *avrBs3, avrBs3-2 (avrBsp)* of *X. vesicatoria* (Bonas et al., 1989, 1993; Canteros et al., 1991), *avrXa5, avrXa7,* and *avrXa10* of *X. oryzae* pv. *oryzae* (Hopkins et al., 1992).

This gene family includes an *avr* gene, *pthA,* of *X. axonopodis* pv. *citri* (Swarup et al., 1992). All members of this family sequenced to date are 95 to 98 percent identical to each other (Hopkins et al., 1992; De Feyter et al., 1993; Bonas et al., 1993; Yang and Gabriel, 1995). By swapping the tandemly repeated regions, the avirulence specificities of *avrB4, avrb6, avrb7, avrBln, avrB101, avrB102,* and *pthA* were shown to be determined by the 102 bp tandem repeats (Yang et al., 1994). Herbers et al. (1992) showed that the avirulence specificity of *avrBs3* is determined by the 102 bp repeats of that gene. Thus the most conspicuous feature of this gene family is the presence of nearly identical, tandemly arranged 102 bp repeats in the central region of the genes.

DNA sequences related to *avrBs3* from *X. vesicatoria* were detected in seven pathovars of *X. campestris* and *X. axonopodis* that cause disease on diverse dicotyledonous plants (Bonas et al., 1989). Homologs of *avrBs3* have been detected in *X. axonopodis* pv. *malvacearum* and *X. axonopodis* pv. *citri* (Swarup et al., 1992). The *X. oryzae* pv. *oryzae* avirulence gene *avrxa5* hybridized with *avrBs3* and hence may be an *avrBs3* homolog (Hopkins et al., 1992). The other two *avr* genes of *X. oryzae* pv. *oryzae, avrXa7* and *avrXa10,* were localized by transposon mutagenesis to regions of the clone that hybridized to *avrBs3* (Hopkins et al., 1992).

Multiple copies of *avrBs3*-related sequences have been detected in *X. axonopodis* pv. *vasculorum, X. translucens* pv. *translucens, X. translucens* pv. *secalis,* and *X. translucens* pv. *undulosa* (Hopkins et al., 1992). An antiserum that was raised against the AvrBs3 protein cross-reacted with six to seven proteins in an extract of *X. oryzae* pv. *oryzae* strain PX086, indicating that several genes which may be expressed in *X. oryzae* pv. *oryzae* encode antigenically similar proteins to *avrBs3* (Knoop et al., 1991).

Physical characterization of *pthA* from *X. axonopodis* pv. *citri* indicated remarkable similarity to *avrBs3* and *avrBsP* of *X. vesicatoria* and to fragments carrying *avrB4, avrb6, avrb7, avrBln3, avrB101,* and *avrB102* cloned from *X. axonopodis* pv. *malvacearum* (Swarup et al., 1992). Multiple Bal I fragments of about 102 bp form the central region of the *pth* gene similar to *avrBs3* and *avrBsp* (Swarup et al., 1992). Southern hybridization revealed that a fragment of DNA carrying *pthA* hybridized at high stringency to the Bam H1 fragments from *avrBs3, avrB4, avrb6, avrb7, avrBln3, avrB101,* and *avrB102*. The sequence of *pthA* is nearly identical to *avrBs3* from position 374 to at least 733 of *avrBs3* and to *avrBsP* from position 1 to at least position 316; this region includes the putative transcriptional and translational start sites of *avrBs3* and *avrBsp* (Knoop et al., 1991; Canteros et al., 1991). Southern hybridization with a *pthA* internal fragment as a probe against total DNA from various xanthomonads revealed multiple fragments of a size similar to *pthA* in *X. axonopodis* pv. *phaseoli, X. axonopodis* pv. *alfalfae, X. axonopodis* pv. *glycines, X. campestris* pv. *asclepiadis, X.*

axonopodis pv. *cyamopsidis, X. translucens* pv. *translucens,* and *X. axonopodis* pv. *vignicola* (Swarup et al., 1992).

Parker et al. (1993) reported that DNA from *X. campestris* pv. *campestris, X. campestris* pv. *raphani, X. axonopodis* pv. *malvacearum, X. campestris* pv. *holcicola,* and *X. vesicatoria* contained sequences hybridizing with *avrXca* of *X. campestris* pv. *raphani.* A cloned gene, *avrBsp,* from one of the avirulent strains of *X. vesicatoria* converted a virulent strain in tomato to avirulent in tomato (Canteros et al., 1991). The avirulence gene cross-hybridized with another avirulence gene from *X. vesicatoria, avrBs3.* The base sequences of the two avirulence genes were almost identical through the 1.7 kb segment of *avrBsp,* with significant differences only in some bases in the repeat region (Canteros et al., 1991).

Conservation of *avr* genes even among different genera of bacterial pathogens has been reported. The *avrA* fragment from *P. syringae* pv. *tomato* hybridized to identical-sized fragments in different pathovars of *X. campestris* (Kobayashi et al., 1990b).

Transfer of some *avr* genes from one bacterial pathogen to another pathogen converts the recipient to show cultivar-specific resistance in host of the transconjugant. When *avrD* from *P. syringae* pv. *tomato* was transferred to *P. savastanoi* pv. *glycinea,* it elicited the HR on soybean cultivars containing the resistance gene *Rpg4* (Keen and Buzzell, 1991). *P. savastanoi* pv. *glycinea* carrying *avrPphC* gene from *P. savastanoi* pv. *phaseolicola* gave plant reactions that were indistinguishable from those carrying *avrC* of *P. savastanoi* pv. *glycinea* in all tested cultivars (Yucel et al., 1994b). The *avrPphB (=avrPph3)* gene from *P. savastanoi* pv. *phaseolicola* made *P. syringae* pv. *pisi* avirulent in all cultivars of pea (Vivian and Mansfield, 1993). The *avrPpiA* gene from *P. syringae* pv. *pisi* converted the virulence of *P. savastanoi* pv. *phaseolicola* in a cultivar-specific manner (Dangl et al., 1992). Parker et al. (1993) cloned a gene, *avrXca,* from *X. campestris* pv. *raphani* strain 1067 which, when transferred into the virulent *X. campestris* pv. *campestris* strain 8004, strongly reduced symptom development and bacterial growth in *Arabidopsis thaliana* 'Columbia' plants. The gene *avrXca* interacted with all *A. thaliana* accessions tested except one, Kas-1, which developed disease symptoms similar to *X. campestris* pv. *campestris* 8004 alone (Parker et al., 1993).

Although there are reports of conservation of *avr* genes among different pathogens, there are many exceptions. Homologs of *avrPphE* of *P. savastanoi* pv. *phaseolicola* could not be detected in *Pseudomonas cichorii, P. syringae* pv. *coronafaciens, P. savastanoi* pv. *glycinea, P. syringae* pv. *maculicola, P. syringae* pv. *pisi,* or *P. syringae* pv. *syringae* (Mansfield et al., 1994). The *avrBs1* locus could be detected only in races of *X. vesicatoria* (Swanson et al., 1988). The *avrPph3* is absent from isolates of *P. savastanoi* pv. *phaseo-*

licola races 1, 2, 5, 6, 7, and 8, *P. cichorii, P. syringae* pv. *coronafaciens, P. savastanoi* pv. *glycinea, P. syringae* pv. *maculicola, P. syringae* pv. *pisi, P. syringae,* pv. *syringae,* and *P. syringae* pv. *tabaci* (Jenner et al., 1991).

Homologs of Avirulence Genes May Also Be Present in Virulent Strains

The *avr* genes may be conserved in both avirulent and virulent strains of many bacterial pathogens. A sequence homologous to *avr10* was present in strains of all races of *X. oryzae* pv. *oryzae* (Kelemu and Leach, 1990). There were no differences in patterns generated by DNA from race 2 (incompatible to rice cultivar Cas 209) and race 6 (compatible to Cas 209) strains. RFLP analysis revealed no polymorphisms that correlated with expression of the *avr10* phenotype. This suggests the presence of recessive or nonfunctional alleles for *avr10* in *X. oryzae* pv. *oryzae* (Kelemu and Leach, 1990).

Gabriel et al. (1986) used RFLP analysis to compare two *X. axonopodis* pv. *malvacearum* strains, one containing three active *avr* genes and another lacking active *avr* genes. Comparison of the hybridization intensities and positions of fragments that hybridized with a cosmid clone containing all three *avr* genes revealed only minor differences between the strains. Since there were no major DNA arrangements over the 32 kb stretch of DNA, alleles of the *avr* genes may be present in the virulent strain. The gene *avrPphE* has been detected within all strains of *P. savastanoi* pv. *phaseolicola* studied (Mansfield et al., 1994). The *avrD* gene has been detected in all *P. savastanoi* pv. *glycinea* isolates (Kobayashi et al., 1990a,b; Yucel et al., 1994a).

Thus *avr* genes may be present in all strains of a pathogen. But some exceptions have also been reported. *avrPphB* was found only in races that are avirulent on bean genotypes with the matching R3 gene (Jenner et al., 1991).

These studies have revealed that different types of *avr* genes may exist in bacterial pathogens. Their functions also may vary.

Some Avirulence Genes Also May Have Pathogenicity Functions

Similar to *hrp* genes, which show both pathogenicity (susceptibility) and hypersensitivity (resistance) functions, some *avr* genes also have dual functions. Some *avr* genes appear to encode a function required for pathogenicity. The *avrBs2* gene from *Xanthomonas vesicatoria* encodes a function required for pathogenicity (Kearney and Staskawicz, 1990). *avrBs2* was present in each of over 500 *X. vesicatoria* isolates analyzed, as well as in isolates from various *Xanthomonas* spp., and mutations in this gene lower bacterial fitness on at least the two different hosts tested (Kearney and Staskawicz, 1990).

Ten *avr* genes have been detected in *X. axonopodis* pv. *malvacearum (Xam)* strain H1005 (Yang et al., 1996). Marker-eviction mutagenesis of *Xam* H1005 and complementation tests revealed that at least seven of the ten *avr* genes exhibited a pleiotropic pathogenicity (water soaking) function on cotton, but only five *avr* genes were needed for full water soaking. The contribution of each gene to pathogenicity was additive, and five of the ten *avr* genes appeared to be redundant (Yang et al., 1996).

The *X. axonopodis* pv. *malvacearum avr* gene, *avrb6,* appears to be more important than any other *avr* genes conferring water soaking on susceptible cotton lines (Yang et al., 1994, 1996). All mutations or mutation combinations involving *avrb6* reduced water-soaking ability on susceptible cotton lines. By contrast, mutations eliminating five of the *avr* genes except *avrb6* did not affect water-soaking symptoms on susceptible cotton lines (Yang et al., 1996). Strains containing *avrb6* (*Xam* H1005 and *Xam*1003/*avrb6*) elicited more water-soaking ability and necrosis and were associated with much more slime oozing from water-soaked areas than strains lacking *avrb6* (XamH1407 and Xam1003) (Yang et al., 1994). *X. axonopodis* pv. *malvacearum* strains carrying *avrb6* always exhibited many secondary infections around the original inoculation site. In contrast, strains lacking *avrb6* rarely exhibited secondary infection (Yang et al., 1994). *avrb6* functions as a pathogenicity gene and increases symptoms of cotton blight, but without eliciting an HR in cotton lines lacking *b6*. *avrb7* as well as *avrb6* were shown to confer enhanced water soaking to another strain on susceptible hosts (De Feyter et al., 1993).

The avirulence/pathogenicity gene *pthA* of *X. axonopodis* pv. *citri* is required to elicit symptoms of citrus canker disease. The *pthA* gene was first identified functionally via its ability to enhance the virulence of a weakly pathogenic *X. axonopodis* pv. *citri (=citrumello)* strain on grapefruit (Swarup et al., 1991, 1992). Mutation of *pthA* in *X. axonopodis* pv. *citri* results in loss of pathogenicity on grapefruit. *pthA* activity requires a minimum uninterrupted stretch of 3.4 kb of DNA (Swarup et al., 1992). In addition to conferring ability to elicit the plant HR on selected plant genotypes when transferred to a variety of xanthomonads, both *pthA* and *avrb6* also confer ability to elicit host-specific pathogenic symptoms: *pthA* confers ability to induce hyperplastic cankers specifically on citrus, and *avrb6* confers ability to induce strong water soaking specifically on cotton (Swarup et al., 1992; Yang et al., 1994). Comparisons of the two genes in selected isogenic *X. axonopodis* background strains revealed that these two identical genes (98.4 percent identical predicted peptide sequences) can determine three very different plant response phenotypes: cankers on citrus, water soaking on cotton, or an HR on many plants (Yang and Gabriel, 1995). Chimeras of *avrb6* and *pthA* revealed that the 102 bp, leucine-rich tandem repeats in the central portion of these genes determined all three plant reaction phenotypes, and that the regions outside of the repeats are functionally interchangeable (Yang et al., 1994).

Although some *avr* genes have pathogenicity functions, several *avr* genes, such as *avrPthE* from *P. savastanoi* pv. *phaseolicola* (Mansfield et al., 1994), *avrBs3* from *X. vesicatoria* (Bonas et al., 1989), *avrRpt2* from *P. syringae* pv. *tomato* (Innes et al., 1993a), and *avrPto* from *P. syringae* pv. *tomato* (Salmeron and Staskawicz, 1993), do not appear to be a significant determinant of pathogenicity.

Some avr Genes Also Have Virulence Functions

Some *avr* genes have been shown to be virulence factors involved in promoting parasitism (Gopalan et al., 1996). The *avr* genes *avrE* and *avrA* from *P. syringae* pv. *tomato* also have virulence functions (Lorang et al., 1994). Mutation of *avrE* in *P. syringae* pv. *tomato* lowered the ability of low doses of bacteria to grow and cause disease symptoms on tomato, although very high titer inoculum was able to cause symptoms. Transconjugants carrying the cloned *avrE* gene in an *avrE* deletion mutant strain were complemented to full virulence at low-dose inoculum. It suggests that *avrE* is required to establish infection in tomato, but that it is not necessary for symptom production if large numbers of bacteria are inoculated into tomato tissue. Further, this virulence function is apparently strain dependent, since mutation of a functional *avrE* allele in another *P. syringae* pv. *tomato* strain had no effect on virulence on tomato (Lorang et al., 1994). These results suggest that *avrE* may be only one of several virulence genes responsible for symptom development. Lorang et al. (1994) also showed a similar, though less pronounced, role for *avrA* virulence on tomato.

Ritter and Dangl (1995) demonstrated that the avirulence gene *avrRpm1*, isolated from the *Arabidopsis* pathogen *Pseudomonas syringae* pv. *maculicola* strain PsmM2, is also required for maximal virulence on the host. Two *avrRpm1* :: Tn3-Spice marker-exchange mutants do not elicit a hypersensitive reaction on *Arabidopsis* resistant accessions Col-O and Oy-O. These mutants neither generate symptoms nor grow in planta after inoculation onto susceptible accessions Nol-O, Fe-1 and Mt-0 (Ritter and Dangl, 1995). These deficiencies could be corrected in a merodiploid containing a wild type *avrRpm1* allele and were not observed following gene-replacement with *avrRpm::Tn-3*-Spice alleles containing insertions just beyond the 3' terminus of the avirulence gene open reading frame (Ritter and Dangl, 1995). It suggests that the *avrRpm1* gene encodes a function necessary for virulence of *P. syringae* pv. *maculicola* M2 on an array of susceptible accessions, as well as being necessary and sufficient for induction of HR on resistant *Arabidopsis* accessions. No obvious effect of Tn3-Spice insertion alleles in, or downstream of, the *avrRpm1* ORF in *P. syringae* pv. *maculicola* M2 on HR induction on either pea cultivars which contain the R2 resistance gene corresponding to *avrRpm1,* or a cultivar which does not. Thus, *avrRpm1* does not function as an *hrp* gene (Ritter and Dangl, 1995) and it functions only as an *avr*/virulence gene. *avrRpm1* is present in only a minority of tested *P. syringae* pv. *maculicola* isolates, and several of the isolates lacking it are pathogenic on *Arabidopsis*. It suggests that *avrRpm1* may not be absolutely required for virulence of the pathogen, and it may be only one of the several factors (Ritter and Dangl, 1995).

Two *X. axonopodis* pv. *malvacearum avr* genes, *avrb6* and *avrb7,* have been shown to have virulence function also (De Feyter and Gabriel, 1991). One of the *X. axonopodis* pv. *malvacearum* race b6 mutants was less virulent and caused reduced water-soaking lesions even in the most susceptible cotton cultivar AC44. When the mutant phenotype was complemented by a 5 kb ECORI fragment containing *avrb6,* its virulence increased. Another mutant *Xam*1003 carrying either *avrb6,* or *avrb7* induced increasing water-soaking symptoms in cotton cultivars (De Feyter and Gabriel, 1991).

Molecular characterization of the *avrBs2* locus from *X. vesicatoria* has revealed that expression of this gene triggers disease resistance in *Bs2* pepper *(Capsicum annuum)* plants and also contributes to virulence of the pathogen (Swords et al., 1996). The *avrBs2* gene encodes a putative 80.1 kDa protein. A divergent *avrBs2* homolog has been cloned from the *Brassica* pathogen *X. campestris* pv. *campestris* (Swords et al., 1996).

Although some *avr* genes have been shown to have virulence functions, several other *avr* genes do not show virulence functions. An *avr* gene, *avrPpiA1,* has been cloned from *P. syringae* pv. *pisi* race 2 (Dangl et al., 1992; Gibbon et al., 1997). Homologs of the gene were present in races 5 and 7 of *P. syringae* pv. *pisi* (Gibbon et al., 1997). The *avrPpiA* is a homolog of *avrRpm1,* which is carried on a plasmid in *P. syringae* pv. *maculicola,* which is required for full virulence toward susceptible accessions of *Arabidopsis thaliana* (Ritter and Dangl, 1995). But virulence function for *avrPpiA* in terms of symptom development or bacterial growth in susceptible cultivars of the host plant, pea, could not be demonstrated. In *P. syringae* pv. *tomato, avrE* has been found to be required for full virulence in strain PT23 but not in strain DC3000 (Lorang and Keen, 1995).

Disease-Specific (dsp) Genes

Some bacterial genes are required for pathogenicity, but not the HR. Genes involved in pathogenicity, but no other phenotype, have been designated *dsp* genes (disease-specific genes). The association of *dsp* genes in pathogenicity of *Erwinia amylovora* has been reported (Steinberger and Beer, 1988; Barny et al., 1990). Barny et al. (1990) could divide the virulence region in *E. amylovora* into an *hrp* cluster and a *dsp* cluster. Bauer and Beer (1991) could identify at least one gene in the cluster of *hrp* genes which was required for pathogenicity, but not the HR. They were unable to clearly differentiate an *hrp* cluster from a *dsp* cluster.

The *dsp* genes have been detected in *Ralstonia solanacearum* GMI 1000 (Boucher et al., 1985). The *dsp* genes appear to be scattered in the genome of strain GMI 1000 (Arlat and Boucher, 1991). By the use of Tn5 mutagenesis, three disease-specific mutants impaired in the ability to wilt tomato were obtained in strain GMI 1000 (Boucher et al., 1985). None of the mutants generated were affected in their HR-inducing ability on tobacco (Arlat and Boucher, 1991). Hence the mutants would have lost only the *dsp* region rather than the *hrp* region.

Arlat and Boucher (1991) screened separately a genomic bank of the wild-type strain GMI 1000 constructed in pLAFR3 with probes corresponding to cloned EcoRI fragments containing Tn5 and the flanking sequences of the three Dsp$^-$ mutants in order to clone the genes altered in the Dsp$^-$ mutants GMI 1299, GMI 1314, and GMI 1330. Each probe led to the isolation of a set of genes carrying overlapping inserts, none of which overlapped with cosmids corresponding to the two other sets. This result suggests that the *dsp* genes altered in the three mutants may be different (Arlat and Boucher, 1991).

A cluster of *dsp* genes seems to be located on a 3 kb region adjacent to the right end of the *hrp* cluster in *R. solanacearum* (Boucher et al., 1987). It modulates aggressiveness toward tomato but does not control HR-inducing ability (Boucher et al., 1987). In *Xanthomonas campestris,* a set of *dsp* genes specifically involved in interaction with the homologous host has been shown to be clustered in all on a 10 kb region (Turner et al., 1985).

X. axonopodis pv. *glycines* causes pustules on susceptible soybean cultivars. Nonpathogenic mutants of *X. axonopodis* pv. *glycines* have been generated with N-methyl-N-nitro-N'-nitrosoguanidine to identify and characterize pathogenicity genes of the bacterium (Hwang et al., 1992). One cosmid clone, which contained a 31 kb insert, complemented mutant NP1. A restriction map of the clone was constructed, and deletion analyses identified a 10 kb fragment that restored pathogenicity of the mutant (Hwang et al., 1992). *X. axonopodis* pv. *glycines* does not have the ability to induce HR on tobacco or tomato plants, and no *hrp* genes have been detected in *X. axonopodis* pv. *glycines*. The genes detected in this bacterium may have only pathogenicity functions (Hwang et al., 1992).

The 10 kb fragment is conserved among *X. vesicatoria, X. axonopodis* pv. *malvacearum, X. campestris* pv. *campestris,* and *X. oryzae* pv. *oryzae* (Hwang et al., 1992). Genes were identified in a library of *X. translucens* pv. *translucens* DNA that restored the pathogenicity of two mutants of *X. campestris* pv. *campestris* (Sawczyc et al., 1989). The pathogenicity genes were identified as global regulatory genes. These genes were present in a range of other pathovars such as *X. axonopodis* pv. *glycines, X. translucens* pv. *graminis, X. campestris* pv. *holcicola, X. campestris* pv. *pisi, X. vesicatoria, X. axonopodis* pv. *vitians, X. campestris* pv. *zinniae,* and *X. oryzae* pv. *oryzae* (Sawczyc et al., 1989).

A large (14 kb) cluster of pathogenicity genes from *Pantoea stewartii,* the maize pathogen, has been cloned (Coplin et al., 1992). These genes were required for both water-soaked lesion formation and wilting on maize seedlings. *Trans* complementation analysis identified three complementation groups in the map order *wtsA, wtsB,* and *wtsC.* These genes are not able to cause *hrp,* and *Pantoea stewartii* is not known to cause an HR in any host or nonhost (Coplin et al., 1992). *Agrobacterium tumefaciens* incites tumors on dicotyledonous plants, and no *hrp* genes have been detected in this pathogen. A set of genes localized in a tumor-inducing (Ti) plasmid is required for pathogenesis. At least eight known virulence genes, *virA-H,* are required for pathogenicity (Stachel et al., 1986).

Host-Specific Virulence (hsv) Genes

The *hrp* genes are general virulence genes that are required for virulence on all hosts. The *hrp* genes are not host specific; their inactivation leads to loss of virulence on all hosts and loss of the nonhost hypersensitive response. Besides *hrp* genes, some bacterial pathogens contain genes which are required for virulence on specific hosts. These genes are designated as host-specific virulence *(hsv)* genes (Waney et al., 1991). These genes are functionally distinct from *hrp* genes; they are not involved in the nonhost HR, and they are not required for virulence on genera other than the one indicated. They are distinct from *avr* genes and there was no evidence of cultivar specificity (Waney et al., 1991).

Xanthomonas translucens pv. *translucens* strain Xct-216.2 causes bacterial leaf streak on barley, wheat, rye, oats, and triticale. Under growth chamber conditions, Xct-216.2 caused water soaking on 17 different cultivars of barley, 14 different cultivars of wheat, and two cultivars each of rye, oats, and triticale. There was no cultivar specificity (Waney et al., 1991). Transposon mutagenesis was done to obtain mutations affecting virulence in the bacteria. Screening of the prototrophic Tn5-*gus*A Xct-216.2 derivatives revealed the presence of prototrophic insertional derivatives that affected virulence on one host and not on the other four hosts. All possible host-specific virulence mutant classes (affected on each individual host genus) were recovered (Table 1.5; Waney et al., 1991).

Growth in planta of Tn5-*gusA* mutant T2 ($HsvB^-$) was affected on barley and not on wheat. Similarly, growth of mutant T18 ($HsvW^-$) was affected on wheat and not on barley. Cosmid clones pUFT 200 carrying *hsvW* loci and pUFT 300 carrying *hsvB* loci were transferred by conjugation to T18 and T2; these clones restored the respective mutants to phenotypic virulence and growth in planta on the appropriate host. The cosmids did not cross-hybridize and did not cross-complement. Thus the two *hsv* loci, *hsvB* and *hsvW,* which are necessary for virulence on either barley or wheat, respectively, appear to be superimposed on a basic ability to parasitize (Waney et al., 1991). These genes are not involved in the nonhost HR, they are not required for virulence on genera other than the one indicated, and there is no evidence of cultivar specificity on any of the 14 wheat and 17 barley cultivars (Waney et al., 1991). Thus these genes may represent a distinct class of virulence genes that may be largely responsible for the host range at the plant genus level.

Similar *hsv* genes have been detected in *Pseudomonas syringae* pv. *tabaci* BR2, which causes wildfire disease on tobacco and bean plants

TABLE 1.5. Symptom Induction on Different Hosts and on a Nonhost Cotton After Inoculation with *Xanthomonas translucens* pv. *translucens* Strain Xct-216.2 and Tn5-*gus*A Mutant Derivatives

	Hosts					Nonhost
Strain	**Barley**	**Wheat**	**Rye**	**Oats**	**Triticale**	**cotton**
Xct-216.2	+	+	+	+	+	HR
T2	–	+	+	+	+	HR
T18	+	–	+	+	+	HR
T3	+	+	–	+	+	HR
T4	+	+	+	–	+	HR
T12	+	+	+	+	–	HR

Source: Waney et al., 1991.

+ = Water soaking; – = no symptoms; HR = hypersensitive response

(Salch and Shaw, 1988). Eight Tn5 insertion mutants were obtained, and they caused typical wildfire disease symptoms on bean plants but were no longer pathogenic on tobacco. A 7.2 kb EcoRI fragment restored pathogenicity to the nonpathogenic mutant. Some genes on the fragment may be related to the *hsv* genes (Salch and Shaw, 1988).

Ralstonia solanacearum strain T2003 was pathogenic to potato but nonpathogenic to groundnut, while *R. solanacearum* strain T2005 was pathogenic to groundnut (Ma et al., 1988). Transferring clones from the strain T2005 genomic library to a T2003 derivative recipient resulted in the identification of two transconjugants that can cause wilting on groundnut, indicating that the host range of the original nonpathogenic (to groundnut) strain T2003 can be extended to groundnut by introducing two cosmid clones from the strain T2005 library. Ma et al. (1988) presented evidence that at least some genes for host specificity (*hsv* genes) in strain T2005 are carried in a 12.5 DNA fragment.

COREGULATION OF hrp, avr, AND OTHER PATHOGENICITY GENES

Some hrp Genes Are Regulatory Genes

Bacterial pathogens containing several *hrp* genes and *hrp* gene clusters are common (Lindgren et al., 1989). The *hrp* genes have dual functions of inducing resistance (hypersensitive reaction) as well as pathogenesis (Boucher et al., 1988). *hrp* genes are conserved among different bacterial genera. For example, two different unrelated phytopathogenic bacteria such as *Xanthomonas campestris* pv. *campestris* and *Ralstonia solanacearum,* which induce different classes of diseases like leaf spot and wilt and have different host specificities, carry similar *hrp* genes (Arlat et al., 1991). It suggests that *hrp* genes may have some common function in different bacteria.

For induction of resistance, all *hrp* genes are required, and deletion of a single *hrp* gene from *hrp* gene clusters results in loss of ability to induce resistance (Bonas et al., 1991; Lidell and Hutcheson, 1994). Besides *hrp* genes, *avr* genes are also required for induction of HR. Similarly, *hrp* genes alone are not sufficient to induce pathogenesis; additional pathogenicity genes are also required (Yang and Gabriel, 1995). Thus *hrp* genes may encode proteins which regulate other genes to induce resistance or susceptibility. *hrp* genes may regulate expression of other *hrp* genes, *avr* genes, and other pathogenicity genes.

Hrp Regulatory Proteins Regulate hrp Genes

Some Hrp Regulatory Proteins Have Sequence Similarity to Response Regulator Proteins of Bacterial Two-Component Sensory Transduction Systems

Two-component sensory transduction systems have been found in many bacterial species (Parkinson and Kofoid, 1992). These systems are parts of complex regulatory networks and cascades (Charles et al., 1992). These systems are made up of a sensor protein and a response regulator protein. This pair of proteins communicate with each other by a conserved mechanism involving protein phosphorylation (Albright et al., 1989). One member of the pair codes for a sensor (modulator) protein that modifies the activity of its partner, a regulator (effector) protein. The sensor proteins are histidine protein kinases that autophosphorylate in direct or indirect response to environmental stimuli, then transfer the phosphate to a specific aspartate residue of the response regulator protein. Phosphorylation of the response regulator protein alters its regulatory function (Charles et al., 1992).

Wengelnik et al. (1996b) isolated and characterized the *hrpG* locus of *Xanthomonas vesicatoria,* the pathogen of pepper and tomato. The pathogen has a 23 kb *hrp* cluster comprising six *hrp* loci, *hrpA* to *hrpF,* which are all required for full pathogenicity (Bonas et al., 1991). Another *hrp* gene, *hrpXv,* has also been reported in the bacteria (Wengelnik and Bonas, 1996). *hrpXv,* which codes for a protein of the Arac family, has been shown to be indispensable for transcriptional activation of the five loci *hrpB* to *hrpF* (Wengelnik and Bonas, 1996). However, induced expression of the *hrpA* locus and *hrpXv* gene itself was independent of HrpXv (Wengelnik and Bonas, 1996; Wengelnik et al., 1996a). The *hrpG* locus plays a key role in *hrp* gene regulation. An *hrpG* mutation no longer allows induction of all other *hrp* loci identified so far *(hrpA, B, C, D, E, F,* and *hrpXv)* under *hrp* gene-inducing conditions (Wengelnik et al., 1996b). Since transcription of the *hrpXv* gene itself is independent of *hrpXv* but depends on *hrpG, hrpG* appears to be the key regulatory gene in the cascade activating *hrp* gene expression in *X. vesicatoria. hrpG* encodes a protein with significant sequence similarity to response regulator proteins of bacterial two-component systems (Wengelnik et al., 1996b). The sensor and regulator proteins in the two-component systems have been classified into subgroups depending upon their predicted domain structure. Accordingly, HrpG belongs to the subgroup containing the *Escherichia coli* proteins Ompr and PhoB and *Agrobacterium tumefaciens* VirG. These proteins have been shown to be phosphorylated, thereby enhancing their binding to specific regulatory sequences involved in transcriptional activation. HrpG shows homology to response regulator proteins of a two-component system, but a corresponding sensor Hrp protein has not yet been isolated (Wengelnik et al., 1996b). The regulatory genes *hrpG* and *hrpXv* are located outside of the large *hrp* gene cluster but next to each other. The two genes are transcribed from a divergent promoter region (Wengelnik et al., 1996b).

One locus at the right end of the *hrp* cluster, designated *hrpSR* in *P. savastanoi* pv. *phaseolicola,* contains two regulatory genes, *hrpS* and *hrpR*. (Lindgren et al., 1989; Grimm et al., 1995). HrpR and HrpS showed high sequence similarities to each other and to other response regulators of the two-component regulatory system. *hrpS* expression is regulated by the *hrpR* gene product. HrpR activates *hrpS* transcription by binding to an activator site. This HrpR binding site was mapped in a fragment which is located 378 to 609 nucleotides upstream of the *hrpS* transcription start site. The *hrpS* transcription site maps 179 nucleotides upstream of the initiation codon ATG (Grimm et al., 1995). HrpR and HrpS are involved in the regulation of the other *hrp* genes (Grimm and Panopoulos, 1989). The *hrpS* promotes the expression of another *hrp* locus, *hrpD* (Grimm and Panopoulos, 1989). There was extensive similarity between the predicted amino acid sequence of HrpS protein and the products of several regulatory genes of enteric and plant symbiotic bacteria such as *Rhizobium* spp., *Klebsiella pneumoniae,* and *Escherichia coli* (Grimm and Panopoulos, 1989). Hence HrpS protein may also be regulatory protein. These proteins constitute the so-called Ntrc family and require the alternate sigma factor (σ^{54}) encoded by *rpoN* as coactivator. Ntrc operates as part of a global control system (the ntr system) which mediates transcriptional activation or repression of many different genes and operons (Magasanik, 1982). The *hrpS* gene has been detected in *Pantoea stewartii, E. herbicola* pv. *gypsophilae* (Nizan et al., 1997), and *E. amylovora* (Wei and Beer, 1995). HrpS has been identified as a transcriptional regulator which contains the σ^{54} interacting domain (Frederick et al., 1993). Horns and Bonas (1996) have cloned the *rpoN* region of the pepper and tomato pathogen *X. vesicatoria*. The *X. vesicatoria rpoN* gene has been shown to be not required for *hrp* gene expression. It suggests that the nature of the *hrp* regulators may be different in *Xanthomonas* and *Pseudomonas*.

Some hrp Regulatory Proteins Act As Alternative Sigma Factors

Sigma factor is a polypeptide subunit of the RNA polymerase. Sigma factors are composed of two functional domains: a core-binding domain and a DNA-binding domain. The major sigma factor is responsible for transcription initiation of most genes. The sigma factor is released from RNA polymerase after initiation of genetic transcription, i.e., is free to program another polymerase molecule; the core enzyme continues to transcribe. Alternative sigma factors have a role in cellular responses to environmental stimuli. One of the *hrp*- encoded proteins, HrpL, from *Pseudomonas syringae* pv. *syringae* appears to be an alternative sigma factor (Xiao et al., 1994; Xiao and Hutcheson, 1994). The *hrpL* cluster is involved in regulation of the other *hrp* genes in *P. syringae* pv. *syringae* (Xiao et al., 1994). *P. savastanoi* pv. *phaseolicola* also has the *hrpL* gene, and the gene is predicted to encode a protein that is 93 percent identical to *hrpL* from *P. syringae* pv. *syringae* (Xiao et al., 1994). Key regulatory roles have been proposed for the different *hrpL* loci in *P. syringae* pv. *syringae, P. savastanoi* pv. *phaseolicola, P. savastanoi* pv. *glycinea,* and *P. syringae* pv. *tomato* (Huynh et al., 1989; Fellay et al., 1991; Rahme et al., 1991, 1992; Innes et al., 1993a; Salmeron and Staskawicz, 1993; Xiao and Hutcheson, 1994; Xiao et al., 1994). Sequences related to HrpL have been detected in *Erwinia herbicola* pv. *gypsophilae* and *Pantoea stewartii* (Nizan et al., 1997). The *hrpL* gene has been detected in *Erwinia amylovora* (Wei and Beer, 1995). It encodes a 21.7 kDa regulatory protein, similar to members of the ECF (extra cytoplasmic functions) subfamily of eubacterial RNA polymerase σ factors. HrpL has been shown to control expression of five independent *hrp* loci, including *hrpN,* which encodes harpin. Expression of *hrpL* is affected by *hrpS,* another regulatory gene of the *hrp* gene cluster of *E. amylovora* (Wei and Beer, 1995). The importance of the alternative sigma factor in regulation of *hrp* genes has been demonstrated by developing an *rpoN* mutant (which lacks σ^{54}) of *P. savastanoi* pv. *phaseolicola,* which became nonpathogenic and unable to elicit hypersensitive response on a heterologous host (Charles et al., 1992). Several *P. savastanoi* pv. *phaseolicola hrp* genes require σ^{54} cofactor for expression (Fellay et al., 1991; Shen and Keen, 1993; Grimm et al., 1995).

hrp Genes Regulate Transcription of avr Genes

The *hrp* genes appear to be needed for transcription of *avr* genes in bacterial pathogens. Lorang and Keen (1995) have reported regulation of the *P. syringae* pv. *tomato avrE* locus, which causes the HR on soybean cultivars. The effect of *hrp* regulatory mutants of *P. syringae* pv. *tomato* on accumulation of another *avr* gene, *avrPto,* measured via RNA blots, was apparently absolute (Salmeron and Staskawicz, 1993). No *avrRpt2* mRNA was observed in *hrp* regulatory mutants of *P. syringae* pv. *tomato* (Innes et al., 1993b). Some *hrp* mutations of *P. syringae* pv. *glycinea* inhibit the transcriptional induction of *avr* genes of the bacterium (Lindgren et al., 1988; Huynh et al., 1989).

A set of plasmid-borne constructs expressing the *P. syringae* pv. *syringae hrp* genes or the *avrB* gene of the bacterium were transformed singly or in combination into *Escherichia coli* cells (Pirhonen et al., 1996). Another plasmid carrying a constitutively expressed *hrpL* gene was included in these strains to enhance expression of the *P. syringae* pv. *syringae hrp* genes in *E. coli*. All soybean cultivars tested were found to be insensitive to *E. coli*–carrying *hrp* genes including the *hrpL* (regulatory gene) or *avrB* gene alone. In contrast, *E. coli* transformants carrying the *hrp* gene cluster and *avrB* were capable of eliciting the HR, but only in those soybean cultivars carrying *Rpg1,* the resistance gene that mediates soybean responses to *P. syringae* strains carrying *avrB*. When inoculated into leaves of soybean cultivars which lack *Rpg1,* a null response similar to that produced by *E. coli*–carrying *hrp* genes or *avrB* alone was observed (Pirhonen et al., 1996). *P. syringae* pv. *maculicola* and *P. syringae* pv. *tomato* transformants carrying *avrB* genes also induced the HR in *Arabidopsis thaliana* accession Col-O carrying *RPM1,* an R-gene mediating recognition of *P. syringae* strains carrying *avrB*. *A. thaliana* accession Col-O was screened for its ability to respond to *E. coli* carrying the *hrp* genes and *avrB*. The HR was observed when an *E. coli* transformant carrying *P. syringae hrp* genes and *avrB* was inoculated into Col-O. *E. coli* strains lacking either the *avrB* gene or the *P. syringae* pv. *syringae hrp* cluster produced a null response (Pirhonen et al., 1996). The nonpathogenic *Pseudomonas fluorescens* also elicited HR in plants under certain conditions. Elicitation of a genotype-specific HR was observed with *avrB*$^+$ *P. fluorescens* in soybean and *Arabidopsis* plants carrying resistance genes *RPG1* and *RPM1,* respectively, but only if the Hrp secretion system, HrpZ harpin, and the appropriate AvrB protein were produced in the same bacterial cell (Gopalan et al., 1996). These observations suggest a strict *hrp* regulatory dependence for transcriptional activation of the *avrB* and *avrPto* genes in *P. syringae* strains.

avrA isolated from *P. savastanoi* pv. *glycinea* race 6, *avrPto* and *avrRpt2* isolated from *P. syringae* pv. *maculicola,* and *avrPph3* isolated from *P. savastanoi* pv. *phaseolicola* were screened for activity in *E. coli*–carrying *P. syringae hrp* genes. Soybean cultivars that reacted hypersensitively to *P. savastanoi* pv. *glycinea* transformants carrying these *avr* genes also exhibited the HR when inoculated with *E. coli* transformants carrying the same *avr* gene in addition to the *hrp* cluster (Pirhonen et al., 1996). The avirulence gene *avrBs3* in *Xanthomonas vesicatoria* is constitutively expressed in the bacteria, but for induction of HR all of the *hrp* loci need to be intact (Bonas et al., 1989; Knoop et al., 1991; Herbers et al., 1992).

The *avr* genes may require specific *hrp* genes for their transcription in bacterial pathogens. *P. syringae avr* genes *avrB* (Huynh et al., 1989; Xiao et al., 1994; Xiao and Hutcheson, 1994), *avrD* (Shen and Keen, 1993), *avrPto* (Salmeron and Staskawicz, 1993), and *avrRpt2* (Innes et al., 1993a) require functional *hrpL* and *hrpS* loci, but not other *hrp* loci for their expression. Transcriptional regulation of *avrRpm1 avr* genes in *P. syringae* pv. *maculicola* was dependent on *hrpL* and *hrpS* (Ritter and Dangl, 1995). Functional *hrpL* and *hrpRS* loci are required for expression of *avrE* transcripts (Lorang and Keen, 1995). The requirement for *hrp* regulators in activation of *P. syringae avr* genes seems to be absolute, lending credence to the model that *avr* genes are part of an hrp-dependent regulon required for pathogenesis.

Conserved regions (so-called harp boxes) associated with regulation by *hrp* genes have been located within promoters of several *avr* genes from pathovars of *P. syringae* and *P. savastanoi* (Jenner et al., 1991; Dangl et al., 1992; Innes et al., 1993a; Salmeron and Staskawicz, 1993; Shen and Keen, 1993). Lorang and Keen (1995) localized the DNA region required for function of the *avrE* locus to the right external border of the *hrp* gene cluster. The *avrE* locus and adjacent DNA are organized into at least four transcriptional units having regulatory as well as physical linkage to the bacterial *hrp* genes. None of the new transcriptional units was required for *hrp* gene function. But putative transcriptional units III and IV, comprising about 9 kb of DNA, were required for *avrE* function when conjugated into *P. syringae* pv. *glycinea* (Lorang and Keen, 1995). The promoter region of *avrPphE,* the avirulence gene from *P. savastanoi* pv. *phaseolicola,* contains a "harp box" motif (Mansfield et al., 1994). The *avrPphE* was linked to *hrpY,* an *hrp* locus identified at the left end of the *hrp* gene cluster (Mansfield et al., 1994).

Promoters of *hrpL* operons contain a conserved consensus sequence similar to the *avr* box motif, G/T GGAACC-N15 or 16-CCAC found upstream of several *P. syringae* and *P. savastanoi avr* and *hrp* genes (Napoli and Staskawicz, 1987; Tamaki et al., 1988; Kobayashi et al., 1990a,b; Jenner et al., 1991; Dangl et al., 1992; Huang et al., 1993; Innes et al., 1993a; Salmeron and Staskawicz, 1993; Shen and Keen, 1993; Yucel et al., 1994a). *hrpL* may interact with this promoter element to activate transcription of *avr* genes (Xiao and Hutcheson, 1994).

Although many *avr* genes are dependent on the *hrp* cluster for their expression, some *avr* genes may not be dependent on the *hrp* cluster. Phenotypic expression of *avrD* in *Escherichia coli* transformants was not dependent on a *P. syringae* pv. *syringae* 61 *hrp* cluster since an HR was elicited by *E. coli* transformants carrying only *avrD* (Pirhonen et al., 1996).

hrp Genes May Regulate Virulence Genes

Some genes, which are outside of the *hrp* cluster and coregulated with *avr* and *hrp* genes, have been reported to be involved in pathogenesis. Lorang et al. (1994) found that a deletion mutation in *avrE* transcriptional unit III of *Pseudomonas syringae* pv. *tomato* strain PT23 greatly reduced virulence in tomato plants. It suggests that genes coregulated with *hrp* genes may function in the pathogenic process.

TRANSCRIPTION OF BACTERIAL PATHOGENICITY GENES IN PLANTA

hrp Genes

All pathogenicity genes of bacterial pathogens *hrp, avr, dsp,* and *hsv* appear to be dormant in bacterial cells, and they are actively transcribed when they come into contact with plants (Rahme et al., 1991; Wei et al., 1992b; Xiao et al., 1992; Bonas et al., 1994). Two separate *hrp* clusters in *Ralstonia solanacearum* have been reported, and increased expression of the two transcriptional units was obtained in response to the presence of plant tissues (Huang et al., 1990b). In *X. vesicatoria, hrp* gene expression was found to be induced in planta (Schulte and Bonas, 1992a,b). Expression of *hrp* in *X. vesicatoria* has been studied using transcriptional *hrp-gusA* fusions and was found to be induced in planta (Wengelnik and Bonas, 1996). An *hrp* gene, *hrpXc,* in *X. campestris* pv. *campestris* was strongly induced in radish leaves (Kamoun and Kado, 1990).

Enhanced expression of *hrp* genes has been reported during pathogenesis of *P. savastanoi* pv. *phaseolicola* in its susceptible host, bean plants (Lindgren et al., 1989; Rahme et al., 1991). Similarly enhanced expression of *hrp* genes has been reported in resistant interactions. *hrp* gene transcriptional unit III of *P. syringae* pv. *syringae* 61 was induced within 3 h after the bacteria were inoculated into tobacco leaves (resistant interaction) (Xiao and Hutcheson, 1991b). The 13 *hrp/hrm* genes of *P. syringae* pv. *syringae* 61 are induced during the induction stage of the HR in incompatible interactions (Heu and Hutcheson, 1991a,b; Huang et al., 1991). A set of chromosomal *hrp-uidA* fusions in *P. syringae* pv. *syringae* 61 by Tn5-*gusA1* mutagenesis of the *hrp/hrm* gene cluster was constructed and transferred into the genome by marker exchange mutagenesis (Xiao et al., 1992). The expression of *hrp-uidA* fusions in *hrp* transcriptional units increased following inoculation into tobacco (incompatible host) leaves. Enhanced expression from these genes, as indicated by accumulation of GUS, was observed within 2 h after infiltration of tobacco leaves. The induction stage of the HR may represent the time required for *hrp/hrm* genes to be expressed and their products to accumulate (Xiao et al., 1992).

Thus *hrp* genes in the bacterial genome are transcribed in both hosts and nonhosts. In some cases *hrp* expression is more in nonhosts. Wei et al. (1992b) found that *Erwinia amylovora hrp* genes were induced more rapidly and to higher levels in tobacco (nonhost) than in pear (host). Arlat et al. (1992) reported that *hrp* genes of *R. solanacearum* GMI 1000 were induced more in incompatible host (tobacco) than in compatible host (tomato) (Table 1.6).

avr Genes

Transcription of *avr* genes of bacteria is rapidly activated in planta (Lindgren et al., 1989; Fellay et al., 1991; Rahme et al., 1991, 1992; Innes et al., 1993a; Salmeron and Staskawicz, 1993; Shen and Keen, 1993; Xiao et al., 1993, 1994; Ritter and Dangl, 1995). The *avr* genes of *Pseudomonas syringae* pv. *tomato avrB, avrC,* and *avrD* are expressed at high levels only in the plant (Huynh et al., 1989). In the case of *avrD*, expression is low when *P. syringae* pv. *tomato* cells are grown in culture medium and increases approximately 100 times when the bacteria are inoculated into soybean leaves (Keen et al., 1990).

The transcription of the *P. syringae* pv. *tomato avrE* and *avrRpt2* in planta has also been reported (Innes et al., 1993a; Salmeron and Staskavicz, 1993; Shen and Keen, 1993; Xiao et al., 1994). A rapid in planta transcriptional induction was observed for *avr B, avrD,* and *avrPto* of *P. syringae* pv. *tomato,* independent of both the resistance gene status of the inoculated plant and the *avr* gene status of the test bacterial strain (Innes et al., 1993b). Similar findings have been obtained in *avrRpm1* of *P. syringae* pv. *maculicola* (Ritter and Dangl, 1995).

TABLE 1.6. Relative Expression of *hrp* Genes of *Ralstonia solanacearum* GMI1000 in Compatible (Tomato) and Incompatible (Tobacco) Hosts

Plant	*hrp* gene activity (expressed as β-galactosidase activity in Miller's units)
Tomato	61
Tobacco	96

Source: Arlat et al., 1992.

Pathogenicity Genes

Pathogenicity genes are also induced by plant signals similar to induction of *hrp* genes by them. *lacZ* was used as reporter gene to investigate *Pantoea stewartii wts* gene expression in planta (Coplin et al., 1992). *lacZ* was used as a reporter gene since wild-type strains of *P. stewartii* cannot utilize lactose unless a functional β-galactosidase gene is introduced into them. Both the *wtsA* and *wtsB lacZ* gene fusions were expressed (180 to 462 units) in planta. Ooze from control corn plants inoculated with the β-galactosidase-deficient parental strain had negligible enzyme activity (<12 units) (Coplin et al., 1992).

For pathogenesis of *Agrobacterium tumefaciens, vir* genes *virA, B, C, D, E, F, G,* and *H* are important. Most of the transcription units are silent in free-living bacteria (Stachel and Nester, 1986), but they are induced in planta (Stachel and Zambryski, 1986; Bolton et al., 1986; Rogowsky et al., 1987; Veluthambi et al., 1987).

PLANT-DERIVED MOLECULES MAY BE INVOLVED IN INDUCTION OF BACTERIAL GENES

hrp Genes

Expression of *hrp* gene transcripts appears to be controlled by many host nutrients and environment. This observation is based on many in vitro experiments. Transcriptional regulation of *hrp* genes is generally suppressed in complex media such as King's medium B, but the transcripts are induced in bacteria grown in minimal media. A similar regulation pattern has been shown for all *hrp* gene clusters investigated to date (Arlat et al., 1991, 1992; He et al., 1993; Rahme et al., 1991, 1992; Xiao et al., 1992; Schulte and Bonas, 1992b; Wei et al., 1992b; Bonas et al., 1994; Lorang and Keen, 1995). *hrp* loci of *R. solanacearum* (Arlat et al., 1992), *P. syringae* pv. *syringae* (Xiao et al., 1992), *X. vesicatoria* (Schulte and Bonas, 1992a), *P. savastanoi* pv. *phaseolicola* (Fellay et al., 1991), and *E. amylovora* (Wei et al., 1992b) were expressed in planta and in defined minimal-salt media almost with equal efficiency.

To determine whether the 13 *hrp* genes of *P. syringae* pv. *syringae* strain 61 are regulated by nutritional conditions, random transcriptional fusions with *uidA* (encodes β-glucuronidase:GUS) were constructed by Tn5-*gusA1* mutagenesis and marker exchange mutagenesis (Xiao and Hutcheson, 1991a). The transcriptional fusions with 12 of the complementation groups were poorly expressed in King's medium B but exhibited detectable expression in planta. A representative *hrp-uidA* fusion could be rapidly induced independently of plant by growth in minimal-salt media. This apparent induction could be repressed by complex nitrogen sources but not by glutamine or ammonium.

In *X. vesicatoria, hrp* gene expression has been studied using transcriptional *hrp-gusA* fusions and was found to be induced in planta (Wengelnik and Bonas, 1996a), in tomato-conditioned medium TMC and in synthetic XVM2 medium (Wengelnik et al., 1996a). Expression is suppressed in complex media (Wengelnik et al., 1996a). These results suggest that the *hrp* genes are regulated by nutritional conditions.

Specific components of minimal media have been reported to induce *hrp* gene expression, and they differ for various bacterial system studies. Carbon sources which induced maximum gene expression varied among several systems (Huynh et al., 1989; Arlat et al., 1991, 1992; Schulte and Bonas, 1992b; Wei et al., 1992b; Xiao et al., 1992). However, sucrose induced gene expression in all cases. Fructose induced *hrp* gene expression to higher levels than other sugars tested (Lorang and Keen, 1995). But there were some gross differences in the activity of fructose and pyruvate in different bacterial pathogens. Fructose, which is the best substrate in *P. savastanoi* pv. *glycinea,* has very little activity in *R. solanacearum;* on the other hand pyruvate, which is very efficient in *R. solanacearum,* ranks lowest in *X. campestris* pv. *campestris* and *P. savastanoi* pv. *glycinea* (Arlat et al., 1991, 1992).

An *hrp* gene transcription in *R. solanacearum* strain GMI 1000 was studied by using the transposon Tn5-B20, which promotes transcriptional gene fusions (Arlat et al., 1992). The gene transcription was assessed by measuring β-galactosidase activity, which was naturally absent in strain GMI 1000. β-galactosidase produced by the mutant GMI 1487 (generated by Tn5-B20 mutagenesis of strain GMI 1000) was measured following growth in minimal medium supplemented with various carbon sources. Bacteria grown in the presence of pyruvate showed the highest activity, whereas growth with fructose and sorbitol gave the lowest activities (Figure 1.3; Arlat et al., 1992).

Glycerol, succinate, and mannitol induced *hrp III* locus of *P. syringae* pv. *syringae* in a minimal-salt medium (Xiao et al., 1992). In *Erwinia amylovora* regulation of several *hrp* genes was linked to nitrogen metabolism (Beer et al., 1991). Complex amino acid sources such as peptone repressed the expression of *P. syringae* pv. *syringae* 61 *hrp* genes (Xiao et al., 1992). Sulfur-containing amino acids are important for *hrp* gene induction in *X. vesicatoria* (Schulte and Bonas, 1992b). But inorganic nitrogen sources, such as ammonium sulfate, as well as individual amino acids, had little ef-

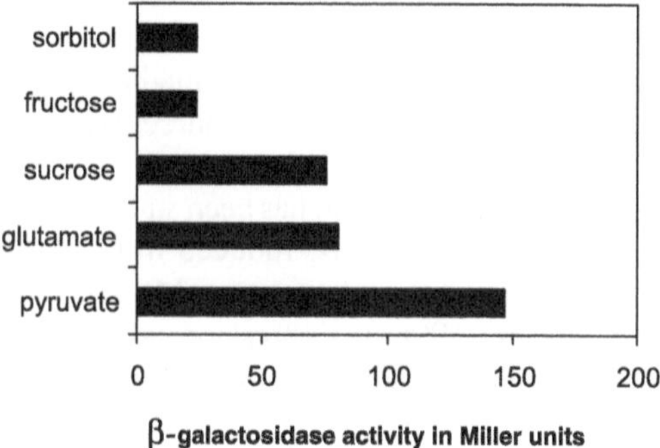

FIGURE 1.3. Effect of different carbon sources on transcriptional activity of *hrp* gene(s) of *Ralstonia solanacearum* mutant GMI1487 grown for 14 h in minimal medium supplemented with various carbon sources (*Source:* Arlat et al., 1992).

fect on induction of *hrp III* in *P. syringae* pv. *syringae* 61-5112 (Xiao et al., 1992).

Phosphate and sodium chloride concentrations are important for *hrp* gene induction in *X. vesicatoria* (Schulte and Bonas, 1992b). High osmotic pressure can repress the expression of several *hrp* genes in *P. savastanoi* pv. *phaseolicola* (Fellay et al., 1991; Rahme et al., 1991). High osmotic pressure inhibited induction of *hrp III* in *P. syringae* pv. *syringae* 61 (Xiao et al., 1992). Low osmolarity and pH are important for *P. savastanoi* pv. *phaseolicola* *hrp* gene induction (Rahme et al., 1992).

Thus changes in host nutrients and environment may induce bacterial *hrp* genes. How do these changes take place? The *hrpXc* gene of *X. campestris* pv. *campestris* is plant-inducible (Kamoun and Kado, 1990). *Hrpxc* mutants of *X. campestris* pv. *campestris* are growth impaired in planta, and can be complemented by coinoculation with wild-type strains, suggesting that this gene may encode for an extracellular component (Kamoun and Kado, 1990). Collmer and Bauer (1994) suggested that an elicitor may be secreted by the bacteria, and this elicitor may release nutrients to the apoplast for bacterial nutrition. The function of elicitors in the release of nutrients will be discussed later in this chapter.

avr Genes

Plant-derived molecules may determine transcription of *avr* genes also in bacteria. Transcription of *Pseudomonas syringae* and *P. savastanoi avr* genes is induced by environmental conditions thought to reflect in planta conditions (Huynh et al., 1989; Innes et al., 1993a; Salmeron and Staskawicz, 1993; Shen and Keen, 1993). Expression of the four putative transcriptional units occurring in the *P. syringae* pv. *tomato avrE* region are induced in bacteria grown in minimal media and not in King's medium B (Lorang et al., 1995). A similar regulation pattern has been shown for *P. syringae avr* genes *avrB* (Huynh et al., 1989), *avrRpt2* (Innes et al., 1993a), *avrPto* (Salmeron and Staskawicz, 1993), and *avrD* (Shen and Keen, 1993). The avirulence gene *avrPphE* of *P. savastanoi* pv. *phaseolicola* was expressed more strongly in minimal media than in nutrient-rich media (Mansfield et al., 1994).

Carbon sources that induced maximum gene expression varied among bacterial pathogens (Huynh et al., 1989). However, sucrose induced gene expression in all cases. Fructose induced *avrE* gene expression in *P. syringae* pv. *tomato* to higher levels than other sugars tested (Lorang and Keen, 1995). *avrB* gene expression in *P. savastanoi* pv. *glycinea* was also induced by fructose (Huynh et al., 1989). Transcription of the *avrRpm1* gene of *P. syringae* pv. *maculicola* is rapidly activated following a shift to minimal media containing either sucrose or fructose as carbon source (Ritter and Dangl, 1995). Huynh et al. (1989) showed that sugars and sugar alcohols optimally induced *lacZ* activity from the *avrB* promoter from *P. savastanoi* pv. *glycinea*. *lacZ* activity could not be detected in the absence of a utilizable carbon source or in the rich media. The *avrPto* gene from *P. syringae* pv. *tomato* was detectably active at extremely low levels in rich media but induced some 800-fold in fructose supplemented minimal media (Salmeron and Staskawicz, 1993). No expression of the *avrRpt2* gene from *P. syringae* pv. *tomato* was seen in rich media, and expression was maximal in a minimal medium supplemented with both fructose and citrate (Innes et al., 1993b). These studies indicate that *avr* gene expression is induced in plants due to secretion of some specific nutrients in the region of infection. How the secretion takes place will be discussed later in this chapter.

Other Pathogenicity Genes

Several pathogenicity genes are also induced by plant-derived molecules. *Agrobacterium tumefaciens* incites tumors on dicotyledonous plants by transferring a segment (T-DNA) of a Ti (tumor-inducing) plasmid into these plants. The T-DNA is flanked on the Ti plasmid by two 25 bp direct repeats. These sequences fully define the T-DNA because any DNA located between these T-DNA is efficiently transferred and integrated into the plant nuclear genome. However, the T-DNA does not encode the products that mediate its transfer. Instead, another region of Ti plasmid, the 35 kbp virulence *(vir)* region including six operons, *virA, virB, virC, virD, virE,* and *virG,* provides most of the *trans* acting functions for T-DNA transit (Stachel and Nester, 1986). Most of the transcription units are silent in free-living bacteria and must be induced for transfer of the T-region to plants. Induction of *vir* gene expression is the control switch for T-DNA transfer to plant cells and occurs only when *A. tumefaciens* is in the presence of wounded susceptible plant cells; these plant cells excrete low molecular weight phenolic compounds that act as *vir*-inducing signal molecules (Stachel et al., 1985, 1986; Spencer and Towers, 1988; Melchers et al., 1989b; Huang et al., 1990a). The induction of the virulence functions requires several factors besides plant signals: acidic pH (Stachel and Zambryski, 1986; Bolton et al., 1986; Rogowsky et al., 1987; Veluthambi et al., 1987), temperature below 28°C, and sucrose or another easily metabolized sugar or glycerol (Alt-Morbe et al., 1988).

The induction of the virulence functions requires the protein products of *virA* and *virG* (Stachel and Zambryski, 1986; Winans et al., 1986; Melchers et al., 1986, 1987; Leroux et al., 1987; Rogowsky et al., 1987). *virA* and *virG* are regulatory genes that belong to a class of two-component regulatory systems used by many bacteria to sense and respond to alterations in environmental conditions. Both these genes are required for the induction of all *vir genes* (Stachel and Zambryski, 1986; Rogowsky et al., 1987). VirA functions as an environmental sensor and VirG functions as a regulator that controls cellular functions, usually transcription activation (Parkinson and Kofoid, 1992). The VirA/VirG pair serve to transcriptionally activate subject operons in response to specific environmental stimuli (Stachel and Nester, 1986; Albright et al., 1989; Gross et al., 1989; Stock et al., 1989). The VirA/VirG regulatory system activates *vir* regulon transcription in response to specific plant phenolic metabolites (Stachel et al., 1985), plant-derived monosaccharides and derivatives (Ankenbauer and Nester, 1990; Cangelosi et al., 1990; Shimoda et al., 1990), and acidic conditions (Stachel et al., 1986; Winans et al., 1988).

VirA is a 91 kDa protein, localized in the membrane (Leroux et al., 1987; Melchers et al., 1987) and is thus the most direct sensor for environmental conditions. It has the crucial role in sensing the plant phenolic-inducing stimuli. VirA is a transmembrane protein with the region between the two hydrophobic sequences residing in the periplasmic space and C-terminal half within cytoplasm (Winans et al., 1989). The C-terminal half of the cytoplasmic portion of the VirA molecule has an ATP-specific autophosphorylating activity (Jin et al., 1990b). The VirA protein autophosphorylates at histidine-474 within the cytoplasmic domain (Huang et al., 1990b; Jin et al., 1990b) and then transfers this phosphate group directly to VirG (Jin et al., 1990a), a

DNA-binding protein that subsequently activates transcription (Jin et al., 1990c; Pazour and Das, 1990). The sensor, VirA, contains at least four domains: periplasmic, linker, kinase, and a C-terminal response regulator homologous domain (Chang and Winans, 1992). The periplasmic domain is responsible for sugar-mediated sensitization of the bacteria to a low concentration of the inducer, and the kinase is the site of autophosphorylation (Cangelosi et al., 1990; Shimoda et al., 1990; Jin et al., 1990b; Huang et al., 1990a; Banta et al., 1994). The extreme C-terminal domain is believed to function as an autoinhibitory domain (Pazour et al., 1991; Chang and Winans, 1992). VirG consists of at least two domains: the N-terminal signal-receiving domain and the C-terminal DNA-binding domain (Winans et al., 1986; Powell and Kado, 1990). The N-terminal domain contains a conserved aspartic acid at position 52 that is phosphorylated by phospho-VirA (Jin et al., 1990a). One of the *vir* gene-inducing compounds identified in plants are phenolic compounds (Stachel et al., 1985; Hohn et al., 1989; Zambryski, 1992). Different phenolic compounds may have different abilities to induce *vir* genes. In *A. tumefaciens* strain KU12, *vir* genes were not induced by phenolic compounds containing 4'-hydroxy, 3'-methoxy, and 5'-methoxy groups such as acetosyringone (Lee et al., 1995). But these phenolic compounds induced *vir* genes of three other strains, C58, A6, and BO542. On the other hand, *vir* genes of strain KU12 were induced by phenolic compounds containing only a 4'-hydroxy group, such as 4'-hydroxy-acetophenone, which did not induce *vir* genes of the other three strains. The *vir* genes of strains KU12, A6, and BO542 were all induced by phenolic compounds containing 4'-hydroxy and 3'-methoxy groups, such as acetovanillone. *virA* locus has been shown to determine which phenolic compounds can function as *vir* gene inducers. VirA protein may directly sense the phenolic compounds for *vir* gene activation (Lee et al., 1995). The protein that binds phenolic compounds has been purified from *A. tumefaciens*. It is a 39 kDa protein with a PI of 4.3. It requires an acidic pH for optimal activity (Kemner et al., 1997).

VirA contains both a two-component kinase module and, at its carboxyl terminus, a receiver module (Chang et al., 1996). This receiver module has been shown to inhibit the activity of the kinase module. Overexpressing the receiver module in *trans* can restore low-level basal activity to a VirA mutant protein lacking the receiver module. Ablation of the receiver module restores activity to the inactive VirA mutant, which has a deletion within a region designated the linker module. This indicates that deletion of the linker module does not denature the kinase module but rather locks the kinase into a phenotypically inactive conformation, and this activity requires a receiver module. Removal of this module causes otherwise nonstimulatory phenolic compounds such as 4'-hydroxy-acetophenone to stimulate *vir* gene expression. It suggests that the receiver module restricts the variety of phenolic compounds that have stimulatory activity.

The periplasmic portion of the VirA protein (region I) is suggested to be the major signal sensing region. However, this region does not sense phenolic compounds (Melchers et al., 1989). This region senses sugar-binding protein ChvE (Cangelosi et al., 1990; Shimoda et al., 1990, 1993). The *chvE* locus is in the bacterial chromosome, and it encodes a sugar-binding protein involved in monosaccharide signaling to the VirA molecule (Huang et al., 1990a).

ChvE is a virulence determinant. It is a periplasmic binding protein that participates in virulence gene induction in response to monosaccharides, which occur in a plant wound environment. The region downstream of the *A. tumefaciens chvE* gene has been cloned and sequenced (Kemner et al., 1997). It revealed the presence of three open reading frames. The first two, together with *chvE,* encoded putative proteins of a periplasmic binding protein-dependent sugar uptake system or ABC-type (ATP binding cassette) transporter. The third open reading frame encoded a protein of unknown function. The deduced transporter gene products were related to bacterial sugar transporters and probably functioned in glucose and galactose uptake (Kemner et al., 1997). These genes were named *ggu-A, ggu-B,* and *ggu-C* for glucose galactose uptake (Kemner et al., 1997). Sugar-bound ChvE may interact with region I of the VirA molecule to make it more sensitive to phenolic compounds (Cangelosi et al., 1990; Shimoda et al., 1990). *Agrobacterium* strains in which *chvE* is nonfunctional are defective in *vir* gene induction (Garfinkel and Nester, 1980; Huang et al., 1990a; Belanger et al., 1997; Heath et al., 1997). It suggests the significant role of monosaccharides in potentiating the *vir* induction.

The periplasmic region of VirA may harbor a repressive function that is modulated by the binding Chv::monosaccharide (Cangelosi et al., 1990; Shimoda et al., 1990; Turk et al., 1993; Banta et al., 1994). The repressive function may be distributed through the periplasmic region and exert its effect in the absence of Chv:: monosaccharide. The majority of the repressive function appears to be concentrated in the N-terminal half of the periplasmic domain (Heath et al., 1997). When the N-terminal half of the periplasmic region is deleted, the repressive effect of the periplasmic region is completely released, allowing *vir* induction and host infection in the absence of ChvE (Heath et al., 1997). Region II of the VirA protein, the N-terminal portion of the cytoplasmic domain, affects the C-terminal kinase activity of the VirA protein. As a domain-connecting bridge, this region is involved in signal transduction from the periplasmic to the kinase domain. It may also be involved in detecting acetosyringone (Jin et al., 1990b).

VirA contains a 117 amino acid residue segment in its extreme C-terminus that is homologous to the N-terminal signal-receiving domain of VirG (Gubba et al., 1995). To study the effect of this domain on *vir* gene expression, Gubba et al. (1995) used site-specific mutagenesis to construct a deletion derivative of VirA, VirAρ, that lacks residues 712 to 828. In the presence of Vir A, expression of *virB* requires the inducer acetosyringone. However, when VirA was substituted expression was observed in the absence of acetosyringone and addition of acetosyringone had little or no effect (Gubba et al., 1995). These results indicate that the deletion of the extreme C-terminal domain of *VirA* leads to a fully constitutive phenotype and this domain functions as a negative regulator of *vir* gene expression. Similar results have been reported by Chang and Winans (1992).

The C-terminal response regulator domain of VirA may function in the prevention of inducer-independent phosphorylation, probably by blocking access to the active site. A likely mode of action of the inducer (acetylsyringone) is to cause a conformational change that will destroy intramolecular interaction between the active site and the VirA C-terminal domain. This loss of interaction may lead to VirA autophosphorylation and subsequent phosphotransfer to VirG (Gubba et al., 1995). The conformational change induced by the inducer may result from its binding to VirA or may be mediated by other proteins (Lee et al., 1992).

These observations are in line with the reports that constitutive VirG mutations, which function independent of both VirA and the plant inducer acetosyringone, can be isolated (Han et al., 1992; Jin et al., 1993; Scheeren-Groot et al., 1994; McLean et al., 1994; Gubba et al., 1995). *virA* mutants that express *vir* genes at high levels in the absence of acetosyringone have been isolated (Ankenbauer et al., 1991). One of the *virA* mutants which was relatively insensitive to acetosyringone contained a single amino acid change in the conserved domain of VirA at a glycine residue near the site of VirA autophosphorylation (Ankenbauer et al., 1991).

Host environmental conditions alter expression of *vir* genes. A high level of *vir* gene induction is obtained at a pH less than 6.0 and a temperature less than 30°C. Very little *vir* gene induction occurred at a pH greater than 6.5 either with (Stachel et al., 1986a; Rogowsky et al., 1987; Winans et al., 1988) or without (Winans, 1990) acetosyringone. Possible explanations for this acidic pH requirement include the inducibility of *virG* by acidic media (Veluthambi et al., 1987; Winans et al., 1988) and maintenance of an acidic conformation of VirA in acidic media which affects the VirA periplasmic domain (Melchers et al., 1989a). Although *virG* expression can be enhanced by acidic pH in the absence of acetosyringone, a number of studies have implicated the VirA protein in the response to low pH in the presence of the inducer (Melchers et al., 1989a; Winans et al., 1989; Jin et al., 1990a; Huang et al., 1990b). Turk et al. (1991) showed that differences in pH sensitivity for *vir* gene expression in octopine-type and nopaline-type *A. tumefaciens* mainly result from differences in their *virA* genes.

Winans et al. (1988) isolated a transposon-generated mutation in the chromosomal gene *chvD* that reduced induction of *virG* upon shift to acidic medium. Neither removal of the periplasmic domain of *virG,* nor mutation of *ChvD* altered the acidic pH optimum for *vir* gene induction (Winans et al., 1988; Melchers et al., 1989a). *virG* is expressed from two promoters (Winans , 1990). Promoter 2 of *virG,* which is activated by low pH in the absence of acetosyringone, is not likely to be solely responsible for the acidic pH requirement for acetosyringone induction of the *vir* genes, because replacement of promoter 2 by a *lacZYA* promoter did not relieve the low pH requirement for *vir* gene induction (Chen and Winans, 1991).

Importance of *virG* in induction of *vir* genes in response to acetosyringone has been demonstrated by inducing multiple (5 to 10) copies of *virG* in *A. tumefaciens* (Liu et al., 1993). The induction of *virB, virD, virE,* and *virG* in a rich medium was greater at alkaline pH than at acidic pH when multiple copies of octopine and agropine-type *virG* genes were present (Liu et al., 1993). Increased copies of a *virG* can confer increased virulence upon the bacterium. Multiple copies of a *virG* gene plus the 3' end of the *virB* gene conferred a supervirulence phenotype on *A. tumefaciens* A348, a non-supervirulent octopine-type *A. tumefaciens* strain (Komari et al., 1986). This enhanced virulence was correlated with an increased expression of *vir* genes (especially that of *virG*). In addition, the virulence of *A. tumefaciens* A281 could be further enhanced by the addition of multiple copies of the entire *virB* and *virG* operons (Jin et al., 1987). The presence of multiple copies of *virG* in *A. tumefaciens* can alter the pH-sensitivity profile of gene induction, suggesting that *virG,* as well as *virA,* may play a role in the pH response to plant signal molecules (Liu et al., 1993).

The plant phenolic signal molecule acetosyringone also induces other *vir* genes of *A. tumefaciens*. The processing of T-DNA from the Ti plasmid is carried out by products of the genes *virD1* and *VirD2* (Yanofsky et al., 1986; Jayaswal et al., 1987; Stachel et al., 1987; Veluthambi et al., 1987, 1988; Albright et al., 1987). Mutations in VirD1 and VirD2 block T-strand production (Stachel et al., 1987). The other *vir* gene, *virE,* is involved in a later step of T-strand transfer (Citovsky et al., 1988). The expression of *virC* increases the levels of T-strand produced (De Vos and Zambryski, 1989).

Transfer of DNA from *A. tumefaciens* into plant cells requires the activities of *virB* genes, *virD2,* and *virE2* genes (Binns et al., 1995). The putative transferred intermediate is a single-stranded DNA (T strand) covalently attached to the VirD2 protein and coated with the single-stranded DNA-binding protein VirE2. The movement of this intermediate out of *Agrobacterium* cells and into plant cells requires the expression of the *virB* operon, which encodes 11 proteins that localize to the membrane system. All the *virB* genes tested are required for the movement of VirE2 (Binns et al., 1995). The 11 predicted gene products of the *A. tumefaciens virB* operon are believed to form a transmembrane pore complex through which T-DNA export occurs. The observed tight associations of VirB9, VirB10, and VirB11 with the membrane fraction support the notion that these proteins may exist as components of multiprotein pore complexes, perhaps spanning both the inner and outer members of *Agrobacterium* cells (Finberg et al., 1995).

All these *vir* genes are induced by acetosyringone (Vernade et al., 1988; Alt-Morbe et al., 1989), probably by the action of the regulatory genes *virA* and *virG*. The transmembrane sensor protein VirA activates VirG in response to high levels of acetosyringone. To respond to low levels of acetosyringone, VirA requires the periplasmic sugar-binding protein Chv and monosaccharides to be released from plant wound sites (Doty et al., 1996).

Another regulatory gene induced by plant signals in *A. tumefaciens* is *ros* (rough outer space) (Close et al., 1985). Mutations in this gene caused elevated levels of transcription of *virC* and *virD*, two loci that augment tumorigenesis on dicotyledonous plants, indicating that *ros* was a negatively acting regulatory gene (Cooley et al., 1991). Using *virC*- and *virD-lacZ* fusions, it was shown that *ros* repressed their transcription and was also autoregulatory, repressing its own transcription (Cooley et al., 1991). The Ros protein bound specifically to a region of dyad symmetry that precedes *virC* and *virD* (D'Souza-Ault et al., 1993). Glucose and iron increased *ros*-mediated repression of transcription (Brightwell et al., 1995).

Another gene, *chvD*, is involved in *virG* expression in response to environmental stress (Winans et al., 1988). The *chvE* locus encodes a sugar-binding protein involved in monosaccharide signaling to the VirA molecule (Huang et al., 1990a). The periplasmic domain of VirA is required for sensing inducing monosaccharides via the ChvE protein, but not for the detection of acetosyringone (Cangelosi et al., 1990; Shimoda et al., 1990). *chvA* and *chvB* may be involved in attachment of *A. tumefaciens* to host cells (Douglas et al., 1985; Altabe et al., 1990; Raina et al., 1995). Other genes of the *A. tumefaciens* chromosome, such as *ivr* and *miaA*, also influence the induction of *vir* genes by acetosyringone (Close et al., 1985; Huang et al., 1990a; Metts et al., 1991; Gray et al., 1992). Thus several genes in *A. tumefaciens* are induced by plant signal molecules, and these genes contribute to recognition of host and pathogen.

The pathogenicity genes *wtsA* and *wtsB* of *Pantoea stewartii* are induced in plants (Coplin et al., 1992). In minimal medium both *wtsA* and *wtsB* were not expressed, but both genes were expressed in the medium amended with 0.1 percent casamino acids or yeast extract or tryptone (Table 1.7; Coplin et al., 1992).

TABLE 1.7. Expression of *wts-lacZ* Gene Fusions in *Pantoea stewartii* Grown in Different Culture Media

	β-Galactosidase (in units)	
Medium	**WtsA : : lacZ**	**WtsB : : lacZ**
DBG (minimal culture medium)	22	9
DBG + casamino acids	1829	1063
DBG + yeast extract	1555	1184
DBG + tryptone	1890	1102

Source: Coplin et al., 1992.

Note: β-Galactosidase levels for the controls were ≤30 units in liquid medium.

SOME PLANT SIGNALS MAY DIRECT SYNTHESIS OF ELICITORS

Some avr Genes Encode Elicitors

Elicitors are signal molecules of pathogens, which trigger defense mechanisms of the host. While a number of elicitors have been isolated from fungal pathogens (Vidhyasekaran, 1997), the research for bacterial elicitors of defense mechanisms has succeeded only in a few cases. Cell-free preparations of bacteria generally lack elicitor activity (Lyon and Wood, 1976). The avirulence genes *avrB* and *avrC* from race O of *P. savastanoi* pv. *glycinea* were found to encode single protein products of 36 and 39kDa, respectively (Tamaki et al., 1988). The *avrC* protein was overproduced in *Escherichia coli* cells and deposited as insoluble inclusion bodies in the cell cytoplasm. The *avrC* protein could be solubilized with urea-octylglucoside treatment, but neither the solubilized protein nor the intact inclusion bodies elicited hypersensitive reaction in soybean leaves (Tamaki et al., 1988). Hence, the *avrC* protein product may not function as an elicitor in the soybean *per se*. The protein product of *avrD* gene of *P. savastanoi* pv. *glycinea* does not elicit the HR when injected into leaves of the appropriate cultivar of soybean (Keen et al., 1990). De novo RNA and protein synthesis by the invading bacterium in the host is required to elicit the HR (Klement, 1982; Sasser, 1982). This requirement during the induction stage, together with the inability to detect elicitor in filtrates or lysates of bacteria grown in culture, suggests that one or more bacterial gene induction events are necessary before elicitor is formed and HR is triggered (Hutcheson et al., 1989). The induction stage may result from the derepression of *hrp* and *avr* genes, which are responsible for induction of resistance and probably susceptibility also.

The first bacterial elicitor has been isolated from *Pseudomonas syringae* (Keen et al., 1990). Although *avrD* gene was first detected in *P. syringae* pv. *tomato,* several *P. syringae* and *P. savastanoi* pathovars harbor DNA sequences with considerable similarity to *avrD* (Kobayashi et al., 1990a). The presence of *avrD* in a collection of *P. syringae* pathovars was tested by performing Southern blots using an *avrD*-specific probe (Yucel et al., 1994a). The *avrD* could be detected in *P. savastanoi* pv. *glycinea, P. savastanoi* pv. *phaseolicola,* and *P. syringae* pv. *lachrymans* (Yucel and Keen, 1994; Keith et al., 1997). All *avrD* alleles encoded proteins with 311 amino acids, precisely same as the previously sequenced *avrD* gene from *P. syringae* pv. *tomato* (Yucel et al., 1994a).

P. syringae pv. *lachrymans* carried two *avrD* alleles on plasmids of different sizes. Allele1 from *P. syringae* pv. *lachrymans* was 95 percent identical to *avrD* from *P. syringae* pv. *tomato* at the amino acid level. The allele2 was 98 percent identical with the *P. savastanoi* pv. *glycinea* allele while the *avrD* gene from *P. savastanoi* pv. *phaseolicola* was 97 percent identical with the *P. savastanoi* pv. *glycinea* allele at the aminoacid level (Yucel et al., 1994a). The *avrD* allele from *P. savastanoi* pv. *glycinea* encodes a protein that has 86 percent amino acid identity with the *avrD* protein from *P. syringae* pv. *tomato* (Kobayashi et al., 1990b). The *avrD*-encoded proteins do not show any elicitor activity. *Escherichia coli* cells carrying constructs of pAVRD produced large amounts of the *avrD*-encoded protein (Keen et al., 1990). This protein was devoid of elicitor activity. Partially purified preparations of the *avrD*-encoded protein from lysed *E. coli* cells also were devoid of any detectable activity (Keen et al., 1990). But culture fluids from *E. coli* expressing *avrD* grown on M9 glucose minimal medium caused necrosis (HR reaction) only on soybean cultivars that were resistant to *P. savastanoi* pv. *glycinea* race 4 carrying the cloned *avrD* gene (Keen et al., 1990). Thus the *avr*-encoded proteins may not be elicitors themselves, but possess enzymatic functions leading to bacterial metabolites that are the actual elicitors.

E. coli cells carrying plasmid constructs of *avrD* caused pronounced HR when infiltrated into primary leaves of only those soybean cultivars previously observed to react hypersensitively to *P. savastanoi* pv. *glycinea* race 4 carrying *avrD* (Keen et al., 1990). The elicitor was purified from *E. coli* cells expressing the *avrD* gene. It was a low molecular weight substance and was active even in the nanomolar range (Keen et al., 1990).

AvrD proteins of *P. syringae* pv. *tomato* and *P. syringae* pv. *lachrymans* allele1 yielded products different from AvrD proteins of *P. savastanoi* pv.

phaseolicola (Yucel et al., 1994b). *E. coli* cells expressing *P. syringae* pv. *tomato* and *P. syringae* pv. *lachrymans* allele 1 *avrD* produced two major elicitor-active acyl glycosides called syringolides 1 and 2 (Keen et al., 1990; Kobayashi et al., 1990a,b; Keen and Buzzell, 1991; Midland et al., 1993; Smith et al., 1993) and they are structurally identical, except that syringolide 1 has an alkyl chain that is two carbons shorter. *E. coli* cells expressing the *avrD* allele from *P. savastanoi* pv. *phaseolicola* yielded only syringolide 1, with no trace of syringolide 2. Two unidentified compounds were also produced by bacteria expressing the *P. savastanoi* pv. *phaseolicola* allele but not the *P. syringae* pv. *tomato* allele (Yucel et al., 1994b). The results suggest that *avrD* protein products may have enzymatic functions that lead to the elicitor-active syringolides as well as structurally related compounds of lesser activities.

The functional *avrD* protein directing synthesis of elicitors has been characterized (Yucel and Keen, 1994). The *avrD* protein from *P. savastanoi* pv. *glycinea* directed the production of extremely small quantities of elicitor when overexpressed in *E. coli* (Keen et al., 1990) while all other *avrD* alleles from *P. syringae* pv. *tomato, P. savastanoi* pv. *phaseolicola,* and *P. syringae* pv. *lachrymans* directed different elicitors when expressed in gram-negative bacteria (Yucel et al., 1994b). Comparison of the nonfunctional *avrD* protein of *P. savastanoi* pv. *glycinea* with the products of four other functional alleles (from *P. syringae* pv. *tomato, P. savastanoi* pv. *phaseolicola,* and *P. syringae* pv. *lachrymans* alleles 1 and 2) identified several amino acids that were unique to the pv. *glycinea* protein (Table 1.8; Yucel and Keen, 1994).

TABLE 1.8. Divergent Amino Acids Found in the Nonfunctional *Pseudomonas savastanoi* pv. *glycinea* avrD Protein in Relation to Other Four Functional Class Proteins

	Amino acid positions			
AvrD allele	**19**	**245**	**280**	**304**
P. savastanoi pv. *glycinea*	Cysteine	Leucine	Alanine	Leucine
P. savastanoi pv. *phaseolicola*	Arginine	Serine	Valine	Serine
P. syringae pv. *tomato*	Arginine	Serine	Valine	Serine
P. syringae pv. *lachrymans 1*	Arginine	Serine	Valine	Serine
P. syringae pv. *lachrymans 2*	Arginine	Serine	Valine	Serine

Source: Yucel and Keen, 1994.

The nonfunctional *avrD* allele of *P. savastanoi* pv. *glycinea* has five unique amino acid substitutions. Oligonucleotide site-directed mutagenesis and recombinant gene constructions were used to determine the role of these amino acids in the dysfunction of *P. savastanoi* pv. *glycinea* AvrD (Yucel and Keen, 1994). The avrD protein of *P. savastanoi* pv. *glycinea* was restored to partial function by oligomutagenesis to create arginine 19, valine 280, and serine 304. A fourth substitution, serine 245, did not affect protein function when in the context of the other three mutations. The valine 280 and serine 304 substitutions did not restore any avrD activity by themselves but, in combination with arginine 19, enabled the protein to direct low but easily detectable levels of elicitor production. The arginine residue was absolutely required for avrD activity since solely altering this residue totally inactivated the avrD protein of *P. savastanoi* pv. *phaseolicola* (Yucel and Keen, 1994). Arginine 19 in the functional avrD may be required for the putative enzymatic activity of the avrD protein (Yucel et al., 1994a).

Two heat-stable proteins, PopA1 and its derivative PopA3, were characterized from the supernatant of *Ralstonia solanacearum* strain GMI 1000 (Arlat et al., 1994). They had HR-like elicitor activities on tobacco (nonhost plant) but no such activity on tomato (host plant). *Petunia* lines responsive to PopA3 and its precursors were resistant to infection by strain GMI 1000, suggesting that *popA* could be an avirulence gene. A *PopA* mutant remained fully pathogenic on sensitive plants, indicating that this gene is not essential for pathogenicity (Arlat et al., 1994).

Elicitors produced by other *avr* genes have not yet been identified. *avrPphE* from *P. savastanoi* pv. *phaseolicola* encodes a hydrophilic protein, and its hydrophilic nature implies cytoplasmic location. Mansfield et al. (1994) suggested that the protein may be enzymic and may be involved in the synthesis of an elicitor.

Some hrp Genes Also Direct Synthesis of Elicitors

Some *hrp* genes are also involved in synthesis of elicitors. The first *hrp*-encoded elicitor characterized is harpin$_{\mathrm{Ea}}$ from *Erwinia amylovora* (Wei et al., 1992a). Mutations in the encoding *hrpN* gene revealed that harpin is required for *E. amylovora* to elicit the HR in nonhost tobacco leaves (Wei et al., 1992a). Another *hrp*-encoded elicitor, Hrp$_{\mathrm{Ech}}$, has been isolated from *Erwinia chrysanthemi*. The pathogen has a wide host range and does not elicit HR. However, mutants of *E. chrysanthemi* lacking pectic enzyme secretion pathway cause a typical HR (Bauer et al., 1994). Elicitation of the HR by *E. chrysanthemi* is dependent on an *hrp* gene *(hrpN$_{Ech}$)* (Bauer et al., 1994). *Echerichia coli* cells expressing *hrpN$_{Ech}$* accumulated HrpN$_{\mathrm{Ech}}$ in inclusion bodies. The protein could be readily purified from cell lysates carrying these inclusion bodies. $HrpN_{Ech}$ suspensions elicited a typical HR in tobacco leaves (Bauer et al., 1995). Comparison of the amino acid sequences of the predicted *hrp*N_{Ea} and $hrpN_{Ech}$ products revealed extensive similarity, particularly in the C-terminal halves of the proteins. The overall identity of the *hrpN* genes and proteins was 66.9 percent (Bauer et al., 1995).

Erwinia carotovora subsp. *carotovora* strain Ecc71 possesses a homlog of *E. chrysanthemi hrpN* known to encode an elicitor of the HR; the corresponding Ecc71 gene was designated $hrpN_{Ecc}$ (Cui et al., 1996). The $hrpN_{Ecc}$ is predicted to encode a glycine-rich protein of approximately 36 kDa (Mukherjee et al., 1997). In *E. coli* strains overexpressing $hrpN_{Ecc}$, the 36 kDa protein has been identified as the $hrpN_{Ecc}$ product. The 36 kDa protein fractionated from *E. coli* elicits the HR in tobacco leaves. The elicitor has been designated Harpin $_{Ecc}$ (Mukherjee et al., 1997). $Harpin_{Ecc}$ has sequence homology with other *Erwinia* harpins, such as those produced by *E. amylovora* and *E. chrysanthemi* (Mukherjee et al., 1997). Another harpin has been detected in *Pantoea stewartii,* the maize pathogen (Ahmad et al., 1996).

Similar elicitors have been isolated from *Pseudomonas syringae* pv. *syringae* (bean pathogen), *P. savastanoi* pv. *glycinea* (soybean pathogen), and *P. syringae* pv. *tomato* (tomato pathogen) (He et al., 1993; Preston et al., 1995). Harpins produced by *Pseudomonas syringae* and *P. savastanoi* pathovars are encoded by the *hrpZ* gene (He et al., 1993). Preston et al. (1995) used an internal fragment of the *P. syringae* pv. *syringae hrpZ* gene to clone the *hrpZ* locus from *P. savastanoi* pv. *glycinea* and *P. syringae* pv. *tomato*. DNA sequence analysis revealed that *hrpZ* is the second ORF in a polycistronic operon.

Harpins produced by *P. syringae* pathovars *syringae* and *tomato* and *P. savastanoi* pv. *glycinea* have been designated $HrpZ_{Pss}$, $HrpZ_{Pst}$, and $HrpZ_{Psg}$, respectively. A comparison of the sequences of the three HrpZ proteins with each other and with HR elicitors from other bacteria indicates that the HrpZ proteins represent a distinct family of elicitors that is conserved among *P. syringae* and *P. savastanoi* pathovars (Preston et al., 1995). The amino acid sequences of the three proteins are sufficiently similar to reveal their relatedness, but they show no significant relatedness to elicitor proteins from other bacteria. $HrpZ_{Pss}$, $HrpZ_{Psg}$, and $HrpZ_{Pst}$ are indistinguishable in several biological and physical properties. They have the same effect on different plants, and they are heat-stable, glycine-rich, and devoid of cysteine and tyrosine (Preston et al., 1995). He et al. (1993) demonstrated with Southern blots and immunoblots that several pathovars of *P. syringae* and *P. savastanoi* contain homologs of the *hrpZ* gene from *P. syringae* pv. *syringae*. It suggests that HrpZ represents a family of elicitor common to all pathogenic strains of *P. syringae* and *P. savastanoi*.

Both *hrpN* and *hrpZ* lie contiguous or within the *hrp* cluster of *E. amylovora* and *P. syringae* pv. *syringae* (Wei et al., 1992; He et al., 1993). There are no genes downstream of the elicitor gene in the *hrpN* operon. In contrast, *hrpZ* lies upstream of at least one other *hrp* gene within an operon (Huang et al., 1991; Xiao et al., 1992). Harpins from *Erwinia* spp., *Pseudomonas syringae,* and *P. savastanoi* appear to be two different classes of harpins. None of the *Erwinia* harpins has greater than 10 percent identity with the *P. syringae* harpins. This contrasts with identities of 30 percent or more between the *Erwinia* harpins. In addition, while the *Erwinia hrpN* genes are organized as monocitronic operons, those of *P. syringae* and *P. savastanoi* are the components of polycistronic operons (Mukherjee et al., 1997).

Some harpin fragments also possess elicitor activity. $\text{Harpin}_{\text{Pss}}$ fragments possessed elicitor activity (He et al., 1993). All fragments of $\text{harpin}_{\text{Pss}}$ elicited the HR in tobacco leaves (Alfano et al., 1996). Elicitor activity was not associated with any consensus sequence (Alfano et al., 1996). Besides harpins, some other elicitors may also be produced by *Pseudomonas syringae.* Although harpin P_{ss} has been reported to function as an elicitor of HR in tobacco, the deletion analyses revealed that harpin production is not required for the initiation of the HR in tobacco (Pirhonen et al., 1996). Transformants expressing *hrp* clusters in which the $hrpZ_2$ gene had been specifically deleted elicited a clearly detectable HR in tobacco, *Arabidopsis thaliana* accession Col-O, or soybean cv. Merit tissue (Pirhonen et al., 1996). It may be that other HR elicitors active in tobacco are encoded by the *P. syringae* pv. *syringae* strain 61 *hrp/hrm* gene cluster (Alfano et al., 1996).

Dual Functions of Harpins

As *hrp* genes are required for pathogenicity and disease resistance, the elicitor harpins produced by the bacterial pathogens may also have dual functions. An hrpN_{Ech} :: Tn5-*gusA1* mutation reduced the ability of a strain of *Erwinia chrysanthemi* to initiate lesions in susceptible chicory *(Cichorinum intybus)* leaves, but did not reduce the size of lesions that did develop (Collmer and Bauer, 1994; Bauer et al., 1995). It suggests that the elicitor $HrpN_{Ech}$ contributes specifically an early stage of pathogenesis.

Mutations in the *hrpN* gene which encodes the harpin, HrpN_{Ea}, revealed that harpin is required for *Erwinia amylovora* to elicit the HR in nonhost tobacco leaves and to incite disease symptoms in highly susceptible pear fruit

(Wei et al., 1992b). These studies indicate that the elicitors encoded by *hrp* genes may induce HR in nonhosts and disease symptoms in hosts.

Other Elicitors

The *hrp* or *avr* genes have not yet been detected in gram-positive bacterial pathogens such as *Clavibacter michiganensis,* probably due to lack of basic tools such as marker exchange methods (Nissinen et al., 1997). Concentrated cell-free culture supernatants from virulent strains of *Clavibacter michiganensis* subsp. *sepedonicus,* the potato pathogen, caused a necrotic reaction on tobacco, whereas cell-free culture suspensions from nonpathogenic strains did not (Nissinen et al., 1997). The elicitor preparations treated with proteinase K no longer elicited a necrotic reaction on tobacco. It suggests that the elicitor may be a protein (Nissinen et al., 1997). Coinfiltration of bacteria with various eukaryotic metabolic inhibitors (α-amanitin-RNA polymerase II inhibitor; cycloheximide-80S ribosome inhibitor; vanadate-ATPase and phosphatase inhibitor; lanthanum-calcium channel blocker) suggested that HR is initiated by the plant and requires active plant metabolism. The elicitor infiltrated into tobacco leaves along with metabolic inhibitors did not trigger any response, whereas the untreated elicitor elicited a necrotic response. Neither bacterial cells nor elicitor gave a rapid necrotic response in host plants (potato and eggplant), although both induced a rapid necrotic response in tobacco (Nissinen et al., 1997). It suggests that the reaction induced by this elicitor preparation is an active plant defense response rather than a toxic response. Cell-free fluid of *Pseudomonas corrugata,* a pathogen of tomato, was capable of eliciting phytoalexin biosynthesis in ladino white clover cells (Gustine et al., 1990).

Specific Plant Signals May Regulate Elicitor Synthesis in Bacteria

Elicitors encoded by *avr* or *hrp* genes could not be isolated from pathogens when they were grown in a nutrient-rich medium. When *Pseudomonas syringae* pv. *tomato* was grown in a nutrient-rich medium, or even in minimal medium, no elicitor could be isolated (Keen et al., 1990). No elicitor could be isolated from *E. coli* carrying *avrD* gene when grown in nutrient-rich medium, but elicitor could be isolated when the bacteria were grown in M9 glucose minimal medium (Keen et al., 1990). *avrD* elicitor could be isolated from culture fluids of *P. syringae* pv. *tomato* with a cloned and overexpressed *avrD* gene (Keen et al., 1990). It suggests that when *P. syringae* pv. *tomato* cells were grown in culture the *avrD* gene is not fully expressed. Expression of *avr* genes increase approximately 100 times when the bacteria are inoculated into plants (Huynh et al., 1989; Keen et al., 1990). Hence, it is possible that elicitor would have been synthesized in planta, and plant signals may enhance expression of the *avr* genes and direct synthesis of elicitors.

All elicitors encoded by *hrp* genes also could be isolated only when the bacterial pathogens were grown in *hrp* gene-inducing minimal media (Preston et al., 1995). They could be isolated from the saprophytic bacterium *E. coli* overexpressing the *hrpN* or *hrpZ* genes, which encoded the harpins (Preston et al., 1995). It suggests that elicitor synthesis is controlled by specific nutrients, however, factors responsible for overexpression of the encoding genes in plants due to specific plant signals directing synthesis of elicitors have not yet been identified.

Synthesis of elicitors in planta has been reported in potato–*Ralstonia solanacearum* interaction (Huang et al., 1989). When cocultivated with incompatible strains of *R. solanacearum,* potato cells died rapidly. No significant death of cells was observed in the compatible combinations. Supernatants from cultures involving incompatible and compatible combinations of bacteria with potato cells were tested for biological activity in the callus assay. No biological activity was detected in the supernatant of bacteria grown in the absence of plant tissues. The factor responsible for the biological activity was identified as a 60-kDa protein. When infiltrated into leaves of *Solanum phureja* (resistant to *R. solanacearum*), the purified 60-kDa protein caused HR within 24 h (Huang et al., 1989).

SECRETION OF ELICITORS FROM BACTERIAL CELLS IN PLANTS

Types of Secretory Pathways

Bacterial plant pathogens produce various proteins due to expression of their *hrp* and *avr* genes. These molecules should be secreted outside of the bacterial cell to activate defense genes of the host or cause disease symptoms. Different types of secretion pathways exist in bacterial cells. Bacteria have two membranes, an inner/cytoplasmic membrane and an outer membrane. Movement of a protein from the cytoplasm across the inner membrane to the periplasm is called *export* while translocation of a protein to the extracellular environment is termed *secretion* (Salmond, 1994).

The proteins produced by bacterial pathogens are secreted by three different pathways. Many bacteria have a type I pathway for the secretion of exoenzymes and toxins which are needed for their virulence. The proteins secreted via this pathway have no classical N-terminal signal sequences that are normally required for export of a protein via the inner membrane Sec or "translocase" complex to the periplasmic space. These proteins do not seem to exist as free periplasmic intermediates. Thus, the proteins are transported from the bacteria cytoplasm directly to the extracellular environment through some form of channel or gated pore formed between the inner and outer membranes. This one-step secretion of proteins requires three accessory proteins, which are membrane-associated. One of these accessory proteins is an ABC protein (containing the *A*TP-*B*inding *C*assette motif) and probably an inner membrane "traffic ATPase" that may hydrolyze ATP to provide energy for some aspect of the translocation process (Wandersman, 1992; Salmond and Reeves, 1993; Pugsley, 1993; Van Gijsegem et al., 1993). This pathway is called a *sec*-independent, one-step, type I pathway.

The second protein secretion pathway is a *sec*-dependent, two-step, general secretory, Out or type II pathway. Pectic enzymes and cellulases of the bacterial pathogens are secreted via this pathway. All proteins targeted through the type II pathway are synthesized with classical N-terminal signal sequences and, when exported, are processed by signal peptidase en route to the periplasm. After export to the periplasm, the proteins may exist as free intermediates. This is the first step, which involves export to periplasm. The second step in the type II secretion pathway is movement of the periplasmic form of the proteins across the outer membrane and into the extracellular environment. The genes responsible for the second step are called *out* genes. Out proteins have been detected in many plant pathogenic bacteria. Out proteins are associated with the cytoplasmic membrane (Salmond, 1994).

The third pathway is novel, the type III secretory pathway which is a *sec*-independent pathway. The animal pathogens *Yersinia enterocolitica* and *Yersinia pestis* secrete several proteinaceous virulence factors, "Yops," to the extracellular environment by this novel pathway. Efficient Yop secretion requires various Ysc (*Y*op *sec*retion) proteins (Michiels and Cornelis, 1991). Most Ysc proteins are associated with the inner membrane (Van Gijsegem et al., 1993). The Yops have no N-terminal classical sequences. Although there is no processing of these proteins, some undefined targeting signals are present in the N-terminal sequences of Yops (Michiels and Cornelis, 1991).

Many hrp Genes Encode Components of Type III Protein Secretion Pathway

Most of the *hrp* genes encode components of a protein secretion pathway that is similar to the type III pathway used by the animal pathogenic bacteria *Yersinia, Shigella,* and *Salmonella* spp. to secrete extracellular proteins involved in animal pathogenesis (Van Gijsegem et al., 1994, 1995; Salmond, 1994). DNA sequences reported for many *hrp*/*hrmA* genes in *Pseudomonas syringae* pv. *syringae* reveal extensive homologies with proteins associated with type III protein secretion pathway in *Yersinia* (Huang et al., 1995).

The *hrpJ4* gene product has been deduced to be an ATPase associated with the inner membrane and is a necessary component of the *hrp*-encoded secretion system of *P. syringae* pv. *syringae* (Lidell and Hutcheson, 1994). *hrpH*[2] gene product is an outer membrane protein that is also essential for the *hrp*-encoded secretion system (Huang et al., 1992; He et al., 1993). These mutations have been shown to produce a secretion minus (Hsc^-) phenotype in *Escherichia coli* transformants (Huang et al., 1992; He et al., 1993; Lidell and Hutcheson, 1994). The HrpI product is a member of the *yersinia* LcrD superfamily of proteins (Huang et al., 1993). HrpJ2 also belongs to the LcrD family (Lidell and Hutcheson, 1994). Proteins in this group have properties of inner membrane proteins and are associated with a sec-independent protein secretion system (Pugsley et al., 1990; Galan et al., 1992; Salmond and Reeves, 1993). Some of the *P. syringae* pv. *syringae* 61 gene products were similar to proteins associated with flagellar biogenesis in *Salmonella typhimurium* and *Bacillus subtilis*. HrpJ2, HrpJ3, HrpJ4, and HrpJ5 of *P. syringae* pv. *syringae* 61 were similar to *Salmonella* proteins Flha, FliG, FliI, and FliJ, respectively. Although the similarities suggest a role for *hrp* genes in flagellar biogenesis, they are not required for export of flagellar components (Lidell and Hutcheson, 1994). The FliI family is involved in the secretion pathway. *hrpU2* also encodes secretory proteins (Lidell and Hutcheson, 1994). Hrp secretion system has been reported also in *P. syringae* pv. *tomato* (Jing and Yang, 1996). Sequences of the *Ralstonia solanacearum hrp* cluster revealed homologies with *Yersinia ysc* genes (Van Gijsegem et al., 1995). Hrp proteins of *R. solanacearum* and virulence determinants of *Yersinia enteolitica* and *Y. pestis* have remarkable sequence similarity (Michiels et al., 1991; Plano et al., 1991). *R. solanacearum hrpI* locus protein is homologous to the *Yersinia* YscJ protein, suggesting a role in the secretion system (Fenselau et al., 1992; Gough et al., 1992). Sequencing of *hrp* genes in *X. vesicatoria* also revealed homologies with components of the type III protein pathway in animal pathogens *Yersinia, Shigella,* and *Salmonella* (Fenselau et al., 1992; Gough et al., 1992; Huang et al., 1992). There were 40 to 50 percent identity and up to 70 percent similarity between Hrp proteins (HrpA1, HrpB3, HrpB6, HrpC2) encoded by *hrpA*, *hrpB,* and *hrpC* loci from *X. vesicatoria* and proteins from *Yersinia*, *Shigella*, and *Salmonella* (Fenselau et al., 1992).

HrpA1 encoded at the left end of the *hrp* gene cluster has been shown to be an outer membrane protein, possibly pore-forming, that belongs to the PulD superfamily of proteins involved in both type II and type III secretion

(Wengelnik et al., 1996a). HrpB3 is predicted to be lipoprotein (Fenselau and Bonas, 1995). HrpB6 could be the energizer of the secretion machinery (Fenselau and Bonas, 1995). HrpB6 is structurally and possibly functionally related to ATPases (Fenselau et al., 1992). HrpB6 may function as an ATPase that is related to a transport apparatus, rather than as part of a proton pump. A number of ATP-binding proteins are involved in specialized secretion systems, such as Prt in *Erwinia* and Hly in *Escherichia coli* (Higgins et al., 1986).

HrpC2 contains eight putative transmembrane domains, and is predicted to be an inner-membrane protein (Fenselau et al., 1992). HrpD1, HrpD2, and HrpD3 also show conservation with type III secretion systems (Huguet and Bonas, 1997). The presence of yet another *hrp* locus, *hrpF,* downstream of *hrpE* has been reported in *X. vesicatoria* (Bonas et al., 1991). The HrpF protein is predicted to be 806 amino acids long, and to have a molecular mass of 86.4 kDa. HrpF is predominantly hydrophobic, lacks a cleavable signal peptide, and contains two hydrophobic domains in its C-terminal region (Huguet and Bonas, 1997). So far no HrpF homologs have been identified in the other gram-negative phytopathogenic bacteria that contain conserved components of type III secretion systems (Huguet and Bonas, 1997). HrpF protein is needed for the *avr* product AvrBs3 recognition by plant cells (Knoop et al., 1991) and it is partially localized in the inner membrane. Hence HrpF may be a component of the Hrp translocation apparatus (Huguet and Bonas, 1997).

Synthesis and export of harpin production by *Erwinia amylovora* depends on a 40 kb chromosome (Bogdanove et al., 1996b). A 10.5 kb portion of the cluster required for harpin secretion contains 11 genes. They were named *hrpN, hrpO, hrpP, hrcQ, hrcS, hrcT, hrcU, hrcT, hrpI, hrcD,* and *hrpU* (Bogdanove et al., 1996 a,b). Regions homologous to *P. syringae* pv. *syringae61 hrpJ, hrpU,* and *hrpH* (which are involved in the secretion system) have been identified in the *Erwinia amylovora hrp* cluster (Laby and Beer, 1992; Wei and Beer, 1994). HrpA, HrpB, HrcJ, HrpD, HrpE, HrpF, and HrpC showed various degrees of similarity to corresponding proteins of *P. syringae* (Kim et al., 1997). HrpA from *R. solanacearum* (Gough et al., 1992) and HrpA1 from *X. vesicatoria* (Fenselau et al., 1992) are similar to the *P. syringae* pv. *syringae* HrpJ2 (Lidell and Hutcheson,1994). The *R. solanacearum* HrpE and the *X. vesicatoria* HrpB6 gene products share similarity to the *P. syringae* pv. *syringae* HrpJ4 product (Lidell and Hutcheson, 1994). HrpA, HrpI, and HrpO from *R. solanacearum* were also highly similar to the HrpA1, HrpB3, and HrpC2 proteins from *X. vesicatoria* (Lidell and Hutcheson, 1994).

Thus *hrp* genes in different bacterial pathogens are involved in protein secretion. They encode highly conserved components of a type III protein secretion system. These genes have been designated as *hrc* (*h*ypersensitive *r*esponse and *c*onserved) (Bogdanove et al., 1996a).

hrp Secretory Proteins Control Secretion of Elicitors

The hrp-encoded elicitors have been shown to require the activities of the *hrp* gene cluster. Mutation studies have revealed that the *hrp*-coded elicitors such as *E. amylovora* $HrpN_{Ea}$, *Pseudomonas syringae* pv. *syringae* HrpZ, and *R. solanacearum* PopA1 proteins are secreted by the Hrp secretion pathway (He et al., 1993; Wei and Beer, 1993; Arlat et al., 1994). Mutation of the *E. chrysanthemi* homolog of an *E. amylovora* gene involved in $HrpN_{Ea}$ secretion abolishes the ability of *E. chrysanthemi* to elicit the HR (Bauer et al., 1995). Since $HrpN_{Ech}$ appears to be the only HR elicitor produced by *E. chrysanthemi,* the effect of the *hrp* secretion gene mutation may be on Hrp_{Ech} (Bauer et al., 1995). There was little difference in the plant interaction phenotype of *E. chrysanthemi* mutants deficient in either $HrpN_{Ech}$ or a putative component of the Hrp secretion pathway (Bauer et al., 1994). Both mutations abolished the activity of the strains to elicit the HR in tobacco, and they both reduced the frequency of successful infections in witloof chicory leaves (Bauer et al., 1995). Attempts to restore the HR phenotype to *E. chrysanthemi* and *E. amylovora hrp* mutants with heterologous $hrpN^+$ subclones failed. Since the *hrp* genes in each subclone successfully complemented *hrpN* mutations in homologous bacteria and expressed in herterologous bacteria, the problem may be most likely the secretion of the harpins by heterologous Hrp systems (Bauer et al., 1995). Thus, the *hrp* secretion system seems to be important in expression of hrp-encoded elicitors.

E. coli cells carrying a *P. syringae* pv. *syringae* 61 *hrp* cluster were capable of eliciting the HR when physiologically intact, whereas *hrpJ4, hrpJ5,* and *hrpU2* mutants required treatments to permeabilize or lyse the cells before they elicited the HR (Lidell and Hutcheson, 1994). Since the ability to elicit the HR has been linked to extracellular $harpin_{Pss}$, these observations indicate that $harpin_{Pss}$ secretion is blocked in *hrpJ4, hrpJ5,* and *hrpU2* mutants. Thus, *hrpJ4, hrpJ5,* and *hrpU2* encode secretory proteins.

TnphoA insertions in *hrpC, hrpE, hrpW, hrpX,* and *hrpY* abolished the ability of *P. syringae* pv. *syringae* 61 to secrete the HrpZ ($harpin_{Pss}$). It indicates that several *hrp* genes may be involved in the secretion of HrpZ and possibly other proteins (Huang et al., 1995). Five transcription units of the *Ralstonia solanacearum hrp* gene cluster are required for the secretion of the HR-inducing PopA1 protein (Van Gijsegem et al., 1995). The total number of Hrp proteins encoded by these five transcription units is 20. Among them, eight belong to protein families involved in type III secretion pathways and in flagellum biogenesis, while two are related solely to proteins involved in secretion systems (Van Gijsegem et al., 1995). A 6.2 kb region of DNA corresponding to complementation groups II and III of the *Erwinia amylovora hrp* gene cluster has been shown to be required for secretion of harpin, an elicitor of the HR (Kim et al., 1997). The sequence of the region revealed 10 open reading frames in two putative transcription units: *hrpA, hrpB, hrcJ, hrpD,* and *hrpE* in the *hrpA* operon (group III) and *hrpF, hrpG, hrc C, hrpT,* and *hrpV* in the *hrpC* operon (group II) (Kim et al., 1997).

hrp Secretory Proteins May Allow Delivery of Different Bacterial Proteins into Plant Cells

Several Avr proteins from pathovars of *Pseudomonas* (AvrB, AvrPto, AvrRpt2) and from *Xanthomonas* (AvrBs3) are recognized inside plant cells (Gopalan et al., 1996; Leister et al., 1996; Scofield et al., 1996; Tang et al., 1996; Van den Ackerveken et al., 1996). Pirhonen et al. (1996) suggested that the *hrp*-dependent secretion system which appears to be functional in *E. coli* may allow delivery of the protein products of *avr* genes directly into plant cells, where they may themselves act as elicitors of the HR. Alternatively, the encoded Avr proteins may have a role in the biosynthesis of elicitors which require *hrp* functions for their delivery (Gopalan et al., 1996).

Protein production encoded by the avirulence gene *avrPphB (= avrPph3)* from *P. savastanoi* pv. *phaseolicola* was assessed by growing it in minimal medium (Puri et al., 1997). Two major peptides of 35 and 28 kDa were detected. The 35 kDa peptide was rapidly processed to the 28 kDa peptide. No macroscopic symptoms developed in bean leaf or pod tissues infiltrated with the peptide. Various extracts prepared from lysates or sonicates of bacterial cells containing high concentrations of the AvrPphB protein also failed to elicit the HR. But coordinated expression of the *hrp* cluster from *P. savastanoi* pv. *phaseolicola* and the avirulence genes *avrPphB* was found to allow an R gene-specific HR in bean. The presence of hrpL in *E. coli* strains harboring constructs containing *avrPphB* allowed induction of the HR in bean phenotype A43 and *Phaseolus acutifolius,* both of which carry the R2 gene, but not in other cultivars (Puri et al., 1997). The results support the proposal that *hrp* genes, many of which appear to be regulated by HrpL, have a role in delivery of signals to the plant.

The nonpathogenic bacteria *Pseudomonas fluorescens* and *Escherichia coli* elicit a genotype-specific hypersensitive response in plants if they possess both the Hrp secretion system and the HrpZ harpin from *P. syringae* pv. *syringae* 61 and a *P. syringae avr* gene whose presence is recognized by a corresponding resistance gene in the plant. Elicitation of a genotype-specific HR was observed with avrB$^+$ *P. fluorescens* in soybean and *Arabidopsis* plants carrying resistance genes *RPG1* and *RPM1,* respectively, and with avrPto$^+$ *E. coli* in tomato plants carrying resistance gene *Pto,* but only if the Hrp secretion system, HrpZ harpin, and the appropriate Avr proteins were produced in the same bacterial cell (Gopalan et al., 1996).

To assess the importance of the *hrp*-encoded secretion system in expression of the *avr* gene *avrB,* experiments were conducted using *Escherichia coli* transformants carrying *hrpJ4: :*Tn *phoA* or *hrpH2: :Tn phoA* (Pirhonen et al., 1996). Phenotypic expression of *avrB* was lost in any of the above *E. coli* transformants. The results suggest that inactivation of genes encoding energy-transducing hrpJ4 or outer membrane components *(hrpH2)* of the hrp-encoded protein secretion system abolished phenotypic expression of *avrB* in the *E. coli* expression system (Pirhonen et al., 1996). These mutants had been shown to abolish *hrp*-dependent secretion of HR-eliciting factors in tobacco (He et al., 1993; Lidell and Hutcheson, 1994). Hence, the *hrp*-encoded secretion apparatus may play an essential role in production of the AvrB phenotype.

Thus several Avr proteins may be members of a larger class of "Hops" (*H*rp-dependent *o*uter *p*roteins) (Alfano et al., 1997). This designation is analogous to the Yops (*Y*ersinia *o*uter *p*roteins) designation for the protein secreted by the prototypical *Yersinia ysc*-encoded type III secretion system. The *ysc*-encoded type III protein systems found in a number of enteric bacteria pathogenic to mammals have been associated with the apparent injection of secreted proteins into mammalian cells (Rosqvist et al., 1994; Parsot et al., 1995; Hutcheson et al., 1996). As with the Yops, *P. syringae* proteins in the Hop class may be targeted to the milieu or to the interior of host cells. Thus, it appears that the Hrp system, while sufficient to deliver Hop proteins targeted to the host cell, has no intrinsic ability to elicit the HR in the absence of such proteins. That is, as with the Yops, the Hops that are targeted to the interior of host cells are the key effectors of the interaction (Alfano et al., 1997). Avr proteins may act within plant cells following secretion by the Hrp pathway (Alfano and Collmer, 1996; Gopalan et al., 1996).

THE ROLE OF hrp AND avr GENES IN THE EARLY RECOGNITION PROCESS IN PLANT-BACTERIAL PATHOGEN INTERACTIONS

Role of hrp and avr Genes in Disease Development

The *hrp* and *avr* genes appear to play important role in early recognition process in plant-bacterial pathogen interactions (Vivian and Gibbon, 1997).

hrp genes are responsible for induction of disease resistance in nonhosts and pathogenicity in host plants (Lindgren et al., 1986, 1988; Chatterjee and Vidaver, 1986; Boucher et al., 1987, 1988). This close association suggests that the factor inducing the HR in nonhosts might contribute to pathogenicity in the host plant. Bacterial pathogens are able to grow in both host and nonhost plants, although multiplication may be less in the latter. *hrp* genes may be required for growth in planta. *Pseudomonas fluorescens* is a saprophyte, but some strains of *P. fluorescens* multiply in planta (Vidhyasekaran et al., 1997). When *P. fluorescens* strain PfG32R was inoculated into leaves of tobacco or *Phaseolus vulgaris* with a low bacterial concentration, the population increased in the first day from 2×10^3 to 1.8×10^4 cfu per leaf disc (Mulya et al., 1996). This strain conserves the regions homologous to *hrpRS* and *hrpF* of *Pseudomonas savastanoi* pv. *phaseolicola* and the regions homologous to *hrpC* and *hrpD* of *P. syringae* pv. *tabaci* (Mulya et al., 1996).

hrp genes in the bacterial pathogens are transcribed in both hosts and nonhosts (Xiao et al., 1992). In some cases *hrp* gene expression is greater in nonhosts. Wei et al. (1992b) found that *Erwinia amylovora hrp* genes were induced more rapidly and to higher levels in tobacco (nonhost) than in pear (host). *hrp* gene expression is induced by nutrients. Leakage of nutrients from plant cells may be important in early host-pathogen interactions.

Whatever the organism studied, most *hrp* mutants are completely nonpathogenic and do not elicit any macroscopically visible response either on host or nonhost. Thus *hrp* gene functions are probably required in an early step in the establishment of the parasitic relationship. In *Ralstonia solanacearum, hrp*-encoded functions probably are not involved in the infection process, since 10 of 11 *hrp* mutants tested retained the ability to naturally infect and colonize tomato plants (Trigalet and Demery, 1986). Following establishment in soil, these mutants could be isolated from the stems of infected tomato plants, although their respective populations in the plants always remained very low compared to those in plants inoculated with a wild-type strain. This impaired ability of *hrp* mutants to multiply within the plant has also been established in other species (Lindgren et al., 1986; Bertoni and Mills, 1987). *hrp* genes appear to support multiplication of the bacteria. The inability of several *hrp* mutants to sustain normal growth in planta has been reported by many workers (Kamoun and Kado, 1990; Bonas et al., 1991; Willis et al., 1991). It does not result from a defect in the basic metabolism of the bacteria since *hrp* mutants grow normally when cultured on minimal media (Lindgren et al., 1986; Boucher et al., 1988). In planta, therefore, growth of *hrp* mutants could be limited by the availability of nutrients. If so, *hrp* genes could play a role in diverting certain plant metabolites from the plant to the bacteria. In support of this hypothesis, it has been shown that *hrp* mutants of *Erwinia amylovora, Pseudomonas syringae,* and *R. solanacearum* each have coordinately lost the ability to induce electrolyte leakage when inoculated on either host or nonhost plants (Baker et al., 1987; Brisset and Paulin, 1991). Leakage of electrolytes could be associated with the leakage of nutrients (Atkinson and Baker, 1987b).

Role of *hrp* genes in the synthesis and secretion of elicitor molecules that could trigger electrolyte leakage in the plant has been reported (Popham et al., 1995; Hoyos et al., 1996). If so, a moderate leakage in the host plant could provide the pathogens with various metabolites originating from the plant cell and thus permit growth of the invading organism. If improperly balanced, such leakage would result in the rapid death of the target plant cells and allow the necrotic response characteristic of the HR.

hrp-encoded elicitors induce extracellular alkalinization (He et al., 1994; Popham et al., 1995). Harpin$_{Pss}$ increased the pH of an external medium when tobacco suspensions were treated with it within a few minutes of treatment (Figure 1.4; Hoyos et al., 1996). The primary function of the Hrp system in pathogenesis appears to be to make apoplastic fluids more suitable for bacterial multiplication by raising the pH (Atkinson and Baker, 1987a,b). Bacteria grow well at natural or alkaline pH. An increase in plant cell wall pH would also be expected to enhance the activity of the *E. chrysanthemi* polygalacturonate trans-eliminase enzymes, which have pH optima of approximately 8.5 (Bauer et al., 1994).

Each bacterial pathogen may have several avirulence genes (Hopkins et al., 1992; Whalen et al., 1993). Each avirulence gene may contribute to virulence to a certain extent. But loss of a single *avr* gene may not result in total loss of virulence. Pathogens may have multiple approaches to host sur-

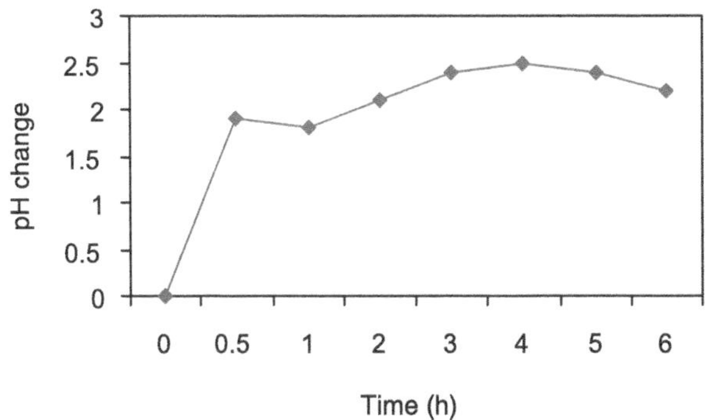

FIGURE 1.4. pH changes in the external medium of tobacco suspension-cultured cells treated with harpin (*Source:* Hoyos et al., 1996).

vival and virulence, so that loss of any one virulence function by mutation might not be detrimental. In support of this hypothesis, Lorang et al. (1994) reported that *P. syringae* pv. *tomato* PT 23 strains carrying individual or multiple *avr* gene mutations caused normal disease symptoms, when relatively high inoculum concentrations (10^7 cfu/ml) were infiltrated into tomato leaves. However, when inoculum concentrations of 10^4 cfu/ml were infiltrated into leaves, strains carrying mutations in the *avrE* locus were considerably less virulent than wild-type *P. syringae* pv. *tomato,* but complete virulence was regained when the mutant strains were complemented by the cloned *avrE* locus (Lorang et al., 1994). Thus the contribution of a particular *avr* gene to survival on plants and virulence may be subtle. Contribution of *avr* genes in virulence of some bacterial pathogens has been reported (Kearney and Staskawicz, 1990; Swarup et al., 1992).

Since *avr* genes are regulated by *hrp* genes, the question arises of whether *avr* genes are gratuitous for survival and pathogenicity of the bacteria. Since bacteria contain multiple *avr* genes, studies on individual *avr* genes may not reveal the actual role of *avr* genes. The complementary effect of all *avr* genes may be important, and such studies are necessary to understand the role of *avr* genes.

Studies made so far reveal that *hrp* and *avr* genes alone may not be sufficient to cause disease. *hrp* genes may code for basic pathogenicity functions, superimposed on which are factors that contribute to disease expression. It is possible that *hrp* genes may induce *dsp* genes, host-selective pathogenicity genes, and virulence genes besides avirulence genes and cause disease symptoms.

The Role of hrp and avr Genes in Inducing Disease Resistance

Plants respond to bacterial pathogen infection by activating certain responses that have been implicated as mechanisms of disease resistance. These responses include the hypersensitive reaction, the production of phytoalexins and pathogenesis-related proteins, and deposition of hydroxyproline-rich glycoproteins (HRGP), lignin, and callose into the plant cell wall. Several plant genes are involved in biosynthesis of these defense chemicals (Vidhyasekaran, 1997). Putative defense protein transcripts include phenylalanine ammonia-lyase (PAL) (Dong et al., 1991; Jakobek and Lindgren, 1993), chalcone synthase (CHS) (Dhawale et al., 1989), chalcone isomerase (CHI) (Jakobek and Lindgren, 1993), chitinase (Godiard et al., 1990), β-1,3-glucanase (Dong et al., 1991), and proteinase inhibitors I and II (Pautot et al., 1991). All of them accumulate in plants due to incompatible bacterial infection. Plants appear to be able to recognize by an unknown mechanism the presence of a potential pathogen and initiate the defense response.

The role of *hrp* genes in induction of these defense genes is indicated, as transcripts for PAL (the key enzyme in synthesis of phenolics, lignin, and phytoalexins), CHS and CHI (the key enzymes for biosynthesis of phytoalexins), and chitinase (Class III PR protein) were detected in bean leaf tissue 6 h after infiltration with *Pseudomonas syringae* pv. *tabaci, P. syringae* pv. *tomato, P. syringae* pv. *pisi,* and *P. syringae* pv. *syringae,* all of which induce a hypersensitive reaction on bean (Jakobek et al., 1993). Two *hsr* (hypersensitivity-related) genes, *hsr515* and *hsr 201,* appeared in tobacco leaves in contact with the avirulent *Ralstonia solanacearum* or with a *Pseudomonas fluorescens* strain carrying the *hrpZ* gene encoding a necrotizing polypeptide, harpin (Czernic et al., 1996). An *hrp* deletion mutant of *X. campestris* pv. *campestris* strain 8420, which grew as well as the wild type, did not induce basic β-1,3-glucanase transcript or any other defense-related transcripts tested (PAL, chitinase, and peroxidase) (Lummerzheim et al., 1993).

Harpin encoded by *hrpN* of *E. amylovora* induces resistance in seven different plants against various diseases caused by fungi, viruses, and bacteria (Wei and Beer, 1996). It induced resistance to southern bacterial wilt of tomato caused by *Ralstonia solanacearum* (Wei and Beer, 1996). The elicitor $HrpZ_{Pss}$ produced by *Pseudomonas syringae* pv. *syringae* induced systemic resistance in cucumber to diverse pathogens including angular leaf spot bacterium, *P. syringae* pv. *lachrymans* (Strobel et al., 1996). An *hrpH* mutant of *P. syringae* pv. *syringae* 61 which is defective in secretion of $HrpZ_{Pss}$ and possibly other protein elicitors failed to elicit systemic resistance. PR-proteins, including β-1,3-glucanase, chitinases, and peroxidase were induced in cucumber plants inoculated with *P. syringae* pv. *syringae* 61$HrpZ_{Pss}$. The induction patterns of PR proteins by $HrpZ_{Pss}$ and *P. syringae* pv. *syringae* 61 were the same (Strobel et al., 1996). The results suggest that the elicitors produced by *hrp* activate defense genes.

When an HR-inducing incompatible pathogen *P. syringae* pv. *syringae* was inoculated on bean leaves, cinnamyl-alcohol dehydrogenase (CAD), cinnamate-4-hydroxylase (C4H), and EL13, a plant defense gene that has high identity with the alcohol dehydrogenase domain of cinnamyl-alcohol dehydrogenase levels, increased at 12 h after inoculation (Buell and Somerville, 1995). Messenger RNAs of anthranilate synthase (ASA) and superoxide dismutase (SOD) were also induced (Buell and Somerville, 1995). CAD, C4H, and EL13 are involved in increased accumulation of phenylpropanoid metabolites such as phenolics and lignin (Dong et al., 1991; Lummerzheim et al., 1993; Wanner et al., 1993). ASA is involved in production of camalexin, an indole-containing phytoalexin (Tsuji et al., 1992). Thus, the HR-inducing incompatible pathogen may induce several defense chemicals.

However, several reports have accumulated to show that *hrp* genes also may not be responsible for induction of defense genes. Several Hrp mutants of bacterial pathogens are known to induce plant defense genes, although they fail to induce an HR on nonhost plants. Jakobek and Lindgren (1993) have demonstrated that transcripts for phenylalanine ammonia-lyase, chalcone synthase, chalcone isomerase, and chitinase accumulated and phytoalexins were produced in bean after infiltration with a wild-type isolate of the incompatible, hypersensitive reaction-inducing bacterium *P. syringae* pv. *tabaci*. *P. syringae* pv. *tabaci* Hrp$^-$ mutants, which do not induce the HR on bean, also induced the accumulation of the transcripts examined and phytoalexin production. The temporal pattern of transcript accumulation seen after infiltration with the Hrp$^-$ mutant was similar to that observed after infiltration with the wild-type strain (Jakobek and Lindgren, 1993). Thus the induction of the transcript accumulation and phytoalexin that were observed was not dependent upon a fully functional set of *hrp* genes in *P. syringae* pv. *tabaci*. Defense transcripts also accumulated in bean 6 h after infiltration with two different Hrp$^-$ mutants of *P. syringae* pv. *glycinea* (Jakobek and Lindgren, 1993). Also, bean leaves preinoculated with *P. syringae* pv. *tabaci* Hrp mutants were more resistant to *P. savastanoi* pv. *phaseolicola* than uninoculated leaves, confirming that the plant responds to the Hrp$^-$ mutants (Lindgren and Jakobek, 1990). These studies show that some genes other than *hrp* may also be involved in induction of disease resistance.

It has been suggested that avirulence genes may be involved in induction of plant defense genes. In *Arabidopsis* plants that recognize the *avrRpt2* gene from *P. syringae* pv. *tomato*, mRNA-encoding PAL accumulated more rapidly in resistant than in susceptible plants (Dong et al., 1991). This pattern of PAL expression was seen also in *Arabidopsis* plants that recognize the *avrB* gene from *P. syringae* pv. *glycinea* (Wanner et al., 1993).

The sulfotransferase (ST) gene in plants functions in synthesis of signals involved in induction of plant defense genes (Lacomme and Roby, 1996). When *Arabidopsis* leaves were inoculated with two *P. syringae* pv. *maculicola* strains differing only by the presence or absence of a cloned *avr* gene *(avrRpm1)*, the strain containing *avrRpm1* induced more ST mRNA than that without the gene. Thus the activation of the ST gene may depend on the bacterially encoded *avr* functions (Lacomme and Roby, 1996).

Several avirulence genes are known to elicit various defense reactions in incompatible hosts (Carney and Denny, 1990; Whalen et al., 1991; Ronald et al., 1992; Fillingham et al., 1992). Some *avr* genes encode elicitors, which elicit defense mechanisms of the incompatible hosts. However, the role of *avr* genes in induction of plant defense genes is also questionable.

Even nonpathogenic bacteria *Pseudomonas fluorescens* Pf101 and *Escherichia coli* DH5a induce defense genes in bean (Jakobek and Lindgren, 1993). *P. fluorescens* strains induce several defense genes in tobacco (Maurhofer et al., 1994), radish (Leeman et al., 1995), cucumber (Liu et al., 1995), and tomato (M'Piga et al., 1997). The nonpathogenic bacteria, which do not have any *avr* gene, may produce signals that are involved with the elicitation of defense transcript accumulation.

These results suggest that besides *hrp* and *avr* genes, some other genes and signals may be involved in inducing disease resistance. Endogenous host elicitors (pectic fragments), bacterial lipopolysaccharides, glucans, and other polysaccharides also induce resistance.

OTHER SIGNAL MOLECULES OF BACTERIAL PATHOGENS

Bacterial Pectic Enzymes May Trigger Plant Defense Genes

Pectic enzymes have been shown to induce a number of plant defense responses, including the production of phytoalexins, phenylalanine ammonia-lyase, and β-1,3-glucanase (Vidhyasekaran, 1997). Pectic enzymes release pectic fragments which induce defense genes and accumulation of pathogenesis-related proteins such as PR-6 (Ryan, 1987), phytoalexins (Nothnagel et al., 1983), hydroxyproline-rich glycoprotein (Roby et al., 1985), and lignin (Robertson, 1986).

Palva et al. (1993) have shown that β-1,3-glucanase (PR2 protein) and chitinases (PR3 proteins) can be induced in tobacco in response to *Erwinia carotovora* subsp. *carotovora* infection and its pectic enzymes. Enzyme export-defective mutants of *E. carotovora* subsp. *carotovora* were unable to induce β-1,3-glucanase. By mini–Tn5-Km and chemical mutagenesis, *RsmA*$^-$ mutants of *E. carotovora* subsp. *carotovora* 71 that produce high basal levels of pectate lyases and polygalacturonase have been isolated (Chatterjee et al., 1995; Cui et al., 1995, 1996). The RsmA$^-$ mutants, but not their parent strains, elicit an HR-like response in tobacco leaves (Cui et al., 1996). Pectinase overproduction by the RsmA$^-$ mutants may induce defense reactions that could culminate in an HR-like response (Cui et al., 1996).

Pectic enzymes and *hrp* genes may independently be involved in elicitation of defense genes. Arlat et al. (1991) observed that Hrp$^-$ mutants of *Xanthomonas campestris* pv. *campestris* retain their ability to produce extracellular pectic enzymes, and a mutant unable to secrete extracellular enzymes still elicits the HR.

Bacterial Lipopolysaccharides May Induce Plant Defense Genes

Lipopolysaccharides (LPS) are constituents of the outer membrane of many bacterial pathogens. Some of the lipopolysaccharides have been shown to elicit plant defense genes. Purified LPS from *X. campestris* pv. *campestris* induce β-1,3-glucanase transcript accumulation in turnip (Newman et al., 1995). This raises the possibility that in plant-pathogen interactions, expression of at least some defense-related genes occurs as a response to LPS released either as a consequence of normal bacterial growth or rapid bacterial death within the plant.

Two forms of LPS which partition into different phases on hot phenol-water extractions of *X. campestris* pv. *campestris* have been reported (Dow et al., 1995). Phenol-phase LPS carries an O-antigen, whereas water-phase LPS comprises lipid A-core oligosaccharide alone (Dow et al., 1995). With 1 μg of water-phase LPS per ml, β-1,3-glucanase transcript accumulation was first detected at 4 h after inoculation, and levels increased steadily up to 24 h (Newmann et al., 1995). Similar results were seen with the phenol-phase LPS forms. The lipid A-inner core structure was required for activity and the O-antigen had no role (Newmann et al., 1995).

LPS from *Ralstonia solanacearum* induces synthesis of new polypeptides in tobacco (Leach et al., 1983). LPS preparations from *P. syringae* pv. *syringae* elicit phytoalexin synthesis in soybean hypocotyls (Barton-Willis et al., 1984). Purified LPS preparations from *P. syringae* pv. *aptata* induce resistance to *P. syringae* pv. *tabaci* in tobacco (Bizarri et al., 1996). Bacterial extracellular polysaccharides also induce defense mechanisms. Extracellular polysaccharides produced by *X. vesicatoria* induce large papillae formation in mesophyll cells of *Capsicum* (Brown et al., 1995) and induce resistance.

Oligosaccharides May Suppress Disease Development

Periplasmic oligosaccharides (OLS) extracted from bacterial pathogens with trichloroacetic acid treatment may act as signal molecules in suppressing disease symptom development (Mazzucchi, 1983). These oligosaccharides may accumulate in bacteria, suspended in the low-molarity congestion fluid during penetration, and subsequently be released under inoculation stress during establishment (Minardi et al., 1989).

The oligosaccharides of *Pseudomonas syringae* pv. *aptata* are a mixture of neutral OLS with a molecular weight of approximately 800-1800, DP (degree of polymerization) 5-11, with linear glucose and mannose chains

(Stefani et al., 1994). The OLS mixture, infiltrated into tobacco leaves, prevented or delayed normosensitive necrosis symptoms caused by a virulent *P. syringae* pv. *tabaci* strain. Pretreatment with OLS inhibited endophytic growth of *P. syringae* pv. *tabaci* in the first five days. In the control tissue, necrosis began to develop on day 3, when the number of endophytic bacteria exceeded the critical threshold. On day 4, the number of bacteria in the treated tissue was already 10 times higher than the critical threshold without causing necrotic disease symptoms (Stefani et al., 1994). It suggests that the OLS may modulate the expression of the *hrp* genes of the pathogen, resulting in suppression of disease development. The regulatory activity of OLS was observed at nonomolar concentrations and an interval of 6 h was necessary between the infiltration of oligosaccharides and inoculation of *P. syringae* pv. *tabaci* for a clear inhibition of normosensitive necrosis (Stefani et al., 1994). These observations suggest that OLS may be signal molecules modulating functions of *hrp* genes.

THE SIGNAL TRANSDUCTION SYSTEM

Ca^{2+}-Dependent Signal Transduction System

On recognition of pathogens, plants synthesize an array of defense-related products to ward off the pathogens. Both *hrp* and *avr* genes, which are involved in recognition, may encode signal molecules, and these signals may be transducted into plant cells for activation of defense genes. Calcium ions are known to transduct extracellular primary signals into intracellular events. Elicitation of defense genes is more effective in the presence of Ca^{2+} in plants (Kohle et al., 1985; Stab and Ebel, 1987).

Calcium can pass through the plant cell plasma membrane. Calcium passively crosses plasma membranes through "channel"-forming proteins embedded in the lipid bilayer of the membrane (Schroeder and Hagiwara, 1989). Calcium channels are commonly present in plant cells (Johannes et al., 1991; Zocchi and Rabotti, 1993). When suspension-cultured tobacco cells were inoculated with *Pseudomonas syringae* pv. *syringae,* an increase in calcium influx was observed (Atkinson et al., 1990). Tobacco cells exhibited a baseline Ca^{2+} influx of 0.02 to 0.06 $\mu mole \cdot g^{-1} \cdot h^{-1}$ and following bacterial inoculation uptake rates increased steadily for 2 to 3 h, reaching 0.5 to 1.0 $\mu mole \cdot g^{-1} \cdot h^{-1}$ (Atkinson et al., 1990).

Stimulation of Ca^{2+} influx by the bacterial pathogen was prevented by lanthanum (La^{3+}), Cd^{2+}, and CO^{2+}. These actions are known to block calcium channels, and the results suggest that the Ca^{2+} influx may be mediated by ion channels (Atkinson et al., 1990). Verapamil and nifedipine, which block certain classes of voltage-dependent calcium channels in plant cells, had minimal effects on the tobacco cell-*P. syringae* pv. *syringae* interaction. It suggests that the Ca^{2+} influx is mediated only by the calcium ion channel. Ca^{2+} appears to be necessary for induction of defense mechanisms. $Harpin_{Ea}$, the elicitor synthesized by *Erwinia amylovora,* and $harpin_{Pss}$, the elicitor synthesized by *Pseudomonas syringae* pv. *syringae,* induce HR in tobacco leaves (the incompatible host of both pathogens). The induction of HR is prevented by inhibitors of calcium influx such as lanthanum chloride and cobalt chloride (He et al., 1993, 1994).

Protein Kinases in Signal Transduction System

Phosphorylation of proteins is an important component in the integration of external and internal stimuli. Plants may use protein phosphorylation as an effective device for responding to endogenous stimuli (Ranjeva and Boudet, 1987). Protein phosphorylation reaction is catalyzed by protein kinase, which is activated by Ca^{2+} (Hetherington and Trewavas, 1984). Increased Ca^{2+} influx observed during the early recognition process in plant-bacteria interactions would have activated protein kinase in plants. Protein kinase C is involved in signal transduction pathways.

Protein kinase C occurs in many plants as a soluble or a membrane-bound enzyme. A shuttle system between cytosol and membrane would allow the phosphorylation of either soluble or membrane-bound proteins and therefore would allow protein kinase C to participate in transmembrane signaling (Ranjeva and Boudet, 1987; Grosskopf et al., 1990; Felix et al., 1991; Conrath et al., 1991).

Recently several disease resistance genes have been cloned. These resistance genes (R) may code for receptors of signal molecules produced by enzymes encoded by bacterial *avr* genes. These signal molecules, which may be perceived by receptors encoded by the corresponding R genes of plant hosts, may induce defense genes. The protein product of the *avr* gene may directly bind to the protein product of an R gene. The predicted peptide sequences of several cloned R genes indicate a function in signal transduction (Martin et al., 1993a,b, 1994; Bent et al., 1994; Fantl et al., 1993; Valon et al., 1993; Whitham et al., 1994, 1996; Mindrinos et al., 1994; Walker, 1994; Grant et al., 1995; Zhou et al., 1995; Song et al., 1995, 1997; Wang et al., 1995; Torli et al., 1996; Dixon et al., 1996; Ronald, 1997; Lawrence et al., 1995; Rommens et al., 1995).

Transcriptional activation of plant defense genes is modulated by phosphorylation (Felix et al., 1991; Yu et al., 1993). The transcriptional activation of plant defense genes can be blocked by inhibitors of protein kinases

(Raz and Fluhr, 1993). It suggests that protein kinases may be involved in phosphorylation cascades leading to plant defense responses (Lamb, 1994). The activated protein kinases and/or transcriptional factors enter the nucleus to activate gene expression (Kerr et al., 1992; Martin et al., 1994).

A disease resistance gene, *Pto,* has been shown to encode a serine/threonine kinase in tomato (Loh and Martin, 1995; Scofield et al., 1996; Tang et al., 1996) and it confers resistance to bacterial speck disease (*P. syringae* pv. *tomato*). A second serine/threonine kinase, *Pto*-interacting 1 *(Pti1),* which physically interacts with *Pto,* was identified (Zhou et al., 1995; Lamb, 1996; Tang et al., 1996; Scofield et al., 1996). *Pto* specifically phosphorylates *Pti1* and *Pti1* does not phosphorylate *Pto*. Expression of a *Pti1* transgene in tobacco plants enhanced the hypersensitive response to *P. syringae* pv. *tabaci* strain carrying the avirulence gene *avrPto* (Zhou et al., 1995). These findings indicate that *Pti* is involved in a *Pto*-mediated signaling pathway, probably by acting as a component downstream of *Pto* in a phosphorylation cascade (Zhou et al., 1995). The involvement of the protein kinase gene *Pti1* which acts in conjunction with the *P. syringae* pv. *tomato* resistance gene *Pto* to activate tomato plant defenses has been reported (Tobias, 1996).

The rice gene *Xa21* confers resistance to many races of *Xanthomonas oryzae* pv. *oryzae,* the bacterial blight pathogen (Song et al., 1995; Wang et al., 1996; Ronald, 1997). It encodes a receptor-like kinase consisting of leucine-rich repeats (LRR) in the putative extracellular domain and a serine/threonine kinase in the intercellular domain similar to tomato Pto kinase. The *Xa21* belongs to a multigene family containing at least eight members (Ronald et al., 1992b; Song et al., 1995, 1997). These members can be classified into two classes. In the presumed extracellular portion of the two classes, the two LRR domains share a low level of identity (59.5 percent). In contrast, the catalytic kinase domains are highly conserved (82 percent identity) (Ronald, 1997). The encoded proteins from the two classes may bind different ligands and/or initiate different signaling pathways. The XA21 LRR domain may bind a polypeptide (signal molecule) produced by the pathogen, and the specific binding may lead to activation of the XA21 kinase with subsequent phosphorylation on specific serine or threonine residues. Phosphorylated residues may then serve as binding sites for proteins that can initiate downstream responses. This reaction may in turn lead to phosphorylation of transcription factors. Upon phosphorylation, the transcription factors can move into the nucleus from the cytosol (Ronald, 1997).

Overexpression of protein kinases may lead to increased disease resistance. The transgenic rice plants expressing the *Xa21* gene show increased resistance to *X. oryzae* pv. *oryzae* (Song et al., 1995; Wang et al., 1996). Similarly, transgenic tomato plants expressing the *Pto* gene show high resistance to *P. syringae* pv. *tomato* (Martin et al., 1993a).

Phospholipase D in Signal Transduction

Phospholipase D is a critical component in signal transduction (Billah, 1993; Divecha and Irvine, 1995). Phospholipase D, which is often the most abundant phospholipase in plants, plays a role in plant signaling systems (Brown et al., 1990; Munnik et al., 1995; Young et al., 1996). Multiple forms of phospholipase D have been identified in plants (Dyer et al., 1994). Phospholipases activate protein kinase C. Phospholipase D is associated with phospholipid metabolism in plants. It hydrolyzes phospholipids generating phosphatidic acid and a free head group, such as choline, which serve as second messengers (Dennis et al., 1991). Phosphatidic acid also can be further metabolized by phosphatidic acid phosphorylase to form diacylglycerol. Diacylglycerol activates protein kinase C.

Phospholipases are calcium-dependent enzymes. Ca^{2+} influx activates phospholipases (Elliott and Skinner, 1986; Bogre et al., 1988; Breviario et al., 1995). Phospholipase D activity is stimulated in tobacco leaves when an incompatible pathogen, *Erwinia amylovora,* was inoculated (Huang and Goodman, 1970).

Phospholipase C in Signal Transduction System

Phospholipase C activity has been reported in plants (Helsper et al., 1986). Phospholipase C plays an important role in the signal transduction system. Phospholipase C appears to be involved in regulating intracellular Ca^{2+} concentrations, which act as a second messenger in the signal transduction system (Drobak and Ferguson, 1985; Berridge and Irvine, 1989; Alexandre et al., 1990). It catalyses hydrolysis of the cell membrane phosphoinositides to release diacylglycerol and inositol phosphates, predominantly inositol 1,4,5-triphosphates (IP_3) (Berridge and Irvine, 1989). Diacylglycerol activates protein phosphorylation and IP_3 releases Ca^{2+} from internal stores (Berridge and Irvine, 1989) and both of them function as second messengers. Inositol phosphates may be involved in elicitor signal transduction (Renelt et al., 1993).

Phosphoinositol phosphates occur in plants (Boss and Massel, 1985). Inositol phospholipid turnover is involved in several signal transduction pathways (Kamada and Muto, 1994a,b). Both phosphoinositides and phosphotidylinositol-specific phospholipases are associated with plant membranes (Pfaffmann et al., 1987). Activities of phosphatidylinositol kinase and phosphatidylinositol-4-phosphate kinase were elevated in vitro by a fungal elicitor treatment (Toyoda et al., 1992). Inositol 1,4,5-triphosphate increases in pea due to fungal elicitor treatment (Toyoda et al., 1993).

When tobacco was inoculated with *P. syringae* pv. *syringae* (an incompatible pathogen), there was increased incorporation of ^{32}Pi into phosphatidic acid and phospholipids (Atkinson et al., 1993). It suggests rapid phosphorylation of diacylglycerol to phosphatidic acid. A rapid breakdown of phosphoinositides was also observed in tobacco-*P. syringae* pv. *syringae* interaction (Atkinson et al., 1993). Increased breakdown of phosphatidylinositol-4-phosphate (PIP) and phosphatidylinositol (PI) was observed. The increased breakdown of PI and PIP may reflect the action of either phospholipase C and/or PI and PIP kinases on these substances (Atkinson et al., 1993). Phosphatidylinositol breakdown was inhibited by bromophenacylbromide, a phospholipase elicitor, and by neomycin, which blocks metabolism of polyphosphoinositides (Atkinson et al., 1993). It suggests a causal role for phosphoinositide metabolism in early signal transduction.

H^+-ATPase in Signal Transduction System

Transport of solutes across the plasma membrane is driven by H^+-ATPase, which produces an electric potential and pH gradient (Spanswick, 1981). Membrane potentials of suspension-cultured tobacco cells ranged from –110 to 160 mV. The diffusion potential was 8–5 mV (Popham et al., 1995). Harpin$_{Ea}$, the elicitor of *Erwinia amylovora,* induced a rapid membrane depolarization to approximately that of the diffusion potential (Popham et al., 1995). Similar immediate membrane depolarization following harpin addition has also been reported in tobacco leaf segments (Popham et al., 1993). The application of harpin$_{Ea}$ or harpin$_{Pss}$ to the bathing solution of tobacco leaf segments causes membrane depolarization within minutes (Hoyos et al., 1996). When cotton cotyledons were inoculated with incompatible bacteria, the membrane potential declined to about – 40 mV, which is substantially below the diffusion potential (Pavlovkin et al., 1986).

These studies indicate that the bacteria may affect the active ATP-dependent portion of the membrane potential. Harpin$_{Ea}$ affects the plasmalemma H^+-ATPase itself (Popham et al., 1995). Fusicoccin, a fungal toxin, stimulates H^+-ATPase (Rasi-Caldogno et al., 1993). Harpin$_{Ea}$ overcame the H^+-ATPase stimulation by fusicoccin (Popham et al., 1995). It suggests that the bacterial elicitor is involved in inhibition of H^+-ATPase.

The H^+-ATPase is active in the dephosphorylated state (Vera-Estrella et al., 1994). Hence the bacterial elicitor may be involved in protein phosphorylation. Protein kinase inhibitor K252a blocks phosphorylation events and alters the harpin-plasmalemma interaction by modifying H^+-ATPase activity (Popham et al., 1995). The harpin$_{Ea}$-induced extracellular alkalinization in tobacco cells requires phosphorylation, which is activated by protein kinase (Popham et al., 1995). The harpin$_{Ea}$ depolarization requires calcium influx and can be delayed by the protein kinase inhibitor K252a (Hoyos et al., 1996). Hence, it is possible that a harpin-membrane interaction signals defense response induction more directly, perhaps via the Ca^{2+} activated phosphorylation required for the harpin-induced depolarization.

K^+/H^+ Exchange Response (XR)

One of the earliest possible plant responses which indicate that recognition has occurred is the K^+/H^+ response (concomitant increases in extracellular pH and K^+) (Atkinson et al., 1985b; Keppler and Baker, 1989; Keppler et al., 1989; Baker et al., 1991). The initiation of a K^+/H^+ response is characterized by specific plasma membrane K^+ efflux, extracellular alkalinization, and intracellular acidification (Atkinson et al., 1990).

Induction of apparent plasmalemma K^+/H^+ exchange response in *Pseudomonas syringae* was associated with both incompatible and compatible bacterium-plant interactions (Atkinson et al., 1985b; Atkinson and Baker, 1987). But the K^+ efflux (Figure 1.5; Baker et al., 1987) and H^+ uptake proceeded rapidly in incompatible interactions (Baker et al., 1987).

Harpin$_{Ea}$ and harpin$_{Pss}$ isolated from *E. amylovora* and *P. syringae* pv. *syringae,* respectively, cause immediate K^+ efflux and extracellular alkalinization in tobacco suspension-cultured cells (Wei et al., 1992a; Baker et al., 1993). Harpin$_{Pss}$ causes K^+ efflux and extracellular alkalinization in suspen-

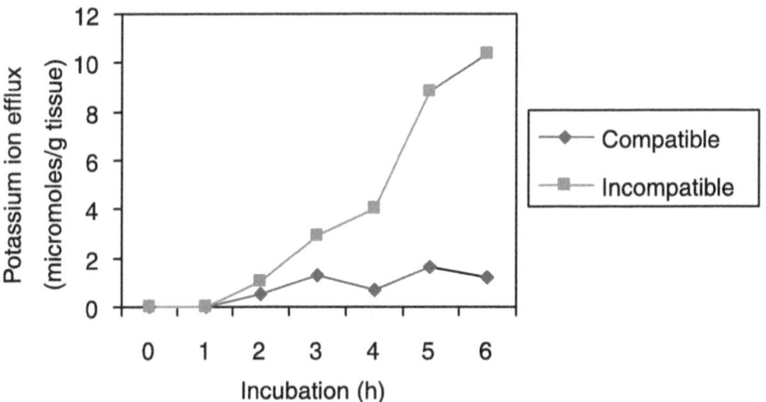

FIGURE 1.5. Monitoring of potassium ion efflux of tobacco suspension cells incubated with compatible pathogen *Pseudomonas syringae* pv. *tabaci* and incompatible pathogen *P. savastanoi* pv. *glycinea* race 4 (*Source:* Baker et al., 1987).

sion-cultured cells of tobacco almost immediately after treatment (Hoyos et al., 1996).

The induction of a K^+/H^+ exchange response appears to be dependent on *hrp* and *avr* genes of the bacterial pathogen (Wei et al., 1992a; Orlandi et al., 1992). *hrp* mutants of *Pseudomonas syringae* pathovars and of *Erwinia amylovora* do not induce the K^+/H^+ exchange response also (Baker et al., 1987; Wei et al., 1992a). Tobacco suspension cells alkalinized the extracellular medium in 2 to 4 h after treatment with wild-type *P. syringae* pv. *syringae,* but not with the harpin mutant ($HrpH^-$) (Hoyos et al., 1996).

The K^+/H^+ exchange response is mediated by Ca^{2+} influx (Atkinson et al., 1990). The protein kinase inhibitor K252a inhibits the K^+/H^+ exchange response (He et al., 1994). It also blocks phosphorylation events and alters membrane diffusion potential mediated by H^+-ATPase (Popham et al., 1995). The K^+/H^+ response does not appear to result from nonspecific membrane damage since extracellular cl^-, the second major cellular ion, does not leak from cells during this period, and an intact plasma membrane demonstrating active ATPase and uptake of Ca^{2+} is required (Atkinson et al., 1990). The K^+/H^+ response may be due to membrane diffusion potential and this response may indicate the involvement of Ca^{2+} influx protein phosphorylation and membrane potential in signal transduction.

Active Oxygen Species As Second Messengers

A sudden burst in active oxygen production characterized by increase in luminol-dependent chemiluminescence has been reported in plants inoculated with bacterial pathogens (Keppler and Baker, 1989; Keppler et al., 1989; Baker et al., 1991, 1993). An immediate increase in chemiluminescence occurred in tobacco suspension cells within 30 min after addition of both compatible (*P. syringae* pv. *tabaci* and *P. syringae* pv. *syringae* B7) and incompatible bacteria (*P. syringae* pv. *syringae* wild type and *P. savastanoi* pv. *glycinea* race 4 and race 6) (Baker et al., 1991). The oxidative burst includes release of superoxide anion (O_2^-) (Keppler and Baker, 1989), singlet oxygen ($^1O_2$) (Keppler et al., 1989), hydroxyl radical (OH^o) (Epperlein et al., 1986), and H_2O_2 (Levine et al., 1996). Production of these active oxygen species has been reported in various plant-bacteria interactions (Keppler and Novacky, 1986, 1987, 1989; Keppler and Baker, 1989). These active oxygen species are involved in signal transduction inducing synthesis of different defense chemicals (Rogers et al., 1988; Van Huytsee, 1987).

H_2O_2 is the important active oxygen species produced during incompatible interactions. Accumulation of H_2O_2 from an oxidative burst cues localized host cell death (Levine et al., 1996). Cell corpse morphology (cell shrinkage, plasma membrane blebbing, and nuclear condensation) characteristic of apoptosis is induced in soybean cells and leaf tissue by avirulent *P. savastanoi* pv. *glycinea*. Similar apoptosis was observed in *Arabidopsis thaliana* inoculated with avirulent *P. syringae* pv. *tomato* (Levine et al., 1996). H_2O_2 has been shown to activate this physiological cell death program, resulting in the generation of large (about 50 kb) DNA fragments and apoptosis. H_2O_2 stimulates a rapid influx of Ca^{2+} into soybean cells, which in turn activates generation of the new 50 kb DNA fragment and apoptosis (Levine et al., 1996). These observations establish a signal for Ca^{2+} downstream of the oxidative burst in the activation of a physiological cell death program.

The importance of H_2O_2 in signal transduction has been demonstrated by using inhibitors of catalase, the enzyme which degrades H_2O_2. 2,6-Dichloroisonicotinic acid (INA) and benzothiadiazole (BTH) induce defense genes like those which encode pathogenesis-related proteins (*PR* genes) and induce enhanced resistance to pathogens. Both these synthetic inducers of defense mechanisms inhibit tobacco catalase (Conrath et al., 1995). Salicylic acid inhibits catalase's H_2O_2-degrading activity and induces *PR* gene expression (Chen et al., 1993b). Salicylic acid, INA, and BTH inhibit the activity of tobacco ascorbate peroxidase, the other key H_2O_2-scavenging enzyme, in vitro (Durner and Klessig, 1995). Antioxidants suppress salicylic acid-, INA-, or BTH-induced *PR-1* gene expression and disease resistance (Conrath et al., 1995; Wendehenne et al., 1998). Transgenic tobacco plants with severely depressed levels of catalase due to either sense cosuppression or antisense suppression were constructed (Chamnongpol et al., 1996; Takahashi et al., 1997). These plants show enhanced disease resistance. It suggests that H_2O_2 is involved in induction of disease resistance.

Salicylic Acid

Salicylic acid has been reported as one of the important signal molecules that act locally in intracellular signal transduction. It is a phenolic compound commonly present in the plant kingdom. It was found in all 34 plant species tested for its content (Raskin et al., 1990). Plants most likely synthesize salicylic acid (*o*-hydroxybenzoic acid) from *trans*-cinnamic acid, the product of the activity of phenylalanine ammonia-lyase (PAL) (Yalpani et al., 1993), which is a key regulator of the phenylpropanoid pathway, which yields a variety of phenolics with structural and defense-related functions.

Two pathways for the formation of salicylic acid from phenylalanine by the action of PAL have been suggested. In the first pathway, *trans*-cinnamic acid (the product of PAL action) is converted first to 2-hydroxycinnamic acid and then through oxidation to salicylic acid. In the alternative pathway, *trans*-cinnamic acid is β-oxidized to benzoic acid and then ortho-hydroxylated to salicylic acid (Ward et al., 1991). Rice shoots converted (^{14}C) cinnamic acid to salicylic acid and the lignin precursors *p*-coumaric acid and ferulic acids, whereas (^{14}C) benzoic acid was readily converted to salicylic acid (Silverman et al., 1995). It suggests that salicylic acid may be synthesized through a benzoic acid pathway.

When cucumber leaves were inoculated with *Pseudomonas syringae* pv. *lachrymans,* induction of synthesis of salicylic acid from both phenylalanine and benzoic acid was observed. Leaf discs from plants inoculated with *P. syringae* pv. *lachrymans* incorporated more (^{14}C) phenylalanine into (^{14}C) salicylic acid than mock-inoculated controls (Meuwly et al., 1995). A specific inhibitor of phenylalanine ammonia-lyase, 2-aminoindan-2-phosphonic acid, completely inhibited the incorporation of (^{14}C) phenylalanine into (^{14}C) salicylic acid, although plants treated with 2-aminoindan-2-phosphonic acid could still produce (^{14}C) salicylic acid from (^{14}C) benzoic acid (Meuwly et al., 1995). It suggests that salicylic acid may be synthesized through a benzoic acid pathway in cucumber.

Salicylic acid accumulates in *Arabidopsis* plants inoculated with *P. syringae* pv. *tomato* (Shah et al., 1997). In cucumber plants inoculated with *P. syringae* pv. *lachrymans,* salicylic acid accumulated (Meuwly et al., 1995). Salicylic acid content increased in 9 and 12 h post-inoculation with *P. syringae* pv. *lachrymans* (Meuwly et al., 1994). Exogenously applied salicylic acid induces expression of pathogenesis-related genes in tobacco (Antoniw and White, 1980; Yalpani et al., 1991) and *Arabidopsis* (Uknes et al., 1993b) and confers resistance to pathogen attack. Salicylic acid-induced resistance to *Erwinia carotovora* subsp. *carotovora* has been reported in tobacco (Palva et al., 1994; Lopez-Lopez et al., 1995). In many plants, following infection a strong correlation between increased levels of salicylic acid and both the expression of salicylic acid-inducible defense-related genes and disease resistance has been reported (Malamy et al., 1990; Metraux et al., 1990).

Salicylic acid may enhance release of H_2O_2 and H_2O_2-derived active oxygen species and induce activities of defense-related genes (Shirasu et al., 1997). Salicylic acid–binding protein has been detected in plants, and it has been identified as catalase (Conrath et al., 1995). Catalase suppresses H_2O_2 activity, and salicylic acid suppresses the H_2O_2-degrading activity of catalase, both in vivo and in vitro (Chen et al., 1993a; Conrath et al., 1995). The other major H_2O_2-scavenging enzyme, ascorbate peroxidase, is also inhibited by salicylic acid (Durner and Klessig, 1995). Elevated levels of H_2O_2 resulting from the inhibition of catalase and ascorbate peroxidase might be directly or indirectly involved in the activation of defense responses (Chen et al., 1993a; Conrath et al., 1995; Dempsey and Klessig, 1995; Durner and Klessig, 1995). H_2O_2 and H_2O_2-derived active oxygen species are known to function in the signal transduction pathway (Du and Klessig, 1997).

G-Proteins

Guanosine triphosphate-binding (GTP-binding) regulatory proteins (G-proteins) are present in plasma membranes of cells of several plants (Ma et al., 1991; Palme et al., 1992; Ishikawa et al., 1995; Seo et al., 1995). G-proteins can cause Ca^{2+} release from membrane vesicles and may be involved in Ca^{2+} movement, which is important in signal transduction (Trewavas and Gilroy, 1991). Stimulation of H^+-ATPase due to a fungal elicitor treatment in tomato cells has been reported due to a membrane-bound phosphatase activated through a G-protein (Vera Estrella et al., 1994). G-proteins can regulate phospholipase also (Freissmuth and Gilman, 1991).

In animals, cholera toxin can activate signaling pathways dependent on heterotrimeric G-proteins. Tobacco plants were transformed with a gene encoding the A1 subunit of cholera toxin. Tissues of transgenic plants expressing cholera toxin showed increased resistance to *P. syringae* pv. *tabaci* (Beffa et al., 1995). In the transgenic plants, high levels of salicylic acid accumulated and the PR proteins PR-1, class II isoforms of PR-2, and PR-3 also accumulated. Contrastingly, there was no induction of class I isoforms of PR-2 and PR-3. Microinjection of cholera toxin also induced PR-1 gene in tobacco plants and not class I isoform of PR-2 (Beffa et al., 1995). These results indicate that cholera toxin-sensitive G-proteins are important in inducing a subset of PR proteins and disease resistance.

What May Be the Sequence of Events in the Intracellular Signal Transduction System?

The signal transduction system involves pathogen signals, calcium ion flux, K^+efflux, protein phosphorylation, membrane potential, active oxygen species, and salicylic acid and other systemic signals. The possible sequence of events in signal transduction has been studied in different plant-pathogen interactions. To investigate whether the protein kinase activation in parsley cells by a fungal elicitor depended on the activity of ion channels that mediate rapid ion fluxes across the cell membrane, parsley cells were incubated with the ion channel blocker anthracene-9-carboxylate (A9C), which inhibits the elicitor-stimulated ion fluxes, thereby blocking all subsequent responses (Ligterink et al., 1997). Under these conditions, the elicitor activation of the protein kinase was completely inhibited, indicating that ion channel activation was also necessary for this reaction. Protein phosphorylation appears to be in the downstream of pathogen signals and ion flux (Ligterink et al., 1997).

The elicitor-induced responses in tobacco suspension cells such as K^+ efflux, active oxygen species production, and membrane depolarization are blocked by a Ca^{2+} channel blocker and K25a, a protein kinase inhibitor (Baker et al., 1993; He et al., 1994; Popham et al., 1995; Hoyos et al., 1996). It suggests that the various cascade of host responses following exposure to a pathogen product resulting in the transcriptional activation of defense-related genes may be in the downstream of Ca^{2+} influx and protein phosphorylation. However, Levine et al. (1996) presented evidence that H_2O_2 stimulates a rapid influx of Ca^{2+} into soybean cells, suggesting a signal function for Ca^{2+} downstream of the oxidative burst.

The elicitor-stimulated production of reactive oxygen species is catalyzed by an NADH (nicotinamide adenine dinuclotide [reduced]) or NADPH (nicotinamide adenine dinucleotide phosphate [reduced]) oxidase that is inhibited by diphenylene iodonium (DPI). In elicitor-treated parsley cells, DPI blocked the oxidative burst, defense gene activation, and phytoalexin accumulation without affecting ion fluxes (Ligterink et al., 1997). It suggests that the oxidative burst may be in the downstream of ion channels in the elicitor signal transduction cascade. Elicitor activation of the protein kinase was not inhibited by DPI, indicating that this kinase may act upstream of the oxidative burst (Ligterink et al., 1997).

Further evidence showing that production of H_2O_2 may be at the downstream of phosphorylation have been provided by Sreeganga et al. (1996). Two transgenic tomato cell suspension cultures (± *Pto* gene) were tested for production of H_2O_2 following independent challenge with two *P. syringae* pv. *tomato* (*avrPto* gene) strains. Only when *Pto* and *avrPto* genes are present in the corresponding organisms, two distinct phases of the oxidative burst were seen, a rapid first burst followed by a slower and more prolonged second burst. In three other plant-pathogen interactions, either no burst or only a first burst was observed, indicating that the second burst is correlated with induction of disease resistance (Sreeganga et al., 1996).

Salicylic acid, another signal molecule, is involved in accumulation of H_2O_2 and other active oxygen species. Induction of H_2O_2 accumulation by avirulent *Pseudomonas savastanoi* pv. *glycinea* in soybean cell suspensions could be blocked by the phenylpropanoid synthesis inhibitor (α-aminooxy-β-phenylpropionic acid) and this response could be rescued by exogenous application of salicylic acid (Shirasu et al., 1997). Salicylic acid is synthesized via a phenylpropanoid pathway. The results indicate that H_2O_2 accumulation may be in the downstream of salicylic acid accumulation.

Salicylic acid may act downstream of pathogen elicitor signals. Salicylic acid administered in the absence of a pathogen is only a weak inducive signal (Shirasu et al., 1997). Salicylic acid had negligible effects when administered at 10 to 100 μM to soybean cell suspensions in the absence of a pathogen. But at these concentrations of salicylic acid, induction of defense gene transcripts was observed in soybean cells in the presence of an avirulent strain of *P. savastanoi* pv. *glycinea* (Shirasu et al., 1997). Endogenous salicylic acid accumulates to concentrations of <70 μM at the site of attempted infection by pathogens, and at this physiological concentration exogenous application of salicylic acid induces defense mechanisms only in the presence of a pathogen (Shirasu et al., 1997). These results suggest that pathogen signals are needed for the action of salicylic acid, and salicylic acid may amplify pathogen signals.

G-proteins are also involved in signal transduction. Activation of G-proteins results in enhanced synthesis of salicylic acid, and hence G-proteins may act upstream of the salicylic acid synthesis pathway (Beffa et al., 1995).

Targeting of Signals to Plant Cell Nucleus

Signals generated by the bacterial genes should enter the nucleus of the plant through second messengers to induce host defense gene expression. Several bacterial *avr* gene products have been found located in the bacterial cytoplasm. Knoop et al. (1991) used an antiserum to demonstrate that most of the AvrBs3 protein was located in the cytoplasm. Brown et al. (1993) showed that the *avrBs3* gene product is located in the cytoplasm of *Xanthomonas vesicatoria* in planta and in bacterial cultures. AvrXa 10 was detected primarily in the cytoplasmic fraction of *X. oryzae* pv. *oryzae,* with less than 12 percent present in the membrane fraction (Young et al., 1994). No AvrXa10 protein was detected in concentrated culture supernatants of *X. oryzae* pv. *oryzae* cells or from intercellular fluids of rice leaves after infiltration with *X. oryzae* pv. *oryzae* (Young et al., 1994). These observations indicate that these Avr proteins are not located outside the bacterial cells.

avrXa10 is expressed constitutively in *X. oryzae* pv. *oryzae* and is not induced to higher levels in planta (Young et al., 1994). *avrBs3, pthA,* and *avrb6* have also been shown to be expressed constitutively in *X. vesicatoria, X. savastanoi* pv. *citri,* and *X. savastanoi* pv. *malvacearum,* respectively (Knoop et al., 1991; Swarup et al., 1992; De Feyter et al., 1993). These results suggest that the *avr* gene may not interact directly with plant cells.

There should be second messengers to carry the signals generated by these *avr* gene products into plant cells.

Host resistance gene (R) products may function as receptors while phytopathogen *avr* products may function as ligands (Gabriel and Rolfe, 1990). The predicted amino acid sequences of PthA, Avrb6, AvrBs3, and AvrXa10 contain tandem repeats that are leucine rich (Yang and Gabriel, 1995). Repetitive domains often function as binding sites (McConkey et al., 1990). Further analysis of amino acid sequences of PthA, Avrb6, AvrBs3, and AvrXa10 revealed the presence of leucine zipperlike heptad repeats closely linked to the amino acid tandem repeats of all these proteins (Yang and Gabriel, 1995).

Transport of proteins into the plant cell nucleus is an active process and the proteins should contain suitable nuclear localization signals (NLS) (Nigg et al., 1991). Most NLSs consist of either the monopartite motif K-R/K-X-R/K (usually adjacent to proline) or a bipartite motif consisting of 2 basic amino acids, a spacer region of any 4 to 10 amino acids, and a cluster of 3 to 5 basic amino acids (Chelsky et al., 1989; Dingwall and Laskey, 1991). NLSs may be present in multiple copies, and the effects of multiple copies are additive (Garcia-Bustos et al., 1991).

Agrobacterium tumefaciens encodes at least two proteins with NLSs that direct the proteins to the plant cell nucleus (Citrvosky and Zambryski, 1993). Several *avr* genes identified in the genus *Xanthomonas,* viz., *avrBs3, avrBs3-2, avrB4, avrb6, avrb7, avrBln, avrB101,* and *avrB102* of *X. campestris, pthA* of *X axonopodis. citri,* and *avrXa5, avrXa7,* and *avrXa10* of *X. oryzae* pv. *oryzae* show three putative NLSs (Yang and Gabriel, 1995). The DNA coding sequences for the C-terminal regions of *avrb6* and *pthA* were independently fused to a β-glucuronidase (GUS) reporter gene (Yang and Gabriel, 1995). When introduced into onion cells, both of these translational fusions were transiently expressed, and GUS activity was specifically localized in the nuclei of transformed cells (Yang and Gabriel, 1995). The results demonstrate that the *Xanthomonas avr/pth* gene fragments encode NLSs.

NLS regions seem to be required for activity for at least 10 members of the *avr* gene family of *Xanthomonas* (Swarup et al., 1991, 1992; Hopkins et al., 1992; De Feyter et al., 1993; Bonas et al., 1993). It suggests that the *Avr* proteins are imported into the plant cells, perhaps by receptor-mediated endocytosis (Horn et al., 1989), and that the proteins can be translocated into the plant nucleus (Yang and Gabriel, 1995).

Only a few disease resistance genes have been cloned from plants. But products of all these genes appear to share striking structural similarities, and they appear to be involved in certain signaling events in plant defense mechanisms (Baker et al., 1997). Analyses of several cloned plant resistance genes revealed the presence of leucine-rich repeats, leucine zippers, and nuclear localization signals (Johnson and McKnight, 1989; Mindrinos et al., 1994).

The RPS2 resistance gene of *Arabidopsis thaliana* confers resistance against *Pseudomonas syringae* pv. *tomato* strains that express avirulence gene *avrRpt2*. This gene has been cloned (Bent et al., 1994). Analysis of the derived amino acid sequence for RPS2 revealed several regions with similarity to known polypeptide motifs. Most prominent among these is a region of leucine-rich repeats (LRRs). The deduced amino acid sequence for RPS2 carries a leucine zipper domain (Bent et al., 1994; Ausubel et al., 1995).

Another resistance gene *RPM1* was identified in *A. thaliana* accession Col-O as conferring resistance to *P. syringae* isolates expressing the *avrRpm1* gene (Dangl et al., 1992). Resistance in *A. thaliana* to the *P. syringae avrB* gene also mapped to the *RPM1* interval (initially termed *RPS3*) (Innes et al., 1993a) and genetic analyses of *A. thaliana* mutants have suggested that *RPM1* conferred resistance to *P. syringae* expressing either *avrRpm1* or *avrB* (Grant et al., 1995). The *RPM1* open reading frame contains a potential six-heptad amphipathic leucine zipper (positions 10 to 51), two motifs of a nucleotide binding site (positions 200 to 208 and 279 to 288) and 14 imperfect LRRs from position 553 (Grant et al., 1995). These features most closely resemble those of the *A. thaliana RPS2* gene (23 percent identity and 51 percent similarity) (Grant et al., 1995). *Xa21* is another disease resistance gene cloned from rice. The cloned *Xa21* gene displays high levels of resistance to *X. oryzae* pv. *oryzae* (Song et al., 1995). The sequence of the predicted protein carries a leucine-rich repeat motif and a serine-threonine kinase-like domain (Song et al., 1995).

The disease resistance gene *N* of tobacco has been cloned (Whitham et al., 1994). It mediates resistance to the viral pathogen tobacco mosaic virus. It encodes a protein of 131.4 kDa with an amino-terminal domain similar to that of the cytoplasmic domain of the *Drosophila* Toll protein and the interleukin-1 receptor in mammals, a nucleotide-binding site (NBS), and four imperfect LRRs. The gene products of *N* and *RPS2* (*Arabidopsis* gene) share common domains and overall sequence similarity (Whitham et al., 1994).

Another resistance gene *Prf* in tomato has been cloned, and it confers resistance to *P. syringae* pv. *tomato* (Salmeron et al., 1996). *Prf* encodes protein containing a putative leucine zipper motif, a leucine-rich repeat domain and a nucleotide binding site (Baker et al., 1997). Other disease resistance genes (showing resistance to fungal pathogens) *L6* and *M* from flax, *RPP5* from *Arabidopsis,* and *I2* from tomato also encode proteins with the LRR-NBS motif (Lawrence et al., 1995; Baker et al., 1997). The tomato resistance genes *Cf-2* and *Cf-9* encode LRR domains (Dixon et al., 1996; Jones et al., 1994). Functional homologs of *Arabidopsis RPM1* exist in pea, bean, and soybean (Dangl et al., 1992). All of them encode proteins with LRR, leucine zipper, and NBS domains. Homolog or homologs of disease resistance (R) locus may be present at the corresponding position in susceptible plants (Martin et al., 1993b; Bent et al., 1994; Mindrinos et al., 1994; Whitham et al., 1994; Jones et al., 1994).

LRRs are contained in a variety of proteins and have been suggested to be involved in protein-protein interactions in a downstream step of the signal transduction pathway (Kobe and Deisenhofer, 1993). They may be involved in ligand binding in a diverse array of proteins (Bent et al., 1994). Leucine zippers and LRRs serve as the sites of protein-protein binding; leucine zippers may directly interact with DNA (Johnson and Mcknight, 1989). The NLS may direct transport of the signals to the plant nucleus (Yang and Gabriel, 1995).

Protein kinase may also be involved in binding the nuclear factors to an elicitor response element (Subramaniam et al., 1997). Treatment of potato tuber discs with specific inhibitors of protein kinase abolished elicitor-induced binding of the nuclear factor PBF-2 to the elicitor response element. This correlated with a reduction in the accumulation of the pathogenesis-related protein PR-10a. In contrast, treatment of potato tuber discs with 12-0-tetradecanoyl-phorbol 13-acetate (TPA), an activator of protein kinase, led to an increase in the binding of PBF-2 to the elicitor response element and a corresponding increase in the level of the PR-10a protein (Subramaniam et al., 1997).

The protein products of disease resistance genes may be cytoplasmic (Mindrinos et al., 1994). If the resistance gene product is a cytoplasmic receptor, either the *avr*-generated ligand (signal) must be a membrane-permeable signal or there must be a primary receptor that converts the *avr*-generated signal into a secondary cytoplasmic signal. Signal transduction in plants may bear similarity to the mammalian signal transduction pathways, which typically involve receptor-binding, transiently increased cellular calcium levels, changes in protein phosphorylation and membrane potential, and de novo protein synthesis (Atkinson et al., 1993).

It is possible that leucine-rich repeats mediate direct interactions between pathogen Avr and plant resistant/susceptibility proteins to form dimers or multimers (Yang and Gabriel, 1995). In at least some cases, the NLS signals may direct transport of the proteins to the plant nucleus. After entering the nucleus, the Avr proteins or protein complexes may act on nuclear transcriptional factors leading to pathogenesis or disease resistance.

SYSTEMIC SIGNAL INDUCTION

Salicylic Acid

Salicylic Acid Induces Defense Gene Activation

The signals generated due to bacterial infection are translocated systemically and the plants synthesize several defense chemicals. Salicylic acid has been implicated as one of the key components in the systemic signal transduction pathway leading to plant resistance to various pathogens (Ryals et al., 1996). Following infection with pathogens, salicylic acid accumulates systemically in plants. When cucumber plants were inoculated with *P. syringae* pv. *syringae,* an incompatible pathogen, salicylic acid accumulated systemically in all leaves (Rasmussen et al., 1991). Uninoculated plants did not contain detectable levels of salicylic acid (Rasmussen et al., 1991). When *Arabiodopsis thaliana* was inoculated with the incompatible pathogen *P. syringae* pv. *syringae,* there was a local and systemic increase in salicylic acid (Summermatter et al., 1995).

When tobacco leaves were fed salicylic acid, pathogenesis-related (PR) protein PR-1 was induced (Yalpani et al., 1991). Induction of PR-1 proteins was positively correlated with leaf salicylic acid content (Yalpani et al., 1991). Salicylic acid application induced nine gene families that encode PR proteins in tobacco (Ward et al., 1991b). These gene families include the PR proteins PR1a, PR-1b, PR-1c, PR-2a, PR-2b, PR-2c, PR-3a, PR-3b, PR-4a, PR-4b, PR-5a, PR-5b, PR-8 (Class III chitinase), and PR-Q', which are all known to be involved in disease resistance (Ward et al., 1991b). Accumulation of salicylic acid in a mutant *Arabidopsis thaliana* plant, which contains lesions in a single accelerated cell death (ACD) gene called *ACD2* and bypasses the need for a pathogen to induce the HR, has been reported (Greenberg et al., 1994). *Arabidopsis* mutants (Cpr1, lsd6, and lsd7) that contain a high level of salicylic acid show induction of PR proteins and enhanced resistance to pathogens (Bowling et al., 1994; Weymann et al., 1995).

Transgenic potato plants exhibiting constitutive expression of a bacterio-opsin proton pump derived from a bacterium *Halobacterium halobium* had increased levels of salicylic acid and overexpressed several pathogenesis-related mRNAs encoding PR-2, PR-3, PR-5, and acidic anionic peroxidase (PR-9) (Abad et al., 1997). The transgenic tobacco plants expressing a bacterio-opsin proton pump contained higher systemic levels of salicylic acid, over-expressed several mRNAs encoding PR proteins, and displayed enhanced resistance to *P. syringae* pv. *tabaci* (Mittler et al., 1995).

Salicylic acid inhibits action of catalase in degrading H_2O_2, which induces disease resistance as a second messenger (Chen et al., 1993b). Transgenic tobacco plants with severely depressed levels of catalase have been developed with antisense suppression (Takahashi et al., 1997). These antisense-suppressed, catalase-deficient tobacco plants exhibited elevated levels of salicylic acid. These plants accumulate PR proteins and show enhanced disease resistance (Takahashi et al., 1997).

Prevention of Accumulation of Salicylic Acid Suppresses Host Defense Mechanisms

Transgenic tobacco and *Arabidopsis* plants in which accumulation of salicylic acid is prevented have been developed by using a bacterial gene *(nahG)* expressing salicylate hydroxylase that degrades salicylic acid to catechol (Gaffney et al., 1993; Lawton et al., 1995). The tobacco plants expressing the *nahG* gene from *Pseudomonas putida* could not accumulate salicylic acid following pathogen infection (Gaffney et al., 1993). Systemic acquired resistance in *Arabidopsis* can be induced by inoculation with *P. syringae* pv. *tomato* or by exogenous salicylic acid, but this response was abolished in transgenic, *nahG*-expressing *Arabidopsis* (Lawton et al., 1995). Preventing the accumulation of salicylic acid in plants through ectopic expression of the *nahG* gene prevents induction of PR genes by salicylic acid and increases susceptibility of these plants to both virulent and avirulent pathogens (Gaffney et al., 1993; Delaney et al., 1994; Lawton et al., 1995). The results suggest the importance of salicylic acid in induction of defense genes.

Signals Upstream of the Salicylic Acid Signal Transduction System

Some systemic signals may act upstream of salicylic acid signal transduction (Rasmussen et al., 1991). Leaf 1 of cucumber plants was detached at 2, 4, 6, 8, and 24 h after inoculation with *P. syringae* pv. *syringae* to more precisely determine the time of appearance of salicylic acid (Rasmussen et al., 1991). Salicylic acid was not detected in phloem exudates of leaf 1 at 2, 4, or, 6 h after inoculation; but could be detected at 8 h after inoculation (Table 1.9; Rasmussen et al., 1991). However the induction of systemic resistance was seen in the plants that had leaf 1 on the plant for 6 h or longer. If salicylic acid is the primary systemically translocated signal, detectable amounts should begin to appear in leaf 1 within this 6 h period; but no detectable amount of salicylic acid accumulated till 6 h. Thus, the systemically

TABLE 1.9. Effect of Detaching Leaf 1 at Intervals After Inoculation with *Pseudomonas syringae* pv. *syringae* on Appearance of Salicylic Acid in Petiole of Leaf 2 of Cucumber Plants

Time leaf 1 detached (h)	Salicylic acid (mM) in leaf 1 petiole phloem exudate at time leaf 1 detached	Salicylic acid (mM) in leaf 2 petiole phloem exudate 24 h after inoculation of leaf 1
2	ND	ND
4	ND	71
6	ND	236
8	31	350

Source: Rasmussen et al., 1991.

ND = Not detected

accumulating salicylic acid does not appear to originate from the inoculated leaf, but rather appears to be induced by yet another systemically translocated signal (Table 1.9; Rasmussen et al., 1991).

Active oxygen species may be involved in the systemic signal transduction pathway. Ozone exposure triggers induced accumulation of $H_2O_{2,k}$, which induced salicylic acid in *Arabidopsis thaliana* (Sharma et al., 1996). H_2O_2 and H_2O_2-generating chemicals induce PR proteins at much higher levels in wild-type tobacco than in *NahG* transgenic plants, where the salicylic acid signal is destroyed (Bi et al., 1995; Neuenschwander et al., 1995). Application of very high concentrations of H_2O_2 was found to stimulate salicylic acid accumulation (Leon et al., 1995; Neuenschwander et al., 1995; Summermatter et al., 1995). Antioxidants suppress salicylic acid-induced PR gene expression (Conrath et al., 1995). These results suggest that H_2O_2 may function upstream of the salicylic acid signal induction pathway.

Downstream Events in the Salicylic Acid Signal Transduction System

Downstream of the salicylic acid signal transduction system, some regulators of the salicylic acid signal transduction pathway may exist. Four mutants of *Arabidopsis* are known that block the transmission of the salicylic acid signal leading to the expression of the PR genes and disease resistance in *Arabidopsis*. The *npr1* mutant does not express the salicylic acid-inducible PR-2 protein (Cao et al., 1994). The *eds5* and *eds53* mutants that are allelic to *npr1* were obtained based on their enhanced susceptibility to *Pseudomo-*

nas syringae pv. *maculicola* (Glazebrook et al., 1996). All these mutants show increased susceptibility to pathogens. In these four mutants the salicylic acid-induced PR genes are not induced by salicylic acid. The recessive nature of these mutations suggests that the corresponding wild-type alleles are positive regulators of the salicylic acid signal transduction pathway.

A *s*alicylic *a*cid-*i*nsensitive *(sai1) Arabidopsis* mutant that is allelic to *npr1* has been identified (Shah et al., 1997). The *sai1* mutant is unable to express the PR-1, PR-2, and PR-5 genes at high levels in response to salicylic acid (Shah et al., 1997). 2,6-Dichloroisonicotinic acid (INA), chemical inducer of plant defense mechanisms (Conrath et al., 1995; Gorlach et al., 1996; Lawton et al., 1996), does not stimulate salicylic acid production but induces defense mechanisms (PR gene induction) in *NahG* transgenic plants that fail to accumulate salicylic acid (Vernooij et al., 1995; Malamy et al., 1996). But INA was unable to induce expression of endogenous PR-1, PR-2, and PR-5 genes in *sai1* mutant plants (Shah et al., 1997). Similar to salicylic acid and INA, BTH (benzothiadiazole), a commercially available activator of induced resistance, also did not induce PR-1 gene expression in *sai1* plants (Shah et al., 1997). These results suggest that INA and BTH may act downstream of salicylic acid (Conrath et al., 1995; Durner and Klessig, 1995; Malamy et al., 1996). These results also suggest that the wild type *SAI1* gene may be an important component in the salicylic acid signal transduction pathway.

sai1 plants accumulate endogenous levels of salicylic acid when infected with *Pseudomonas syringae* pv. *tomato* containing the *avrRpt2* gene (Shah et al., 1997). It rules out the possibility that the mutant phenotype of *sai1* plants is due to a defect in metabolism of salicylic acid. Thus it appears that *sai1* mutation is in a bonafide component of the salicylic acid signal transduction pathway. The wild-type *SAI1* gene may function as a positive regulator of the salicylic acid signal transduction pathway.

H_2O_2 and H_2O_2-derived active oxygen species may act downstream of the salicylic acid signal transduction pathway (Chen et al., 1993a,b; Conrath et al., 1995; Du and Klessig, 1997). But H_2O_2 alone may not be sufficient to induce PR genes. Some transgenic tobacco plants expressing the *nahG* gene (which encodes salicylate hydroxylase) produced little, if any, PR-1 protein, even though they had very low levels of catalase (allowing H_2O_2 production) (Du and Klessig, 1997). Severely reduced catalase activity appears to be insufficient for the activation of certain defense responses in the absence of elevated levels of salicylic acid (Du and Klessig, 1997). Salicylic acid appears to be important in the systemic signal transduction system. The importance of salicylic acid in systemic signal transduction has been reported in many plants (Gaffney et al., 1993; Delaney et al., 1994; Chen et al., 1995; Lawton et al., 1995; Hunt and Ryals, 1996; Fauth et al., 1996; Kessmann et al., 1996; Mur et al., 1996; Shirasu et al., 1996; Ryals et al., 1996; Keller et al., 1996b; Hunt et al., 1997).

Systemic Signal Induction Pathways Independent of Salicylic Acid May Be Present in Plants

Salicylic acid may not be involved in all systemic signal transduction pathways, although it may be an important component in many plant-pathogen interactions. Salicylic acid content did not increase after inoculation with the avirulent pathogen *Pseudomonas syringae* strain D20 or with the rice pathogens *Magnaportha grisea* and *Rhizoctonia solani* (Silverman et al., 1995). Exogenous application of salicylic acid strongly induced the PR-1a mRNA in transgenic tobacco plants expressing a peroxidase gene, *Shpx6b,* of the legume *Stylosanthus humilis*. But *Shpx6b* was not induced by salicylic acid in the transgenic tobacco plants (Curtis et al., 1997). The thionin (Epple et al., 1995) and defensin (Penninckx et al., 1996) genes of *Arabidopsis thaliana* were not induced by salicylic acid. But all these genes were induced by methyl jasmonate. There may be a second pathway independent of salicylic acid. A salicylic acid insensitive *(sai1) Arabidopsis* mutant has been identified. Exogenous application of salicylic acid did not induce PR-1 gene expression in *sai1* plants (Shah et al., 1997). But infection with *P. syringae* pv. *tomato* containing the *avrRpt2* gene (incompatible) resulted in the accumulation of PR-1 mRNA (Shah et al., 1997). It suggests presence of a second pathway, independent of SAI1 for the induction of the PR-1 gene. Similar results have also been reported by Delaney et al. (1995) and Glazebrook et al. (1996) with mutant alleles of *nim1* and *npr1,* respectively. In these mutants also, the PR genes are not induced by salicylic acid.

The transgenic *Arabidopsis* plants, which are unable to accumulate salicylic acid due to the expression of the *nahG* gene (that encodes salicylate hydroxylase which degrades salicylic acid), accumulate PR-1 transcripts after infection with *P. syringae* pv. *tomato* containing the *avrRpt* gene at levels higher than those seen in mock-infected leaves (Delaney et al., 1994; Lawton et al., 1995). When tobacco plants were infected with *Erwinia carotovora* subsp. *carotovora* (incompatible interaction), PR genes were induced (Vidal et al., 1997). When transgenic tobacco plants expressing *nahG* genes that overproduce a salicylate hydroxylase inactivating salicylic acid were inoculated with the bacteria, similar induction of PR genes was observed (Vidal et al., 1997). These results provide evidence for the existence of a second pathway for PR gene induction that is independent of salicylic acid. Induction of plant defense genes by *Erwinia* and salicylic acid seems to be by two distinct pathways.

Methyl Jasmonate and Jasmonic Acid

Methyl jasmonate and jasmonic acid (JA) are naturally occurring compounds in plants (Yamane et al., 1981; Meyer et al., 1984; Anderson, 1985; Farmer and Ryan, 1990; Falkenstein et al., 1991; Siedow, 1991; Staswick, 1992; Creelman and Mullet, 1995; Nojiri et al., 1996; Schweizer et al., 1997a). Jasmonic acid and its methyl ester have been shown to originate from linolenic acid. De novo synthesis of jasmonate may be controlled by the availability of free linolenic acid substrate (Farmer and Ryan, 1992). Lipoxygenase (LOX) converts linolenic acid into jasmonic acid or its precursor 12-*oxo*-phytodienoic acid (Sembdner and Parthier, 1993). Biosynthesis of jasmonic acid in elicitor-treated tobacco cells is suppressed when the cells were preincubated with eicosatetraynoic acid (ETYA), an inhibitor of LOX (Rickauer et al., 1997). It suggests that jasmonic acid is synthesized through the LOX pathway.

LOX enzyme activity was associated with disease resistance (Keppler and Novacky, 1987; Croft et al., 1990, 1993; Rickauer et al., 1990; Ohta et al., 1991; Koch et al., 1992). Enhanced LOX activity might result in the activation of the octadecanoid signaling pathway with jasmonic acid as the central component (Sembdner and Parthier, 1993). Methyl jasmonate and/or jasmonic acid induce many defense genes encoding phenylalanine ammonia-lyase (Gundlach et al., 1992), phytoalexins (Tamogami et al., 1997), proline-rich cell wall protein (Creelman et al., 1992), and PR-6 protein (Farmer and Ryan, 1990). It enhances elicitation of active oxygen species (Kauss et al., 1994). Jasmonic acid is an enhancer of defense responses (Xu et al., 1994; Kauss et al., 1994; Graham and Graham, 1996; Nojiri et al., 1996; Schweizer et al., 1997a,b).

2,6-Dichloroisonicotinic acid, a synthetic inducer of plant defense mechanisms, may act through a signaling network comprising the octadecanoid pathway (Schweizer et al., 1997b). INA increased jasmonic acid content in rice leaves (Figure 1.6; Schweizer et al., 1997b).

Exogenously applied jasmonic acid enhanced INA-induced accumulation of PR1 and ASP31 proteins in rice leaves (Schweizer et al., 1997b). Transcripts of two closely related peroxidase isogenes, *Shpx6a* and *Shpx6b,* of the legume *Stylosanthes humilis* were rapidly induced following treatment with methyl jasmonate (Curtis et al., 1997). Jasmonic acid application induced the accumulation of a number of PR gene products in rice (Schweizer et al., 1997a). Accumulation of PR1-like proteins was reduced in plants treated with tetcyclacis, an inhibitor of jasmonate biosynthesis (Schweizer et al., 1997a). Thus jasmonic acid may play an important role in signal transduction.

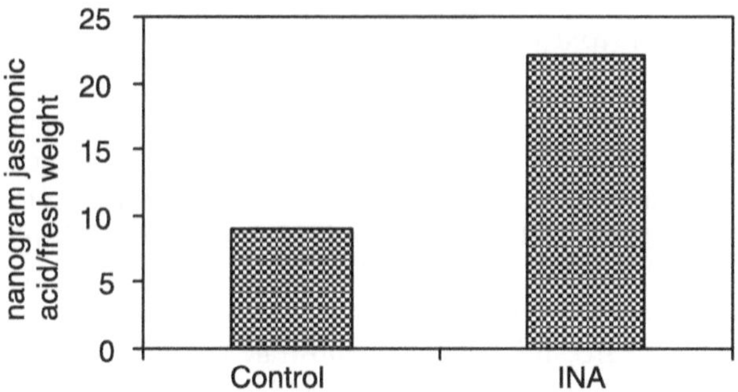

FIGURE 1.6. Induction of jasmonic acid in rice leaves after soil drench application of INA (100 ppm) (*Source:* Schweizer et al., 1997b).

However, jasmonic acid alone may not be sufficient to induce resistance. An elicitor induced sesquiterpene cyclase gene in tobacco cells; but methyl jasmonate did not induce it (Rickauer et al., 1997). Although the octadecanoid signaling pathway was inducible by INA in rice, resistance induction by a low dose of INA was not associated with elevated levels of endogenous jasmonic acid (Schweizer et al., 1997b). Exogenously applied jasmonic acid enhanced INA-induced resistance and accumulation of the protein ASP 31 in rice leaves. But JA by itself could not induce accumulation of this protein (Schweizer et al., 1997b). It suggests that some other signal may have synergistic effect with jasmonic acid in inducing synthesis of defense-related proteins.

Ethylene

Increased biosynthesis of ethylene has been observed due to pathogen attack (Toppan and Esquerre-Tugaye, 1984). Ethylene is known to regulate defense genes in plants (Ecker and Davis, 1987; Boller, 1988; Ishige et al., 1991). Ethylene caused osmotin mRNA accumulation in tobacco. After 24 h of ethylene treatment, five-day-old tobacco seedlings containing an osmotin promoter-β-glucuronidase (GUS) fusion gene exhibited a fivefold increase in GUS activity (Xu et al., 1994). The basic PR-1 transcript is readily induced by ethylene in tobacco (Eyal et al., 1992). But acidic PR-1 transcript levels did not respond to ethylene (Lotan and Fluhr, 1990a). However, it was induced with α-aminobutyric acid, which depends on ethylene evolution for elicitation (Lotan and Fluhr, 1990a). It suggests that the ethylene is necessary but not sufficient for induction of acidic type PR-1 expression. There may be another signal which may synergistically act with ethylene to induce acidic PR-1 protein in tobacco. Ten PR proteins have been detected in tomato. Nine of them are induced by ethylene while the tenth PR protein, C_{10}, was not induced by it (Granell et al., 1987). These results suggest that different signals may be necessary to induce different defense genes.

Role of Salicylic Acid, Ethylene, and Methyl Jasmonate in Coordinating Defense Gene Responses

Signal molecules may differ in inducing different defense genes. Sometimes a combination of signal molecules may be needed to induce a single defense gene. The osmotin gene was induced by ethylene while salicylic acid did not induce the osmotin gene in tobacco (Xu et al., 1994). The osmotin gene was not induced by methyl jasmonate also (Xu et al., 1994). Methyl jasmonate induced the PR protein PRm in maize, but could not induce another maize PR protein MPI (Cordero et al., 1994).

Signal combinations may synergistically hyperinduce plant defense genes. The osmotin promoter in tobacco was not responsive to treatment with methyl jasmonate alone. But when methyl jasmonate was applied in combination with ethylene, the activity of the osmotin promoter was induced dramatically beyond that seen with ethylene or methyl jasmonate in whole tobacco seedlings containing an osmotin promoter-β-glucuronidase (GUS) fusion gene (Table 1.10; Xu et al., 1994).

Besides osmotin, the PR-1b gene also was induced synergistically by ethylene + methyl jasmonate in tobacco (Xu et al., 1994). Salicylic acid induced the accumulation of PR-1b protein in tobacco, and the combination of methyl jasmonate and salicylic acid induced PR-1 protein to accumulate severalfold more. Although methyl jasmonate in combination with ethylene or salicylic acid hyperinduced osmotin and PR-1b mRNA accumulation, osmotin protein accumulated to very high levels only when methyl jasmonate

TABLE 1.10. Synergistic Effect of Signal Molecules in Tobacco

Treatment	GUS activity in tobacco seedlings (nmoles of methyl umbelliferone per milligram of protein per hour)
Control	80
Methyl jasmonate	119
Ethylene	239
Ethylene + Methyl jasmonate	4820

Source: Xu et al., 1994.

was combined with ethylene, and PR-1b protein did so only when methyl jasmonate was combined with salicylic acid (Xu et al., 1994). These results suggest that maximal accumulation of more than one defense protein requires multiple signals. Multiple signal transduction systems may exist in plants (Schweizer et al., 1997a).

Sulfotransferase in the Signal Transduction System

Flavonoids are a diverse group of secondary metabolites found in plants (Hahlbrock and Scheel, 1989). Some of the flavonoids function as signal molecules (Long, 1989; Lynn and Chang, 1990). Sulfation is an important reaction in the biotransformation. The enzymes which catalyze this reaction are the sulfotransferases: they allow the transfer of a sulfonate group from PAPs (3'-phosphoadenosine-5'-phosphosulfate) to an acceptor compound to produce a sulfate ester. The flavonol sulfotransferases have been identified in plants and shown to transfer a sulfate group to flavonols (Varin and Ibrahim, 1992; Varin et al., 1992). A class of flavonoids esterified with sulfate groups has been reported to be ubiquitous in plants (Barron et al., 1988). Sulfated flavonoids may act as regulators or signals in response to environmental stimuli (Lacomme and Roby, 1996). Sulfate metabolites such as Nod factors in the interaction between *Rhizobium* and alfalfa have been shown to be involved in signaling processes (Truchet et al., 1991).

Arabidopsis leaves were inoculated with incompatible pathogens (*X. campestris* pv. *campestris* 147, *P. syringae* pv. *maculicola* M2 and M4/*avr Rpm1*) or compatible pathogens (*X. campestris* pv. *campestris* 8004 and *P. syringae* pv. *maculicola* M4) (Lacomme and Roby, 1996). Infection with both compatible and incompatible pathogens resulted in a strong and rapid increase in the sulfotransferase mRNA levels. The presence of the *avr* gene in *P. syringae* pv. *maculicola* M4 induced more sulfotransferase than when the gene was lacking (Lacomme and Roby, 1996). Signal molecules such as methyl jasmonate and salicylic acid when applied to *Arabidopsis* leaves strongly increased sulfotransferase mRNA levels. These observations suggest that sulfotransferase may participate in the synthesis/decoration of signals involved in the molecular communication between the bacteria-plant interactions (Lacomme and Roby, 1996).

IS CELL DEATH INVOLVED IN THE SIGNAL TRANSDUCTION PATHWAY?

Induction of disease resistance was related to the hypersensitive response with phenotypic expression of accelerated cell death in many host-pathogen interactions. In *Arabidopsis, a*ccelerated *c*ell *d*eath *(acd2)* mutants that form spontaneous lesions show enhanced disease resistance (Dietrich et al., 1994; Greenberg et al., 1994; Weymann et al., 1995). Similar lesioned phenotypes have been reported in tomato (Langford, 1948) and wheat (Wolter et al., 1993), and they show enhanced disease resistance. A correlation between spontaneous cell death and disease resistance is also evident in transgenic tobacco expressing a bacterial proton pump (Mittler et al., 1995), the A1 subunit of cholera toxin (Beffa et al., 1995), or a ubiquitin variant (Becker et al., 1993). These observations suggest that cell death may be involved in the signal transduction pathway.

Transgenic *Arabidopsis* plants which are unable to accumulate salicylic acid due to the expression of the salicylate hydroxylase *(nahG)* gene were crossed with *Arabidopsis* mutants expressing the accelerated cell death gene *(acd2)* (Hunt et al., 1997). Progeny from the crosses did not show enhanced disease resistance. A high level of PR-1 expression in *lsd2* was substantially reduced by the expression of *nahG*. PR-2 and PR-5 also accumulate to substantially lower levels in *lsd* mutants expressing *nahG* (Hunt et al., 1997). However, these progeny retained the spontaneous cell death phenotype. Hence, cell death may not be necessary for induction of disease resistance. The cell death phenomenon may occur prior to salicylic acid accumulation in the signal transduction pathway. The loss of disease resistance in *nahG*-expressing *lsd* mutants indicates that induction of disease resistance depends upon salicylic acid rather than cell death.

Pseudomonas syringae pv. *tabaci Hrp*$^-$ mutants, which do not induce the hypersensitive cell death, induced the accumulation of transcripts for phenylalanine ammonia-lyase, chalcone synthase, chalcone isomerase, and chitinase and phytoalexins in bean (Jakobek and Lindgren, 1993). *P. savastanoi* pv. *phaseolicola hrpD*$^-$ strain does not induce hypersensitive cell death, but induces the formation of large paramural papillae in lettuce mesophyll cells adjacent to bacterial colonies. Deposition of hydroxyproline-rich glycoprotein (HRGP), phenolics, and callose in the plant cell wall was also induced by the *hrp*$^-$ strain (Bestwick et al., 1995). These results suggest that cell death may not be involved in signal transduction.

In fact, synthesis of defense chemicals requires active metabolism of the host plant. The cell necrosis elicited by the bacterial elicitors takes several hours to develop and requires active metabolism of living plant cells. Pectic enzymes produced by *Erwinia chrysanthemi* killed plant cells before hypersensitive necrosis could develop, and no activation of antibacterial defense responses occurred in the plants (Bauer et al., 1994).

HOW PATHOGENS AVOID OR OVERCOME HOST DEFENSE MECHANISMS INDUCED BY THE SIGNAL TRANSDUCTION SYSTEM

Suppressors

Active resistance in plants involves recognition of the pathogen by its host. Defense-related genes are activated in both compatible and incompatible interactions. To date, only one gene exhibiting a truly specific pattern of expression during incompatible interactions has been isolated from plants: the tobacco gene *hsr 203J* (Pontier et al., 1994). Otherwise the defense gene mRNA accumulates during both compatible and incompatible interactions. Plant pathogens evolve to overcome resistance by evading host recognition and response. Plant defense suppressor genes have been extensively described in fungi (Van Etten et al., 1989; Vidhyasekaran, 1997). But such genes have remained poorly understood in bacteria even though there is circumstantial evidence for their existence (Sequeira, 1976). *Xanthomonas campestris* pv. *campestris* is a pathogen on cruciferous plants. Kamoun and Kado (1990) identified an *hrpXc* locus in a 2.1 kb fragment of *X. campestris* pv. *campestris* genome. All tested strains of *X. campestris* pv. *campestris* contained sequences homologous to *hrpXc* located in an 11 kb EcoRI fragment (Kamoun et al., 1992). By using marker-integration mutagenesis with plasmids containing a 338 bp fragment internal to *hrpXc, HrpXc* mutants were constructed in 10 isolates of *X. campestris* pv. *campestris*. Loss of *hrpXc* by mutation resulted in pathogenicity, and the mutants gained the ability to cause the hypersensitive response in their respective host plants (Kamoun et al., 1992). It suggests that *hrpX* confers a means to evade this host response.

In contrast to wild isolates, *X. campestris* pv. *campestris hrpXc* mutants appeared unable to grow either in the xylem or the mesophyll of cauliflower and cabbage. It appears, therefore, that *hrpXc* is essential for pathogenicity and growth of *X. campestris* pv. *campestris* on crucifers (Kamoun et al., 1992). Further *hrpXc* mutants in *X. oryzae* pv. *oryzae,* the rice bacterial blight pathogen, have been constructed by marker-integration mutagenesis (Kamoun et al., 1992). These mutants were unable to cause leaf blight on rice or induce a hypersensitive response on pepper and *Datura* (nonhosts of *X. oryzae* pv. *oryzae*), suggesting that *hrpXc* is highly conserved and may be essential for phytopathogenicity of all *Xanthomonas.*

Several *hrpXc* mutants of *X. campestris* pv. *campestris* induce a vascular hypersensitive response on crucifers, which is similar to the response induced by incompatible *X. campestris* pv. *campestris* races. Interestingly,

HR induction was suppressed by coinoculation of the mutants with their wild-type counterparts, and the HrpXc mutant's in planta growth deficiency was subsequently rescued in coinoculated leaves (Kamoun et al., 1992). This observation suggests that wild-type compatible *X. campestris* pv. *campestris* may have the ability to suppress plant defense responses in order to grow and invade plant tissue, whereas incompatible strains induce the HR. The *hrpXc* gene may encode a product that directly or indirectly functions in suppressing plant defense response to *X. campestris* pv. *campestris*.

The function of HrpX protein was predicted by sequencing *hrpXc* and *hrpXo* from *X. campestris* pv. *campestris* and *X. oryzae* pv. *oryzae*, respectively (Oku et al., 1995). The predicted amino acid sequences of the protein encoded by these respective genes revealed similarities to HrpB protein of *Ralstonia solanacearum*, which has sequence identity to the transcriptional activator VirF of *Yersinia enterocolitica* and Arac of *Escherichia coli*. A putative DNA-binding domain present in the carboxyl terminal half of HrpX is highly conserved among HrpB, VirF, and AraC (Oku et al., 1995). It suggests that HrpX may regulate *Xanthomonas* virulence genes.

Each of the *hrp* genes may also encode suppressors. Strains of *Xanthomonas vesicatoria* with insertion mutations in each of the clustered *hrp* loci *hrpA, hrpB, hrpC, hrpD, hrpE,* and *hrpF* were obtained (Brown et al., 1995). All *hrp* mutants caused resistance reaction in mesophyll cells of a susceptible cultivar of *Capsicum annuum*. They induced localized formation of large papillae. Similar deposits also accumulated after inoculation with the saprophytic strain of *X. campestris,* which lacks the entire *hrp* cluster. By contrast, only minor cell wall alterations were observed in response to inoculation with a wild-type strain with all *hrp* loci in the susceptible *Capsicum* cultivar (Brown et al., 1995). It suggests that each of the clustered *hrp* loci suppresses defense mechanisms in the susceptible plants.

When bean leaves were infiltrated with incompatible pathogens such as *Pseudomonas syringae* pv. *tabaci, P. syringae* pv. *angulata, P. savastanoi* pv. *glycinea, P. syringae* pv. *tomato, P. syringae* pv. *pisi,* and *P. syringae* pv. *syringae,* transcripts for PAL, CHS, CHI, and chitinase were induced. In contrast, transcripts for PAL, CHS, and CHI were not present after infiltration with the bean pathogen *P. savastanoi* pv. *phaseolicola* (Jakobek et al., 1993). The induction of chitinase transcripts was significantly delayed in *P. savastanoi* pv. *phaseolicola*–infiltrated bean leaves when compared with those infiltrated with incompatible *P. syringae* isolates. Probably *P. savastanoi* pv. *phaseolicola* suppresses the accumulation of transcripts for PAL, CHS, and CHI. Prior infiltration with *P. savastanoi* pv. *phaseolicola* suppressed the typical transcript accumulation and phytoalexin production that occurs in bean after infiltration with *P. syringae* pv. *tabaci* (Jakobek et al., 1993). Significantly, the suppressor activity was lost when *P. savastanoi* pv. *phase-*

olicola cells were heat killed or treated with protein synthesis inhibitors, indicating that active metabolism is a prerequisite for suppressor activity (Jakobek et al., 1993).

When *Arabidopsis thaliana* ecotype Pr-O was inoculated with an incompatible pathogen, *P. syringae* pv. *syringae,* several defense genes such as cinnamyl alcohol dehydrogenase (CAD), cinnamate-4-hydroxylase (C4H), EL13 plant defense gene (EL13), and anthranilate synthase (ASA1) were induced (Buell and Somervilla, 1995). But when this *Arabidopsis* ecotype was inoculated with the pathogen *Xanthomonas campestris* pv. *campestris,* expression of all these defense genes was suppressed (Buell and Somerville, 1995).

Preinfiltration of tobacco leaves with *Erwinia carotovora* subsp. *carotovora* at low concentration (10^5 cfu/ml) prevented the appearance of water soaking and necrosis upon reinoculation at the same site with the incompatible pathogen *P. syringae* pv. *pisi.* The ability of preinoculated cells to inhibit the HR-like response was apparent by 12 h after inoculation, and by 24 h production of the response was completely suppressed (Cui et al., 1996). It suggests that at low concentration of bacterial cells suppressor activity is predominant.

The gene *rsmA* of *E. carotovora* subsp. *carotovora* strain 71 suppresses the synthesis of the cell density (quorum) sensing signal N-(3-oxohexanoyl)-L-homoserine lactone (Asita et al., 1996). When *rsmA* of *E. carotovora* subsp. *carotovora* was introduced into *E. amylovora,* elicitation of HR in tobacco leaves by *E. amylovora* was suppressed. It suppressed the production of disease symptoms in apple shoots (Asita et al., 1996). It suggests that the gene *rsmA* suppresses function of *hrp* genes in *E. amylovora*. Homologs of *rsmA* have been detected in *E. amylovora, E. carotovora* subsp. *atroseptica, E. carotovora,* subsp. *betavasculorum, E. carotovora* subsp. *carotovora, E. chrysanthemi, E. herbicola,* and *Pantoea stewartii* (Asita et al., 1996).

An element in the avirulence gene locus may act as a suppressor. Transposon insertion mutants of *X. campestris* pathovars have been isolated that induce hypersensitive reactions on normally susceptible plants, suggesting that bacterial pathogens may produce suppressors (Daniels et al., 1984; Kamoun et al., 1992). The bacterial pathogen *X. vesicatoria* mutates to overcome genetically defined resistance to pepper *(Capsicum annuum)* by the transposon-induced mutation of *avrBs1,* a bacterial gene that provokes the plant's resistance response (Kearney et al., 1988). The transposon-induced mutants of *X. vesicatoria* completely overcame the resistance and induced fully water-soaked lesions when inoculated onto pepper plants. The *avrBs1* locus of each mutant was examined and compared to the wild-type gene from *X. vesicatoria*. Southern analysis revealed that 1.2 kb of DNA was inserted into different regions of *avrBs1* in each mutant. Each mutation was accompanied by an insertion of 1.2 kb into *avrBs1*. The DNA inserted into each case had an identical restriction map, suggesting that the independent mutations were caused by the same element. The element was 1,225 bp long and displayed the structures of a typical bacterial insertion sequence (Kearney et al., 1988). These results suggest the suppressor may be present in the *avr* gene locus itself.

Lipopolysaccharides

Lipopolysaccharides (LPS) or protein-LPS complex (pr-LPS) are constituents of the outer membrane of the bacteria. They are spontaneously sloughed off from cell walls during the log-growth phase of the bacteria (Rothfield and Pearlman-Kothencz, 1969; Mazzucchi and Pupillo, 1976; Minardi et al., 1989). These LPS or protein-LPS complexes from a number of bacteria are known to prevent the hypersensitive response caused by avirulent plant-pathogenic bacteria (reviewed by Sequeira, 1983; Vidhyasekaran, 1988).

Pr-LPS complexes were purified from a virulent strain of *P. syringae* pv. *tabaci*. These pr-LPS complexes, preinjected into tobacco leaves, prevented hypersensitive necrosis evolved by seven incompatible pseudomonads (Mazzucchi et al., 1979). Cellular suspensions of *Erwinia chrysanthemi* infiltrated into tobacco leaves completely prevented the hypersensitive necrosis induced by incompatible pathogen *Pseudomonas syringae;* during the growth phase of *E. chrysanthemi* pr-LPS complexes were released (Mazzucchi and Pupillo, 1976).

Infiltration of tobacco leaves with EDTA extracts from the outer membrane of *Pseudomonas syringae* pv. *morsprunorum* (a pathogen of cherry) strain C28 (Hignett and Quirk, 1979) can prevent the development of the hypersensitive response to subsequent inoculation with incompatible necrotizing pathogens. The EDTA extract of outer membrane of *P. syringae* pv. *morsprunorum* strain C28 contained 30 percent lipopolysaccharides, 8 percent phospholipid, and 16 percent protein (Hodson et al., 1995b). The activity of the LPS is potentiated by its associated cations such as Ca^{2+} and Mg^{2+} (Hodson et al., 1996).

Preinoculation of pepper cultivar ECW10R with LPS from *X. campestris* pv. *campestris* strain 8004 suppresses HR induced by the same bacteria (Newman et al., 1997). The effect of *X. campestris* LPS was not dependent on the strain or on its interaction with pepper ECW10R. LPS from *X. vesicatoria* strain 71-21 (which is virulent on pepper ECW10R) and *X. campestris* pv. *raphani* and armoracine (which are both avirulent on pepper ECW10R) all prevented the HR induced by *X. campestris* pv. *campestris* strain 8004 (Newman et al., 1997). It suggests nonspecificity of LPS in prevention of HR.

LPS from *Ralstonia solanacearum* prevents HR in tobacco (Graham et al., 1977). Hydrolysis of the core-lipid A region destroys the activity indicating the importance of lipid-A region in suppression of HR (Graham et al., 1977). Contrastingly the lipid A moiety of LPS from *X. campestris* pv. *campestris* was totally ineffective in preventing the HR; the core oligosaccharide of the LPS was effective in suppressing the HR (Newman et al., 1997). It suggests that different components of the LPS may have specificity in suppression of HR.

For suppression of HR, a period of 10 to 12 h was required between inoculations with LPS and the challenge inoculation (Newman et al., 1997). It indicates that active metabolism of the host is required for the action of LPS. However, Newman et al. (1994) provided evidence that LPS may have a role in survival and growth of the bacteria in planta rather than suppressing activation of defense genes. An LPS mutant of *X. campestris* induced β-1,3-glucanase, and the level of transcript accumulation was nearly 75 percent of that seen with the wild type, although the bacterial numbers were up to three orders of magnitude lower and symptoms were markedly reduced. It was suggested that the role of LPS may be to mask the bacterial presence to prevent induction of defense-related genes. Close contact with the host cell has been shown to be necessary to induce defense genes by the bacterial pathogens (Hutcheson et al., 1989), and LPS may prevent contact with the host cell by the bacterial cell.

The possibility of suppression of defense genes by LPS cannot be ruled out, as Newman et al. (1994) have tested only β-1,3-glucanase, and the other possible defense genes may be suppressed by LPS. The time lag required for the action of LPS (Mazzucchi et al., 1979; Newman et al., 1997) suggests this possibility.

POSSIBLE ROLE OF THE SIGNAL TRANSDUCTION SYSTEM IN EVASION OF THE HOST RECOGNITION BY PHYTOPATHOGENIC BACTERIA DURING PATHOGENESIS

The molecular basis of evasion of host recognition and response by bacterial pathogens is not yet known. Host resistance gene products may play an important role in susceptibility or resistance in plants as disease resistance genes control recognition of invading pathogens and subsequent activation of defense responses. Classical genetic studies have revealed the presence of numerous resistance genes in plants. Individual resistance genes may be highly specific in function, being effective only against particular strains of a viral, bacterial, or fungal pathogen. The function of disease resistance genes is still obscure, and only very few disease resistance genes have been cloned so far. But products of all bacterial disease resistance genes share striking structural similarities, and they appear to be involved in the signal transduction system (Dangl et al., 1992; Martin et al., 1993a; Bent et al., 1994; Song et al., 1995; Grant et al., 1995; Baker et al., 1997).

Can we conclude that due to absence of resistance genes in the susceptible varieties, the defense mechanism is not activated and the pathogen causes disease? The disease resistance genes seem to be conserved in many plants. Functional homologs of *Arabidopsis RPM1* exist in pea, bean, and soybean (Dangl et al., 1992). Homologs of the disease resistance locus have been suggested to be present at the corresponding position in susceptible plants (Whitham et al., 1994; Jones et al., 1994; Mindrinos et al., 1994). Activation of the signal transduction system in both susceptible and resistant interaction has been observed to be similar, if not identical. All signal molecules such as salicylic acid, jasmonic acid, ethylene, Ca^{2+}, protein kinases, phospholipases, and H^{+}-ATPase are commonly present in the plant kingdom. In fact, activation of defense genes which are responsible for biosynthesis of defense-related products (phenolics, phytoalexins, callose, lignin, HRGP, and PR proteins) has been observed in both resistant and susceptible interactions (Pontier et al., 1994). Then what may be the difference between resistant and susceptible interactions?

The speed and amount of expression of the signal transduction system appears to be different between resistant and susceptible interactions. Protein kinases are the enzymes involved in protein phosphorylation. Higher expression of protein kinases induces resistance to *Xanthomonas oryzae* pv. *oryzae* in rice (Wang et al., 1996). Phospholipase C appears to be involved in regulating intracellular Ca^{2+} concentrations which act as a second messenger in signal transduction system (Alexandre et al., 1990). Phospholipase C transcript accumulation peaked at two days in resistant rice-*X. oryzae* pv. *oryzae* interactions. In susceptible interactions there was a delayed increase in phospholipase C, and the transcript levels never reached those observed for the peak levels in resistant interactions (Young et al., 1996).

X. oryzae pv. *oryzae* strains isogenic for *avrXa10* and rice cultivars near-isogenic for the corresponding resistance gene *(Xa10)* were used to investigate changes in phospholipase D, the enzyme which activates protein kinases which are involved in the signal transduction system, and its distribution in rice tissues undergoing resistant and susceptible responses (Young et al., 1996). In the susceptible interaction, phospholipase D mRNA gradually increased over four days and then was barely detected at day 5; at this time, the lesions had spread up and down the leaves and were tan and desiccated. The expression pattern of phospholipase D mRNA was different for the resistant interactions in that the mRNA peaked earlier (two days) in resistant interactions than in susceptible interactions (four days). By five days after inoculation, the mRNA had decreased to the levels at day 0 in all treatments (Young et al., 1996).

Gold-labeled antibodies specific for phospholipase D immunodecorated the plasma membranes of rice mesophyll cells in both susceptible and resistant interactions. In susceptible interactions, phospholipase D was evenly distributed in the plasma membranes of cells. In contrast, in the resistant interactions, the particles were clustered at sites in which bacteria were associated with the responding cell (Young et al., 1996). Probably this change in distribution occurs under conditions of increased enzyme turnover, and the translated enzyme may be targeted to sites at which bacteria are associated with the hypersensitively responding plant cells.

The K^+/H^+ exchange response, which is mediated by Ca^{2+} influx, protein kinases, and H^+-ATPase (all involved in the signal transduction system) is associated with both incompatible and compatible interactions, but the K^+/H^+ exchange proceeds rapidly in incompatible interactions (Baker et al., 1987). The sudden outburst in active oxygen species involved in signal transduction was observed in both susceptible and resistant interactions in tobacco (Baker et al., 1991). Incompatible bacteria induced more sulfotransferase than the compatible bacteria in *Arabidopsis* (Lacomme and Roby, 1996). Induction of higher amounts of salicylic acid and ethylene results in disease resistance (Xu et al., 1994; Dempsey et al., 1997). Induction of defense genes is also seen in both susceptible and resistant interactions. But in susceptible interactions, induction of plant defense genes is delayed (Dong et al., 1991).

No detectable level of β-1-3-glucanase transcript was present in the total RNA of healthy turnip *(Brassica campestris)* plants (Newman et al., 1994). In the compatible interaction with *Xanthomonas campestris* pv. *campestris* strain 8004, accumulation of β-1-3-glucanase transcript was first detectable at low levels at 12 h after inoculation and reached a maximum at 24 h. In contrast, *X. campestris* pv. *raphani* strain 1946 and *X. campestris* pv. *armoraciae* strain 1930, which are both incompatible with turnip cultivar Just Right, caused a much earlier accumulation of β-1-3-glucanase transcript, which could be detected as early as 4 h after inoculation in both cases (Newman et al., 1994).

Similar differential induction of other defense-related genes in bean in compatible and incompatible interactions with races of *Pseudomonas savastanoi* pv. *phaseolicola* (Voisey and Slusarenko, 1989; Meier et al., 1993) and in *Arabidopsis* in compatible and incompatible interactions with strains of *P. syringae* pv. *maculicola* has been reported (Dong et al., 1991). *X. campestris* pv. *campestris* strain 8004 is compatible with *Arabidopsis thaliana,* whereas *X. campestris* pv. *campestris* strain 147 is incompatible. In the incompatible interaction there was induction of a basic β-1-3-glucanase transcript at 48 h after inoculation, whereas the compatible strain gave a very weak induction of this transcript at 48 h after inoculation (Lummerzheim et al., 1993). In the bean cultivar Red Mexican, chitinase transcript levels increased between 3 and 6 h after inoculation with an incompatible isolate of *P. syringae* pv. *phaseolicola,* whereas during the interaction with a compatible isolate, chitinase transcript accumulation was delayed to 20 to 24 h (Voisey and Slusarenko, 1989).

Thus delay in expression of defense genes appears to be a common phenomenon during pathogenesis of bacterial pathogens. How does the delay in expression of defense genes occur? Suppressors which delay the expression of defense genes have been isolated from fungal pathogens (Yamada et al., 1989; Shiraishi et al., 1991). The activation of PAL activity had very similar kinetics in elicitor-treated and elicitor-plus suppressor-treated tissues, except for a delay of about 6 h as observed in the pattern of pisatin biosynthesis in pea (Yamada et al., 1989). The activity of an ATPase in the pea plasma membranes, which is markedly suppressed by 6 h, recovers within 9 h after the start of the treatment with the suppressor from *Mycosphaerella pinodes,* the pea pathogen (Shiraishi et al., 1991). The inhibition of the ATPase causes a temporary suppression of expression of the PAL and chalcone synthase (CHS) genes that are responsible for production of pisatin (Yoshioka et al., 1992). Similar suppressors may be present in bacterial pathogens also (Jakobek et al., 1993). However, no such suppressors have been isolated from bacterial pathogens so far. Delayed release of elicitors from fungal cells into host tissues has also been suggested to be responsible for delayed induction of host defense mechanisms in compatible host-fungal pathogen interactions (Bostock, 1989). The secretion of bacterial elicitors into host tissues has not yet been demonstrated. Thus several molecular biological studies are still needed to understand the recognition mechanism in plant-bacterial pathogen interactions.

CONCLUSION

The interactions of plants and bacterial pathogens appear to be carefully regulated, complex biological relationships. During the interactions, plant-derived molecules function as signals that induce expression of specific bacterial genes; the products of these bacterial genes, in turn, induce changes in plant gene expression. When bacterial pathogens establish physical contact with plants, different types of genes are activated in the bacteria in response to plant signals. Four groups of bacterial genes appear to be involved in host recognition. The first group of genes, *hrp* genes, is required for both disease development in susceptible plants and disease resistance in nonhost plants. Several *hrp* genes are required for a single bacterial pathogen to induce disease as well as disease resistance. *hrp* genes are conserved among several bacterial pathogens. In general, *hrp* regions of one bacterial pathogen can be introduced into another bacteria and *hrp* expression can be demonstrated. Even unrelated phytopathogenic bacteria, one causing leaf spot and another causing wilt and having different host specificities, carry similar *hrp* genes. It suggests that *hrp* genes may have some common function in different bacteria.

For induction of resistance, all *hrp* genes are required and deletion of a single *hrp* gene from *hrp* gene clusters results in loss of ability to induce resistance. But *hrp* genes alone are not sufficient to induce resistance with phenotypic expression of HR. Avirulence *(avr)* genes are also required for induction of HR. Similarly, *hrp* genes alone are not sufficient to induce pathogenesis. Additional pathogenicity genes are also required. Hence, *hrp* genes may encode proteins which regulate other genes to induce resistance or susceptibility.

Not all *hrp* genes are regulatory genes. Some of the *hrp* genes among *hrp* gene clusters alone show regulatory function. But these regulatory *hrp* genes may regulate expression of other *hrp* genes as well as *avr* genes and other pathogenicity genes. Some *hrp* genes encode elicitors. These elicitors have dual functions. They induce HR in nonhosts and incite symptoms in susceptible plants. Experimental removal of an *hrp* gene results in loss of ability of a bacterial pathogen to elicit disease resistance on nonhost plants and loss of pathogenicity on the susceptible host. Most of the *hrp* genes encode components of a protein secretion pathway. The *hrp* secretory proteins control secretion of elicitors. The *hrp* secretory proteins may allow delivery of different bacterial proteins into plant cells.

Pathogen genes that are critical in determining the host range of the pathogen are called avirulence *(avr)* genes. The avirulence gene, when carried by a normally virulent strain, may induce cultivar-specific resistance in natural host and nonhost species. The cloned *avr* gene converts a virulent strain of a bacterial pathogen to avirulence in a cultivar-specific manner. Experimental removal of *avr* genes makes the bacteria cause disease on otherwise resistant cultivars and unlike the *hrp* genes, such strains lacking the *avr* genes are usually not impaired in pathogenicity on susceptible hosts.

Several *avr* genes may be present in a single bacterial pathogen. The *avr* gene locus may be conserved among many plant pathogens. Transfer of *avr* genes from one bacterial pathogen to another pathogen converts the recipient to show cultivar-specific resistance in the host of the transconjugant. Homologs of *avr* genes may also be present in virulent strains. Some avirulence genes also have pathogencity functions, while some *avr* genes also have virulence functions. *avr* genes may encode proteins which possess enzymatic functions leading to synthesis of elicitors. *avr* genes may require specific *hrp* genes for their transcription in bacterial pathogens. *avr* genes may be part of an *hrp*-dependent regulon required for pathogenesis. Conserved regions, "harp boxes," associated with regulation by *hrp* genes, are located within promoter of several *avr* genes. Virulence genes may also be coregulated with *avr* and *hrp* genes. Disease-specific pathogenicity *(dsp)* genes (without causing HR on nonhosts) and host-specific virulence *(hsv)* genes (without cultivar specificity) are also present in some bacterial pathogens.

All pathogenicity genes of bacterial pathogens *hrp, avr, dsp,* and *hsv* are dormant in bacterial cells and they are actively transcribed when they come into contact with plants. Different host nutrients (sugars, nitrogen sources, minerals, phenolics) and host environment (pH) determine transcription of the pathogenicity genes. Some *hrp* and *avr* genes direct synthesis of elicitors and specific plant signals may regulate elicitor synthesis in bacteria. Secretion of these elicitors into plant cells depends upon a protein secretion pathway and most of the *hrp* genes encode components of the secretion pathway. The *hrp* secretory proteins control secretion of elicitors. Lipopolysaccharides, oligosaccharides, and pectic fragments released by pectic enzymes of bacterial pathogens also act as elicitors. These elicitors act as signal molecules and induce an array of defense-related products in host and nonhost plants. Signals generated by the bacterial genes should enter the nucleus of the cell through second messengers to induce host defense gene expression.

The transduction of these signals into plant cells for activation of defense genes involves a conserved signal transduction pathway. Calcium ions transduct extracellular primary signals into intracellular events. Phosphorylation of proteins is important in integration of external and internal stimuli. Protein phosphorylation is catalyzed by protein kinase, which is activated by Ca^{2+}. Phospholipases C and D are also involved in signal transduction. H^{+}-ATPase involved in membrane depolarization is a part of the signal transduction system. Active oxygen species and flavonoids also function in signal transduction.

Analyses of several cloned plant disease resistance genes revealed the presence of leucine-rich repeats (LRR), leucine zippers, and nuclear localization signals (NLS). Several *avr* genes also encode LRRs, leucine zippers and NLS regions. LRRs may serve as the sites of protein-protein binding, leucine zippers may directly interact with DNA, and the NLS may direct transport of the signals to the plant nucleus. After entering the nucleus, the Avr proteins or protein complexes may act on nuclear transcriptional factors leading to pathogenesis or disease resistance.

The signals generated due to bacterial infection are translocated systemically, and the plants synthesize several defense chemicals. Salicylic acid, jasmonic acid, and its methyl ester and ethylene are involved in systemic signal transduction. A synergistic effect of these signals has also been reported. Defense-related genes are activated in both compatible and incompatible interactions by the signals. In compatible interactions, suppressors may be present which may suppress defense mechanisms of the host. Some *hrp* genes encode synthesis of suppressors. An element in the *avr* gene locus may act as a suppressor. The bacterial lipopolysaccharides also may suppress the defense genes in susceptible interactions.

Studies conducted so far reveal the molecular basis of an early recognition process in plant-bacterial pathogen interactions. Bacterial pathogens are different from other bacteria (saprophytes or animal pathogens) in that they are able to make growth in both host and nonhost plants. This basic character appears to be encoded by *hrp* and *avr* genes of the bacterial pathogens, and transcription of these genes is activated by plant signals. Both *hrp* and *avr* genes encode elicitors which induce leakage of nutrients from plants, which are needed for growth of bacterial pathogens in plants. *hrp*-encoded elicitors induce extracellular alkalinization, and the primary function of the Hrp system appears to make apoplastic fluids more suitable for bacterial multiplication by raising the pH.

Besides *hrp* and *avr* genes, *dsp, hsv,* and virulence genes of the bacteria may be important in causing disease in susceptible hosts and in inducing disease resistance in nonhosts. *hrp* genes are needed for induction of both disease and disease resistance. Thus *hrp* genes may be regulatory genes. They regulate *avr* genes to induce resistance and *dsp, hsv,* and virulence genes to induce disease. Some of the *avr* and *hrp* genes are also involved in disease development by suppressing defense genes, probably producing suppressors. Plant disease resistance genes may be involved in recognition of bacterial pathogens. Disease resistance genes also seem to be conserved in plants. Functional homologs of a resistance gene seem to be present in many unrelated plants. Homologs of resistance genes appear to be present in susceptible cultivars. Products of all cloned resistance genes share striking similarities and they appear to be involved in the signal transduction system. The signals induce host defense genes, which are quiescent in normal plants. Defense genes are induced in both hosts and nonhosts.

Thus all bacterial pathogens (belonging to *Pseudomonas, Xanthomonas, Erwinia,* and *Ralstonia*) have *hrp* genes; all of them are quiescent in the bacteria and are induced by plant signals. Similarly, all defense genes in plants are quiescent and are induced by bacterial signals. The signal transduction systems appear to determine the early recognition process in plant-pathogen interactions. On recognition of bacterial pathogens, defense mechanisms of the host are activated. Molecular mechanisms of activation of defense genes are described in the following chapters.

Chapter 2

Host Defense Mechanisms: The Cell Wall—the First Barrier and a Source of Defense Signal Molecules

THE FIRST BARRIER TO BACTERIAL INFECTION IN PLANTS

Plant-pathogenic bacteria do not possess penetration structures, such as penetration pegs commonly found in fungi (Vidhyasekaran, 1997). Hence they are unable to exert mechanical or physical forces to penetrate intact epidermal cells (Huang, 1986). The bacteria enter their hosts through stomata, hydathodes, lenticels, trichomes, or wounds. A stoma usually consists of a pair of guard cells with a pore between them. The epidermal cells adjacent to the guard cells frequently differentiate into subsidiary cells. The space formed by the guard cells, the subsidiary cells, and the neighboring cells constitutes the substomatal cavity. Stomata occur primarily on aerial parts of plants, particularly on leaves and young stems. Hydathodes have structures similar to stomata except the guard cells associated with hydathodes do not function to regulate aperture opening. Hydathodes usually occur on the marginal teeth or serrations of leaves or at the leaf tips. The water droplets secreted by the hydathodes are brought to the surface by the terminal tracheids of the veins and pass through the intercellular spaces of the loosely packed parenchyma, called epithem. Lenticels usually occur in the periderm of stems and roots, often beneath a stoma in the original epidermis. Lenticels generally develop from a stoma or a group of stomata. During the transformation of stomata into lenticels, cells in the first two subepidermal cell layers around the substomatal cavity divide in both inward and outward directions. This results in the formation of a mass of loosely arranged parenchyma cells with large intercellular spaces (Huang, 1986). Trichomes are epidermal projections, and they are fragile and readily collapse under slight pressure. In many cases, bacterial pathogens enter through wound sites rather than natural openings (Huang, 1986).

The plant tissues are made up of symplast and apoplast. The symplast constitutes the sum total of living protoplasm of a plant. The total mass of living cells of a plant constitutes a continuum, the individual protoplasts being ultimately connected throughout the plant by plasmodesmata. The apoplast constitutes the total nonliving cell wall continuum that surrounds the symplast (Buvat, 1989). It also constitutes a continuous permeable system through which water and solutes may freely move (Crafts and Crisp, 1971).

In the area that separates two cells, there is a median substance called middle lamella. On either side, layers derived from one or the other cell are juxtaposed. The first is called the primary layer or primary cell wall; it is applied against the middle lamella. Toward the interior of the cell cavity, other layers can be found, the whole structure producing the secondary cell wall (Buvat, 1989). The xylem system constitutes a specialized phase of the apoplast. Xylem is characterized by cells that after differentiation generate vessels or tracheids (Buvat, 1989).

Bacterial pathogens, which enter through natural openings or wounds, multiply in the apoplast (intercellular spaces), particularly in the middle lamella and xylem tissues (O'Connell et al., 1990; Benhamou, 1991; Benhamou et al., 1991; Brown et al., 1993; Mansfield et al., 1994; Grimault et al., 1994; Boher et al., 1995; Bestwick et al., 1995; Dai et al., 1996; Dreier et al., 1997). The plant cell wall is considered as the extracellular matrix secreted into intercellular spaces (Bolwell, 1993) and hence the cell wall may be the first barrier the bacterial pathogens encounter in the intercellular spaces.

STRUCTURE OF THE PLANT CELL WALL

The plant cell wall constitutes a barrier to microbial invasion of cells. The cell wall is a biphasic structure, consisting of a rigid skeleton of cellulose microfibrils held together (and apart) by a gel-like matrix. The matrix makes up about two-thirds of the wall's dry weight and is built up of several noncellulosic polysaccharides and glycoproteins (Fry, 1986). The cell wall matrix is in a dynamic state (Bolwell, 1993). The primary cell wall is at first poor in cellulose and rich in pectic compounds, polysaccharides, and hemicelluloses. It is thickened by the addition of amorphous polysaccharide material and cellulose fibrils, first deposited on the middle lamella. The middle lamella is mainly pectic, and it contains calcium and magnesium pectates (Buvat, 1989). Cross-linking in pectins includes Ca^{2+} bridges, other ionic bonds, H-bonds, glycosidic bonds, ester bonds, and phenolic coupling (Fry, 1986).

Hemicellulosic polysaccharides are also predominant in plant cell walls. Cellulose microfibrils are spun into a matrix consisting of both pectinaceous and hemicellulosic polysaccharides (Bolwell, 1993). In the primary wall of the dicot, xyloglucans are the major hemicelluloses while in monocots xylans substituted with a number of sugars are the main hemicelluloses (Bolwell, 1993).

The plant cell wall contains proteins, hydroxyproline-rich glycoproteins, proline-rich proteins, and glycine-rich proteins. It also contains phenolics. The phenolics of the wall constitute a diverse group of molecules. Wall phenolics are based upon the phenylpropanoid unit and found as both conjugated acids and lignin alcohols. In secondary walls, *p*-coumaryl-, coniferyl-, and sinapyl-alcohols are polymerized into lignin. The cell wall contains suberin, and minerals such as calcium and silica (Bolwell, 1993). Calcium enhances the structural integrity of cell walls (Jones and Lunt, 1967). The cell wall contains enzymes such as acid phosphatases and peroxidases (Buvat, 1989). The xylem is a rigid assemblage of plant polysaaccharides composed of cellulose, hemicellulose, pectic substances, lignin, and protein (Benhamou, 1991).

Studies conducted so far have revealed that the cell wall has a complex ordered structure, with a mosaic of microdomains still largely unexplored (Liners and Van Cutsen, 1991; Vian et al., 1992, 1996; Bolwell, 1993; Roberts, 1994; Roy et al., 1994). Therefore, the bacterial pathogen may face a complex network of different polymers that require specific enzymes for their degradation (Temsah et al., 1991).

PECTIC POLYSACCHARIDES

Complexity of Pectic Substances

The primary cell walls are characterized by a high content of pectic polysaccharides. Plant pectic polysaccharides comprise the most complex polysaccharides known (Carpita and Gibeaut, 1993). Pectins are the group of polysaccharides that are associated with D-galactosyluronic acid residues (Bolwell, 1993). Of these, homogalacturonan, rhamnogalacturonan-I, and a substituted galacturonan, rhamnogalacturonan-II, are commonly detected. Most plant walls contain this pectin family in variable ratios. Homogalacturonan is a chain of 1,4-linked α-D-galactosyluronic acid residues. The carboxyl groups of the galacturonosyl residues of the cell wall pectic polysaccharides are known to be highly methyl esterified. The degree of esterification of the carboxyl groups varies depending on the source of the pectic polymers in different plants. There are regions which are highly methyl esterified as well as regions which are relatively free of methyl esters (Darvill et al., 1980). Pectins are regarded as α-1,4-galacturonans with various degrees of methyl esterification, and the terms "pectic acid" and "pectinic acid" refer to the nonesterified and partially esterified forms, respectively (Stoddardt, 1984).

Rhamnogalacturonan-I is a family of polysaccharides in which rhamnosyl residues are glycosidically linked to galacturonosyl residues (Bolwell, 1993). The rhamnosyl residues are closely associated with galacturonosyl residues in that both are integral components of the same polysaccharide chain (Darvill et al., 1980).

Arabans have been isolated from the cell walls of many dicotyledonous plants. A number of complex pectic polysaccharides have been demonstrated to contain arabinosyl residues (Darvill et al., 1980). Several pectic polysaccharides have been demonstrated to contain 3- and 6-linked galactosyl residues. Two distinct rhamnogalacturonans splice into polygalacturonic acid, and several kinds of natural sugar side chains are attached to many of the rhamnosyl units (Carpita and Gibeaut, 1993). A considerable portion of the arabinosyl and galactosyl residues appear to be components of araban and of galactan side chains, which are covalently attached to the rhamnogalacturonan backbone. The primary cell walls may contain arabinogalactans. Arabinogalactans are composed of interior chains of 1, 3-β-linked-D-galactopyranosyl residues to which are attached side chains. The pectic materials may contain separate arabinans and galactans (Stoddardt, 1984). Other pectic domains, rhamnogalacturonan-II contain at least 12 different monosaccharides (McNeil et al., 1984). On hydrolysis, rhamnogalacturonan yields the rarely observed cell sugars 2-O-methyl fucose, 2-O-methyl xylose, and apiose. Apiose-containing galacturonans have been reported to be present in plant tissues (Darvill et al., 1980).

Cross-linking of polyuronides with calcium has been reported (McMillan et al., 1993b). Most of the calcium in plant tissues is located in the cell walls, where it is bound not only by electrostatic interactions with carboxylic groups of the polygalacturonans but also by coordinate linkages with hydroxylic groups of diverse polysaccharides (Demarty et al., 1984).

Pectic Polysaccharides May Act As a Physical Barrier

Bacterial pathogens produce colonies within intercellular spaces or xylem vessels (Cason et al., 1978; Hildebrand et al., 1980; Al-Mousawi et al., 1982; Al-Issa and Sigee, 1982; Brown and Mansfield, 1988, 1991; Benhamou, 1992; Dreier et al., 1997). These pathogens in the intercellular spaces and xylem vessels may face the complex pectic polysaccharides and unless the pectic polysaccharides are degraded by pectic enzymes, they may not be able to breach the cell well barrier. *Erwinia carotovora* subsp. *carotovora,* the tobacco pathogen, is able to multiply in tobacco leaves. A transposon-induced mutant of this bacterium affected in exoenzyme production (Exp$^-$) was unable to multiply in the leaves (Palva et al., 1993). The mutant was virtually negative for all exoenzymes tested and specifically for pectic enzymes.

Pseudomonas viridiflava is a soft rot pathogen, and a mutant of the strain which lacked the ability to produce pectic enzymes was developed by site-directed mutagenesis. The mutant produced 70- to 100-fold less pectic enzyme than the wild type and failed to cause any symptoms in its market vegetable hosts (Liao et al., 1988, 1992). *Ralstonia solanacearum* is the causal agent of bacterial wilt disease of many important crop species. Marker exchange mutagenesis was used to inactivate the structural gene for a pectic enzyme in a mutant strain of *R. solanacearum*. The resultant mutant was very much reduced in virulence (Roberts et al., 1988; Schell et al., 1988). Similar reduced virulence has been observed in mutants of *E. chrysanthemi* (Payne et al., 1987; Ried and Collmer, 1988) and *E. carotovora* subsp. *carotovora* (Saarilahti et al., 1992; Pischik et al., 1996) which are deficient in pectic enzymes.

Disease resistance has been correlated with the complexity of pectic substances in plant cell walls in many host-pathogen interactions. Potato varieties which have a highly methylated pectin are resistant to *E. chrysanthemi*. The stem cell walls of the potato genotype resistant to black leg disease caused by *Erwinia carotovora* subsp. *atroseptica* are characterized by a higher content of highly methylated and branched water-soluble pectins (Marty et al., 1997). Resistance of cell wall pectic substances to maceration by pectic enzymes produced by erwinias has been suggested to be the cause of resistance in potato. Resistance to *Erwinia* spp. in potato appears to be closely related to the degree of pectin esterification of the stem and tuber tissue of the different potato cultivars (McMillan et al., 1993a). Methylation of the carboxyl groups of galacturonic acid and the formation of bridges between rhamnogalacturonan chains offer resistance to *Erwinia* spp. and they confer the gelling property of pectic substances (Cooper, 1983). Pagel and Heitfuss (1989) found that potato tuber rotting susceptibility of six cultivars to erwinias was inversely related to cell wall pectin esterification. Cross-linking of polyuronides with calcium confers resistance to enzymatic breakdown of pectin by erwinias (McGuire and Kelman, 1986).

Pectin degradation has also been demonstrated during bacterial pathogenesis. *Clavibacter michiganensis,* the tomato pathogen, quickly spreads along the xylem vessels of the host. Swelling, shredding, and partial wall dissolution were typical features in areas adjacent to sites of bacterial accumulation (Benhamou, 1991). Pectin-like molecules and galactose residues were the major components of the fibrillar or amorphous material accumulating at those sites where bacteria were actively growing (Benhamou, 1991).

These studies suggest that the pectic barrier is to be degraded for bacterial pathogenesis.

Complexity of Pectic Enzymes

Bacterial pathogens produce pectic enzymes to breach the pectic barrier. Since the bacterial pathogens have to breach the complex pectic barrier, they have to produce different types of pectic enzymes with varying specificity to the different types of pectic substances. Pectic substances of plant cell walls contain mostly polymerized galacturonic acid which is usually present also as the methyl ester in varying proportions (McMillan et al., 1993b). Pectins are regarded as α-1,4-galacturonans with various degrees of methyl esterification. The term pectic acid (or polygalacturonic acid) refers to the nonesterified forms while the pectinic acid (or pectin) refers to partially esterified forms (Stoddardt, 1984).

The unesterified uronic acid molecules containing up to 75 percent of methoxyl groups are known as pectinic acid, and those with more than 75 percent of methyl groups are known as pectins. The difference between pectinic acid and pectin is narrow, and generally the pectinic acids are called pectins (Vidhyasekaran, 1993). The enzyme which catalyses the conversion of pectin by de-esterification of the methyl ester group with production of methanol is called pectin methyl esterase (PME or PEM). PME removes the methoxyl group from rhamnogalacturonan chains (McMillan et al., 1993b). PME is more active on methylated oligogalacturonates than on pectin.

The pectic enzymes vary in their mechanism to split α-1,4-glycosidic bonds (hydrolytic or lytic) and in the type of cleavage (random or terminal). Exopolygalacturonase (exo PG or Peh) cleaves polygalacturonic acids in a terminal manner, releasing monomeric products (galacturonic acids), while endopolygalacturonase (endo PG) cleaves polygalacturonic acids in a random manner releasing oligogalacturonic acid. Exo and endo PGs cleave the α-1,4-glycosidic bond in the substrate by hydrolysis. Exo- and endo-pectin methylgalacturonases (PMG) act similarly to polygalacturonases, but the preferred substrate is pectin instead of polygalacturonic acid (Vidhyasekaran, 1997).

Pectate lyases (PL or Pel) and pectin lyases (PTE) induce lytic degradation of the glycosidic linkage, resulting in an unsaturated bond between carbons 4 and 5 of a uronide moiety in the reaction product. The action of *trans*-eliminases results in oligouronides that contain an unsaturated galacturonyl unit. Pectate and pectin lyases cleave the α-1,4-glycosidic bond by β-elimination. Both exo (terminal) and endo (random) types of reactions are seen in the *trans*-eliminases.

The activity of pectate lyase, which preferentially attacks nonesterified rhamnogalacturonan chains, is enhanced when methoxyl groups had been removed by PME (McMillan et al., 1993). Similarly, endo PG preferentially hydrolyzes nonesterified pectic substances (McMillan et al., 1993a). PME facilitates the action of polygalacturonases by producing poorly methylated pectin. Low methylated pectins are the true substrate for PG (Pozsar-Hajnal and Polcasek-Racz, 1975). Calcium is commonly located in the plant cell walls. In general, endo PG activities are reduced and endo-PL activities are stimulated by an increase in the amount of calcium (Moran et al., 1968; Garibaldi and Bateman, 1971). Thus, different cell wall substrates activate various types of pectic enzymes in different manners.

Production of various pectic enzymes by bacterial pathogens belonging to various genera *Pseudomonas, Xanthomonas, Erwinia, Ralstonia, Burkholderia, Agrobacterium,* and *Clavibacter* has been reported (Liao et al., 1989; Denny et al., 1990; Beimen et al., 1992; Schell et al., 1988; Herlache et al., 1997; Rong et al., 1994; Alfano et al., 1995; Liao et al., 1996; Bauer and Collmer, 1997; Gonzales et al., 1997; Grimault et al., 1997; Tardy et al., 1997).

CELLULOSE

Cellulose is one of the major components of the plant cell wall. While pectin is the main constituent of the middle lamella, cellulose is the main constituent of xylem tissue (Roberts et al., 1988). Xylem degradation has been documented in many bacterial pathogen-infected tissues (Nelson and Dickey, 1970; Benhamou, 1991). Several bacterial pathogens are known to produce cellulolytic enzymes (Krataka, 1987; Roberts et al., 1988; Bequin, 1990).

Cellulose is composed of glucose units in the chain configuration, connected by β-1,4-glycosidic bonds. Cellulase is a general term for a synergistic system of three major categories of enzymes that hydrolyze the β-1,4-glucosidic bonds; they are endoglucanase (endo β-1,4-glucanase), exoglucanase (β-1,4 -cellobiohydrolase), and β-glucosidase. They are classified according to their mode of action and substrate specificity. Endoglucanase cleaves β-glucosidic bonds in amorphous regions of cellulose, creating sites for exoglucanase to cleave cellobiose from the nonreducing ends (Coughlan and Mayer, 1992). β-Glucosidase would hydrolyze the resultant cellobiose to glucose, preventing cellobiose buildup and exoglucanase inhibition.

Endoglucanase has been demonstrated in *Ralstonia solanacearum* races 1 and 2 (Schell, 1987). *Clavibacter michiganensis* subsp. *michiganensis* produces an endoglucanase (Meletzus et al., 1993). *C. michiganensis* subsp.

sepedonicus produces exoglucanase, endoglucanase, and β-glucosidase (Baer and Gudmestad, 1995). A β-1,4-endoglucanase-encoding gene from *C. michiganensis* subsp. *sepedonicus* was cloned into a plasmid. When this construction was electroporated into competent cells of a cellulase-deficient mutant, it restored cellulase production at almost wild-type levels (Laine et al., 1996). *C. michiganensis* subsp. *insidiosus* produces cellulases (Krataka et al., 1987). *Xanthomonas campestris* pv. *campestris* produces endoglucanase (Gough et al., 1988; Tang et al., 1991).

Erwinia chrysanthemi produces cellulases (Lojkowska et al., 1995) and EGZ is the major endoglucanase secreted by *E. chrysanthemi* (Py et al., 1991). The EGZ consists of a catalytic N-terminal domain linked to a C-terminal cellulose-binding domain by a Ser Thr-rich linker (Brun et al., 1995). *E. carotovora* subsp. *carotovora* produces cellulase (Murata et al., 1994). The cellulase CelV, a β-1,4-endoglucanase, is responsible for at least 95 percent of the detectable carboxymethyl cellulase activity of *E. carotovora* subsp. *carotovora* (Cooper and Salmond, 1993; Walker et al., 1994). In common with most other cellulases, CelV consists of several functionally distinct domains: at the amino terminus there is a classical sequence, followed by a large catalytic domain that shows homology to catalytic domains of EGZ of *E. chrysanthemi* (Gilkes et al., 1991). The carboxy-terminal domain shows homology to cellulose-binding domains of several *Bacillus* cellulases (Jauris et al., 1990). Production of cellulase has been reported also in *E. carotovora* subsp. *atroseptica* and *E. carotovora* subsp. *betavasculorum* (Murata et al., 1994; Cui et al., 1995). All cellulase-producing bacteria produce endoglucanase in combination with an exoglucanase and/or β-glucosidase (Beguin, 1990).

HEMICELLULOSE

Hemicelluloses are covalently linked to pectic polysaccharides and noncovalently bound to cellulose fibrils of the cell walls. The important plant cell wall hemicelluloses are xylan, araban, and galactan. Xylans are the most abundant hemicelluloses, constituting up to 40 percent of the primary cell walls of some plants (Aspinall, 1980). *Clavibacter michiganensis* subsp. *michiganensis* produces xylanase to degrade xylans (Beimen et al., 1992). The enzyme has been purified and characterized. Its molecular weight was 42 kDa and the primary protein was 413 amino acids with a leader peptide (31 amino acid) that was cleaved during secretion to the bacterial periplasm (Keen et al., 1996). Production of hemicellulases by other bacterial pathogens has not yet been studied.

CELL WALL PROTEINS

Plant cell walls contain up to 10 percent protein. The cell wall protein is exceptionally rich in hydroxyproline. There are five plant cell wall protein classes. These include extensins, proline-rich proteins (PRPs), arabinogalactan proteins (AGPs), lectins, and glycine-rich proteins (GRPs). Each of them, with the exception of GRPs, contains hydroxyproline-rich glycoproteins (HRGPs). They are rich in hydroxyproline and serine and they usually contain the repeating pentapeptide motif Ser-Hyp4. PRPs represent another class of plant cell wall proteins, and all of the PRPs are characterized by the repeating occurrence of Pro-Pro repeats. They contain approximately equimolar quantities of proline and hydroxyproline. Hydroxyproline and arabinose are major constitutents of lectins. The serine-hydroxyproline-rich glycopeptide domain of lectins bears a striking biochemical resemblance to the extensins. Arabinogalactan proteins are HRGPs that are highly glycosylated. The protein moiety of AGPs is typically rich in hydroxyproline, serine, alanine, threonine, and glycine. They contain Ala-Hyp repeats. GRPs lack hydroxyproline and contain Gly-X repeats, where X is most frequently Gly but can also be Ala or Ser. GRPs have nucleotide sequence similarity to the extensins (Showalter, 1993).

Several bacterial pathogens produce proteases. *X. campestris* pv. *campestris* (Dow et al., 1993), *X. campestris* pv. *armoraciae* (Dow et al., 1993), *P. syringae* pv. *tomato* (Bashan et al., 1986), *E. carotovora* (Murata et al., 1994; Walker et al., 1994), and *E. chrysanthemi* (Dahler et al., 1990) produce proteases. The pathogens may produce multiple proteases. *E. carotovora* subsp. *carotovora* isolate177 produces a metalloprotease which degrades lectin (Heilbronn et al., 1995). The metalloprotease was different from the two extracellular proteases produced by other isolates of *E. carotovora* (Kyostio et al., 1991). *E. carotovora* subsp. *carotovora* produces a number of extracellular proteases (Delepelaire and Wandersman, 1989). A locus, *pat-1,* has been identified in a plasmid of *C. michiganensis* subsp. *michiganensis* (Dreier et al., 1997). The nucleotide sequence of the *pat-1* region revealed a single open reading frame (ORF1). The computer search for possible homologies between *C. michiganensis* subsp. *michiganensis* ORF1 and other proteins indicated significant homologies to serine proteases of various bacteria. Thus ORF1 protein is a serine protease. However, proteolytic activity of the ORF1 gene product could not be demonstrated (Dreier et al., 1997). It is possible that ORF1 may be expressed in planta and this possibility is yet to be assessed.

BACTERIAL GENES ENCODING EXTRACELLULAR ENZYMES

Genes encoding various pectic enzymes have been cloned from various phytopathogenic bacteria (Liao et al., 1996; Bauer and Collmer, 1997). The gene encoding for the pectate lyase (Pel) enzyme from *X. axonopodis* pv. *malvacearum* has been cloned (Liao et al., 1996). Analysis of nucleotide sequence of the *X. axonopodis* pv. *malvacearum* 1.8 kb *pel* fragment revealed an ORF consisting of 1,131 necleotides. This ORF was predicted to encode a pre-Pel consisting of 377 amino acids (Liao et al., 1996). The *pel* gene cloned from *X. axonopodis* pv. *malvacearum* was used to probe genomic digests prepared from 12 strains of xanthomonads. The *pel* homologs were detected in four strains of *X. axonopodis* pv. *malvacearum,* two strains each of *X. vesicatoria* and *X. campestris,* one strain of *X. axonopodis* pv. *glycines,* and *X. axonopodis* pv. *phaseoli*. A single hybridization band of about the same intensity was detected in the genomic digest of each strain, indicating that *pel* genes are well conserved in all xanthomonads (Liao et al., 1996).

The *pel* gene of *X. vesicatoria (pel-XV1)* has been cloned (Beaulieu et al., 1991) and it shared homology to the DNA of *X. axonopodis* pv. *citri (=X. campestris* pv. *citrumelo), X. axonopodis* pv. *dieffenbachiae,* and *X. campestris* pv. *campestris* (Beaulieu et al., 1991). However, it shared no homology to DNA of *E. chrysanthemi* (Beaulieu et al., 1991). The *pel* gene has been cloned from *X. campestris* pv. *campestris* (Dow et al., 1989).

The *pme* gene encoding PME has been cloned from *Ralstonia solanacearum* (Clough et al., 1994). The polygalacturonase gene (called *pehA*) was located about 15 kb from end of the Vir2 cluster of *hrp* genes of *R. solanacearum* strain GM11000 (Allen et al., 1991). The *pehA* gene has been located in 200 kb plasmid in *Burkholderia cepacia* (Gonzalez et al., 1997). The *pehA* gene is located chromosomally in both *E. carotovora* subsp. *carotovora* (Willis et al., 1987) and *R. solanacearum* (Allen et al., 1991). An amino acid sequence showed several conserved regions in the *pehA* gene products of *E. carotovora* subsp. *carotovora, R. solanacearum,* and *B. cepacia* (Gonzalez et al., 1997). This histidine residue in the GHGXSIGS sequence (residues 353 to 360), the three aspartic acid residues in NTD (residues 303 to 305), TGDD (residues 324 to 327), and the RIK sequence (residues 395 to 397), and the tyrosine residue at position 431 all have been implicated in the activity of polygalacturonase enzyme from different sources (Bussink et al., 1991; Stratilova et al., 1996). *Agrobacterium vitis* also produces a polygalacturonase and the gene encoding the enzyme *pehA* has been cloned (Herlache et al., 1997). DNA sequencing of the *Agrobacterium vitis pehA* gene revealed a predicted protein with an Mr of 58,000 and significant similarity to the polygalacturonases of *E. carotovora* and *R. solanacearum* (Herlache et al., 1997). The amino acid sequence of *pehA* gene product from *A. vitis* also shows these conserved sequences (Herlache et al., 1997). The conserved sequences in the enzyme appear to be required for activity. The gene *pgl,* which encodes a predicted protein with 67 to 72 percent conserved amino acid homology to polygalacturonases from several bacterial species, has been cloned from *Agrobacterium tumefaciens* (Rong et al., 1994).

A pectate lyase gene, *pelS,* has been cloned from *P. syringae* pv. *lachrymans* (Bauer and Collmer, 1997). DNA sequencing of the *pelS* locus revealed an open reading frame encoding a predicted protein of 40.3 kDa (Bauer and Collmer, 1997). The cloned *pelS* was used to probe *P. syringae, P. savastanoi* pathovars, and other bacteria. The *pelS* probe hybridized with DNA from *P. syringae* pv. *tabaci, P. savastanoi* pv. *phaseolicola, P. savastanoi* pv. *glycinea, P. fluorescens* (= *P. marginalis*), *P. viridiflava,* and *X. campestris* pv. *campestris* but not with *P. syringae* pv. *pisi, P. syringae* pv. *syringae, P. syringae* pv. *tomato, P. syringae* pv. *populans, E. chrysanthemi,* or *Rastonia solanacearum* (Bauer and Collmer, 1997).

PelS encoded by the gene *pelS* showed strong similarity to members of the *E. chrysanthemi* PelE family (Bauer and Collmer, 1997). PelS was 79 percent similar to *P. viridiflava* PelV, 78 percent similar to *P. fluorescens* PelF, 76 percent similar to *X. axonopodis* pv. *malvacearum* PelX, and 42 percent similar to *E. chrysanthemi* PelE (Bauer and Collmer, 1997). *pel* genes have been cloned from *P. marginalis* and *P. viridiflava* (Liao et al., 1992; Nikaidou et al., 1993). DNA sequence analysis has revealed the *P. marginalis* and *P. viridiflava* Pels to be members of the *E. chrysanthemi* PelADE family (Liao et al., 1996). By using the 1.2 kb DNA fragment of *P. viridiflava, pel* homologs were detected in *P. syringae* pv. *lachrymans, P. savastanoi* pv. *phaseolicola, P. syringae* pv. *tabaci,* and *X. axonopodis* pv. *malvacearum* (Liao et al., 1992). Sequences homologous to the *pel* genes from *P. viridiflava* were not detected in *Erwinia chrysanthemi, E. carotovora* subsp. *carotovora,* and *P. syringae* pv. *tomato* (Liao et al., 1992).

Several *pel* genes have been cloned from *E. chrysanthemi*. These *pel* genes are organized in two clusters on the bacterial chromosome. The *pelA, pelD,* and *pelE* genes are organized in one cluster, whereas the *pelB* and *pelC* genes are on the second one (Boccara et al., 1988). The sequences of *pelD* and *pelE* structural genes of *E. chrysanthemi* show high homology (Van Gijsegem, 1989). The upstream noncoding sequences of the *pelD* and *pelE* genes are divergent, but three domains are conserved in the promoter region of the *pelD* and *pelE* genes (Reverchon et al., 1989; Van Gijsegem, 1989). A gene adjacent to *pelC, pelZ,* was cloned from *E. chrysanthemi* (Pissavin et al., 1996). It encoded a protein of 420 amino acids with an endo-pectate lyase activity. The *pelZ* gene was widespread in different strains of

E. chrysanthemi. A gene homologous to *pelZ* existed in *E. carotovora* subsp. *atroseptica* adjacent to the cluster containing the pectate lyase-encoding genes *pel1, pel2,* and *pel3* (Pissavin et al., 1996). A new *pel* gene, *pelL,* has been cloned from *E. chrysanthemi*. It encodes a novel, asparagine-rich, high-alkaline enzyme that is similar in primary structure to PelX (derived from *X. axonopodis* pv. *malvacearum*) and in enzymological properties to PelE (Alfano et al., 1995). A gene coding for PME, *pemB,* was also cloned from *E. chrysanthemi* strain 3937 gene library (Shevchick et al., 1996). *pemB* could be detected in four other *E. chrysanthemi* strains but not in *E. carotovora* strains (Shevchick et al., 1996).

The major extracellular endopectate lyase gene from *E. carotovora* subsp. *carotovora* has been cloned (Roberts et al., 1986). The gene for a periplasmic Pel enzyme has been isolated from *E. carotovora* subsp. *carotovora* SCR 1193 (Keen et al., 1987). Endo-Pel and exo-Pel genes of *E. carotovora* subsp. *carotovora* are related to genes mediating production of the PelBC family of extracellular *Erwinia* enzymes and to the family of intracellular pectate lyases (Yang et al., 1992).

A polygalacturonase gene, *peh1,* has been located chromosomally in *E. carotovora* subsp. *carotovora* (Willis et al., 1987). The *pel3* gene has also been detected in the chromosome. There is a tight linkage between *pel3* and *peh1* within the chromosome of *E. carotovora* subsp. *carotovora* strain 71 (Liu et al., 1994). The 1041 bp *pel3* ORF and the 1206 bp *peh1* ORF are separated by a 579 bp sequence. The genes are transcribed divergently from their own promoters (Liu et al., 1994). *pel3* appears not to have significant homology with the *pel* genes belonging to *pelBC, pelDE,* or periplasmic *pel* families (Liu et al., 1994).

The DNA segment that elicits Pel production in *E. carotovora* subsp. *carotovora* revealed a 1,122 bp open reading frame corresponding to *pel1* which could encode a polypeptide of 374 amino acid residues (Chatterjee et al., 1995). Four *pel* genes, *pel1, pel2, pel3,* and *pelZ,* have been cloned from *E. carotovora* subsp. *atroseptica* (Pissavin et al., 1996).

The gene encoding an endoglucanase (cellulose-degrading enzyme) has been cloned from *Ralstonia solanacearum* and it was designated *egl* (Roberts et al., 1988; Schell et al., 1988; Denny et al., 1990). A β-1,4 endoglucanase-encoding gene from *Clavibacter michiganensis* subsp. *sepedonicus* has also been cloned (Laine et al., 1996). EGZ is the major endoglucanase secreted by *E. chrysanthemi*. The gene encoding the endoglucanase EGZ has been cloned and it was designated *celZ* (Reverchon et al., 1994). The cellulase CelV, a β-1,4 endoglucanase, is the major enzyme produced by *E. carotovora* subsp. *carotovora,* and the *celV* gene has been cloned and sequenced (Cooper and Salmond, 1993).

The gene *(xynA)* encoding an endoxylanase has been cloned from *E. chrysanthemi* (Keen et al., 1996). Protease genes *(prt)* have been cloned from *Erwinia* spp. and *X. campestris* pv. *campestris* (Allen et al., 1986; Tang et al., 1991; Wandersman et al., 1987).

BACTERIAL GENES REGULATING PRODUCTION OF EXTRACELLULAR ENZYMES

Two-Component Regulatory Protein System

Bacterial pathogens respond to various plant signals by using a conserved signal transduction mechanism that activates a specific set of genes inducing production of extracellular enzymes (Stock et al., 1995). A signal transduction mechanism for this adaptation is the two-component regulatory protein system. The majority of two-component regulatory protein systems comprise two different proteins: a histidine protein kinase sensor protein and a response regulator protein that subsequently becomes phosphorylated by the sensor protein (Parkinson and Kofoid, 1992; Stock et al., 1995). A subset of the two-component regulatory systems in which a single protein contains both histidine kinase sensors and response regulator domains exist in some bacteria (Frederick et al., 1997). The sensor kinase proteins in two-component systems are believed to undergo autophosphorylation in response to a signal molecule or signal molecules produced by host plants (Parkinson and Kofoid, 1992). The transmembrane sensor proteins contain an ATP-binding site and a conserved histidine residue that becomes the site of autophosphorylation upon activation. The phosphorylated form of the sensor protein then acts as a kinase and transfers the phosphoryl group to a conserved aspartic acid residue in the receiver domain of the cytoplasmic response regulator protein. Upon phosphorylation, the response regulator protein activates transcription of a set of genes (Parkinson and Kofoid, 1992; Stock et al., 1995).

LemA from *P. syringae* pv. *syringae,* RpfC from *X. campestris* pv. *campestris,* RepA from *P. viridiflava,* and PheN from *P. tolassi* belong to two-component sensor-regulator proteins (Hrabak and Willis, 1990; Frederick et al., 1997).

lemA, repA, and repB Genes

The *lemA* gene is required by *P. syringae* pv. *syringae* for disease lesion formation on bean *(Phaseolus vulgaris)* (Willis et al., 1990). It regulates protease production in the bacterium (Hrabak and Willis, 1992, 1993). At least two gene loci (*repA* and *repB*) are involved in the regulation of Pel in *P. viridiflava* (Hrabak and Willis, 1992). DNA sequence analysis of the *repA* gene revealed that it has 87 percent similarity in nucleotide sequence to the *lemA* gene of *P. syringae* pv. *syringae* (Liao et al., 1994). The function of the *repA* and *lemA* genes appears to be similar and interchangeable. The plasmid carrying the *P. syringae* pv. *syringae lemA* gene is capable of restoring the enzyme-producing and disease-causing ability in Rep^- mutants of *P. viridiflava* (Liao et al., 1994).

The *lemA/repA* gene family is involved in the regulation of a wide variety of biological or pathological activities in bacteria and appears to be widely distributed in fluorescent pseudomonads. LemA has been shown to regulate *pelS* expression in *P. syringae* pv. *lachrymans* (Bauer and Collmer, 1997). The *lemA* locus is well conserved within pathovars and strains of *P. syringae* and within *P. aeruginosa* (Liao et al., 1994). So far the *lemA/repA* gene has been identified only in fluorescent pseudomonads. No *repA/repB* homologs could be detected in genomic digests of *Erwinia* and *Xanthomonas* strains (Liao et al., 1994).

The *lemA* gene of *P. syringae* pv. *syringae* encodes the sensor kinase of a bacterial two-component signal transduction system. The *gacA* gene has been identified as encoding the response regulator of the *lemA* regulon (Rich et al., 1994; Kitten and Willis, 1996). The *gacA* gene encodes a protein of 221 amino acids that contains the conserved residues characteristic of response regulators and a helix-turn-helix domain in the carboxy-terminal region of the protein of target promoters (Rich et al., 1994). A locus was identified that restores extracellular protease production to a *lemA* insertion mutant of *P. syringae* pv. *syringae*. The locus encoded the *P. syringae* homologs of translocation initiation factor IF3 and ribosomal proteins L20 and L35 of *Escherichia coli* and other bacteria. Deletion of both the L35 and L20 genes resulted in loss of protease restoration, whereas disruption of either gene alone increased protease restoration. Hence, overexpression of either L20 or L35 is sufficient for protease restoration (Kitten and Willis, 1996).

vsr and phc Genes

Two separate two-component systems encoded by *vsrAD* and *vsrBC* appear to positively and negatively regulate production of a number of extracellular enzymes in *Ralstonia solanacearum* (Huang et al., 1993; Clough et al., 1994). Another component of the regulatory network is encoded by *phcA* (Brumbley and Denny, 1990; Brumbley et al., 1993). Mutation of *phcA* dramatically changed the levels of most of the extracellular enzymes produced by *R. solanacearum.* There was a 50-fold reduction in β-1,4-glucanase (EG) production coincident with reduced transcription of *egl* in the mutant (Huang et al., 1989; Brumbley and Denny, 1990). A 15-fold decrease in PME activity and a 10-fold increase in endo PG activity have also been reported in the mutants (Brumbley and Denny, 1990). The deduced amino acid sequence of PhcA (Brumbley et al., 1993) suggests that it is a member of the LysR family of transcriptional regulators, implying that PhcA is a global regulator. A new locus, designated *phcB,* has been detected in *R. solanacearum.* Like a *phcA* mutant, *phcB* mutant produces dramatically less endoglucanase and has increased endo PG (Clough et al., 1994). *phcB* may also play a role in the complex regulatory network.

rpf Genes

A cluster of *rpf* genes involved in positive regulation of synthesis of extracellular enzymes in *X. campestris* pv. *campestris* has been reported (Tang et al., 1991). Seven of *rpf* genes that are required for production of Pel, protease, endoglucanase, and amylase have been identified (Tang et al., 1991). The functions of the *rpfC* gene of *X. campestris* pv. *campestris* and the *repA* gene of fluorescent pseudomonads *(P. viridiflava)* appear to be similar. However, a *repA/repB* homolog could not be detected in a genomic digest of *X. campestris* pv. *campestris* (Liao et al., 1994).

A genetic locus, *rpfA* (for *r*egulator of *p*athogenicity *f*actors), has been identified in *E. carotovora* subsp. *carotovora.* The mutant lacking this gene locus was deficient in extracellular protease and cellulase activity, although it produced normal levels of pectate lyase and polygalacturonase (Frederick et al., 1997). Sequencing of the *rpfA* region identified an open reading frame of 2,787 bp. The predicted 929 amino acid polypeptide shared high identity with several two-component sensor-regulator proteins (Frederick et al., 1997). Analysis of the RpfA sequence shows that the conserved histidine and aspartic acid residues are present at position 301 in the putative sensor domain and at position 725 in the putative response regulator domain, respectively (Frederick et al., 1997). RpfA may act as a transcriptional activator for *cel* and *prt* gene expression in *E. carotovora* subsp. *carotovora.*

AepA, aepH, and rex Genes

A gene locus designated *aepA* (*a*ctivator of extracellular *e*nzyme *p*roduction) is required for the production of extracellular enzymes such as Pel, polygalacturonase (Peh), cellulase, and protease by *E. carotovora* subsp. *carotovora* (Murata et al., 1991). The mutants of *E. carotovora* subsp.

carotovora were pleiotropically defective in the production of Pel, Peh, cellulase, protease, and phospholipase (Beraha and Garber, 1971; Murata et al., 1991; Pirhonen et al., 1991). A one-step revertant that regained the ability to produce all of those enzymes was also isolated (Beraha and Gerber, 1971). The nucleotide sequence of a 2.7 kb *aepA*$^+$ segment revealed an ORF of 1395 bp (Liu et al., 1993). *aepA* has been predicted to encode a protein of 465 amino acid residues with a molecular mass of approximately 51 kDa and pI of 6.52. AepA has been shown to activate *pel1* transcription (Liu et al., 1993). In the absence of a functional *aepA*, the level of enzymatic activities (Murata et al., 1991) and *pel1* expression (Liu et al., 1993) was low. A low level of expression of *pel, peh, cel,* and *prt* may occur in the absence of AepA. The full expression of these genes in the presence of inducing signals requires AepA. *E. carotovora* subsp. *carotovora* strain SCR 1193 carrying the *aepA*$^+$ plasmid produced Pel, Peh, Cel, and Prt at six, two, six, and three times higher levels, respectively, than the levels in the strain carrying the cloning vector (Table 2.1; Liu et al., 1993). *AepA* has been shown to activate transcription of *pel1* (Murata et al., 1991; Liu et al., 1993).

Homologs of *E. carotovora* subsp. *carotovora* strain 71 *aepA* occur in other *E. carotovora* subsp. *carotovora* strains and *E. carotovora* subsp. *atroseptica* strains (Liu et al., 1993). The *aepA* gene shows very little or no homology with other prokaryotic regulatory genes and is different from the *repA/lemA* gene (Liao et al., 1994). The *aep* locus contains the *aepA* gene and a small ORF, *aepH,* encoding a 47 amino acid protein (Murata et al., 1994). a*epH* has been found to be an *a*ctivator of *e*xtracellular *p*rotein production (Murata et al., 1991; Liu et al., 1993). Multiple copies of *aepH* activate the production of Pel, Peh, Cel, and Prt in *E. carotovora* subsp. *carotovora* and *E. carotovora* subsp. *atroseptica* (Murata et al., 1994).

Another locus, *rex,* also acts as an activator of extracellular enzyme production. The smaller fragment of the *rex* locus able to complement the *rex* mutations is 250 bp long and shows extensive sequence identity with the

TABLE 2.1. The Effect of *aepA*$^+$ Plasmid on the Levels of Extracellular Enzymatic Activities in *Erwinia carotovora* subsp. *carotovora*

Plasmid of bacterial strain	**Specific activity (units/mg protein) of extracellular enzymes**			
	Pel	**Peh**	**Cel**	**Prt**
With *aepA*	183	413	110	56
Without *aepA*	27	164	20	19

Source: Liu et al., 1993.

aepH region (Salmond et al., 1994). When these small *aepH* or *rex* DNA fragments were transferred into the wild-type *E. carotovora* strains, they caused exoenzyme hyperproduction. The *rex* fragment can also restore exoenzyme production in other downregulated mutants such as *expI*. The 250 bp *rex* fragment encodes no protein but would titrate out a putative repressor (Salmond et al., 1994). The role of the *aep* or *rex* region is still not clear.

exp Gene

Another regulatory locus, *exp,* has been detected in *E. carotovora* subsp. *carotovora* (Pirhonen et al., 1991). The Exp$^-$ mutants show a pleiotropic defect in production and secretion of pectic enzymes, cellulase, and protease (Pirhonen et al., 1991). The Exp$^-$ phenotype resembles the Out$^-$ mutants (which will be described later in this chapter) in that the exoenzymes are retained in the periplasm. However, in Out$^-$ mutants, enzyme production is not affected.

Expression of both the *PelB* and *PehA* genes were shown to require a functional *expI* gene (Heikinheimo et al., 1995). The *expI* codes for an autoinducer synthase needed for the production of a small diffusible signal molecule (*Erwinia* autoinducer, EAI) similar to *Vibrio (=Photobacterium) fischeri* autoinducer *N*-(3-oxohexanoyl) homoserine lactone (OHHL or HSC). The EAI is suggested to form a complex with a DNA binding protein, which would mediate the effect of the autoinducer by activating the transcription of the target genes (Heinkinheimo et al., 1995).

The gene *expI* is related both structurally and functionally to the *luxI* gene product of *Photobacterium fischeri* (Jones et al., 1993; Pirhonen et al., 1993; Salmond et al., 1995). The *luxI* synthesized autoinducer HSL regulates the production of Pel, Cel, Peh, and Prt (Jones et al., 1993; Pirhonen et al., 1993). The exp$^-$ mutants have been shown to lack the freely diffusible signal, the EAI (Pirhonen et al., 1993; Jones et al., 1993). Extracellular enzyme production in *E. carotovora* and *E. chrysanthemi* appears to be controlled by *exp* and *aep* (Pirhonen et al., 1991, 1993; Murata et al., 1991) and by another locus, *rex* (Jones et al., 1993).

crp Gene

Pectate lyase production in *Erwinia* species is subjected to cyclic AMP (cAMP)-controlled catabolite repression. High concentrations of unsaturated digalacturonate exert cAMP-reversible self-catabolite repression on pectate lyase production. The gene encoding catabolite activator proteins *(crp)* has been cloned from *E. chyrsanthemi*. *crp* mutants are deficient in pectate lyase production (Hugouvieux-Cotte-Pattat, 1996). It suggests that *crp* is needed for pectate lyase production.

rdg and rec Genes

Induction of pectin lyase (Pnl) production in *E. carotovora* subsp. *carotovora* strain 71 requires a functional *recA* gene and *rdg* locus. The *rdg* locus contains two regulatory genes, *rdgA* and *rdgB* (Liu et al., 1994). The activation of Pnl production in *E. carotovora* subsp. *carotovora* occurs via a unique regulatory circuit involving *recA, rdgA,* and *rdgB* (Liu et al., 1997). In a similar Pnl-inducible system reconstituted in *Escherichia coli,* the *rdgB* product activated the expression of *pnlA,* the structural gene for pectin lyase. Transcription of *pnlA* followed that of the *rdgB* in *E. carotovora* subsp. *carotovora rdg* product (an 11 kDa polypeptide) is required for *pnlA* expression, and *rdgB* encodes a transcriptional factor which specifically interacts with *pnlA* promoter/regulatory region (Liu et al., 1997).

pehR Gene

A locus called *pehR* has been suggested to encode a *trans*-acting positive regulator of polygalacturonase production in *Ralstonia solanacearum* (Allen et al., 1991). Saarilahti et al. (1992) have characterized a *pehA* (encoding polygalacturonase)-specific regulator, called *pehR,* from *E. carotovora* subsp. *carotovora.*

Negative Regulator Genes

rsmA Gene

Negative regulator genes have also been detected in bacterial pathogens. *rsmA* (*r*epressor of *s*econdary *m*etabolites), a negative regulator gene, has been identified in several *Erwinia* spp. (Cui et al., 1995; Chatterjee et al., 1995). A 183 bp open reading frame encodes the 6.8 kDa Rsm A (Cui et al., 1995). *rsmA* suppressed the production of Pel, Peh, Cel, and protease (Prt) (Cui et al., 1995). *rsmA* and *rsmA*-like genes are expressed in *E. carotovora* subsp. *carotovora, E. carotovora* subsp *atroseptica, E. carotovora* subsp. *betavasculorum, E. chrysanthemi, E. amylovora, E. herbicola,* and *Pantoea (= Erwinia) stewartii* (Asita et al., 1996). *rsm* appears to be a global regulator of several secondary metabolites. A low-copy plasmid carrying *rsmA* of *E. carotovora* subsp. *carotovora* strain 71 caused suppression of flagellum formation in *E. carotovora* subsp. *carotovora* and antibiotic production in

E. carotovora subsp. *betavasculorum*. In *E. amylovora, rsm* of *E. carotovora* subsp. *carotovora* suppressed the elicitation of the HR in tobacco leaves and the production of disease symptoms in apple shoots, in addition to repressing motility and extracellular polysaccharide production (Asita et al., 1996).

An RsmA$^-$ mutant of *E. carotovora* subsp. *carotovora* was isolated, and it overproduced extracellular enzymes (Chatterjee et al., 1995). In the mutant Peh is mostly constitutive; Pel, Cel, and Prt production is still inducible with plant extract (Chatterjee et al., 1995). The high basal levels of *pel-1, pel-3,* and *peh-1* mRNAs in an RsmA$^-$ mutant demonstrated that overproduction of the pectolytic enzymes is due to the stimulation of transcription (Chatterjee et al., 1995). The Rsm$^-$ mutant, like its parent, produces HSL, the sensing signal required for extracellular enzyme production (Chatterjee et al., 1995). The Rsm$^-$ mutant produces extracellular enzymes even in the absence of HSL. HSL-deficient strains were constructed by replacing *hsl,* a locus required for HSL production. While the basal levels of Pel, Peh, Cel, and Prt were comparable in the RsmA$^-$ mutant and its HSL derivative, these enzymes were barely detectable in the HSL derivative of the Rsm$^+$ parent strain (Chatterjee et al., 1995). It suggests that *rsmA* acts independently of HSL. However, it has been shown that *rsmA* suppresses HSL synthesis in *E. carotovora* subsp. *carotovora, E. carotovora* subsp. *atroseptica, E. carotovora* subsp. *betavasculorum,* and *E. chrysanthemi* (Cui et al., 1995). *rsmA* reduces the levels of transcripts of *hslI* required for HSL biosynthesis (Cui et al., 1995).

kdgR Gene

The *kdgR* gene encodes the KdgR protein, which acts as a transcriptional repressor of a set of pectolytic genes in *Erwinia chrysanthemi* (Condemine et al., 1986; Condemine and Robert-Baudouy, 1987, 1991; Reverchon et al., 1991; Nasser et al., 1992, 1994). In *E. chrysanthemi,* all genes involved in pectin degradation *(pelA, pelB, pelC, pelD,* and *pelE)* are specifically controlled by the KdgR repressor and induced in the presence of a pectin catabolic product, 2-keto-3-deoxygluconate (KDG) (James and Hugouvieux-Cotte-Pattat, 1996). The global control exerted by the catabolic activator protein (CAP) and the specific regulation mediated by the kdgR repressor appear to be important in regulation of transcription of the pectinase genes of *E. chrysanthemi* (James and Hugouvieux-Cotte-Pattat, 1996). Comparison of the regulatory regions of the KdgR-controlled genes revealed the existence of a conserved motif proposed as a KdgR-binding site (KdgR box) (Condemine and Robert-Baudouy, 1991; Hugouvieux-Cotte-Pattat and Robert-Baudouy, 1992). KdgR binding prevents gene expression. KdgR is a 306 amino acid protein with a molecular mass of 35,029. The purified KdgR binds to a synthetic KdgR box (Nasser et al., 1992). KDG, 5-keto-4-deoxygluconate (DKI), and 2,5-diketo-3-deoxygluconate (DKII) are the inducing molecules interacting with the KdgR. In vitro, KDG can dissociate the KdgR operator complex (Nasser et al., 1994). All the inducing molecules in KDG analogues contain the motif COOH-CO-CH_2-CHOH-C-C in a pyranic cycle (Nasser et al., 1991). This motif is also found in DKI and DKII, but a direct interaction of these two compounds with KdgR has not been demonstrated (Hugouvieux-Cotte-Pattat et al., 1996). *pemB* expression in *E. chrysanthemi* was inducible in the presence of pectin and was controlled by the negative regulator KdgR (Shevchik et al., 1996). Binding of KdgR to the regulatory regions of the genes it controls has been proven in vitro for *pelA, pelB, pelC,* and *pelE* (Nasser et al., 1992, 1994). However, the regulatory regions of some genes like *pemA* regulated in vivo by *kdgR* are not able to bind purified KdgR protein in vitro (Nasser et al., 1994). It suggests that KdgR binding may require an additional regulatory protein(s).

In *E. chrysanthemi* KdgR mutants, pectate lyases are still inducible in the presence of pectin derivatives, suggesting the existence of other regulatory factors in *E. chrysanthemi* (Hugouvieux-Cotte-Pattat et al., 1992). *rsmA, pecT, pecS,* and *pecM* are the other known negative regulators of pectic enzymes in *E. chrysanthemi* (Reverchon et al., 1994; Surgery et al., 1996; Hugouvieux-Cotte-Pattat et al., 1996). *exp, aep,* and *rex* genes are also involved in regulation of pectic enzymes in *E. chrysanthemi* (Murata et al., 1991; Pirhonen et al., 1991, 1993; Jones et al., 1993).

pecT, pecS, and pecM Genes

pecT encodes a 316 amino acid protein with a size of 34761 Da that belongs to the LYSR family of trancriptional activators. PecT represses the expression of pectate lyase genes *pelC, pelD, pelE, pelL,* and *kdgC,* activates *pelB,* and has no effect on the expression of *pelA* or the pectin methylesterase genes *pemA* and *pemB* (Surgery et al., 1996). PecS is another protein that negatively regulates production of pectic enzymes. The *pecS* gene encodes a protein of 166 amino acids with a calculated molecular mass of 19,287 (Reverchon et al., 1994). PecS, a cytoplasmic protein homologous to other transcriptional regulators, can bind in vitro to the regulatory regions of pectinase and cellulase genes in *E. chrysanthemi* (Praillet et al., 1996; Hugouvieux-Cotte-Pattat et al., 1996). *pecS* regulates negatively the expression of the five *pel* genes of *E. chrysanthemi* (Lojkowska et al., 1995). Regulation of *pelL* expression appeared to be independent of the KdgR repressor, which controls all the steps of pectin catabolism (Lojkowska et al., 1995). *pecM* is another gene that regulates negatively the expression of *pel, pem,* and *celZ* genes in *E. chrysanthemi* (Reverchon et al., 1994). Both *pecS* and *pecM* have the same spectrum of action. However, the level of derepression of the controlled genes is higher in the *pecS* mutant (15-fold) than in the *pecM* mutant (4-fold) (Reverchon et al., 1994). The *pecM* gene encodes a protein of 297 amino acids with a calculated mass of 31,913. PecM is anchored in the bacterial inner membrane, whereas PecS is located in the cytoplasm. PecM may be involved in the sensing and transduction of the external stimulus (Reverchon et al., 1994).

BACTERIAL GENES REGULATING SECRETION OF EXTRACELLULAR ENZYMES

Out Genes

Secretion of pectinases from bacterial cells occurs in a common two-step process. The pectinases are synthesized with a signal sequence that is cleaved during transfer across the inner membrane by the general Sec machinery (Salmond, 1994). The type II or Sec-dependent secretion system is used by diverse gram-negative bacterial pathogens for secretion of extracellular enzymes. Several genes regulate the transcription of genes involved in synthesis of extracellular enzymes in bacteria. Similarly, several genes have been identified in the bacterial pathogens which regulate the secretion of these synthesized enzymes. A cosmid clone isolated from *E. chrysanthemi* strain EC16 enabled *Escherichia coli* to secrete heterologously expressed *E. chrysanthemi* Pels (Lindeberg et al., 1996). Sequencing in a region required for secretion revealed the presence of 12 genes, *outC-M* and *outO*. Each *out* gene was required for secretion of *E. chrysanthemi* PelE from *E. coli* with the exception of *outH* (Lindeberg et al., 1996). *E. carotovora* has 13 *out* genes *(outCDEFGHIJKLMNO),* whereas the equivalent *outN* gene appears to be absent from the *E. chrysanthemi* cluster (Salmond, 1994). No homologs of *outB, outC,* and *outT* genes of *E. chrysanthemi* have been found in *E. carotovora* (Salmond, 1994). Pel1 secretion was conferred on the *E. chrysanthemi* Out system by the presence of OutC-M, S, and B from *E. carotovora.* Only *outC* and *outD* from *E. carotovora* did not confer secretion of *pel1* on the *E. chrysanthemi* Out system (Lindeberg et al., 1996). It suggests the presence of species-specific recognition factors in the Out system of the two *Erwinia* species.

Secretion from the periplasmic space to the outer medium of Pels, Pem, and endoglucanase EGZ in *E. chrysanthemi* occurs through the *out* gene cluster, a group of 15 genes organized in five operons *(outS, outB, outT, outCDEFGHIJKLM,* and *outO)* (Condemine et al., 1992; He et al., 1992; Lindeberg and Collmer, 1992). Genes coding for proteins similar to those of the Out secretion system have been identified in several protein-secreting bacteria (Condemine et al., 1992; Pugsley, 1993). OutT has been characterized as a 12 kDa protein and it has no homology with any known protein (Condemine et al., 1992).

Out$^-$ mutants of *E. carotovora* subsp. *carotovora* and *E. chrysanthemi* were defective in secretion of pectic enzymes and cellulases (Andro et al., 1984; Chatterjee et al., 1985; Pirhonen et al., 1991). A number of the Out apparatus components possess domains in the cytoplasm and/or the periplasm with potential for protein-protein interactions which facilitate the secretion of periplasmic enzyme intermediates across the outer membrane to the external milieu (Thomas et al., 1997). OutD could form a channel-like structure in the outer membrane (Shevchik et al., 1997).

Genes regulating synthesis of extracellular enzymes and secretion may be tightly regulated in bacterial pathogens (Condemine and Robert-Baudouy, 1995). *kdgR* is a repressor of the genes involved in synthesis of extracellular enzymes, and *outT* and *outC* operons are also regulated in vivo by *kdgR* (Condemine et al., 1992). Although a putative KdgR binding sequence was found in regulatory regions of these two operons, bandshift experiments have shown that only the *outT* regulatory region could bind the purified KdgR protein in vitro (Nasser et al., 1994). Therefore, regulation of *outC* by *kdgR* could not be explained by binding of KdgR preventing transcription of the gene. Regulation of the *outC* operon expression by *KdgR* requires the *outT* gene product, suggesting that OutT could be a transcriptional activator of *outC* (Condemine and Robert-Baudouy, 1995).

pecS

The gene *pecS,* which regulates transcription of *pel* and *celZ* in *E. chrysanthemi,* also negatively controls the *outC* operon in a manner independent of the regulation by OutT (Condemine and Robert-Baudouy, 1995). Residues within the C-terminus of PehA (endopolygalacturonase) of *E. carotovora* subsp. *carotovora* also may have a role in secretion, possibly through stabilization of a structure needed for proper exposition of the targeting motif (Palomaki and Saarilahti, 1995).

dsbA and *dsbC*

Two genes, besides *out* genes, have been detected in *E. chrysanthemi*, which are involved in secretion of pectate lyases and cellulase EGZ. They were designated *dsbA* and *dsbC* (Shevchik et al., 1995). They encode for periplasmic proteins with disulfide isomerase activity. In the absence of *dsbA*, pectate lyases and cellulase EGZ are rapidly degraded in the periplasm of an *E. chrysanthemi* mutant. Moreover, pectate lyases expressed in an *Escherichia coli dsbA* mutant were very unstable. Contrastingly, stability of pectate lyases was unaffected in a *dsbC* mutant. Hence, *DsbA* and *DsbC* could have different specificities (Shevchik et al., 1995).

Most of the studies on secretory mechanisms have been made in *Erwinia* spp. Secretion of extracellular enzymes across the outer membrane in *Xanthomonas campestris* pv. *campestris* also has been studied. The last open reading frame of an *Xps* gene cluster, designated *xpsD*, is required for the secretion of extracellular enzymes. *xpsD* encodes a protein of 759 amino acid residues. A consensus N-terminal lipoprotein signal peptide was revealed from its deduced amino acid sequence. XpsD was fatty-acylated (Hu et al., 1995).

SECRETION OF PROTEASES

While pectinases and cellulases of *Erwinia* spp. are secreted by the type II or Sec-dependent pathway, proteases appear to be secreted by a one-step, Sec-independent pathway (Reeves et al., 1993). The *E. chrysanthemi* proteases have no classical N-terminal signal sequences that are normally required for export of a protein via the inner membrane Sec complex to the periplasmic space (Salmond, 1994). The proteases require three accessory proteins (PrtD, E, and F) for secretion. The PrtD protein is an ABC protein (containing the *A*TP-*B*inding *C*assette motif). The PrtE protein is located mainly in the inner membrane, and PrtF is located in the outer membrane (Salmond, 1994). An extracellular metalloprotease, PrtG, is secreted by *E. chrysanthemi* through a signal peptide-independent secretion pathway (Ghigo and Wandersman, 1994). The PrtG secretion signal is COOH-terminal and located in the last 56 residues of PrtG. COOH-terminal exposition of the last four amino acids play a role in the secretion of PrtG (Ghigo and Wandersman, 1994).

THE SIGNALING SYSTEM IN INDUCTION OF BACTERIAL EXTRACELLULAR ENZYMES

Plant Cell Wall Signals Regulate Production of Bacterial Extracellular Enzymes

Plant cell walls appear to contain signal molecules to induce synthesis of extracellular enzymes in bacterial cells. The role of putative plant factor(s) in inducing extracellular enzymes has been reported in many bacterial pathogens. The basal level production of polygalacturonase (PG) was very low when *Ralstonia solanacearum* was grown in rich medium (Allen et al., 1991). When the bacteria were grown in minimal medium supplemented with intercellular fluids from tobacco leaves, there was a fivefold increase in PG activity. About 100-fold increase in PG activity was observed when the bacteria grew directly in tobacco leaves (Allen et al., 1991). In nutrient media, *P. syringae* pv. *lachrymans* produced only less PelS. But the *pelS* gene was strongly expressed in planta (Bauer and Collmer, 1997).

Erwinia carotovora subsp. *carotovora* produces extracellular enzymes Pels, Cel, Peh, and Prt, and the extracellular levels of these enzymes are extremely low when the bacterium is grown in salts-yeast extract-glycerol (SYG) medium. But when it is grown in SYG medium supplemented with celery extract, the production of these enzymes is highly induced (Chatterjee et al., 1995). Plant signals activate the expression of *pel-3* and *peh-1* genes in *E. carotovora* subsp. *carotovora* (Liu et al., 1994). A consensus integration host factor (IHF) binding sequence upstream of *pel-3* appears physiologically significant, since *pel-3* promoter activity is higher in an *Escherichia coli* IHF^+ strain than in an IHF^- strain (Liu et al., 1994).

Potato tuber slices were indirectly inoculated with *E. carotovora* subsp. *carotovora* suspension by using an inert polysulfone membrane to separate the tuber from the bacteria (Yang et al., 1989). This system permitted quantitative isolation of bacteria free of plant cells. Increases in Pel mRNA levels were observed at 4 h after inoculation in bacterial cells free of plant cells and maximum pel activity was seen at 24 h. It suggests that plant signals induce pectic enzymes in bacterial cells (Yang et al., 1989).

The presence of enzyme-inducing factor, particularly in plant cell walls, has been demonstrated by Choi and Han (1996). No pectate lyase activity was detected in cultures of the bacterium *Erwinia rhapontici* in minimal salts glycerol medium or in minimal salts polygalacturonate medium (MSP) containing plant extracts. Pel activity was, however, detected in MSP media amended with cell walls of Chinese cabbage, lettuce leaves, potato tubers, celery petioles, onion bulbs, and carrot roots. The Pel-inducing plant factors were water insoluble and heat labile (Choi and Han, 1996).

The Nature of Plant Signals

The nature of the plant signal molecule, which induces extracellular enzymes in bacteria, has been assessed (Rong et al., 1990, 1991, 1994). Crude extracts from carrot roots induced the *pgl* locus of *Agrobacterium tumefaciens*. The *pgl* locus encoded a predicted protein with homology to known polygalacturonases from several bacterial species (Rong et al., 1991). The inducing activity in carrot root extracts was sensitive to digestion by pectinase, a purified polygalactuonase and is, therefore, most likely derived from the pectic portion of plant cell walls (Rong et al., 1990). Rong et al. (1994) have partly characterized the carrot root extract. Incubation of purified carrot extract with pectate lyase (a cloned PelE protein from *Erwinia chrysanthemi*) abolished the signaling activity of the root extract. Chemical reduction of uronic acids in the carrot extract or fractionated polygalacturonic acid solution to their respective neutral sugars also abolished "inducing" activity (Rong et al., 1994). The purified carrot extract was rhamnogalacturonan, requiring both arabinose and galacturonic acid for activity. The inducer may be a signal molecule from the plant as it shows inducer activity even at very low galacturonate concentrations of 5 to 10 μM (Rong et al., 1994). The pectic fragment from plant cell walls, thus, appears to be an important signal molecule.

The degraded products of pectic substances in plant cell walls may also serve as inducers of pectic enzymes of the bacterial pathogens. The degraded products include galacturonic acid, saturated digalacturonate (DG), unsaturated digalacturonate (uDG), 2-keto-3-deoxygluconate (KDG), 2,5-diketo-3-oxygluconate (DKII), and 5-keto-4-deoxygluconate (DKI) (Collmer et al., 1982). The pectic substances may be initially degraded by constitutive pectic enzymes of bacterial pathogens, and then the degraded products may induce several pectic enzymes. Pectate lyases (Pel) and polygalacturonases (PG) depolymerize the polygalacturonate of host plant cell walls into saturated and unsaturated uronate oligomers, predominately dimers and trimers (Forrest and Lyon, 1990). The products of pectate degradation are taken up by the bacteria and converted intracellularly by oligogalacturonide lyase (OGL) into DKI, DKII, and galacturonic acid (Chatterjee et al., 1985; Condemine et al., 1986). DKI and DKII induce an extensive production of pectolytic enzymes (Hugouvieux-Cotte-Pattat et al., 1983). uDG is the final product of the action of pectate lyase (Weber et al., 1996). uDG functions as an inducer for expression of bacterial pectinolytic enzymes (Weber et al., 1996).

Pel of *E. chrysanthemi* produces predominantly an unsaturated oligogalacturomic acid with a degree of polymerization (DP) of 2. It does not produce any DP3 unsaturated oligogalacturonic acid during the first 200 min of reaction (Preston et al., 1992). Contrastingly, PelB and PelC isozymes of *E. chrysanthemi* produced predominantly the DP3 unsaturated oligogalacturonic acid from plant cell walls (Baker et al., 1990; Preston et al., 1992). The Pel isozyme I purified from *E. carotovora* subsp. *carotovora* produces DP2, 3, and 4 unsaturated oligogalacturonic acids (Forrest and Lyon, 1990). *Pseudomonas viridiflava* Pel produced DP2 unsaturated oligogalacturonic acid quickly (Hotchkiss et al., 1996).

Short oligogalacturonates, galacturonic acid, dimer, and trimer may induce synthesis of different pectic enzymes in the bacteria. These conclusions are based on several studies, and a typical example is provided by Yang et al. (1992). A membrane-separated system to facilitate isolation of *E. carotovora* subsp. *carotovora* cells apart from soft-rotted potato tubers was developed to examine in planta expression of the bacterial genes encoding various pectolytic enzymes (endo-Pel, exo-Pel, and endo-PG). Exo-Pel had the highest basal levels of mRNA accumulation at time zero; endo-Pel mRNA was induced at 3 h after inoculation, peaking at 6 h. Accumulation of endo-PG was not significant until 6 h and did not reach its maximum until 12 h after inoculation. The results suggest that exo-Pel is the constitutive enzyme followed by endo-pel and endo-PG. Potato tuber slices were preinoculated directly with *Escherichia coli* expressing the *E. carotovora* subsp. *carotovora* exo-Pel using the membrane-separated system. Preinoculation with *E. coli* expressing *E. carotovora* subsp. *carotovora* exo-Pel resulted in significant increases in mNA levels for endo-Pel and endo-PG besides exo-Pel over *E. coli* alone (Yang et al., 1992). These results indicate that exo-Pel–produced reaction products (unsaturated dimers and trimers) activate other pectic enzyme genes during pathogenesis.

Activation of pectate lyases by a monomer of galacturonic acids in planta has been demonstrated. Potato tubers contain factors that activate extracellular enzymes. The activating factor in the tuber sap extract has been identified as a monomer of galacturonic acid (McMillan et al., 1993a). Of the two enzymes produced by *E. carotovora* subsp. *atroseptica* which degrade polygalacturonic acid, only Pel is activated by tuber sap extract. PG is constitutively produced whereas Pel is induced by breakdown products of polygalacturonic acid (McMillan et al., 1993b). The activators might be released from cells of tubers by the activity of PG, which, being a constitutive enzyme, is produced early during the infection process. Endo-Pel cleaves polygalacturonic acid by β-elimination, giving rise to two galacturonide chains; one has a nonreducing end (4-5-unsaturated galacturosyl) and the other a reducing end (galacturonic acid). In contrast, exo-PG hydrolyzes polygalacturonic acid to galacturonic acid.

These studies reveal that bacterial pathogens produce different types of pectolytic enzymes to break down plant cell walls. Both smaller pectic fragments (monomers, dimers, and trimers) and larger fragments (oligomers) are produced. The compounds are formed extracellularly and their rapid assimilation by the bacteria may play a crucial role in determining disease development. A subset of molecules released due to the depolymerization of pectate are taken up by the pathogen, after which they are catabolized to yield inducers for the production of pectolytic enzymes (Chatterjee et al., 1985; Barras et al., 1994). Thus the uptake system for galacturonic acid, dimers, and trimers may determine the induction of disease.

The uptake systems for galacturonic acid (GA) and 2-keto-3-deoxygluconate (KDG) have been studied in *E. chrysanthemi* (Condemine and Robert-Baudouy, 1987b; San Francisco and Keenan, 1993). The transport of these molecules has been shown to be energy dependent and activated by growth of the bacteria in the presence of galacturonic acid and digalacturonic acid. The *exuT* gene(s) encode the uptake system for galacturonic acid (Freeman and San Francisco, 1994) while the *kdgT* gene encodes the KDG permease (Condemine and Robert-Baudouy, 1987b; Allen et al., 1989).

The uptake system for the dimer molecule, digalacturonic acid, in *E. chrysanthemi* strain EC16 was studied (San Francisco et al., 1996). Glucose-grown cells showed negligible uptake activity, suggesting repression of the digalacturonic acid uptake system. Galacturonic acid-grown cells consistently showed lower uptake activity than cells grown on the dimer or polymer of galacturonic acid (San Francisco et al., 1996). This indicates incomplete induction of the uptake system by galacturonic acid. Inducers of the pectin-degrading enzymes DKI, DKII, and KDG are generated due to the intracellular catabolism of digalacturonic acid (and possibly trimers and oligomers), whereas intracellular catabolism of galacturonic acid results only in the formation of 2-keto-3-deoxygluconate (Reverchon et al., 1991). It suggests that digalacturonic acid molecules are taken up by the bacteria prior to their intracellular conversion into inducers for pectinase production (Nasser et al., 1992, 1994).

The metabolic inhibitors, potassium cyanide and 2,4-dinitrophenol, reduced digalacturonic acid uptake activity (Table 2.2; San Francisco et al., 1996). Both these compounds interfere with electron transport and the maintenance of an energized membrane. Thus, the uptake of digalacturonic acid appears to be proton motive force dependent.

Unsaturated digalacturonic acid (udGA), one of the primary products of pectate degradation by pectate lyases, resembles dGA except for 4,5-unsaturation at the nonreducing end. Uptake of dGA by *E. chrysanthemi* cells was competitively inhibited by udGA (San Francisco et al., 1996).

TABLE 2.2. Effect of Metabolic Inhibitors on Digalacturonic Acid Uptake by *Erwinia chrysanthemi*

Addition	Percent uptake of digalacturonic acid
None	100
Potassium cyanide	39
2,4-Dinitrophenol	15

Source: San Francisco et al., 1996.

This suggests that the two molecules are taken up by the same system. This uptake system may be involved in the release of inducers of pectic enzymes.

A methanol-like factor released from degradation of pectin in plant cell walls has been suggested to be involved in induction of extracellular enzymes. When *Escherichia coli* carrying a *pme* (pectin methyl esterase) gene from *Ralstonia solanacearum* was grown on pectin-supplemented medium in lid-agar plates, it stimulated many virulence genes including *egl* coding for endoglucanase of an *R. solanacearum* mutant lacking the locus *phcB* via the vapor phase (Clough et al., 1994). *R. solanacearum* produces PME that may release enough methanol from hydrolysis of pectin to activate virulence genes. Virulence of an *R. solanacearum* mutant that lacked a *phcB* locus coding for endoglucanase and several extracellular proteins was restored when it was coinoculated with the low-virulence, extracellular factor-producer *R. solanacearum* strain, suggesting that all important factors were restored to normal by the extracellular factor. The extracellular factor was produced by *R. solanacearum* de novo. The chemical nature of the extracellular factor produced is not known, but it may be a vapor and may be a methanol-like factor having an OCH3 group (Clough et al., 1994). Thus several types of plant pectic molecules may be signal molecules, and all of them may be released by action of different enzymes of bacterial pathogens. The production of pectic enzymes in a proper sequence is needed for production of the signal molecules.

Signals for Activation of Genes for Secretion of Enzymes

The bacterial genes coding for proteins involved in the secretion system appear to be coregulated with the regulatory genes responsible for production of enzymes. The expression of two of the five operons involved in the secretion of Pels, Pem, and endoglucanase enzymes from the bacterial periplasmic space to the outer medium, *outT* and *outC,* is negatively activated by *kdgR* in *Erwinia chrysanthemi* and thus both *outT* and *outC* belong to the *kdg* regulon (Condemine et al., 1992). *OutT* has been detected in the *pel* cluster (Condemine et al., 1992). Regulation of the *outC* operon expression by *kdgR* requires the *outT* gene product (Condemine and Robert-Baudouy, 1995). The gene *pecS,* which regulates *pel* and *pelZ* gene expression in *E. chrysanthemi,* controls the *outC* operon (Condemine and Robert-Baudouy, 1995). All plant signals that affect pectate lyase production also modulate Out protein synthesis (Condemine and Robert-Baudouy, 1995). Hence, the signals and environment which regulate both production and secretion of the enzymes may be tightly linked.

PLANT CELL WALL COMPONENTS INVOLVED IN DEFENSE MECHANISMS AGAINST BACTERIAL PATHOGENS

Pectin Esterification

Several studies have indicated that bacterial pathogens rely mostly on their ability to produce pectic enzymes, cellulases, proteases, and hemicellulases to infect host tissue (Meletzus et al., 1993; Pischik et al., 1996; Dreier et al., 1997; Gonzalez et al., 1997). Cell wall structures offer resistance by repressing synthesis of these enzymes. The pathogens have to overcome this barrier to produce these enzymes. Pectin esterification in plant cell walls has been correlated with bacterial disease resistance. The cortex of tubers of all potato genotypes tested was significantly more resistant to rotting than the medulla when inoculated with *E. carotovora* subsp. *carotovora.* The resistance of both the cortex and medulla of genotype 4680 was significantly greater than those of its sibling 4708. Both tuber and stem rotting appeared to be more closely related to the degree of pectin esterification of the stem and tuber tissue of the different potatoes (Table 2.3; McMillan et al., 1993b).

Pagel and Heitfuss (1989) found that potato tuber rotting susceptibility of six cultivars to *Erwinia carotovora* subsp. *atroseptica* was inversely related to cell wall pectin esterification. The stem cell walls of the genotypes belonging to *Solanum tuberosum* subspecies resistant to black leg disease caused by *E. carotovora* subsp. *atroseptica* showed a higher content of highly methylated and branched water-soluble pectins. Numerous pectin methyl esterase isoforms were extracted from all cell walls, but an acidic form, present in the cell walls of the susceptible genotypes, could not be detected in the resistant cell walls (Marty et al., 1997). The isoform might, therefore, be involved in the *in muro* demethylation of the pectins. Pectate lyases are the major enzymes produced by *Erwinia* spp. which are involved in soft rot and black leg disease development. These enzymes do not attack

TABLE 2.3. Relationship Between Pectin Esterification of Potato Tuber and Stem and Intensity of Rotting Caused by *Erwinia carotovora* subsp. *atroseptica*

Potato genotype	Tuber rot*		Stem rot **	Pectin esterification %		
	Medulla	Cortex		Tuber	Cortex	Stem
4708	12.9	1.5	5.0	27.9	53.0	42.0
468	0.6	0.3	0.0	69.4	94.1	69.3

Source: McMillan et al., 1993b.

*Tuber rot was assessed in terms of lesion diameter (mm).
**Stem rot was rated as 0-5 depending upon intensity of infection.

esterified rhamnogalacturonan chains but prefer polygalacturonic acid (Evtushenkov and Fomichev, 1996). Polygalacturonases, which are the other important enzymes of *Erwinia,* also do not hydrolyze esterified pectic substances. Hence, potato cultivars with high esterified pectic substances may be resistant to *Erwinia* spp. (McMillan et al., 1993b). Although *Erwinia* spp. produce pectin methyl esterase (PME) (Tsuyumu and Chatterjee, 1984), which can deesterify the pectic substances, its activity in planta was negligible. Hence, pectin esterification may be involved in *Erwinia* disease resistance. Pectin esterification may be an important host defense mechanism when the pathogen is unable to produce abundant PME and/or the host plant is not able to produce suitable pectin methyl esterase to convert pectin into polygalacturonic acid. However, several bacterial pathogens are known to produce PME (Hugouvieux-Cotte-Pattat et al., 1996; Shevchick et al., 1996) and many host plants also produce PME (Vidhyasekaran, 1972; Marty et al., 1997). Although PME is produced by several bacteria, the enzyme should be produced either constitutively or at the very early stage of infection to make available polygalacturonic acid to Pels and Pehs for further degradation of cell wall pectic substances. The sequence of production of these enzymes may determine disease resistance or susceptibility.

Cell Wall Calcium

Cell walls contain several minerals and among them calcium is predominant. Resistance of cell wall pectic substances to maceration by pectic enzymes produced by bacterial pathogens has been suggested to be due to cross-linking of polyuronides with calcium (Jauneau et al., 1994). Potato tuber calcium concentration has been related to resistance to rotting caused by various *Erwinia* spp. (McGuire and Kelman, 1986). Cortex of potato tuber shows greater resistance than medulla to rotting by *E. carotovora* subsp. *atroseptica*. Calcium concentration was higher in cortex than in medulla in all potato genotypes tested (McMillan et al., 1993b).

Calcium alone may not be responsible for disease resistance. Calcium levels of cortex and medulla of the different potatoes were similar irrespective of whether the potatoes were susceptible or resistant to *E. carotovora* subsp. *atroseptica*. Calcium concentration in tubers of potato cultivars grown under similar conditions did not differ, although their relative susceptibility to rotting and blackleg was different. Probably both methylation of the carboxyl groups of galacturonic acid and cross-linking of polyuronides may synergistically contribute for resistance (Cooper, 1983).

Differential response of pectic enzymes to Ca^{2+} has also been reported. Ca^{2+} is stimulatory for the gene encoding for a pectate lyase, *pelB*, while it is inhibitory to the gene encoding for a polygalacturonase, *pehA*, in *E. carotovora* subsp. *carotovora* (Heikinheimo et al., 1995). In general, endo-polygalacturonase activities are reduced and endo-pectate lyases are stimulated by an increase in the amount of calcium (Moran et al., 1968; Garibaldi and Bateman, 1971). Henrissat et al. (1995) demonstrated that the Pel protein superfamily has a highly conserved amino acid sequence that is likely a catalytic site, which binds calcium. The hydrolytic enzymes may have different catalytic sites (Hotchkiss et al., 1996).

Calcium may also act as a signal molecule and high calcium content may suppress synthesis of pectic enzymes. Enhanced resistance to *E. carotovora* has been observed in high-calcium content plants (Flego et al., 1997). An increase in extracellular calcium to mM concentrations represses *pehA* gene (encoding PehA endopolygalacturonase) expression of this pathogen. Inhibition of in planta production of PehA resulted in enhanced resistance to *E. carotovora* infection. Ectopic expression of *pehA* from a calcium-insensitive promoter allowed *E. carotovora* to overcome this calcium-induced resistance (Flego et al., 1997). The results indicate that plant calcium acts by modulating the genes encoding pectic enzymes of the pathogen.

Apoplast Nitrogen

High nitrogen application induces host susceptibility to bacterial diseases (Vidhyasekaran, 1993). Low nitrogen significantly reduces Pel synthesis in *E. chrysanthemi* (Hugouvieux-Cotte-Pattat et al., 1992). Low nitrogen nutrition represses *OutT, OutC, OutB,* and *OutS* expression in *E. chrysanthemi* (Table 2.4; Condemine and Robert-Baudouy, 1995). It sug-

TABLE 2.4. Effect of Nitrogen on the Transcription of *outT-lacZ, outC-lacZ, outB-uidA,* and *outS-uidA* Fusions in *Erwinia chrysanthemi*

Growth condition	**β-galactosidase or β-glucuronidase activity of bacteria with the following genotype**			
	OutA-lacZ	***OutC-lacZ***	***OutB-uidA***	***OutS-uidA***
High nitrogen	132	159	1750	4250
Low nitrogen	27	26	280	1160

Source: Condemine and Robert-Baudouy, 1995.

Note: β-galactosidase or β-glucuronidase activities are expressed in nanomoles of *o*-nitrophenol or *p*-nitrophenol formed per minute per milligram of bacteria.

gests that nitrogen nutrient in the intercellular fluid may determine the virulence of the bacteria causing cell wall disintegration.

Apoplast Iron

Iron may play a role in planta in modulating the expression of bacterial pectate lyases (Pels). Iron limitation induces *pelB, pelC,* and *pelE* transcription in *Erwinia chrysanthemi* (Sauvage and Expert, 1994). *Erwinia* cells multiply in the intercellular fluid of plants, which is very poor in available iron. A low iron level induces pectate lyase synthesis (Hugovieux-Cotte-Pattat et al., 1996). In response to iron deprivation, *E. chrysanthemi* synthesizes two siderophores, chrysobactin and achromobactin. Chrysobactin is a stronger iron ligand than achromobactin. The *fct* gene encoding the *f*errichrysobactin outer membrane *t*ransport function and *cbs* genes involved in the four primary steps of *c*hrysobactin *b*iosynthesi*s* define an operon, *fct cbs CEBA* (Franza and Expert, 1991), that is strongly repressed by iron (Expert et al., 1992; Masclaux and Expert, 1995). In addition, *E. chrysanthemi* expresses a second iron acquisition system, dependent on the siderophore achromobactin (Mahe et al., 1995), that is induced under conditions of iron deficiency weaker than those required for chrysobactin depression. Mutations affecting ferriachromobactin permease, an ABC transporter encoded by the *cbr ABCD* operon, give rise to accumulation of achromobactin in the external medium, thereby decreasing iron availability to the bacterial cells (Mahe et al., 1995). The result is the derepression by chrysobactin production under conditions normally repressive. *Cbr* (*c*hryso*b*actin *r*egulatory gene) mutants of *E. chrysanthemi* display delayed symptoms on African violets *(Saintpaulia ionatha)* (Sauvage and Expert, 1994). The production of Pel was stimulated by low iron levels in the wild-type strain of *E. chrysanthemi* but not in a *cbr* mutant (Sauvage and Expert, 1994). The low level of iron in intercellular fluids slowed down the growth and spread of the *cbr* mutants (Enard et al., 1988; Neema et al., 1992; Masclaux and Expert, 1995).

Pel activity appeared to increase gradually, and at 24 h after inoculation a significant increase in activity was observed only in wild isolate of *E. chrysanthemi* and not in the *cbr* mutant when these bacteria were inoculated on *Saintpaulia ionantha*. A significant decrease of PelD and PelE activities was noticed in *cbr* mutant cells (Masclaux et al., 1996). Achromobactin-mediated iron uptake is affected in the *cbr* mutant. Because this mutation resulted in derepression of the chrysobactin-mediated iron transport pathway, the mutant is probably less susceptible to iron deprivation than wild-type cells are when entering the host. Thus resistance to iron deprivation may result in a delay in Pels production, leading to delayed symptoms. Wild-type cells produce chrysobactin when they are starved intracellularly for iron, and iron starvation leads to induction of *pelD, pelB,* and *pelC* in *E. chrysanthemi* wild isolate (Masclaux et al., 1996). If any of the functions of the chrysobactin are impaired by insertional mutation, the mutant fails to incite a systemic disease in *S. ionantha* (Hugouvieux-Cotte-Pattat et al., 1996). These studies indicate that concentration of iron in the host intercellular fluids may increase or delay the symptom development due to infection by *Erwinia chrysanthemi.*

BACTERIAL EXTRACELLULAR ENZYMES INDUCE HOST DEFENSE MECHANISMS

Bacterial exocellular enzymes induce several host defense mechanisms. Tobacco seedlings [tissue-culture (axenic cultured) plants] were sprayed with the pectic enzymes of *E. carotovora* subsp. *carotovora* and after two days inoculated with the virulent *E. carotovora* subsp. *carotovora* strains. Exoenzyme treatment induced enhanced resistance to *E. carotovora* subsp. *carotovora* (Palva et al., 1993). Polygalacturonase (PehA) treatment could afford good protection against the bacterial pathogen. Pectate lyase treatment delayed symptom development (Table 2.5; Palva et al., 1993).

Pathogenesis-related (PR) proteins are the important defense proteins in plants and β-1,3-glucanase belongs to PR-2 group of proteins (Vidhyasekaran, 1997). Induction of β-1,3-glucanase mRNA has been observed in cell cultures of *Arabidopsis thaliana* treated with culture filtrates containing endo-Pel of *E. carotovora* subsp. *carotovora* (Davis et al., 1984; Davis and Ausubel, 1989; Yang et al., 1992). β-1,3-Glucanase mRNA accumulated rapidly in tobacco seedlings sprayed with culture filtrate from the wild-type

TABLE 2.5. Pectic Enzyme-Induced Protection of Tobacco Against *Erwinia carotovora* subsp. *carotovora*

	Disease symptom development (days)*			
Pretreatment	**1**	**2**	**3**	**4**
Pectate lyase	–	+	++	+++
Polygalacturonase	–	–	+	+
No treatment	+	+	++	+++

Source: Palva et al., 1993.

*Disease intensity ranging from – (no maceration of plants) to +++ (plant material is totally macerated)

E. carotovora strain SCC 3193 (Palva et al., 1993). In contrast, no accumulation of mRNA could be observed in plants treated with culture filtrates from the exoenzyme-deficient mutant (Exp^-) SCC 3065 (Palva et al., 1993). It suggests that exoenzymes are needed for induction of β-1,3-glucanase. The endopolygalacturonase (PehA) appeared to be a strong inducer of the plant β-1,3-glucanase mRNA accumulation (Palva et al., 1993). Endopectate lyases *PelA, B, C,* and *D* also induced β-1,3-glucanase. In contrast to the strong accumulation of the β-1,3-glucanase mRNA by the different pectate-degrading enzymes, no accumulation of β-1,3-glucanase mRNA could be observed when plants were treated with cellulase (CelS). These results suggest that the major pectate-degrading enzymes of *E. carotovora* subsp. *carotovora* are also the main elicitors of plant defense response (Palva et al., 1993).

Pretreatment of tobacco leaf tissue with low levels of purified *E. chrysanthemi* PelC induces resistance against challenge with compatible pathogen *P. savastanoi* pv. *phaseolicola* (Baker et al., 1986). The polygalacturonase isolated from bean leaves infected with *P. savastanoi* pv. *phaseolicola* has been shown to be a potent elicitor of bean phytoalexins (Longland et al., 1992). The culture filtrates of *E. carotovora* induced pterocarpan phytoalexins in soybean cotyledons. The culture filtrates contained two pectate lyases. Increased amounts of Pel resulted in increased elicitor activity inducing phytoalexins (Davis et al., 1984).

Potato tubers under reduced moisture conditions show resistance response to *E. carotovora* subsp. *carotovora* while they are susceptible at anaerobic conditions (Perombelon and Lowe, 1980; Lyon, 1989; Davis et al., 1990). Phenylalanine ammonia-lyase (PAL) mRNA and enzyme activity were elevated in response to both compatible and incompatible interactions

(Yang et al., 1989). Potato tuber slices were inoculated directly with *Escherichia coli* expressing *Erwinia carotovora* subsp. *carotovora* pectic enzyme genes from plasmids (Yang et al., 1992). Inoculation with *E. coli* alone did not increase PAL mRNA levels; but plasmid pDR1 in *E. coli* encoding *E. carotovora* subsp. *carotovora* endo-Pel caused an increase in potato PAL mRNA. Plasmid pDR30 in *E. coli,* encoding endo-Peh and exo-Pel did not induce accumulation of PAL mRNA (Yang et al., 1992). It suggests that specific pectic enzymes alone trigger defense mechanisms of the host.

E. carotovora subsp. *carotovora* endo-pel increased expression of a number of plant defense genes including PAL and 4-coumarate : coenzyme A ligase (4CL) genes (Davis and Ausubel, 1989). The pectate lyase induced phytoalexin accumulation in soybean (Davis et al., 1984, 1986). Bauer et al. (1994) have shown that *E. chrysanthemi* can trigger a hypersensitive response when the production of Pels is disrupted.

Thus several studies have revealed that some specific pectic enzymes may be involved in induction of host defense mechanisms. Pectic fragments released by the action of pectic enzymes have been shown to be elicitors of defense genes (Davis and Hahlbrock, 1987; Collinge and Slusarenko, 1987; Farmer et al., 1991; Yang et al., 1992; Vidhyasekaran, 1997). The commercial preparation of sodium polypectate was treated with the endopolygalacturonase of *E. carotovora* subsp. *carotovora,* PehA, and the partially hydrolyzed pectate was used to spray tobacco seedlings. Peh-A treated polypectate was active as an elicitor of β-1,3-glucanase mRNA accumulation. The untreated preparations of sodium polypectate did not show β-1,3-glucanase–inducing activity (Palva et al., 1993). It suggests that the endopolygalacturonase of *E. carotovora* subsp. *carotovora* releases some pectic fragments which function as elicitors.

A pectate lyase of *E. carotovora* released heat-stable elicitors from soybean cell walls and the heat-stable elicitor-active material solubilized from soybean cell walls by the Pel was composed of at least 90 percent w/v uronosyl residues (Davis et al., 1984). Oligogalacturonates elicit numerous physiological responses in plants (Darvill and Albersheim, 1984; Ryan, 1988; Ryan and Farmer, 1991; Darvill et al., 1992). Oligogalacturonates activate genes encoding phytoalexins (Nothnagel et al., 1983; Davis et al., 1984; Forrest and Lyon, 1990), PR proteins (Davis and Ausubel, 1989), proteinase inhibitors (Bishop et al., 1984) and genes inducing rapid localized cell death (Keen, 1992). Polygalacturonates induce synthesis of phytoalexins (Nothnagel et al., 1983), protease inhibitors (Bishop et al., 1984), enzymes of the phytoalexin pathway and pathogenesis-related proteins in plants (Davis and Hahlbrock, 1987; Davis and Ausubel, 1989). Oligogalacturonates can induce reinforcement of cell walls (Bruce and West, 1989; Bradley et al., 1992). Oligogalacturonates induce responses such as rapid and reversible plasmalemma depolarization, K^+ efflux, and an influx of Ca^{2+} (Thain et al., 1990; Mathieu et al., 1991; Messiaen et al., 1993).

Several studies have shown that oligogalacturonates with a degree of polymerization (DP) of 7-20 produced by the action of pectic enzymes elicit defense responses in plants (Nothnagel et al., 1983; Bishop et al., 1984; Davis and Hahlbrock, 1987; Davis and Ausubel, 1989; Yang et al., 1992). Short oligogalacturonates (DP < 4) showed no or weak elicitor activity (Thain et al., 1990; Mathieu et al., 1991; Yang et al., 1992; Messiaen et al., 1993). Oligogalacturonates with a degree of polymerization of less than 6 have been reported to be inactive as elicitors (Farmer et al., 1991; Darvill et al., 1992).

Some recent studies have revealed that shorter oligogalacturonates also act as elicitors (Moloshok et al., 1992; Weber et al., 1996). The exoenzyme Pel3 of *E. carotovora* subsp. *atroseptica* strain C18 forms oligogalacturonates, predominantly tri- and dimers, from pectins (Bartling et al., 1995a,b). Weber et al. (1996) demonstrated that unsaturated digalacturonates produced by Pel enzyme action induced defense reactions against *Erwinia* soft rot in potato tissue. The unsaturated digalacturonate (uDG) induced cell death and membrane damage (increase in ion leakage) in potato similar to that induced by *E. carotovora* subsp. *atroseptica* (Table 2.6; Weber et al., 1996). It indicates that exogenous uDG was sufficient to induce cell death. Accelerated cell death is known to induce resistance by liberating phenolics and phenolases in potato (Tomijama et al., 1967). Another compound, 5-keto-4-deoxyuronate (DKI), which is produced when incubated with uDG by the action of oligogalacturonide lyase (OGL) secreted by *E. carotovora* subsp. *carotovora,* also induced cell death similar to uDG. Potato tissue incubated with DKI inhibited the soft-rot development caused by *E. carotovora* subsp. *atroseptica* similar to uDG. Incubation with galacturonic acid had no effect, reflecting that a transeliminative degradation of digalacturonates could be involved in the induction of plant defense responses against *E. carotovora* subsp. *atroseptica* (Weber et al., 1996).

TABLE 2.6. Induction of Cell Death and Ion Leakage in Potato Tissues Incubated with *Erwinia carotovora* subsp. *atroseptica* or Digalacturonate

Treatment	Cell death (%)	Increase in ion leakage (conductivity in µS)
E. carotovora	76	3140
Unsaturated digalacturonate	20	816

Source: Weber et al., 1996.

The Pel3 of *E. carotovora* subsp. *atroseptica* forms dimers and trimers from pectins (Bartling et al., 1995a,b). Transgenic potato plants expressing the heterologous Pel3 enzyme were developed (Wegener et al., 1996). The Pel3 released by wounding the tuber tissue of transgenic plants caused an increased resistance to maceration by *E. carovotora* and its enzymes. A strong induction of *pal* (phenylalanine ammonia-lyase) gene transcription was observed in the transgenic plant with high inhibitory effect against *Erwinia* maceration. There was no detectable PAL induction in the susceptible transformant expressing only small amounts of Pel3 in its tubers (Wegener et al., 1996). These observations suggest that the pectic fragments (dimer/trimer) released by Pel3 may act as signal molecules. Moloshok et al. (1992) found digalacturonic acid to be the most active oligomer to induce synthesis of a protease inhibitor in tomato seedlings. Oligogalacturonates of DP > 3 elicit H_2O_2 (Legendre et al., 1993). Weber et al. (1996) suggested that both short and larger oligogalacturonides may act as elicitors. Oligogalacturonide elicitors of DP > 3 may be active in micromolar concentrations while uDG may be active at millimolar concentrations in inducing resistance.

PECTIC FRAGMENTS INDUCE VIRULENCE GENES IN BACTERIA AND DEFENSE GENES IN PLANTS

Pectic fragments of different degrees of polymerization (DP) have been suggested to act differently. The incomplete digestion of pectic substances by pectate lyases (optimal activity at pH 8-10) at the nonoptimal physiological pH of plant tissues (pH 6 or lower) results in release of elicitor-active oligogalacturonides (DP ranging from 7 to 20) which induce host defense genes under optimum conditions. These enzymes produce short oligogalacturonates (dimers and trimers) which are effective in inducing various pectic enzymes of the bacteria (Stack et al., 1980; Collmer and Keen, 1986). However, the short oligomers have also been shown to induce host defense mechanisms (Weber et al., 1996). Rong et al. (1994) demonstrated that polygalacturonate oligomers of DP 6 to 17 from carrot roots induced *picA* (*p*lant-*i*nducible *c*hromosome) locus in *Agrobacterium tumefaciens* (Figure 2.1).

The *picA* locus activates genes needed for bacterial infection. Thus pectate fragments may induce both susceptibility and resistance in plants. It is not possible unless the pectic fragments have some specific structural differences among them to induce resistance or susceptibility depending upon their structures. Oligonucleotide site-directed mutations were introduced into the *pelC* gene of *E. chrysanthemi* EC16, which directed single or double amino acid changes. Three different mutations at lysine 172 with greatly

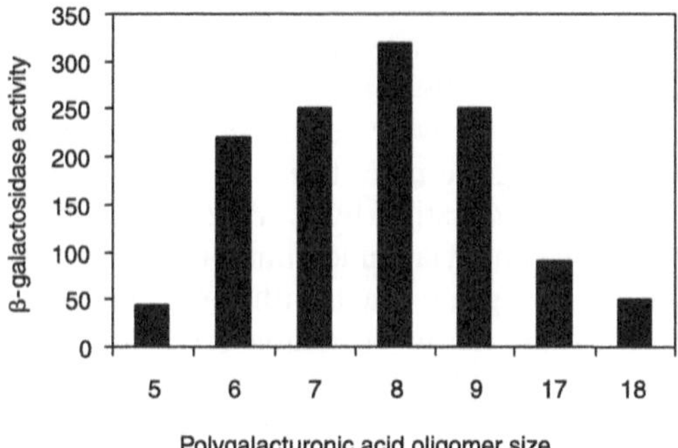

FIGURE 2.1. β-galactosidase activity of *Agrobacterium tumefaciens* incubated with individual galacturonic acid oligomers (*Source:* Rong et al., 1994).

reduced pectinolytic activity but as much elicitor activity as the wild-type protein were obtained (Kita et al., 1996). PelE macerated plant tissue 10 times more efficiently than PelC but showed equal activity in the elicitor assay (Kita et al., 1996). The results indicated that factors other than pectinolytic activity per se are involved in elicitor activity.

Similarly, the *A. tumefaciens picA* locus-inducing activity from carrot roots may not be a simple oligouronide (Rong et al., 1994). Although polygalacturonate oligomers of specific sizes were able to induce β-galactosidase activity of the *picA* : : *lacZ* fusion, the peak activities were much lower than those induced by carrot extracts. The carrot extracts purified by Sephadex chromatography contained only trace amounts of polygalacturonic acid oligomers. It suggests that the inducing activity is not a simple oligogalacturonide. The carrot extract was approximately 100-fold more active per uronic acid content (Rong et al., 1994). The carrot extract may contain complex pectin-derived oligomeric signals that are far more active than are simple oligogalacturonides. Further purification of oligogalacturonates may be needed to assess the exact role of pectic fragments released from plant cell walls by the action of pectic enzymes.

PECTIC ENZYMES VARY IN INDUCING RESISTANCE OR SUSCEPTIBILITY

Involvement of specific bacterial pectic enzymes in host plant resistance by releasing host pectin fragments has been reported in several host-pathogen interactions. A typical example is the interaction of endo- and exo-pectate lyases of *E. carotovora* subsp. *carotovora* and potato soft rot disease resistance. Endo-Pel of *E. carotovora* subsp. *carotovora* releases oligogalacturonates of DP of more than 7 which elicit defense mechanisms in potato tuber tissue (Yang et al., 1992). Exo-Pel of *E. carotovora* subsp. *carotovora* produces unsaturated dimers and trimers (Roberts et al., 1986b) which do not induce defense responses. It may cleave elicitor-active large oligogalacturonides to non–elicitor-active small oligogalacturonides. Exo-Pels are involved in producing reaction intermediates as inducers to mediate their own induction as well as other pectate lyases (Yang et al., 1992). When potato tubers were inoculated with the bacteria, both the enzymes were produced. Significant levels of endo-Pel mRNA were maintained through 24 h under conditions incompatible with soft rot (aerobic conditions) but not under conditions compatible with rot (anaerobic conditions). At 24 h endo-Pel could not be detected in a compatible interaction (Yang et al., 1992). It suggests that endo-Pel may be available to produce oligogalacturonides acting as inducers of host defense responses in incompatible interactions. Exo-Pel mRNA accumulation was rapid under compatible conditions but was absent early in the interaction under incompatible conditions (Yang et al., 1992). Exo-Pel may induce pectic enzymes, which cause tissue maceration (soft rot). PAL mRNA involved in phenylpropanoid metabolism continued to increase through 12 h after *E. carotovora* subsp. *carotovora* inoculation under incompatible conditions but not under compatible conditions (Yang et al., 1989). The endo-Pel increased expression of a number of plant defense genes including PAL and 4-coumarate: coenzymeA ligase (4CL) genes (Davis and Ausubel, 1989). Thus induction of the bacterial endo-Pel results in resistant reactions while induction of exo-Pel results in susceptible reactions.

The cell wall factors may modulate the production of extracellular enzymes, and a delicate balance may be maintained so that the enzymes may induce susceptibility or resistance in host plants. The extracellular enzymes induce synthesis of defense chemicals in plants or induce host tissue disintegration depending upon the type of signals generated by the plant cell wall. Bacterial pectic enzymes have dual and apparently opposing functions: macerating plant tissues and triggering defense responses (Hahn et al., 1989). The elicitor activity depends on the type and amount of enzyme involved in the interaction, and the signaling mechanism involved may determine the type of reaction.

POLYGALACTURONASE-INHIBITING PROTEINS

Some proteins detected in plant cell walls have been shown to inhibit polygalacturonases, resulting in production of pectic fragment elicitors, which induce host defense mechanisms. Polygalacturonase inhibitory proteins (PGIPs) have been isolated from a number of dicotyledonous plants (Bergman et al., 1994; Yao et al., 1995; Caprari et al., 1996; Pressey, 1996; Cervone et al., 1997). Among monocotyledons, onion *(Allium cepa)* and leek *(Allium porum)* contain polygalacturonase inhibition activity (Favaron et al., 1993, 1997). PGIPs are proteins structurally related to several resistance gene products recently cloned in plants (Bent, 1996) and belong to a superfamily of leucine-rich repeat (LRR) proteins specialized for recognition of nonself molecules and rejection of pathogens. PGIPs and resistance gene products may function as integrated components of a cell surface apparatus, part of the "plant immune system" in which the role of each component is defined by both its structure and regulation (De Lorenzo and Cervone, 1997).

PGIP is encoded by a family of genes (Toubart et al., 1992; Frediani et al., 1993; Favaron et al., 1997). The PGIPs which have been purified to date differ in organ-specific accumulation and in some functional characteristics (Salvi et al., 1990; Toubart et al., 1992; Stotz et al., 1993; Johnston et al., 1993; Favron et al., 1994; Yao et al., 1995). The different PGIP genes encode PGIPs with distinct regulation and distinct specificity, i.e., the ability to interact with and inhibit polygalacturonases from different pathogens (Desiderio et al., 1997). PGIPs purified from bean, tomato, and pear show differential specificity toward polygalacturonases of different pathogens (Hoffman and Turner, 1984; Johnston et al., 1993; Desiderio et al., 1997).

The PGIPs have been shown to inhibit only fungal polygalacturonases. Their role in inhibition of bacterial polygalacturonases is not yet known. There is a report that PGIP from raspberry did not inhibit exo-polygalacturonase of *E. carotovora* subsp. *atroseptica* (Johnston et al., 1993). But different PGIPs have been shown to inhibit polygalacturonases of pathogens in a different manner (Sharrock and Labavitch, 1994). For example, the product of one *pgip* gene, PGIP-1 (Toubart et al., 1992) is unable to inhibit a polygalacturonase from *Fusarium moniliforme,* and only partially inhibits polygalacturonase from *Fusarium oxysporum* f.sp. *lycopersici, Botrytis cinerea,* and *Alternaria solani* (Desiderio et al., 1997). Transgenic tomato plants expressing high levels of PGIP-1 did not exhibit enhanced resistance to these fungi (Desiderio et al., 1997). Contrastingly, expression of pear PGIP improves resistance in transgenic tomatoes to the fungal pathogens (Powell et al., 1994). These results suggest that more intensive studies are needed to assess the role of individual PGIPs in inhibiting bacterial polygalacturonases. The PGIP encoding genes have been shown to be inducible by treatment with biotic and abiotic elicitors (Bergmann et al., 1994; Nuss et al., 1996). Hence, isolation of PGIPs after elicitor treatment or incompatible bacterial inoculation may reveal more information. Intensive research is needed in this area as PGIPs can be exploited for bacterial disease management, and success in control of fungal diseases by PGIPs already has been reported (Powell et al., 1994).

CELL WALL MODIFICATIONS AND BACTERIAL DISEASE RESISTANCE

Papillae Formation

The cell wall is in a dynamic state, and when a bacterial pathogen tries to disintegrate a cell wall to facilitate its spread, the cell wall reinforces itself by rapidly modifying its structure. The most common response is formation of cell wall appositions called papillae (Morgham et al., 1988; Dai et al., 1996). The ability of plant cell walls to respond in a highly localized manner to the presence of bacterial pathogens has been demonstrated by electron microscopy in several host-pathogen interactions. The formation of cell wall apposition or papillae adjacent to the bacterial colony has been demonstrated in bean inoculated with *P. savastanoi* pv. *phaseolicola* (Brown and Mansfield, 1988), in tobacco inoculated with *P. syringae* pv. *pisi* (Politis and Goodman, 1978), in pepper inoculated with *X. vesicatoria* (Brown et al., 1993), and in cotton inoculated with *X. axonopodis* pv. *malvacearum* (Dai et al., 1996).

Papillae formation was associated with bacterial disease resistance. The first response of lettuce leaves to an incompatible pathogen *P. savastanoi* pv. *phaseolicola* was the apparent convolution of the plasma membrane next to bacterial cells, and by 3 to 5 h after inoculation, the accumulation of lightly stained fibrillar material was observed within the convoluted membrane next to bacterial cells (Bestwick et al., 1995). Progressive thickening and increase in complexity of the paramural deposits occurred within 8 h after inoculation. Swelling of the endoplasmic reticulum (ER) and an increase in numbers of variously sized smooth vesicles were observed at sites of paramural deposition (Bestwick et al., 1995). Papillae were found only in the cotton plants resistant to *X. axonopodis* pv. *malvacearum*.

The role of papillae in restricting bacterial multiplication is not known. However, in papillae many toxic substances such as phenolics, flavonoids, and H_2O_2 may accumulate (Peng and Kuc, 1992; Dai et al., 1996) or the papillae may strengthen the cell wall by incorporating callose (Dai et al., 1996) and hydroxyproline-rich glycoproteins (Bestwick et al., 1995).

Tyloses

Tyloses are commonly induced in vascular tissues of plants resistant to bacterial pathogens. Tyloses are balloonlike outgrowths. The origin and type of tylose cell wall is not known. In tomato the protective layer (neoformation lining the secondary wall) is described as the tylose cell wall (Bishop and Cooper, 1984). The tylose cell wall may be directly derived from the contact cell primary wall (Wallis and Truter, 1978; Grimault et al., 1994). Contact cells responsible for tylose production are known to be involved in plant defense against pathogens (Grimault et al., 1994).

In tomato cultivars resistant to *Ralstonia solanacearum,* tyloses occluded the colonized vessels and the contigous ones, limiting bacterial spread (Grimault et al., 1994). In susceptible wilting plants, no tyloses were observed in colonized vessels, and the bacterial spread was not limited. But tyloses were detected in many noncolonized vessels, and these tyloses may be responsible for wilting in the susceptible cultivars (Grimault et al., 1994). It is not known why tyloses accumulate in colonized vessels in resistant interactions and not in susceptible interactions. Probably the compatible bacteria may produce suppressors to suppress the tylose formation. Further studies are needed to elucidate this phenomenon.

Callose Deposition

When bacteria infect host tissues, callose deposition is observed in primary cell walls, middle lamellae, and newly formed papillae (Bestwick et al., 1995; Boher et al., 1995; Brown et al., 1996; Kpëmoua et al., 1996; Dai et al., 1996). Callose is a polymer of β-1,3-glucans and it is synthesized from glucose by an enzyme, callose synthase (=β-1,3-glucan synthase) (Lawson et al., 1989; Dugger et al., 1991). The purified β-1,3-glucan synthase from different plants has been shown to contain several polypeptides suggesting the existence of a multi-submit enzyme complex (Delmer et al., 1991; Wu et al., 1991). Callose synthesis is activated by Ca^{2+} (Kohle et al., 1985; Kauss, 1987). β-1,3-glucan synthase is located in the plasma membrane (Schmele and Kauss, 1990). The activity of the plasma membrane-located β-1,3-glucan synthase is strictly dependent on Ca^{2+} (Kauss et al., 1991). β-1,3-glucan synthase is latent in epidermal cells of healthy plants and is activated by attempted invasion by pathogens (Schmele and Kauss, 1990).

A membrane perturbation leads to elevated cytoplasmic Ca^{2+} (Atkinson et al., 1990; Ohana et al., 1992) and bacterial invasion causes membrane perturbation (Popham et al., 1995). Callose biosynthesis does not involve transcriptional activation of plant defense genes and may be induced by elicitor signals that cause changes in the plasma membrane (Kauss et al., 1990).

Besides Ca^{2+}, a β-glucoside is involved in callose synthesis (Kauss et al., 1990). β-furfuryl-β-glucoside (FG) is a specific endogenous activator of plant β-1,3-glucan synthase (Ohana et al.,1992). FG is localized within the vacuole in healthy plants and hence is inaccessible to the active site of β-1,3-glucan synthase, which is localized on the cytoplasmic side of the plasma membrane (Fredrikson and Larsson, 1989). However, lowering of cytoplasmic pH leads to release of FG from vacuole and under this condition more FG is detected in the cytoplasmic compartment of cells (Ohana et al., 1993). Bacterial invasion leads to membrane perturbation and lowers cytoplasmic pH (Atkinson et al., 1990). The combination of low pH and supplementation with high levels of external Ca^{2+} leads to dramatic stimulation of callose deposition (Ohana et al., 1993). Bacterial invasion may increase Ca^{2+} and FG in cytoplasm, which may lead to synthesis of callose in bacteria-infected tissues. However, this possibility has not been studied in plant-bacteria interactions.

Callose deposition in papillae has been observed in resistant interactions (Brown et al., 1996; Davis et al., 1996). Callose deposition was found in the phloem of cassava plants resistant to *X. axonopodis* pv. *manihotis* (Kpëmoua et al., 1996), and papillae containing callose deposits were found only in the bacterial blight-resistant infected cotton plants (Dai et al., 1996). In the compatible cotton-*X. axonopodis* pv. *malvacearum* pathosystem, paramural material enriched with callose was seen at the edge of the necrotic area (Dai et al., 1996).

Callose appears to strengthen cell wall barrier to prevent bacterial spread. All papillae that are formed in different plants appear to contain callose (Bayles et al., 1990; Benhamou, 1995).

HRGPs

Deposition of HRGPs in Cell Walls

Besides callose, hydroxyproline-rich glycoproteins (HRGPs) also accumulate in cell wall appositions. In bean and melon inoculated with bacteria, deposition of HRGPs within papillae was observed (O'Connell et al., 1990). The earliest changes in the cell wall of lettuce leaves inoculated with an incompatible pathogen, *P. savastanoi* pv. *phaseolicola,* involved the incorporation of HRGPs (Bestwick et al., 1995). Extensin-like HRGPs accumulated over the fibrillar-granular material formed around bacterial cells in tomato inoculated with *Clavibacter michiganense* (Benhamou, 1991). A large induction of extensin-like HRGPs was seen in turnip *(Brassica campestris)* leaves inoculated with an incompatible strain of *X. campestris* pv. *raphani* (Davies et al., 1997b).

Three types of changes in HRGPs have been noticed. New HRGPs may be synthesized and/or HRGPs may be insolubilized and/or intermolecular cross-linking of HRGPs with pectin, phenolics, and lignin may occur in plants which are under stress.

New HRGPs May Be Synthesized

HRGPs are encoded by multigene families expressing a variety of mRNAs (Showalter et al., 1991). Seven HRGP genes have been detected in the sunflower genome (Adams et al., 1992). Three HRGP mRNA species have been identified in tomato (Showalter et al., 1985; Zhou et al., 1992). Three cDNA clones (pDc11, pDc12, and pDc16) have been isolated from cDNA library of carrot. cDNA pDc11 encodes extensin while pDc12 and pDc16 encode a proline-rich protein (Chen and Varner, 1985). Cereal HRGPs are coded by simple gene families (Stiefel et al., 1990; Raz et al., 1991; Caelles et al., 1992; Guo et al., 1994).

Davies et al. (1997b) provided evidence that bacterial inoculation resulted in two new HRGP proteins which were absent in healhy turnip *(Brassica campestris)* plants. Two HRGPs (gp45 and gp120) have been detected in healthy turnip leaves. But when turnip leaves were inoculated with an incompatible strain of *X. campestris* pv. *raphani,* two new HRGP proteins, gp160 and gpS, were induced. In compatible interaction with *X. campestris* pv. *campestris* strain, no induction of these glycoproteins was observed. Reductions in the levels of the constitutive glycoprotein gp45 were seen, particularly in the compatible interaction (Davies et al., 1997b). The newly induced gpS comprised several components. There were three distinct fractions of gpS, gpS-1, gpS-2, and gpS-3. The gpS-3 fraction was rich in hydroxyproline and serine with substantial levels of tyrosine and lysine. In contrast, the gpS-1 and gpS-2 fractions had much lower levels of hydroxyproline but were both rich in serine and glycine (Table 2.7; Davies et al., 1997b). The gpS-3 fraction may be an HRGP of the extensin family while gpS-1 may be an arabinogalactan protein, an HRGP of another family (Davies et al., 1997b).

HRGPs May Be Insolubilized

When pathogens attack plants, HRGPs become insolubilized in plant cell walls (Chen and Varner, 1985). The insolubilization may be caused by link-

TABLE 2.7. Variations in Structure of Three Glycoproteins Induced

	mol %		
Structure	**gpS-1**	**gpS-2**	**gpS-3**
Amino acids			
Hydroxyproline	5.2	4.0	19.8
Glycine	10.1	12.6	4.7
Tyrosine	2.5	2.9	9.4
Lysine	6.4	3.1	9.1
Serine	12.6	15.4	13.9
Leucine	7.5	5.0	2.9
Monosaccharides			
Arabinose	60.6	92.4	96.1
Xylose	20.6	trace	trace
Galactose	18.8	7.6	3.9

Source: Davies et al., 1997b.

ing HRGPs by isodityrosine residues and by a diphenyl ether linkage (Fry, 1982). Treatment of soybean cells with H_2O_2 promoted the loss of two sodium dodecyl sulfate (SDS)-extractable proteins, a proline-rich protein (PRP), and an HRGP from cell walls. These two proteins could be still detected in the cell walls by using antisera of these proteins. It suggests that the two glycoproteins remain in the cell wall, but H_2O_2 induces a change in form that makes these proteins no longer extractable by SDS (Bradley et al., 1992). The insolubilization of plant cell wall proteins would have been caused by peroxidase requiring hydrogen peroxide generated via the oxidative burst (Fry, 1986; Everdeen et al., 1988; Bradley et al., 1992; Brownleader et al., 1993; Brisson et al., 1994). Increase in peroxidases has been reported in plants inoculated with incompatible bacteria (Young et al., 1995).

Cross-Linking of HRGPs

HRGPs are cross-linked with different cell wall constituents. Extensin is covalently linked to pectin (Showalter, 1993). The positively charged lysine and protonated histidine residues of extensin may be involved in ionic interactions with the negatively charged uronic acids of pectins (Showalter and Varner, 1989; Zhou et al., 1992). There may be HRGP cross-linking sites in pectin (Oi and Mort, 1990). HRGPs may provide nucleation sites for lignin-like polymer deposition (Whitemore, 1978).

In plants invaded by different vascular wilt bacteria, two types of material that could be distinguished easily by their texture and electron density were found to occur in infected cells and colonized vessels (Wallis, 1977; Mansfield and Brown, 1986; Benhamou, 1991). In tomato plants inoculated with *Clavibacter michiganense,* pectin-like molecules are the major components of the fibrillar or amorphous material accumulating at those sites where bacteria are actively growing (Benhamou, 1991). HRGPs accumulated over the fibrillar-granular material formed around bacterial cells (Benhamou, 1991). HRGPs together with pectic compounds may play a protective role during the infection process via intermolecular cross-linking (Oi and Mort, 1990). Lignin-like polymer and other phenolics accumulated in cell wall sites where HRGPs accumulated in lettuce leaves inoculated with an incompatible pathogen *P. savastanoi* pv. *phaseolicola* (Bestwick et al., 1995). Increased cross-linking of HRGPs may lead to strengthening of the cell wall to localize pathogens (Esquerre-Tugaye et al., 1979; Showalter et al., 1985). The increased HRGP cross-linking may lead to a more impenetrable cell wall barrier, thus impeding pathogen infection. Cell walls which undergo ultrarapid HRGP cross-linking, due to pathogen invasion, are tougher than cell walls of uninoculated healthy tissues (Showalter, 1993).

Some HRGPs May Agglutinate Bacterial Cells

Some of the HRGPs are known to agglutinate bacterial cells. Leach et al. (1982) reported that an HRGP from potato was capable of agglutinating strains of the bacterial wilt pathogen, *Ralstonia solanacearum.* Agglutination and hence immobilization of phytopathogenic bacteria by cell surface HRGPs may also contribute to plant defense (Mellon and Helgeson, 1982; Swords and Staehelin, 1993).

Cell Wall-Bound Phenolics

Different phenolic compounds have been detected in plant cell walls and have been shown to be involved in bacterial disease resistance. Phenolics were found to be deposited in the plant cell wall when lettuce leaves were inoculated with the incompatible bacterium *P. savastanoi* pv. *phaseolicola* (Bestwick et al., 1995). Flavonoids are the important plant cell wall-bound phenolic compounds. Anthocyanin compounds, the most conspicuous class of flavonoids, play a vital role in cotton resistance against *X. axonopodis* pv. *malvacearum* (Jalali et al., 1976; Holton and Cornish, 1995). When cotton cotyledons were infiltrated with *X. axonopodis* pv. *malvacearum,* flavonoids accumulated from 9 h after infiltration in the resistant variety (Reba B50) but no such flavonoid accumulation was observed in the susceptible variety (Acala 44) (Table 2.8; Dai et al., 1996). Other phenolics could be detected in both resistant and susceptible cotton cultivars while lignin was not detected in cotyledons of both resistant and susceptible cotton cultivars after infiltration with *X. axonopodis* pv. *malvacearum* (Table 2.8; Dai et al., 1996). The flavonoids which are synthesized in the cytoplasm (Snyder and Nicholson, 1990) were translocated toward the paramural areas and in newly formed papillae where they predominantly accumulated in primary cell walls and middle lamellae (Dai et al., 1996). In resistant interactions, flavonoid synthesis was activated in mesophyll cells close to infection sites (Dai et al., 1996).

The flavonoids were fixed to polysaccharides including cellulose, pectin, and callose (Dai et al., 1996). Flavonoids are known to inhibit several *Xanthomonas* species (Wyman and VanEtten, 1978).

TABLE 2.8. Accumulation of Defense Chemicals in Cotyledons of Cotton Cultivars Resistant or Susceptible to *Xanthomonas axonopodis* pv. *malvacearum* After Infiltration with the Bacteria

		Reaction according to the time course of infection (h) after infiltration					
Defense chemicals	**Cotton cultivar**	**0**	**6**	**9**	**15**	**24**	**48**
Flavonoids	R	–	–	+	+	+	+
	S	–	–	–	–	–	–
Phenolics	R	–	+	+	+	+	+
	S	–	±	±	±	±	+
Lignin	R	–	–	–	–	–	–
	S	–	–	–	–	–	–

Source: Dai et al., 1996.

R, Resistant; S, Susceptible; +, presence; –, absence; ±, trace

Lignin Deposition

Deposition of lignin in plant cell walls has been observed in many plant-bacterial interactions (Reimers and Leach, 1991; Guo et al., 1993; Milosevic and Slusarenko, 1996; Kpëmoua et al., 1996). Lignin is a polymer formed by the random condensation of phenylpropanoid units (Grisebach, 1981). Lignin monomers are produced from phenylalanine by a branch of phenylpropanoid metabolism. The first step of the phenylpropanoid pathway is the deamination of phenylalanine by phenylalanine ammonia-lyase (PAL) to yield *trans*-cinnamic acid. Hydroxylation of cinnamic acid by cinnamic acid-4-hydroxylase (CA4H) produces *p*-coumaric acid. Further hydroxylation by *p*-coumaric hydroxylase (CH) results in the synthesis of caffeic acid or 5-hydroxyferulic acid. Methylation of these acids by *O*-methyl transferases (OMT) produces ferulic and sinapic acids, respectively. The three acids (*p*-coumaric, ferulic, and sinapic acids) are coupled to CoA and reduced to the corresponding alcohols. The end products of this pathway are the three monolignols, viz., *p*-coumaryl, coniferyl, and sinapyl alcohols (Vidhyasekaran, 1988). Polymerization of these monolignols is catalyzed by peroxidases (Gross, 1978). The plants possess a number of different peroxidase isozymes (Flott et al., 1989; Harrison et al., 1995; Baga et al., 1995; El-Turk et al., 1996). Many of them have been found to be localized to the cell walls (Espelie et al., 1986; Lagrimini et al., 1987). The cell wall-bound peroxidases fall into two subgroups, anionic and cationic. Both the acidic and basic peroxidases have been found to be located in the plant cell walls (Kim et al., 1988; Hu et al., 1989). Both anionic peroxidases (Lagrimini et al., 1990; Lagrimini, 1991) and cationic peroxidases (Abeles and Biles, 1991) are involved in lignification. Cell wall-bound peroxidases are involved not only in the oxidative polymerization of hydroxylated cinnamyl alcohols but also in the generation of hydrogen peroxide necessary for lignification. The polymerization of the three cinnamyl alcohols is mediated by the peroxidase-H_2O_2 system (Goldberg et al., 1987; Peng and Kuc, 1992).

Rapid lignification appears to be an important defense mechanism in plants. Lignification was observed in both phloem and xylem of infected cassava plants resistant to *X. axonopodis* pv. *manihotis*. This reaction was observed at a higher intensity in resistant plants than in susceptible plants (Kpëmoua et al., 1996). When rice leaves were inoculated with an incompatible strain of *X. oryzae* pv. *oryzae,* lignin-like compounds accumulated in rice leaves. Similar accumulation of lignin is not seen in compatible interaction (Reimers and Leach, 1991).

Among the various peroxidase isozymes, only a few appear to be involved in induction of lignification and disease resistance. Three peroxidase genes, POS 22.3, POX 8.1, and POX 5.1, were identified from a cDNA library that was constructed from leaves of plants undergoing a resistant reaction to *X. oryzae* pv. *oryzae*. Only two peroxidase genes, POX 8.1 and POX 22.3, were predominantly expressed during resistant interaction (Chittoor et al., 1997). These two genes also were expressed during susceptible interactions, but induction was delayed compared with resistant interactions. POX 5.1 was induced equally well in mock inoculations with water or bacteria (Chittoor et al., 1997). Another peroxidase gene, *POXgX9*, was not induced in leaves after pathogen challenge (Chittoor et al., 1997).

In resistant or incompatible interactions between rice and *X. oryzae* pv. *oryzae*, increases in the activities of three intercellular peroxidases (two anionic PO-A1 and PO-A2 and one cationic PO-C1) have been correlated with the accumulation of lignin-like compounds and reduction in bacterial multiplication in the leaves (Reimers and Leach, 1991; Reimers et al., 1992; Guo et al., 1993). Activity of the cationic peroxidase PO-C1 increased more dramatically with the onset of resistance when compared with the two anionic peroxidases (Reimers et al., 1992). In the compatible interactions, the increase in PO-C1 activity was observed within 16 to 48 h. But PO-C1 did not accumulate until 72 h after infection in susceptible or compatible interactions (Reimers et al., 1992; Hopkins et al., 1992; Guo et al., 1993; Leach et al., 1994). The PO-C1 has been purified (Young et al., 1995).

PO-C1 was shown by immunoelectron microscopy to accumulate within the apoplast of mesophyll cells and within the cell walls and vessel lumen of xylem elements of plants undergoing incompatible interactions (Young et al., 1995).

PO-C1 accumulated faster and to higher levels in resistant than in susceptible host/pathogen interactions (Young et al., 1995). The timing of the accumulation of PO-C1 activity correlates with lignin deposition, and the isozyme has been shown to utilize the lignin precursor coniferyl alcohol as a substrate in vitro (Reimers and Leach, 1991). It suggests that PO-C1 may be a lignin peroxidase (Young et al., 1995).

When primary leaves of 'Red Mexican' bean plants (*Phaseolus vulgaris* cv. Red Mexican) were inoculated with *P. savastanoi* pv. *phaseolicola* race 1 (incompatible) and race 3 (compatible) by syringe infiltration, lignin deposition was observed in the inoculated leaf area in the incompatible interaction but not in the compatible interaction at 40 h after inoculation (Milosevic and Slusarenko, 1996). Peroxidase activity was also located in the same zone in the incompatible interaction. Peroxidase activity increased in this zone at 12 h after inoculation. The increase in peroxidase activity in the incompatible interaction could largely be attributed to the appearance of a novel anionic peroxidase from 18 h after inoculation onward (Milosevic and Slusarenko, 1996). The same anionic peroxidase also appeared in the compatible interaction, but only weakly, and first at 48 h after inoculation. The novel peroxidase may be involved in the lignification observed in the bacteria-inoculated zone (Milosevic and Slusarenko, 1996).

Cationic peroxidases also may be involved in lignification and induction of disease resistance. A strong correlation between a cationic peroxidase and lignification has been shown in rice leaves inoculated with *X. oryzae* pv. *oryzae*. When leaves of the rice cultivar Cas209 were inoculated with an avirulent strain of *X. oryzae* pv. *oryzae* (strain PXO86) and exposed to 24 h darkness after inoculation, water-soaking lesions typical of susceptible reaction appeared instead of a characteristic hypersensitive reaction (Guo et al., 1993). Duration of light exposure had no apparent effect on compatible interactions. Increase in the activity of a cationic peroxidase and accumulation of lignin-like polymers were observed in the resistant reaction while little cationic peroxidase was detected and lignin did not accumulate in susceptible reactions (Table 2.9; Guo et al., 1993). The results suggest that lignification may be involved in disease resistance and the cationic peroxidase may be involved in the increased lignification process. Increased expression of some cationic isoforms and subsequent increase in lignification have been reported in bean leaves inoculated with an incompatible strain of *P. savastanoi* pv. *phaseolicola* (Milosevic and Slusarenko, 1996).

Although lignification of plant cell walls has been observed in many resistance reactions, lignification may not serve as physical barrier against bacterial pathogens because bacterial plant pathogens do not penetrate host cells. However, lignified materials may prevent bacterial spread by blocking movement between the epithem and xylem vessels. The lignin biosynthetic process itself may be an important component of the defense response against bacteria, because bacteria may be inhibited by toxic phenolic compounds which are associated with lignification (Venere, 1980; Horino and Kaku, 1989; Guo et al., 1993).

Peroxidases

Several peroxidases have been found in xylem of *Helianthus*, strawberry, tomato, French bean, cucumber, rice, pear, and apple (Biles and Abeles, 1991; Magwa et al., 1993; Young et al., 1995). Peroxidase activity increases in a number of resistant plant-bacteria interactions (Reimers et al., 1992; Young et al., 1993). Peroxidase activities increased in rice leaves inoculated with *X. oryzae* pv. *oryzae*. The changes were more pronounced in the resistant lines than in the susceptible lines (Wu and Zeng, 1995). Peroxidase activity increased 4 h after inoculation of the resistant cotton line with *X. axonopodis* pv. *malvacearum* (Dai et al., 1996). Besides lignification, (Grisebach, 1981; Campa, 1991) peroxidases have a number of physiologi-

TABLE 2.9. Relationship Between Bacterial Blight Resistance and Activation of Cationic Peroxidase and Lignification in Rice Leaves

X. oryzae pv. *oryzae* strain	Treatment	Disease reaction	Cationic peroxidase (intensity in grades)	Lignification (intensity in grades)
Incompatible strain PXO86	24 h light	Resistant	+++++	++++
	24 h darkness	Susceptible	++	+
Compatible strain PXO99	24 h light	Resistant	++	+
	24 h darkness	Susceptible	+	+

Source: Guo et al., 1993.

cal functions that may contribute to resistance, including oxidation of hydroxy-cinnamyl alcohols into free radical intermediates (Gross, 1980), phenol oxidation (Schmid and Feucht, 1980), polysaccharide cross-linking (Fry, 1986), and cross-linking of extensin monomers (Everdeen et al., 1988). Increased peroxidase has been correlated with deposition of phenolic materials into plant cell walls during race-specific resistant responses (Graham and Graham, 1991; Reimers and Leach, 1991). Construction of cross-links due to peroxidases has been suggested to limit pathogen ingress and spread in these interactions (Tiburzy and Reisener, 1990). Venere (1980) demonstrated that the increase in peroxidase activity in cotton leaves during the incompatible *X. axonopodis* pv. *malvacearum* interaction was accompanied by a decline in the number of bacteria recovered and the accumulation of brown materials in the inoculated site. It was suggested that oxidation of phenolics such as catechin by peroxidase may be a factor in restricting growth of the bacterial pathogen in blight-resistant cotton.

Peroxidases are responsible for membrane oxidative burst during incompatible and resistant interactions (Vidhyasekaran, 1997). Plant peroxidases may also generate antimicrobial phenolics (Kobayashi et al., 1994). Ionically wall-bound peroxidases may be associated with the high level of phenolic esters incorporated into the cell wall of *Medicago* embryogenic culture (Hrubcova et al., 1994). Several phenolics have been reported to be converted into oxidized or polymerized compounds by activated peroxidases (Patzlaff and Barz, 1978; Takahama, 1988; Morales et al., 1993). In resistant cotton-*X. axonopodis* pv. *malvacearum* interactions, flavonoids accumulated in cell walls and paramural papillae, peroxidase activity may be involved in flavonoid incorporation into cell wall polysaccharides, and this peroxidase activity was not involved in synthesis of lignin, suberin, catechin, and condensed tannins (Dai et al., 1996).

The polymerization of phenolics within cell walls is also believed to result from peroxidase (Hahlbrock and Scheel, 1989; Graham and Graham, 1991; Bolwell, 1993). The polymerization of phenolics was highly localized within lettuce cells adjacent to bacterial colonies (Bestwick et al., 1995).

Insolubilization of HRGPs is also induced by the formation of intermolecular isodityrosine cross-links mediated by a specific peroxidase (Fry, 1986; Everdeen et al., 1988; Brownleader et al., 1993). HRGPs are rapidly insolubilized by activation of peroxidase (Bradley et al., 1992) in bacteria-infected tissues (Bestwick et al., 1995). Anionic peroxidases have been shown to be involved in suberization (Kolattukudy et al., 1987). Peroxidases play an important role in forming rigid cross-links between cellulose, pectin, HRGP, and lignin (Lagrimini et al., 1987). Peroxidase is involved in cross-linking pectic polysaccharides with phenolic acids in cell walls (Espelie and Kolattukudy, 1985) and cross-linking of extensin with feruloylated polysaccharides is also activated by peroxidase (Fry, 1986).

Cross-linking of several components of plant cell walls by peroxidase may be important in bacterial disease resistance. In cassava plants resistant to *X. axonopodis* pv. *manihotis,* deposition of callose, pectic and lignin-like materials, suberization, accumulation of phenolic-like molecules, and incorporation of flavonoids were observed in xylem and phloem cell walls. Similar reactions were not observed in susceptible plants (Kpëmoua et al., 1996). In the cell walls of lettuce plants, HRGPs, phenolics, and lignin-like polymers were found to be deposited simultaneously when the plants were inoculated with the incompatible bacterium *P. savastanoi* pv. *phaseolicola.* Cross-linking of these molecules by peroxidases has been suggested to be responsible for disease resistance (Bestwick et al., 1995).

Suberization and Gum Deposits

In the incompatible interactions, suberization was also observed. Suberin is an insoluble polymeric material attached to the cell walls. Suberin consists of a phenolic matrix attached to the cell wall and aliphatic components attached to the phenolic matrix. The aliphatic domains may be embedded in a layer of soluble waxes. Cross-linking of phenolics forms a polymeric matrix, and this matrix is made hydrophobic by attachment of aliphatic domains and by deposition of highly nonpolar waxes into this layer (Kolattukudy, 1981). The formation of the aromatic matrix is the first step in suberization. This matrix is catalyzed by peroxidases (Espelie et al., 1986). Suberization has been observed in inoculated cassava plants resistant to *X. axonopodis* pv. *manihotis*. It was not observed in infected susceptible plants (Kpëmoua et al., 1996).

Numerous gum deposits with lumpy appearance were observed in xylem vessels in a resistant tomato variety after inoculation with *R. solanacearum*. These deposits were absent in inoculated susceptible tissues (Grimault et al., 1994). Thus cell wall thickening appears to be a common phenomenon in plants expressing resistance to bacterial pathogens.

CONCLUSION

The plant cell wall appears to be the first dynamic defense barrier for bacterial infection. Fungal pathogens possess penetration structures to penetrate the plant cell wall barrier, but bacteria do not have any penetration structure and hence do not penetrate the cell wall. The bacteria enter their hosts through wounds and/or natural openings. However, bacterial pathogens multiply and spread through the apoplast, which is made up of primary and secondary cell walls. The plant cell wall is in a dynamic state and it is an extracellular matrix secreted into intercellular spaces. Several components are continuously synthesized and secreted into the cell wall layer. Hence, structure of the cell wall varies across time. When bacterial pathogens invade the cell wall, several molecular events take place to ward off the pathogens. The complex structure of the dynamic cell wall is still unexplored and the complex ordered structure has a mosaic of microdomains, which may induce or suppress the bacterial multiplication and spread.

The primary cell walls contain high amounts of pectic polysaccharides, cellulose, hemicelluloses, and proteins. These barriers have to be broken by the bacterial pathogens for their multiplication and spread in the apoplast (middle lamella and xylem vessels). Bacterial pathogens synthesize and secrete several extracellular enzymes to degrade the cell wall barrier. Genes encoding various enzymes and genes regulating secretion of extracellular enzymes have been cloned from various phytopathogenic bacteria. These genes are quiescent and they respond to various plant signals by using a signal transduction mechanism, which is a two-component regulatory protein system. This system comprises a sensor protein and a response regulator protein. The sensor protein undergoes autophosphorylation in response to plant signal molecules. The phosphorylated form of the sensor protein transfers the phosphoryl group to the response regulator protein, which upon phosphorylation activates transcription of the bacterial genes encoding extracellular enzymes and those genes involved in secretion of enzymes. Other positive and negative regulatory genes that regulate synthesis and secretion of extracellular enzymes also have been detected in bacterial pathogens.

Plant pectic fragments appear to be key plant signals that induce the bacterial enzymes. These fragments induce susceptibility or resistance depending upon the size of the fragments. Short oligogalacturonates induce synthesis of different pectic enzymes while large oligomers induce plant defense responses. Some smaller oligomers have also been shown to be involved in inducing disease resistance. Due to action of bacterial pectic enzymes, both smaller and larger pectic fragments are produced extracellularly. Their rapid assimilation by the bacteria may play a crucial role in determining the outcome of either disease or resistance responses by the host plant. It is still not known whether the same bacterial pathogen releases short oligomers in a susceptible host and large oligomers in a resistant host. Studies in these areas will be highly fruitful in developing technology for crop disease management. Some toxic chemicals such as phenolics may be synthesized in the resistant host, and these phenolics may inhibit production of pectic enzymes resulting in larger oligomers, which may trigger host defense mechanisms. Polygalacturonase-inhibitor proteins have been detected in plant cell walls; however, their role in inhibiting bacterial pectic enzymes is not yet studied.

The extracellular enzymes do not seem to be needed for bacterial pathogenesis. Even the role of individual enzymes in virulence of bacterial pathogens has not been established in many cases. A collaborative action of the range of secreted degradative enzymes may be involved in bacterial pathogenesis. Further, a correct sequence of production and secretion of these enzymes is needed for pathogenesis. Fine-tuning of various bacterial genes, probably by plant signals, may lead to disease development. Any alteration in the fine-tuning may result in disease resistance. How do these alterations take place in the incompatible interactions? Intensive studies are needed to unfold this complex phenomenon, which are lacking at present. Initial investigations have revealed that pectin esterification, cross-linking of polyuronides with calcium, and levels of nitrogen and iron in intercellular fluid may be involved in fine-tuning the production of enzymes.

When bacterial pathogens try to distintegrate the cell wall for their multiplication and spread, the cell wall reinforces itself by forming appositions. These appositions strengthen the cell wall by incorporating callose, hydroxyproline-rich glycoproteins, lignin, suberin, and several toxic compounds such as phenolics, particularly flavonoids and H_2O_2, may accumulate. Gum deposits have also been found in infected plant cell walls. When the bacterial pathogens invade the cell wall, new HRGPs are synthesized and/or HRGPs are insolubilized. Intermolecular cross-linking of HRGPs with pectin, phenolics, and lignin also occurs. Flavonoids are incorporated into cellulose, pectin, and callose. Several peroxidases accumulate in plant cell walls, and these peroxidases are involved in lignification, suberization, phenol oxidation, free radical formation, phenol synthesis, and polysaccharide cross-linking. All these reactions occur in resistant interactions, and similar reactions are not seen in susceptible interactions. Even if these reactions occur, they are very much delayed. It is not yet known why such reactions occur predominantly in resistant interactions. A recent study showed that *hrp* genes may not be involved in activation of cell wall defense mechanisms (Bestwick et al., 1995). Early release of elicitors from a pathogen's cell wall has been suggested as the possible cause for such early cell wall reinforcement in fungal pathogen-plant interactions (Vidhyasekaran, 1997). Similar studies in bacterial pathogen-plant interactions are lacking. Both endogenous (plant origin) and exogenous (bacterial origin) elicitors may be involved in this process.

Intensive research efforts are needed to unravel this important phenomenon, and these studies will pave the way to develop effective bacterial disease management.

Chapter 3

Active Oxygen Species

MECHANISMS OF PRODUCTION OF ACTIVE OXYGEN SPECIES

Plant defense responses against bacterial pathogens often involve rapid production of active oxygen species (AOS) which is called the oxidative burst (Adam et al., 1989, 1995; Keppler et al., 1989; Levine et al., 1994; May et al., 1996; Hammond-Hosack et al., 1996). In its ground state, molecular O_2 (dioxygen) is relatively unreactive. But it is capable of giving rise to lethal reactive excited states as active oxygen species. Utilization of O_2 proceeds most readily via a complete stepwise, four-electron reduction to water during which partially reduced relative intermediates are generated (Scandalios, 1993). The reactive species of reduced dioxygen include the superoxide radical (O_2^-), H_2O_2, and the hydroxyl radical (°OH). H_2O_2 is a nonradical and hence the collective term active oxygen species is used in the literature in preference to oxygen radicals in order to include nonradical forms such as H_2O_2. Singlet oxygen ($^1O_2$) is the physiologically energized form of dioxygen. All of these activated oxygen species are extremely reactive, and hence they are also called reactive oxygen species (ROA). They are cytotoxic in all organisms (Scandalios, 1993).

The AOS are intermediates that result from successive one-electron steps in the reduction of molecular O_2. There are several routes to AOS production in plants (Sutherland, 1991; Vianello and Macri, 1991; Elstner and Osswald, 1994). AOS are routinely generated at low levels by plant cells due to electron transport in chloroplasts, mitochondria, and enzymes in other cell compartments involved in the reduction-oxidation pathway. The first reaction during the pathogen-induced oxidative burst is believed to be the one-electron reduction of molecular oxygen to form a superoxide anion (Mehdy, 1994). O_2^- bears an unpaired electron and is routinely generated, in low concentrations, by the electron transport system (Rich and Bonner, 1978). O_2^- is also produced by a number of enzymes that participate in oxidation-reduction processes. Enzymes such as xanthine oxidase, aldehyde oxidase, and other flavin dehydrogenases are capable of generating superoxide anion as a catalytic product (Fridovich, 1986). An NADPH-dependent oxidase catalyses the single electron reduction of oxygen to form O_2^-, using NADPH as the reductant (Doke, 1985; Sutherland, 1991; Segal and Abo, 1993; Tenhaken et al., 1995; Dwyer et al., 1995; Desikan et al., 1996):

$$2\,O_2 + \text{NADPH} \rightarrow 2\,O_2^- + \text{NADP}^+ + \text{H}^-$$

The NADPH oxidase may reside in the plasma membrane (Mehdy, 1994; Doke et al., 1994). The activation of NADPH oxidase causes an abrupt rise in oxygen consumption, termed the respiratory burst (Chanock et al., 1994). The O_2^- production is inhibited by compounds such as diphenyleneiodonium that inhibit the NADPH oxidase (Nurnberger et al., 1994; Auh and Murphy, 1995) suggesting the importance of NADPH oxidase in superoxide anion production. NADPH oxidase-activated O_2^- production has been reported in human systems also (Segal and Abo, 1993; Tenhaken et al., 1995; Dwyer et al., 1995; Desikan et al., 1996; Groom et al., 1996). One of the key components of the human NADPH oxidase is gp91 phox, a 91 kDa membrane-bound glycoprotein, which together with a smaller component, gp22 phox, forms cytochrome b558 (Parkos et al., 1987). A gene, *rbohA,* which encodes a predicted protein with pronounced similarity to gp91 phox has been detected in rice (Groom et al., 1996). The existence of this gene suggests that plants generate O_2^- via an NADPH oxidase (Groom et al., 1996). *rbohA* may be a component of an NADPH oxidase that contributes toward O_2^- production.

Another oxidase that could be the source of the oxidative burst is an NADPH-dependent peroxidase that is associated with the external surface of the plasma membrane (Sutherland, 1991; Vera-Estrella et al., 1992). Peroxidase generation of O_2^- in plant tissues has been reported (Vera-Estrella et al., 1992). Xanthine oxidase activity produces O_2^- (Montalbini, 1992). Xanthine oxidase is a reductase supplying electrons to NAD^+ to produce NADH (Halliwell and Gutteridge, 1989). O_2^- arises as a by-product of many oxidoreductase enzymes when electrons leak from the reaction and reduce molecular oxygen, e.g., ferridoxin-$NADP^+$-reductase (Halliwell and Gutteridge, 1989).

H_2O_2 is produced from O_2^-. In aqueous solutions, the superoxide anion undergoes spontaneous dismutation to produce H_2O_2:

$$2\,O_2^- + 2\,H^+ \rightarrow H_2O_2 + O_2$$

Several enzyme activities can lead to the production of H_2O_2 inside and at the surface of plant cells (Halliwell and Gutteridge, 1989; Vianello and

Macri, 1991). Superoxide dismutase (SOD) activity produces H_2O_2 as a direct consequence of the disproportionation of superoxide anions (Apostol et al., 1989). The dismutation of two superoxide anions produces H_2O_2 (Tolbert, 1982). A membrane-associated NADH-dependent redox enzyme might be involved in the reactions, which generate H_2O_2 (Apostol et al., 1989). An NADPH-dependent oxidase activity, which generates O_2^-, in conjunction with SOD may lead to H_2O_2 production (Milosevic and Slusarenko, 1996). The H_2O_2 may result from NAD(P)H oxidation by peroxidase (Gross et al., 1977; Halliwell, 1978; Mader and Amber-Fisher, 1982). Peroxidase activity may contribute to the production of H_2O_2 and other activated oxygen species (Vianello and Macri, 1991; Vera-Estrella et al., 1992; Mehdy, 1994; Bestwick et al., 1995). Xanthine oxidase activity produces both O_2^- and H_2O_2 (Montalbini, 1992). Some enzymes, e.g., urate oxidase and glycollate oxidase, can produce H_2O_2 without going via O_2^- (Halliwell and Gutteridge, 1989).

Salicylic acid treatment of plants causes an increase in H_2O_2 concentrations in vivo (Delaney et al., 1994). Salicylic acid inhibits catalase, which can remove H_2O_2 (Milosevic and Slusarenko, 1996). Catalases are present as multiple isoforms in plants, and isolation of cDNA clones from several plant species has shown that catalases exist as small gene families (Scandalios, 1994). Expression analysis of catalases in plants showed that each of the genes is associated with a specific H_2O_2-producing process (Ni and Trelease, 1991; Willekens et al., 1994a). Xanthine oxidase and peroxidase also reduce the level of catalase and hence increase the production of H_2O_2 (Milosevic and Slusarenko, 1996).

The °OH radical is one of the strongest oxidizing agents known. O_2^- is able to react with other intermediates to produce °OH (Sutherland, 1991).

$$O_2^- + H_2O_2 + H^+ \rightarrow O_2 + H_2O + {}^\circ OH$$

O_2^- can also act as a reducing agent for transition metals such as Fe^{3+} and Cu^{2+}. These metals may be reduced even if they are complexed with proteins or low molecular weight chelators. The consequence of metal reduction is that it can lead to the H_2O_2-dependent formation of hydroxyl radicals (Mehdy, 1994).

$$O_2^- + Fe^{3+} \rightarrow O_2 + Fe^{2+}$$
$$Fe^{2+} + H_2O_2 \rightarrow Fe^{3+} + OH^- + {}^\circ OH$$

Superoxide and H_2O_2 can react in a "Haber-Weiss" reaction to generate the hydroxyl radical (Halliwell and Gutteridge, 1989; Scandalios, 1993).

Singlet oxygen ($^1O_2$) production requires energy, and an activation energy of approximately 22 kcal/mol is required to raise molecular O_2 from its ground state to its first singlet state (Scandalios, 1993). Lipoxygenases (Baker and Orlandi, 1995) and oxalate oxidase (Zhang et al., 1995) also have been proposed as alternative sources of AOS production.

SIGNALS FOR INDUCTION OF ACTIVE OXYGEN SPECIES IN BACTERIA-INFECTED PLANTS

Several studies have indicated that active oxygen species may be involved in bacterial disease resistance. Bacteria also are known to produce elicitors and these elicitors may act as signal molecules to induce active oxygen species. Adam et al. (1989) showed that the failure of the *hrpD* mutant of *P. savastanoi* pv. *phaseolicola* to cause a macroscopic HR in tobacco was associated with the loss of ability to induce O_2^- generation and extensive lipid peroxidation. Harpin, the protein product of the *hrpN* gene from *Erwinia amylovora,* stimulated active oxygen production in tobacco and acted as an elicitor of the HR (Baker et al., 1993). Harpin-producing *E. amylovora* induced AOS in tobacco cell suspensions, but these AOS were not induced by *E. amylovora* transposon mutants that do not produce harpin (Baker et al., 1993). Moreover, addition of cell-free extracts of *Escherichia coli* expressing the harpin gene increased AOS levels within 5 to 10 minutes (Baker et al., 1993). The autoclaved *Pseudomonas corrugata* cells elicited the oxidative burst (Devlin and Gustine, 1992).

Plant cell wall-derived elicitors released by bacterial infection (as described in Chapter 1) may also act as signal molecules in inducing AOS. Legendre et al. (1993) have shown an oligogalacturonide-induced oxidative burst. Elicitor receptors may be associated with heterotrimeric GTP-binding or G proteins. Agents known to interact with heterotrimeric G proteins were shown to promote generation of AOS in soybean cell cultures either in the presence or absence of elicitor (Legendre et al., 1992).

Active oxygen species generation in several plant species appears to depend on increased intracellular Ca^{2+} and protein kinase activation (the importance of Ca^{2+} and protein kinase in signal transduction is discussed in Chapter 1). A Ca ionophore, A23187, induced active oxygen formation in spruce cell suspensions (Schwacke and Hager, 1992). Some oxidized fatty acids are calcium ionophores (Serhan et al., 1981). A Ca^{2+} channel blocker, La^{3+}, inhibited harpin-induced AOS production in tobacco (Baker et al., 1993). The protein kinase inhibitors staurosporine and K252a inhibited elicitor-induced increases in AOS in tobacco (Baker et al., 1993) and spruce cell cultures (Schwacke and Hager, 1992).

Active oxygen species themselves may act as second messengers in the signal transduction system (Epperlein et al., 1986; Vick and Zimmerman, 1987; Slusarenko et al., 1991; Levine et al., 1994). The role of H_2O_2 and O_2^- in signal transduction is discussed in Chapter 1. AOS stimulates a rapid Ca^{2+} influx (Levine et al., 1996) and the role of Ca^{2+} influx in signal transduction is discussed in Chapter 1. Salicylic acid and jasmonic acid are also induced by AOS, and their role as systemic signals has also been described previously.

BACTERIAL INFECTION LEADS TO PRODUCTION OF ACTIVE OXYGEN SPECIES IN PLANTS

When tobacco plants were infected with an incompatible strain of *P. syringae* pv. *syringae,* increased O_2^- generation was observed (Adam et al., 1989). An increase in AOS (both O_2^- and H_2O_2) was observed within 30 min after inoculation with *P. syringae* pv. *tabaci* or *P. syringae* pv. *syringae* in tobacco cell suspensions (Keppler et al., 1989). A rapid production of H_2O_2 was observed in soybean cells incubated with *P. savastanoi* pv. *glycinea* (Levine et al., 1994). *Erwinia amylovora* induced AOS production in tobacco cell suspensions within 2 h after inoculation (Baker et al., 1993). When incompatible bacteria *Pseudomonas corrugata* were added to suspension-cultured cells of white clover *(Trifolium repens),* AOS production was observed within 10 min (Devlin and Gustine, 1992). Baker et al. (1991) found that both incompatible and compatible bacteria caused an increased production of AOS within 30 min after addition to tobacco cell suspensions. Leaf tissue and cell cultures produced an oxidative burst within minutes of exposure to pathogenic bacteria, involving the generation of O_2^-, °OH, and H_2O_2 (Keppler and Baker, 1989; Keppler et al., 1989). Xanthine oxidase activity which results in production of O_2^- and H_2O_2 increased in incompatible (avirulent race 1) *P. savastanoi* pv. *phaseolicola*/bean *(Phaseolus vulgaris)* interaction at 24 and 48 h after inoculation with the bacteria. But in the compatible combination, the enzyme activity increased at 48 h after inoculation (Milosevic and Slusarenko, 1996).

The oxidative burst consists of two phases of AOS production in many host-pathogen interactions. Phase I AOS production is rapid, transient, and nonspecific, occurring almost immediately after challenge with both compatible and incompatible pathogenic bacteria. Phase II AOS production, in contrast, occurs much later, is larger and long-lived, and is specifically stimulated only by incompatible bacteria resulting in disease resistance reaction. When soybean cv. Williams 82 cells were inoculated with compatible or incompatible strains of *P. savastanoi* pv. *glycinea,* a rapid but weak and transient burst of H_2O_2 production was observed. No further H_2O_2 production was observed in cells inoculated with the compatible strain, but in contrast, about 3 h after inoculation with the incompatible strain of *P. savastanoi* pv. *glycinea* there was a second and massive oxidative burst, sustained for several hours (Levine et al., 1994). H_2O_2 has been shown to be involved in hypersensitive cell death in soybean cells inoculated with an incompatible strain of *P. savastanoi* pv. *glycinea* and similar cell death was not observed in soybean cells inoculated with a compatible strain (Levine et al., 1994). It suggests that the second burst of H_2O_2 may be involved in disease resistance. Almost similar results have been reported in tobacco cell suspensions inoculated with a pathogen of tobacco *(P. syringae* pv. *tabaci)* and a nonpathogen of tobacco *(P. syringae* pv. *syringae)* (Keppler et al., 1989).

An early (step 1) transient increase in AOS (both H_2O_2 and O_2^-) occurred in both compatible and incompatible combinations in tobacco inoculated with *P. syringae* pv. *tabaci* and *P. syringae* pv. *syringae*. After this initial increase, levels of AOS, both H_2O_2 and O_2^-, continued to decline in the compatible interaction. In incompatible combination, O_2^- levels increased between 2 and 4 h and H_2O_2 levels increased from 3.5 h (Keppler et al., 1989). Baker et al. (1991) also found that both compatible and incompatible bacteria caused an increased production of AOS within 30 min after addition to tobacco cell suspensions. However, only bacteria that elicited the hypersensitive reaction stimulated a second, long-term production, beginning at 3 h and continuing until at least 6 h after inoculation.

The oxidative burst, consisting of two successive O_2^- and H_2O_2 bursts, has been shown to occur upon pathogen attack, and the second oxidative burst has been exclusively characterized in incompatible plant-pathogen interactions (Doke et al., 1994). The short-lived oxidative burst and long-term production of AOS may be of significance for determining the outcome of host-pathogen interactions (Apostol et al., 1989; Bradley et al., 1992; Davis et al., 1993; Devlin and Gustine, 1992; Levine et al., 1994).

ACTIVE OXYGEN SPECIES MAY INDUCE LIPID PEROXIDATION

The oxidative burst may function as a first line of defense in resistant plants by directly attacking the pathogen during the earliest stages of infection. The highly reactive active oxygen molecules initiate a lipid peroxidation chain reaction in the plasmalemma (Keppler and Novacky, 1986, 1989; Rogers et al., 1988; Adam et al., 1989). Lipid peroxidation appears to be characteristic of incompatible interactions. Lipid peroxidation was not detected in *Phaseolus vulgaris* leaves infected with the compatible pathogen *P. savastanoi* pv. *phaseolicola* (Croft et al., 1990). Keppler et al. (1986) measured increased lipid peroxidation in cucumber cotyledons infected with *P. syringae* pv. *lachrymans* (compatible interaction) but the level of increase was much less than that in the incompatible interaction. Tobacco plants infected with *P. syringae* pv. *syringae* (incompatible pathogen) increased lipid peroxidation along with increased O_2^- generation (Adam et al., 1989). The bacterial infection was associated with a decreased double bond index of fatty acids and a decrease in phospholipids and galactolipids (Adam et al., 1989). The increased saturation of fatty acids in the polar lipids is the consequence of lipid peroxidation. The phospholipid and galactolipid breakdown was also detected during the development of the hypersensitive (resistant) reaction. Phosphatidic acid accumulation was found at the same time as phospholipid breakdown (Adam et al., 1989).

Lipid peroxidation seems to be the consequence of the accumulation of active oxygen species (Dhindsa et al., 1981; Elstner, 1982). O_2^-, $^1O_2$, and °OH radicals are able to cause lipid peroxidation of polyunsaturated fatty acids since such compounds are particularly vulnerable to attack by these molecular species (Elstner, 1982; Halliwell, 1982; Halliwell and Gutteridge, 1986). Lipid peroxides may be generated by nonenzymic reaction of active oxygen species (Farmer and Ryan, 1992).

Increased O_2^- generation and lipid peroxidation were first detected in tobacco inoculated with *P. syringae* pv. *syringae* (incompatible pathogen) (Adam et al., 1989). Subsequently, decrease in the amounts of phospholipids and galactolipids was observed. There were reductions in the degree of unsaturation of polar acyl lipids and in the free sterol pool. The incompatible bacteria induced membrane deterioration in the plant cell and induced symptoms of the hypersensitive response (HR) (Adams et al., 1989). Application of different types of free radical scavengers, such as ascorbic acid, reduced glutathione, and α-tocopherol reduced all these activities and delayed the development of the HR (Adams et al., 1989). The autocatalytic chain of lipid peroxidation is controlled by different types of enzymic and nonenzymic free radical scavenging systems (Dhindsa et al., 1981; Elstner, 1982; Halliwell, 1982; Abeles, 1986). α-Tocopherol directly inhibits the chain reaction of lipid peroxidation (Halliwell, 1982; Lindsay et al., 1985; Lielder et al., 1986). It suggests that a major role of O_2^- generation may be lipid peroxidation.

Higher lipid peroxidation was also observed in incompatible bacteria-treated plant cell suspensions (Keppler and Baker, 1989). Addition of O_2^- scavengers SOD and tiron delayed the transient increase in lipid peroxidation in tobacco suspension cells treated with incompatible bacteria *P. syringae* pv. *syringae* (Keppler and Baker, 1989), suggesting that O_2^- may be responsible for increased lipid peroxidation.

INCREASES IN ACTIVE OXYGEN SPECIES LEAD TO ACTIVATION OF LIPOXYGENASE

Lipoxygenase (LOX) (linoleate : oxygen oxidoreductase) is a dioxygenase catalyzing the hydroperoxidation of fatty acids containing a *cis, cis*-1,4-pentadiene structure, e.g., linoleic, linolenic, and arachidonic acids (Hildebrand, 1989; Siedow, 1991). LOX is classified as 5-, 12-, or 15-LOX accordng to where in the carbon chain of arachidonic acid the hydroperoxy function is introduced (Bostock et al., 1992). Large numbers of LOX isozymes have been reported in plants (Eskin et al., 1977; Hildebrand et al., 1988; Ohta et al., 1991). Several LOX genes have been cloned from various plants (Shibata et al., 1987, 1988; Ealing and Casey, 1988; Siedow, 1991). Increase in LOX activity may be a response to fatty acids released by active oxygen species attack of membrane lipids during lipid peroxidation. LOX activity increased in plants inoculated with both incompatible and compatible bacterial pathogens (Keppler and Novacky, 1987). There was an increase in lipoxygenase activity in bean cotyledons infiltrated with the pathogen *P. syringae* pv. *lachrymans* or with an incompatible pathogen, *P. syringae* pv. *pisi* (Keppler and Novacky, 1987).

There are several reports of earlier and greater induction of LOX enzyme activity in resistant than in susceptible reactions in several pathosystems (Slusarenko and Longland, 1986). Earlier induction of LOX enzyme activity in *P. syringae* pv. *syringae* (incompatible pathogen) than in *P. syringae* pv. *tomato* (compatible pathogen)-inoculated tomato leaves was observed (Figure 3.1; Koch et al., 1992).

LOX mRNA was induced by 3 h and enzyme activity began to increase between 6 and 12 h and had reached maximum levels by 24 to 48 h in tomato leaves inoculated with the incompatible pathogen *P. syringae* pv. *syringae.* But in tomato leaves inoculated with the pathogen *P. syringae* pv. *tomato,* LOX mRNA was induced later and enzyme activity changed only marginally in the first 24 h, then increased steadily up to 72 h reaching the level seen in the resistant reaction (Koch et al., 1992). The results suggest that early induction of LOX may lead to disease resistance. The early increase in LOX may lead to membrane damage, leading to electrolyte leakage.

ACTIVE OXYGEN SPECIES PRODUCTION LEADS TO CELL MEMBRANE DAMAGE

Normal plant cell membrane structure and functions are altered during the incompatible interactions. Dramatic changes in leaf tissue membrane permeability, as measured by electrolyte leakage, have been reported in

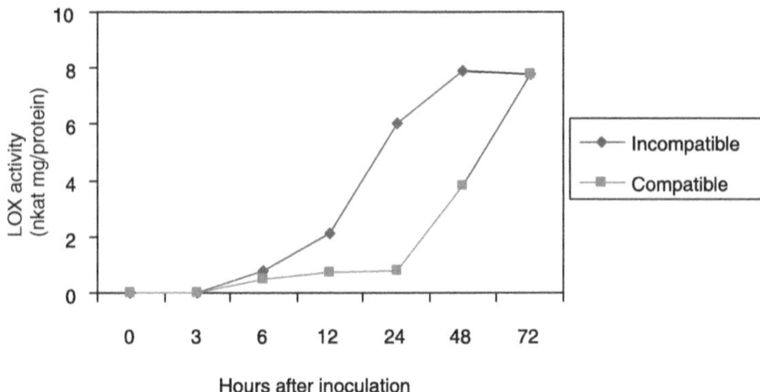

FIGURE 3.1. Changes in LOX enzyme activity in tomato leaves inoculated with *Pseudomonas syringae* pv. *syringae* (incompatible pathogen) and *P. syringae* pv. *tomato* (compatible pathogen) (*Source:* Koch et al., 1992).

those interactions (Goodman, 1968; Cook and Stall, 1968). Plant cells lose electrolytes as a consequence of irreversible modifications of the plasma membrane (Ohashi and Shimomura, 1982). A decline in membrane potential during the resistant interactions has been reported (Keppler and Novacky, 1986; Pavlovkin et al., 1986). In the attempted infection of cotyledons of an incompatible host (cotton) by *P. syringae* pv. *tabaci,* a depolarization of the host plasma membrane diffusion potential has been observed (Pavlovkin et al., 1986).

Membranes may deteriorate as a result of peroxidation of unsaturated fatty acids in their phospholipids (Halliwell and Gutteridge, 1989). Peroxidation of membrane lipids occurs as a result of increased levels of AOS or increased LOX activity (Wolff et al., 1986; Keppler and Novacky, 1987; Adam et al., 1989; Croft et al., 1990). Increased AOS generation was implicated in membrane damage of cells undergoing incompatible pathogen-induced HR (Keppler and Novacky, 1987). Increased LOX precedes electrolyte leakage in leaves of *Phaseolus vulgaris* undergoing incompatible bacteria-induced HR (Keppler and Novacky, 1986; Croft et al., 1990).

Electrolyte leakage increased in tobacco cell suspensions treated with both the incompatible pathogen *P. syringae* pv. *syringae* and the compatible pathogen *P. syringae* pv. *tabaci,* but to a much lower level in *P. syringae* pv. *tabaci*-treated cell suspensions than in *P. syringae* pv. *syringae*-treated cell suspensions (Keppler and Baker, 1989). Lipid peroxidation due to active oxygen species or lipoxygenase may result in electrolyte leakage from plant cells. Accumulation of maleindialdehyde, which is the end product of lipid peroxidation, slightly preceded the increase in electrolyte leakage in bacterially induced HR in plants (Keppler and Novacky, 1986). Lipid peroxidation damage of the plasmalemma leads to leakage of cellular contents, rapid dessication, and cell death (Scandalios, 1993). Early cell death leads to disease resistance. In the hypersensitive response, a lowering in the level of membrane unsaturation was observed (Keppler and Novacky, 1989), consistent with the oxidation and subsequent removal of unsaturated fatty acids from the bilayer.

The disruption of the plant cell membrane may release a signal from the membrane or induce an influx of signal molecules into the plant cell (Epperlein et al., 1986; Rogers et al., 1988; Apostol et al., 1989). Lipid peroxides generated by lipid peroxidation may serve as precursors in the synthesis of jasmonic acid, which is a well-known systemic signal molecule (Farmer and Ryan, 1992). The role of jasmonic acid in signal transduction has been discussed in Chapter 1.

ACTIVE OXYGEN SPECIES MAY DIRECTLY KILL BACTERIAL PATHOGENS

AOS may directly kill bacterial pathogens (Keppler and Baker, 1989; Peng and Kuc, 1992). Cells of the bacteria *X. hortorum* pv. *pelargonii* (= *X. campestris* pv. *pelargonii*), *P. savastanoi* pv. *phaseolicola, P. syringae* pv. *pisi,* and *Erwinia amylovora* were killed when they were exposed to AOS (Kiraly et al., 1997). Both catalase which degrades H_2O_2 and superoxide dismutase which scavenges O_2^- protected the bacteria from AOS (Table 3.1; Kiraly et al., 1997). It suggests that both O_2^- and H_2O_2 are involved in killing the bacteria.

Cells of *P. savastanoi* pv. *phaseolicola* present in the intercellular space of bean leaves were reduced in culturability by exposure to AOS (Kiraly et al., 1997). O_2^- scavengers added with *P. syringae* pv. *tabaci* inhibited the O_2^- production, lipid peroxidation, and increased recovery of bacteria in tobacco (Keppler and Baker, 1989). These observations suggest direct killing of bacterial pathogens by AOS. The amount of H_2O_2 released in plant cells due to pathogen attack or elicitor treatment (Legendre et al., 1993) was roughly equivalent to that present in activated mammalian phagocytes as part of their chemical arsenal (Morel et al., 1991; Mehdy, 1994). The killing effect of AOS may be responsible for reduction of bacterial disease symptom development. Treatment that led to the production of AOS almost completely suppressed disease symptom development in bean leaves inoculated with *P. savastanoi* pv. *phaseolicola* (Kiraly et al., 1997). Lower levels of antioxidant enzymes in different bean cultivars were correlated with greater

TABLE 3.1. The Killing Action of the Xanthine-Xanthine Oxidase System on Bacteria and the Protective Effect of Superoxide Dismutase (SOD) and Catalase

	Percent inhibition of bacterial growth		
Bacteria	**Xanthine-xanthine oxidase system only**	**Xanthine-xanthine oxidase system+SOD**	**Xanthine-xanthine oxidase system+catalase**
P. syringae pv. *pisi*	95	41	43
P. savastanoi pv. *phaseolicola*	99	37	35
X. campestris pv. *pelargonii*	95	31	45
E. amylovora	63	12	14

Source: Kiraly et al., 1997.

Note: Xanthine-xanthine oxidase is an active oxygen species-producing system.

resistance to pathogen infection (Buonaurio et al., 1987). Decreases in superoxide dismutase in rice-*X. oryzae* pv. *oryzae* resistant interactions have been reported (Wu and Zeng, 1995).

BACTERIAL PATHOGENS MAY TOLERATE TOXICITY OF ACTIVE OXYGEN SPECIES

In compatible plant-bacteria interaction, the bacteria are able to multiply in the intercellular space. Even in incompatible interactions, the bacteria are able to make initial growth (see Chapter 1). The capacity of infecting bacteria to multiply in plant tissues may be due to their ability to resist the toxic effects of AOS produced in plant tissues due to bacterial infection (Kiraly et al., 1991; Sutherland, 1991; Klotz and Hutcheson, 1992). Bacterial pathogens, particularly *Xanthomonas* spp. growing in their natural environment in plants, exhibit an initial rapid growth phase, after which the number of bacteria remains constant for many days while disease symptoms develop (Hopkins et al., 1992; Lummerzheim et al., 1993). During this stationary growth period in plants, bacteria must overcome plant defense responses, one of which may be increasing accumulation of AOS (Sutherland, 1991).

In order to protect themselves from the potentially deleterious effects of AOS, the bacteria seem to have evolved a whole battery of defensive antioxidant enzymes. The first line of defense appears to be mediated by protective enzymes such as catalase and superoxide dismutase, which scavenge H_2O_2 and O_2^-, respectively, and in combination catalyze their reduction to nontoxic compounds. *X. oryzae* pv. *oryzae* develops stationary phase resistance to stresses due to H_2O_2 and O_2^- (Vattanaviboon et al., 1995) by producing catalase.

Catalase is the major bacterial H_2O_2 protective enzyme. It catalyzes the breakdown of H_2O_2. Multiple catalase isozymes are present during different stages of bacterial growth, and induction of specific isozymes in response to oxidative stresses has been observed (Klotz et al., 1995). In most bacteria, H_2O_2 is a potential catalase inducer (Bol and Yasbin, 1990; Bishai et al., 1994). Superoxide anions are known to induce some catalases (Greenberg and Demple, 1989). A superoxide generator (paraquat) has been shown to be a potent inducer of *X. oryzae* pv. *oryzae* catalase (Chamnongpol et al., 1995a).

Superoxide anions also have been shown to be a potent inducer of catalase in other bacteria such as *X. oryzae* pv. *oryzicola, X. axonopodis* pv. *phaseoli,* and *X. axonopodis malvacearum* (Chamnongpol et al., 1995a). But in *X. translucens* pv. *translucens (= X. campestris* pv. *translucens)* and *X. vesicatoria,* O_2^- was a poor catalase inducer (Chamnongpol et al., 1995a). Ascorbate also induced the bacterial catalase (Chamnongpol et al., 1995a). Ascorbate has been shown as a potent catalase inducer in many bacteria (Loewen and Triggs-Raine, 1984; Bishai et al., 1994). Catalase induction involved de novo enzyme synthesis due to increased *kat* gene expression (Greenberg et al., 1990; Demple, 1991). The catalase induction in *X. oryzae* pv. *oryzae* was due to increasing synthesis of an 80 kDa monofunctional catalase (Chamnongpol et al., 1995b).

Superoxide dismutases (SOD) are metalloproteins that catalyze the dismutation of O_2^-. Structural genes for SOD have been cloned from a number of bacteria and have been found to encode three types of enzymes, differentiated by their metal cofactors, which can be either iron, manganese, or copper plus zinc (Benov and Fridovich, 1994). MnSOD is encoded by *sodA* while FeSOD is encoded by *sodB*. CuZnSOD, encoded by *sodC,* occurs primarily in the chloroplasts of plants, but a few bacteria also contain this enzyme (Benov and Fridovich, 1994). Chamnongpol et al. (1995b) reported that the major SOD enzyme of *X. oryzae* pv. *oryzae* is an MnSOD.

All *Xanthomonas* strains tested were highly resistant to O_2^-, probably due to activity of SOD (Chamnongpol et al., 1995a). The *sod* gene from *X. campestris* pv. *campestris* strain 8004 was upregulated following introduction of the bacteria into turnip (compatible host) and pepper (incompatible host) (Smith et al., 1996). The increase in SOD in both compatible and incompatible interactions suggests that the pathogen's SOD may not be a limiting factor for its growth in a host.

Superoxide anions may not be toxic to bacterial pathogens. But even low concentrations of O_2^- induced catalase, which will degrade H_2O_2 (Chamnongpol et al., 1995a). Even low concentrations of H_2O_2 retard growth of several *Xanthomonas* spp. (Chamnongpol et al., 1995a). Hence, superoxide anions may protect the bacteria against other AOS. However, it has been shown that during the stationary growth phase of the bacteria, levels of the oxidative stress protective enzymes catalase and SOD decrease, indicating that AOS may inhibit further growth of bacterial pathogens (Chamnongpol et al., 1995a).

Some bacteria, such as *X. oryzae* pv. *oryzae,* develop stationary phase resistance to AOS (Vattanaviboon et al., 1995). Stationary phase cells of *X. oryzae* pv. *oryzae* are much more resistant to killing concentration of H_2O_2 and O_2^- than cells from early log and mid-log phase. The stationary phase stress resistance phenotype did not require de novo protein synthesis (Vattanaviboon et al., 1995). It suggests that growth of compatible bacterial pathogen in the susceptible host may be either due to preformed AOS-degrading enzymes or it would have developed tolerance to the AOS at later stages of growth.

Bacterial pathogens produce extracellular polysaccharide (EPS) slime. Slime nonproducing or poorly producing strains of *P. savastanoi* pv. *phaseolicola* and *Erwinia amylovora* were more sensitive to the killing effect of AOS-producing systems than the slime-producing wild strains (Kiraly et al., 1997). These observations suggest that EPS slime protects bacteria from AOS. EPS may function as a protective layer that could be effective against the "natural resistance" factor of infected host plants (Leigh and Coplin, 1972). Rudolph et al. (1987) reported on the enhancement of bacterial multiplication in susceptible as well as in resistant bean leaves by supplementing inocula with EPS-solutions derived from *P. savastanoi* pv. *phaseolicola* and *X. axonopodis* pv. *malvacearum.* Takahashi and Doke (1984) have also reported similar observations with *X. axonopodis* pv. *citri* in citrus leaves.

ANTIOXIDANTS OF THE HOST MAY PROTECT BACTERIAL PATHOGENS AGAINST ACTIVE OXYGEN SPECIES

AOS generation induced by pathogens appears to be only transient (Schwacke and Hager, 1992; Baker et al., 1993; Legendre et al., 1993). The AOS-generating activity is downregulated by various endogenous reducing agents and antioxidant enzymes in the host. These include cytosolic and peroxisomal forms of catalase (Dhindsa et al., 1981; Law et al., 1983; Pearce et al., 1997), SOD, ascorbate peroxidase, GSH reductase, and glutathione peroxidase (Lielder et al., 1986; Thompson et al., 1987).

Catalase degrades H_2O_2 while SODs scavenge the O_2^-. The catalases and SODs are the most efficient antioxidant enzymes. Their combined action converts the potentially dangerous superoxide anion and H_2O_2 to water and molecular oxygen. SOD may function in the elimination of $^1O_2$ and °OH radicals (Matheson et al., 1975). Peroxidases (ascorbate peroxidase and glutathione peroxidase) scavenge H_2O_2 (Van Montagu et al., 1994).

Nonenzymatic antioxidants include ascorbate, α-tocopherol, GSH, cysteine, hydroquinones, mannitol, flavonoids, some alkaloids, and β-carotene (Siegel and Halpern, 1965; Drolet et al., 1986; Larson, 1988; Pearce et al., 1997). Ascorbic acid reacts rapidly with O_2^- and °OH and helps to remove the H_2O_2 by the ascorbate peroxidase reaction (Halliwell, 1982; Abeles, 1986). Oxy-radical–bearing fatty acyl chains may partition into the hydrophilic surface of the bilayer so that ascorbic acid interacts directly with them (Lielder et al., 1986). α-Tocopherol directly inhibits the chain reaction of lipid peroxidation (Lielder et al., 1986).

Catalase is a tetrametric heme protein that catalyses the dismutation of H_2O_2 into water and oxygen (Halliwell and Gutteridge, 1989).

$$2\ H_2O_2 \rightarrow 2\ H_2O + O_2$$

Plant catalases are predominantly peroxisomal enzymes, and most of them contain a carboxy terminal consensus for peroxisomal import (Scandalios, 1994). Catalases are present as multiple isoforms in plants. They exist as small gene families (Scandalios, 1994). Each of these genes is associated with a specific H_2O_2-producing process (Ni and Trelease, 1991; Willekens et al., 1994a). It is suggested that H_2O_2 itself may induce its own scavenging mechanisms by activating production of catalase (Prasad et al., 1994; Willekens et al., 1994a; Scandalios, 1994).

Catalase suppresses H_2O_2 production in plants inoculated with bacterial pathogens. When catalase was added to white clover *(Trifolium repens)* suspension-cultured cells inoculated with incompatible pathogen *P. corrugata,* H_2O_2 production was suppressed (Devlin and Gustine, 1992). Bacterial infection may induce the synthesis of catalase. A catalase, Cat2St, has been detected in potato (Niebel et al., 1995). When potato plants were inoculated with the compatible pathogens *Erwinia carotovora* and *Corynebacterium sepedonicum,* a strong induction of Cat2St was observed in infected roots. Cat2St mRNA levels increased upon compatible bacterial infection (Niebel et al., 1995). It suggests that this catalase is pathogen-induced isoform. The induction of *Cat2St* was systemic due to infection with *E. carotovora* or *C. sepedonicum* (Niebel et al., 1995). The Cat2St belongs to class II catalases and the *Cat2St* gene has been shown to be induced by various abiotic stresses such as ozone, UV light and SO_2 (Willekens et al., 1994b), and AOS (Willekens et al., 1994a). Thus any stress may induce catalase. Catalase downregulates AOS-induced defense activities. Some catalases bind salicylic acid (Chen et al., 1993).

Glutathione-S-transferase (GST) metabolizes the products of membrane lipid peroxidation and other products of cellular oxidative stress (Berhane et al., 1994). GST transcripts are induced by lower concentrations of exogenous H_2O_2. It suggests that H_2O_2 may be a diffusible signal for induction of protectant genes against AOS (Levine et al., 1994). Oxidative stress is known to induce antioxidant systems (Scandalios, 1993). In incompatible interactions, increased oxidative stress is seen. This stress may induce accumulation of antioxidants in incompatible interactions.

Glutathione reductase reduces oxidized glutathione (Halliwell and Gutteridge, 1989) and may protect plant pathogenic bacteria against AOS. Increase in the enzyme activity was observed in an incompatible interaction. The enzyme activity increased significantly between 18 and 24 h postinoculation with an incompatible race of *P. savatanoi* pv. *phaseolicola* in bean (Milosevic and Slusarenko, 1996). It suggests that the increase in glutathione reductase activity is only a response to the increased oxidative conditions occurring in the HR lesion. In susceptible interactions, such an increase in glutathione reductase activity is not seen (Milosevic and Slusarenko, 1996).

Coordinate expression of various antioxidant enzymes has been reported. Introduction into tobacco plants of a chimeric gene that encodes a Cu/Zn SOD from pea results in a threefold increase in total SOD activity (Allen et al., 1994). Levels of ascorbate peroxidase activity were more than threefold greater in SOD^+ plants than in nonexpressing control plants (SOD^-). Large increases in MnSOD mRNAs were also seen in leaves of SOD^+ plants. These results indicate that expression of various SOD and ascorbate peroxidase isoforms are coordinately regulated in tobacco plants. The pathogen induces AOS production and the AOS, in turn, induces antioxidant systems (Scandalios, 1993). Elicitor treatment of alfalfa and bean cells induces synthesis of antioxidants such as glutathione (Edwards et al., 1991). In soybean cells, the AOS-generating activity was downregulated or desensitized by subsequent exposures to the elicitor (Legendre et al., 1993). The induced antioxidants would have suppressed the AOS-generating activity.

THE POSSIBLE ROLE OF ACTIVE OXYGEN SPECIES IN DISEASE RESISTANCE

AOS are toxic to bacterial pathogens. But bacterial pathogens have potential to tolerate this toxicity. They produce enzymes to detoxify the AOS. When bacterial pathogens infect the host tissue, AOS production is elicited.

But the AOS induce production of antioxidant systems in plants, which in turn downregulates AOS production. Thus accumulation of AOS appears to be only transient, and it is not known whether the quantity of AOS produced in plant tissues is sufficient to kill the bacteria. However, an oxidative burst seems to be characteristic of several incompatible bacteria-plant interactions. It is tempting to assign a role for AOS in reducing bacterial multiplication in resistant plant tissues. But an in vivo effect of AOS in killing bacterial cells has not yet been demonstrated.

AOS may indirectly contribute to various plant defense mechanisms. H_2O_2 is known to be involved in cross-linking cell wall hydroxyproline-rich glycoproteins (Bradley et al., 1992; Brisson et al., 1994). Formation of intermolecular isodityrosine cross-links is mediated by H_2O_2 (Gross et al., 1977; Halliwell, 1978; Mader and Amber-Fisher, 1982). H_2O_2 is involved in lignification (Gross, 1980; Moerschbacher et al., 1990; Peng and Kuc, 1992). Hydroxyl radicals have been shown to be involved in elicitation of phytoalexins in legumes (Epperlein et al., 1986). H_2O_2 and O_2^- are known to direct phytoalexin accumulation (Montillet and Degousse, 1991; Sharma and Mehdy, 1992). Treatment of bean cell suspensions with H_2O_2 induced the accumulation of mRNAs encoding phenylalanine ammonia-lyase, chalcone synthase, and chalcone isomerase (the enzymes involved in synthesis of phenolics and phytoalexins) and a basic chitinase (a pathogenesis-related protein) (Mehdy, 1994). The role of phenolics, phytoalexins, and PR-proteins in bacterial disease resistance are discussed in later chapters. AOS may be involved in signal transduction (Levine et al., 1994). AOS are involved in synthesis of salicylic acid and jasmonic acid (see Chapter 1).

CONCLUSION

AOS are rapidly and abundantly produced when bacterial pathogens invade host plants. O_2^-, H_2O_2, °OH, and $^1O_2$ are the important AOS produced in plant tissues due to pathogen stress. There are several routes to AOS production in plants. The enzymes that are involved in oxidation-reduction processes may generate AOS. NADPH-dependent oxidase, NADH-dependent peroxidase, xanthine oxidase, superoxide dismutase, glycollate oxidase, urate oxidase, oxalate oxidase, and lipoxygenases have been shown to have a role in AOS production. Elicitor molecules produced by bacterial pathogens may signal the oxidative burst. Endogenous elicitors (of host origin) may also signal AOS production. Ca^{2+} influx and phosphorylation may be part of the signal transduction pathway to induce AOS.

The oxidative burst has been reported in both susceptible and resistant interactions. But in resistant interactions the oxidative burst occurs in two distinct phases; the second one is larger and long-lived. The oxidative burst results in lipid peroxidation in plant cell plasma membranes. It may cause membrane deterioration and result in a leakage of electrolytes characteristic of the hypersensitive reaction. Lipoxygenase activity also increases, leading to cell membrane deterioration and electrolyte leakage. The disruption of the plant cell membrane may release a signal from the membrane or induce an influx of signal molecules into the plant cell.

Although AOS are toxic to bacteria, many bacteria have developed mechanisms to protect them from toxic AOS. The bacterial elicitor molecules signal the production of antioxidants, which detoxify AOS. It is still not known whether AOS can kill the bacteria in planta. However, AOS can be an important defense mechanism, and several defense genes can be activated by AOS. The finding that transgenic tobacco plants expressing depressed levels of catalase showed enhanced disease resistance (Takahashi et al., 1997) suggests the importance of AOS in disease resistance.

Chapter 4

Inducible Plant Proteins

INTRODUCTION

Plants respond to bacterial pathogen attacks by transcriptionally activating a number of genes coding for proteins which are thought to help the plant ward off the invader (Ahl et al., 1981; Bashan et al., 1986; Metraux and Boller, 1986; Meins and Ahl, 1989; Voisey and Slusarenko, 1989; Castresana et al., 1990; Marco et al., 1990; Godiard et al., 1990, 1991; Pautot et al., 1991; Smith and Metaux, 1991; Dong et al., 1991; Kim and Yoo, 1992; Reimmann et al., 1992; Uknes et al., 1992, 1993; Reimmann and Dudler, 1993; Lummerzheim et al., 1993; Brisson et al., 1994; Meuwly et al., 1994; O'Garro and Charlemange, 1994; Pontier et al., 1994b; Newman et al., 1994, 1995; Buell and Somerville, 1995; Alonso et al., 1995; Kim and Hwang, 1995; Swoboda et al., 1995; Hoffland et al., 1996; Lee and Hwang, 1996; Strobel et al., 1996; Stromberg, 1996; White et al., 1996; Chittoor et al., 1997; Schweizer et al., 1997; Vidal et al., 1997). It is generally assumed that a wide variety of the induced proteins play different roles in the defense of plants against bacterial pathogens (Boller et al., 1983; Roberts and Selitrennikoff, 1988; Molina and Garcia-Olmedo, 1997; Van Loon, 1999). The development of disease resistance has been correlated with the accumulation of the inducible plant proteins in many plant-bacterial pathogen interactions.

The induced proteins accumulate more in resistant plant-bacteria interactions than in the susceptible ones (Marco et al., 1990; Castresana et al., 1990; Alonso et al., 1995). Early induction of these proteins in resistant interactions has also been reported (Voisey and Slusarenko, 1989; Dong et al., 1991; Pautot et al., 1991; O'Garro and Charlemange, 1994; Newman et al., 1995). Transgenic plants constitutively overexpressing these proteins show increased resistance to bacterial pathogens (Beffa et al., 1995; Huang et al., 1997; Molina and Garcia-Olmedo, 1997). These studies indicate the importance of induced proteins in bacterial disease resistance.

Several signal molecules are involved in induction of these proteins (Grillo et al., 1995; Bergey et al., 1996; Ryals et al., 1996; Hammond-

Kosack and Jones, 1996; Wobbe et al., 1996; O'Donnell et al., 1996; Sato et al., 1996; Du and Klessig, 1997; Durner et al., 1997; Subramanim et al., 1997; Shulaev et al., 1997; Takahashi et al., 1997; Wasternack and Parthier, 1997; Vidhyasekaran, 1997, 1998). Specific signal transduction pathways leading to gene activation have been studied (Matton et al., 1993; Dammann et al., 1997; Fukuda, 1997; Subramaniam et al., 1997; Penninckx et al., 1998).

The inducible defense proteins have been of particular interest to molecular biologists because of their potential for genetically regulating defensive genes to develop disease-resistant plants. The major challenge at present is to identify the signal transduction events that lead to induction of these genes and exploit them for disease management. The role of inducible proteins in bacterial disease resistance, the mechanism of induction of these proteins, and the complexity of signal transduction systems are discussed at the molecular level in this chapter.

NOMENCLATURE OF PATHOGEN-INDUCIBLE PLANT PROTEINS

Plants respond to pathogen attack by synthesizing a set of proteins. Pathogen-inducible plant proteins comprise three major groups:

1. Pathogenesis-related (PR) proteins which comprise proteins that are newly induced by pathogens
2. Proteins that are constitutive but increasingly induced by pathogens which comprise enzymes of phenylpropanoid metabolism, phytoalexin biosynthetic pathway, and octadecanoid pathway
3. Hydroxyproline-rich glycoproteins (HRGPs)

PR proteins have been defined as proteins encoded by the host plant but induced only in pathological situations (Antoniw et al., 1980). Induction of PR proteins was first reported in tobacco due to infection with a viral pathogen (Gianinazzi et al., 1970; Van Loon and Van Kammen, 1970). PR proteins have been shown to be induced not only by viruses, but also by fungi (Niderman et al., 1995), bacteria (Reimann and Dudler, 1993), and viroids (Vera and Conejero, 1988; Ruiz-Medrano et al., 1992). Besides these pathogens, insect pests (Ryan, 1990), and nematodes (Rahimi et al., 1996) also induce PR proteins. In addition to pathogens, some saprophytes are also known to induce PR proteins. Saprophytic bacteria, such as some strains of *Pseudomonas fluorescens,* induce PR proteins (Maurhofer et al., 1994;

M'Piga et al., 1997). Several endophytic bacteria induce PR proteins (White et al., 1996).

Biological products from bacteria and fungi also induce PR proteins. Elicitors (glycoproteins, proteins, carbohydrates, fatty acids) isolated from fungi (Matton and Brisson, 1989) and proteinaceous elicitors isolated from bacteria (Strobel et al., 1996) induce PR proteins in plants. Toxins isolated from fungi (Dixelius, 1994) and pectic enzymes (Ryan, 1987), cellulases (Chang et al., 1995), and xylanases (Lotan and Fluhr, 1990) isolated from fungi and bacteria induce PR proteins. Various heavy metals and toxic chemicals are inducers of PR proteins (De Tapia et al., 1986; Rowland-Bamford et al., 1989; Bowles, 1990; Tronsmo et al., 1993; Karenlempi et al., 1994; Kim and Huang, 1997). Polyacrylic acid (Gianinazzi and Kassanis, 1974), salicylic acid (Pierpoint et al., 1990; Mouradov et al., 1993; Jung et al., 1993), methyl salicylate (Shulaev et al., 1997; Seskar et al., 1998), methyl jasmonate/jasmonic acid (Mouradov et al., 1993; Schweizer et al., 1997; Curtis et al., 1997), 2,6-dichloroisonicotinic acid (INA) and benzothiadiazole (BTH) (Shah et al., 1997; Du and Klessig, 1997), mannitol (Pierpoint et al., 1981), and probenazole (3-allyloxy-1,2-benzisothiazole-1,1-dioxide) (Midoh and Iwata, 1996) are known to induce PR proteins. Several growth regulators, such as ethylene (Mouradov et al., 1993), kinetin (Memelink et al., 1987), indoleacetic acid (de Loose et al., 1988; Clarke et al., 1998), and abscisic acid (Singh et al., 1987; Zhu et al., 1993a; Grillo et al., 1995) induce PR proteins in various plants. Oligouronides also induce PR proteins (Farmer et al., 1991; Ryan, 1992).

Several environmental factors also induce PR proteins. Temperature (White et al., 1983), light (Asselin et al., 1985), ozone (Ernst et al., 1992), and ultraviolet rays (Brederode et al., 1991) have been shown to influence induction of PR proteins. Osmotic stress induces PR proteins in some plants (Singh et al., 1987). Any injury or mechanical wounding induces several PR proteins (Matton and Brisson, 1989; Godiard et al., 1990; Cordero et al., 1994).

Thus the so-called PR proteins can be induced by several stresses besides pathogens. Hence, Van Loon et al. (1990, 1994) defined PR proteins as plant proteins that are induced by various types of pathogens as well as stress conditions. However, even this definition needs modifications. The PR proteins have been detected in healthy tissues without any stress in many plants. Several PR proteins appeared in the leaves of potato plants maturing in the glasshouse without any stress application (Pierpoint et al., 1990). The PR proteins β-1,3-glucanase and chitinase are induced in leaves of unstressed potato plants (Garcia-Garcia et al., 1994). PR proteins appear in leaves of healthy tobacco plants during flowering (Fraser, 1981; Neale et al., 1990). Some PR proteins appear constitutively in bean leaves (Clarke et al., 1998).

In some plants, PR proteins, which appear in leaves only after a stress, naturally occur in other parts of plants such as roots (Keefe et al., 1990; Koiwa et al., 1994), flowers (Lotan et al., 1989; Neale et al., 1990; Memelink et al., 1990; Constabel and Brisson, 1995), pollens (Breiteneder et al., 1989), stigma (Constabel and Brisson, 1996), and seeds (Hoj et al., 1989; Jacobsen et al., 1990; Kragh et al., 1990; Casacuberta et al., 1991; Leah et al., 1991; Hejgard et al., 1991, 1992; Zu et al., 1992). Many of the PR proteins appear constitutively in cultured plant cells (Crowell et al., 1992; Koiwa et al., 1994). Thus some proteins, induced by pathogens in one type of plant organ such as leaf, have been found to be constitutive components in other organs. Van Loon et al. (1994) also called these proteins PR proteins.

In general, induction of proteins by pathogens has been observed in leaves rather than in any other organs. However, exceptions have also been reported. A PR protein, PR-10a, accumulated in potato tubers, stolons, stems, and petioles after infection by a pathogen; but it did not accumulate in leaves after infection (Constabel and Brisson, 1996). Another PR protein, MPI, accumulates in germinating maize embryos upon fungal infection, but not in vegetative tissues (Casacuberta et al., 1991; Cordeo et al., 1994). In barley, PR proteins accumulated in roots infected with a pathogen (Vale et al., 1994). More than 30 PR proteins accumulate in roots of Norway spruce *(Picea abies)* after infection with a fungal pathogen (Sharma et al., 1993). PR proteins accumulated in pepper stems after infection (Huang et al., 1991).

Some PR proteins, which are induced proteins in some varieties, occur constitutively in other varieties. The hybrid produced from *Nicotiana glutinosa* and *Nicotiana debneyi* makes PR proteins constitutively, while in *N. glutinosa* and *N. debneyi* the PR proteins were induced only after infection by pathogens (Pierpoint et al., 1992). Even in the same plant, PR proteins appear in lower old leaves without any stress (Keefe et al., 1990), while these proteins could not be detected in young leaves near the top of the tobacco plant (Felix and Meins, 1986; Memelink et al., 1990; Neale et al., 1990). In these young leaves, the PR proteins are induced by pathogens (Kauffmann et al., 1987; Meins and Ahl, 1989; Brederode et al., 1991; Stintzi et al., 1991; Ward et al., 1991).

Van Loon et al. (1994) suggested that to call a protein a PR protein, the protein has to be newly expressed upon infection. This seems to be the most important criterion to define a PR protein (Van Loon, 1999). However, many reports have appeared indicating that these PR proteins can be detected even in uninfected control tissues, when Western blot analyses were made using antisera or when cDNA probes were used (Eyal et al., 1992; Beerhues and Kombrink, 1994; Dixelius, 1994; Lawrence et al., 1996). These PR proteins may be present in uninfected tissues in amounts too small for detection on gels by general protein stains. Hence, the definition of PR proteins had to be modified as PR proteins are the proteins which are readily detected in infected tissues but not in uninfected ones (Van Loon, 1999). Some proteins are exclusively induced during disease development and such proteins are not induced in resistant interactions (Wolpert and Dunkle, 1983; Traylor et al., 1987; Baga et al., 1995). Such proteins are not considered PR proteins (Van Loon, 1999). PR proteins should be expressed more in resistant interactions.

The second group of pathogen-inducible proteins consists of constitutive proteins, which are increasingly induced by pathogens. Several enzymes involved in synthesis of phenolics, lignin, callose, and phytoalexins are included in this group. Enzymes involved in the octadecanoid pathway are also included in this group. HRGPs constitute the third group of inducible proteins; their constitutive level is generally low in healthy plants but is induced severalfold in response to pathogen ingress. In most cases, structure of HRGPs is modified due to infection through intra- and intermolecular oxidative cross-linking of HRGP monomers (Esquerre-Tugaye et al., 1999).

There are some proteins which are homologous to PR proteins in healthy tissues (Chen et al., 1996). They have amino acid (nucleotide) sequences similar to already characterized PR proteins. They may also have enzyme activity similar to some PR proteins. However, they are never induced by pathogens. Such proteins are called PR-like proteins (Van Loon et al., 1994).

OCCURRENCE OF PR PROTEINS IN VARIOUS PLANTS

Numerous PR proteins have been detected in rice, wheat, maize, sorghum, oat, barley, tomato, potato, bean *(Phaseolus vulgaris),* broad bean *(Vicia faba),* azuki bean *(Phaseolus lunatus),* jack bean *(Canavalia ensiformis),* pea, tobacco *(Nicotiana tabacum, N. debneyi, N. langsdorfii, N. plumbaginifolia, N. rustica, N. sylvestris, N. tomentosiformis),* cucumber, radish *(Raphanus sativus),* pumpkin *(Cucurbita pepo),* okra *(Abelmoschus esculentus),* sugar beet *(Beta vulgaris),* chickpea *(Cicer arietinum),* soybean *(Gycine max),* pepper *(Capsicum annuum),* sunflower *(Helianthus annuus),* carrot *(Daucus carota),* peanut *(Arachis hypogaea),* cotton (*Gossypium* spp.), grapevine (*Vitis* spp.), alfalfa *(Medicago sativa),* celery *(Apium graveolens),* rubber *(Hevea brasiliensis),* oranges (*Citrus* spp.), oilseed rape *(Brassica napus), Brassica nigra, Asparagus officinalis,* and *Lablab purpureus* (Vidhyasekaran, 1997; Neuhaus, 1999; Leubner-Metzger and Meins, 1999).

It appears that PR proteins can be detected in almost all plants. Large numbers of PR proteins may be present in each plant. At least 33 PR proteins have been described in tobacco (Stintzi et al., 1993). Twenty PR proteins have been detected in sugar beet (Fleming et al., 1991). More than 30 PR proteins have been identified in Norway spruce *(Picea abies)* (Sharma et al., 1993).

CLASSIFICATION OF PR PROTEINS

Criteria for Classification

The structure of PR proteins varies widely. Hence, it is difficult to classify them. Van Loon et al. (1994) tried to classify PR proteins into families on the basis of their nucleotide sequence or predicted sequence of amino acids, serological relationship, and/or enzymatic or biological activity. PR proteins are grouped into classes identified by their migration on native polyacrylamide gel electrophoresis (PAGE), and they have also been classified using specific antisera, mRNA probes, and sequence data (Stintzi et al., 1993).

Some times PR proteins that could not be detected by PAGE could be detected using cDNA clones as probes. *Nicotiana langsdorfii* appears to lack the characteristic electrophoretically mobile acidic PR proteins of the PR-1 family typical to tobacco (Gordon-Weeks et al., 1997). However, a cDNA clone of PR-1a from tobacco used as a probe in Northern blots of tobacco mosaic virus-infected *N. langsdorfii* RNA detected a band of the expected size of mRNA encoding PR-1 proteins (Gordon-Weeks et al., 1997).

Antisera alone cannot be used to classify the PR proteins. A PR-1 protein of tobacco, PR-1g, could not be detected by an antibody to PR-1a (Gordon-Weeks et al., 1997). Antisera raised to PR-1b from tobacco cultivar' Samsun-NN' do not detect the tobacco PR-1g. But PR-1g could be detected by an antibody raised to the tomato PR-1 protein, P14a (Niderman et al., 1995). Antisera raised to acidic PR-1 proteins do not react to basic PR-1 proteins (Cornelissen et al., 1987). But exceptions also occur. A basic PR-1 protein of tobacco, PR-1h, reacts to an antibody of the acidic PR-1a (Gordon-Weeks et al., 1997). Two acidic β-1,3-glucanases are antigenically related to the basic isoforms in bean leaves (Awade et al., 1989). The bean chitinases lack serological relatedness to tobacco basic class I and class II chitinases (PR-3 proteins) (Neuhaus, 1999).

Biological activity of the PR proteins alone cannot be taken as criteria for classifying them into a family. PR-3, PR-4, PR-8, and PR-11 families show chitinase activity, but they strongly differ in sequence and/or substrate preference (Legrand et al., 1987; Metraux et al., 1989; Melchers et al., 1994; Ponstein et al., 1994; Van Loon et al., 1994; Van Loon, 1999). A bean PR protein, Saxa PR-4, belongs to PR-5 family because of its structural resemblance to tobacco PR-5 protein, but it possesses chitinase activity (Awade et al., 1989). Similarly, another pinto bean PR protein, PR-4d, is structurally homologous with tobacco PR-5 proteins, but functionally different showing β-1,3-glucanase activity (Sehgal and Mohamed, 1990; Sehgal et al., 1991). The PR-10 proteins have a ribonuclease-like structure, but a PR-10 protein from parsley, PR-1, did not display ribonuclease activity (Moiseyev et al., 1994; Van Loon, 1999).

Thus structure of PR proteins is considered most important for classifying PR proteins into families. The PR proteins are classified into families based on shared sequence homology. The isoelectric points of the PR proteins are considered to classify some of the PR proteins into subclasses (Koiwa et al., 1994).

PR-1 Proteins

Acidic Proteins

PR-1 proteins are the major group of PR proteins detected in the plant kingdom. Several PR-1 proteins have been detected in tobacco, tomato, rice, barley, maize, parsley, *Arabidopsis,* and in many other plants belonging to Gramineae, Solanaceae, Chenopodiaceae, and Amaranthaceae (Gordon-Weeks et al., 1997; Schweizer et al., 1997; Vidhyasekaran, 1997). The PR-1 proteins remain soluble in acidic buffers (pH 3.0), whereas most other plant proteins are degraded under these conditions (Van Loon, 1976). They have low molecular weight (14-16 kDa). They are present in acidic and basic isoforms (Payne et al., 1989; Linthorst, 1991; Eyal et al., 1992).

At least six acidic PR-1 proteins have been detected in different tobacco cultivars and species (Antoniw et al., 1980; Pfitzner and Goodman, 1987). They have been designated PR-1a, PR-1b, PR-1c, PR-1d, PR-1e, and PR-1f. While PR-1a, PR-1b, and PR-1c have been detected in *N. tabacum* cv. Samsun NN, PR-d has been described in *N. tabacum* cv. White Burley (Antoniw and White, 1980). PR-1e has been detected in *N. sylvestris* (Ahl et al., 1985) and PR-1f has been found in *N. tomentosiformis* (Ahl et al., 1985). They are serologically related. Most of them have been thoroughly characterized both at protein and DNA levels (Antoniw et al., 1980; Cornelissen et al., 1986, 1987; Van Loon et al., 1987; Pftizner and Goodman, 1987; Cutt et al., 1988; Eyal et al., 1992). The 138 amino acid mature PR-1a protein is synthesized on membrane-bound ribosomes as higher molecular weight precursor containing an N-terminal 30 amino acid hydrophobic signal peptide, which is cleaved to yield 15 kDa mature protein (Carr et al.,

1985). The coding sequences of two cDNAs for PR-1a and PR-1b of tobacco share 93 percent homology, and the deduced amino acid sequences of PR-1a and PR-1b precursors, which are synthesized as larger precursors containing signal peptides, are 91 percent homologous. The homology of mature PR-1a and PR-1b regions is higher than that of larger precursors, 94 percent in the nucleotide sequence and 93 percent in the amino acid sequence, whereas that of the signal peptide regions is 80 and 90 percent, respectively (Matsuoka et al., 1987).

PR proteins similar to tobacco PR-1 proteins have been detected in tomato. It was named P14. Three distinct P14 proteins have been identified (Camacho-Henriquez and Sanger, 1982; Fischer et al., 1989; Vera et al., 1989; Joosten et al., 1990). Serological cross-reactions and amino acid sequence comparisons showed that the three proteins are members of the PR-1 group of proteins. P14a and P14b showed high similarity to a previously characterized P14, whereas P14c was found to be very similar to a basic-type tobacco PR-1, PR-1g (Niderman et al., 1995). A PR-1 protein serologically related to tobacco PR-1 has been detected in cowpea (Nassuth and Sanger, 1986). Gillikin et al. (1991) has detected an acidic PR-1 from maize.

Tobacco cv. Samsun NN contains at least eight genes encoding acidic PR-1 proteins (Cornelissen et al., 1986, 1987; Pfitzner and Goodman, 1987; Pfitzner et al., 1988; Oshima et al., 1987, 1990). At least six of them are expressed at the mRNA level (Matsuoka et al., 1987). There may be at least six PR-1 genes in the barley genome (Bryngelsson et al., 1994; Mouradov et al., 1994). The PR-1 gene family in parsley consists of approximately three to six genes (Somssich et al., 1988). In *Arabidopsis,* a gene encoding PR-1 has been identified (Uknes et al., 1993b).

Basic Proteins

cDNAs and genomic clones corresponding to basic proteins, which are approximately 65 percent similar to the acidic PR-1a, PR-1b, and PR-1c proteins, have been characterized from TMV-infected tobacco leaves (Payne et al., 1989). The basic protein has been designated PRB-1b (Eyal et al., 1992). Niderman et al. (1995) have isolated a 17 kDa basic PR-1 protein and named it PR-1g. Another basic PR-1 protein, PR-1h, has been detected in tobacco Xanthi-nc (Gordon-Weeks et al., 1997). Generally, antibodies against the tobacco acidic PR-1 proteins did not react with the basic proteins (Cornelissen et al., 1987). But PR-1h could be detected by an antibody to acidic PR-1a (Gordon-Weeks et al., 1997). The basic PR-1 gene family consists of at least four genes, of which at least two are expressed at the mRNA level (Cornelissen et al., 1987). The gene encoding PRB-1b *(prb-1b)* was cloned, sequenced, and characterized (Eyal et al., 1992).

Two homologs of the tobacco PR-1 have been detected in barley. They are HvPR-1a and HvPR-1b (Bryngelsson et al., 1994). Both these proteins were basic proteins. They show pI values of approximately 10.5 and 11.0, respectively (Bryngelsson et al., 1994). The molecular weight of these proteins was about 15 kDa. With the exception of one amino acid, the partial amino acid sequences of HvPR-1a and HvPR-1b are identical to internal sequences of the polypeptides. These derived polypeptides are each 164 amino acids long and both have putative N-terminal leader sequences of 24 amino acids (Bryngelsson et al., 1994). Another basic PR protein, Prb-1, has been isolated from barley (Mouradov et al., 1994). Three basic PR proteins (designated P2, P4, and P6) of approximately 15 kDa have been reported in tomato (Joosten et al., 1990). P4 and P6 (15.5 kDa, pI 10.9 and 10.7, respectively) were serologically related to each other and to the PR-1 protein family of tobacco. P4 and P6 represent two differently charged, highly basic isomers of P14 (Joosten et al., 1990). The structure of cDNA clones encoding P14 isomers of P4 and P6 has been described (Van Kan et al., 1992). Protein P4 is a highly related isomer, with only five amino acid substitutions compared to P6. The proteins P4 and P6 may be encoded by at least two, but less than five, copies per genome (Van Kan et al., 1992). P6 has been renamed PR-1b (Wubben et al., 1993). A basic PR-1 protein has been identified in maize seeds (Casacuberta et al., 1991).

The function of PR-1 proteins is not yet known. However, it has been suggested that PR-1 protein may be involved in cell wall thickening and may offer resistance to the spread of pathogens in the apoplast (Benhamou et al., 1991).

PR-2 Proteins

PR-2 proteins show β-1,3-glucanase activity. β-1,3-Glucanases (glucan endo-1,3-β-glucosidases) catalyze endo-type hydrolytic cleavage of the 1,3-β-D-glucosidic linkages in β-1,3-glucans. β-1,3-Glucan, the substrate for the enzyme β-1,3-glucanase, is widespread in plant tissues and is associated with the formation of callose, leaf and stem hairs, root hairs, pollen grains, ovules, and wound parenchyma cells (Abeles et al., 1970). As such, endo β-1,3-glucanases are abundant proteins widely distributed in plant species (Meins et al., 1992). Constitutive expression of several β-1,3-glucanases in healthy plants has been reported. They accumulate in high concentrations in the roots and in the epidermis of lower leaves of healthy plants (Keefe et al., 1990). High expression of β-1,3-glucanase in flowers of healthy tobacco plants has been reported (Lotan et al., 1989; Neale et al., 1990). A glycosylated β-1,3-glucanase, which is specifically expressed in tobacco flowers, has been identified (Ori et al., 1990). β-1,3-glucanases have been reported to accumulate in plant seeds during maturation and germination (Swegle et al., 1989; Hoj et al., 1989; Jacobsen et al., 1990). Two β-1,3-glucanases have been detected constitutively in all tomato cultivars (Lawrence et al., 1996). One β-1,3-glucanase has been detected constitutively in cell suspension cultures of barley *(Hordeum vulgare)* (Kragh et al., 1991). β-1,3-glucanases have been implicated in several physiological and developmental processes in normal healthy plants, such as cell division (Waterkeyn, 1967), microsporogenesis (Worrall et al., 1992), pollen germination (Roggen and Stanley, 1969), fertilization (Lotan et al., 1989), and seed germination (Vogeli-Lange et al., 1994).

Several β-1,3-glucanases are also considered as PR proteins. In young leaves near the top of the plant, β-1,3-glucanases are normally absent (Felix and Meins, 1986; Memlink et al., 1990; Neale et al., 1990). They are induced in leaves of several plants after infection by pathogens (Kauffmann et al., 1987; Meins and Ahl, 1989; Ward et al., 1991). The PR-2 protein β-1,3-glucanase exists in multiple forms in a number of plant species, including tobacco (Kauffmann et al., 1987), potato (Pan et al., 1989), tomato (Van Kan et al., 1992), wheat (Sock et al., 1990), pepper (Hwang et al., 1991), bean (Awade et al., 1989), rice (Schweizer et al., 1997), Norway spruce (Sharma et al., 1993), sugar beet (Rousseau-Limouzin and Fritig, 1991), chickpea (Vogelsang and Barz, 1993), pea (Mauch et al., 1988a), soybean (Yi and Hwang, 1996), and maize (Nasser et al., 1990). The various β-1,3-glucanases (PR-2 proteins) of the genus *Nicotiana* have been classified into three structural classes based on amino acid sequence identity. The class I β-1,3-glucanases are basic proteins localized in the cell vacuole. They include a 33 kDa PR-2e subgroup of tobacco *(N. tabacum)* PR proteins, and Gn1 and Gn2 of *N. plumbaginifolia*. The class II β-1,3-glucanases are acidic proteins secreted into the extracellular space. They include tobacco PR-2a (PR-2), PR-2b (PR-N), and PR-2c (PR-O). The class III β-1,3-glucanases are also acidic proteins secreted into the extracellular space. The 35 kDa PR-2d (= PR-Q') is the sole representative of tobacco class III β-1,3-glucanase and differs in sequence by at least 43 percent from the class I and class II enzymes (Leubner-Metzger and Meins, 1999). Tobacco contains at least eight genes encoding extracellular acidic β-1,3-glucanases and five or six genes encoding intracellular basic β-1,3-glucanases (Linthorst et al., 1990; Ohme-Takagi and Shinshi, 1990). In the tetraploid tobacco, 13 to 14 different β-1,3-glucanase genes are active when infected with TMV (Linthorst et al., 1990).

Tomato produces at least three β-1,3-glucanases. One extracellular, slightly acidic 35 kDa protein was characterized both at the protein and the cDNA level (Van Kan et al., 1992). The acidic β-1,3-glucanase is encoded by a single gene in the tomato genome (Van Kan et al., 1992). A basic 35 kDa protein with high homology to vacuolar, basic tobacco β-1,3-glucanases has been detected in tomato. A C-terminal extension with a potential glycosylation site, which is involved in vacuolar targeting of the basic tobacco β-1,3-glucanase, is also present in the C-terminal part of the precursor protein of the basic 35 kDa β-1,3-glucanase of tomato (Van Kan et al., 1992). This area also contains the dipeptide sequence FG (at position 34), which was suggested to be a cleavage signal for removal of the C-terminal extensions from precursors of vacuolar PR proteins (Payne et al., 1990). The intracellular basic 35 kDa β-1,3-glucanase may be encoded by two or three genes (Van Kan et al., 1992). Two class III β-1,3-glucanases have also been detected in tomato. Tom PR-Q'a is an acidic isoform 86.7 percent identical to tobacco PR-Q' and Tom-PR-Q'b is a basic isoform 78.7 percent identical to tobacco PR-Q' (Leubner-Metzger and Meins, 1999).

Two acidic β-1,3-glucanases were detected in mercuric chloride-treated bean leaves (Awade et al., 1989). A basic 36 kDa β-1,3-glucanase was also detected in bean (Vogeli et al., 1988; Mauch and Staehelin, 1989; Awade et al., 1989). Both acidic and basic isoforms of the enzyme were antigenically related (Awade et al., 1989). In sugar beet, three β-1,3-glucanases of molecular mass of approximately 35 kDa (two acidic and one basic enzyme) were detected (Rousseau-Limouzin and Fritig, 1991). Approximately 22 electrophoretic forms of β-1,3-glucanases have been detected in infected wheat plants (Sock et al., 1990). At least four acidic, two neutral, and two basic isozymes of β-1,3-glucanases could be detected in infected potato tubers (Pan et al., 1989). Several induced β-1,3-glucanases have been detected in potato plants (Rahimi et al., 1996). Two β-1,3-glucanases (PR-2 and PR-Q) have been detected in *Brassica napus* and *B. nigra* (Dixelius, 1994). In rice, several induced PR-2 (29 to 36 kDa) have been reported (Simmons et al., 1992; Zhu et al., 1993; Xu et al., 1996; Schweizer et al., 1997). Two β-1,3-glucanases have been identified in roots of Norway spruce *(Picea abies)* (Sharma et al., 1993). Many β-1,3-glucanases were induced in pepper *(Capsicum annuum)* stems after infection or mercuric chloride treatment (Kim and Hwang, 1997).

PR-3 Proteins

PR-3 proteins show chitinase activity. Several chitinases have been reported in almost all plant species studied. Chitinases are endo β-1,4-

glucosaminidases. They hydrolyze the β-glycosidic bond at the reducing end of glucosaminides, which can be parts of various polymers, such as chitin, chitosan, or peptidoglycan (from bacterial cells) (Neuhaus, 1999). Plant chitinases have been classified into seven classes based on sequence homology and the presence or absence of a chitin-binding domain (CBD). Class I contains CBD while class II lacks the CBD of class I. Class III chitinases are structurally different from other chitinases. Class IV chitinases differ from class I by several internal deletions within both chitin-binding and catalytic domains (loops 1, 3, and 4). Class V chitinases contain two CBDs. Class VI chitinases are characterized by the presence of a truncated CBD and a long proline-rich spacer. Class VII possesses a catalytic domain homologous to class IV, but lacks the CBD (Neuhaus, 1999). Besides these chitinases, a new type of chitinase with the most related sequences belonging to bacterial chitinases has been identified (Heitz et al., 1994; Melchers et al., 1994). Another class of chitinases constitutes hevein and Win proteins, which contain a CBD and PR-4 proteins which lack the CBD (Linthorst et al., 1991).

PR-3 proteins consist of chitinases belonging to class I to class VII except class III (Van Loon, 1994). Class III proteins, which are structurally very much different from other classes, are considered a separate family of PR proteins and called PR-8. The chitinases, which are similar to bacterial chitinases, have been grouped as another family and called PR-11. The PR-11 proteins consist of class I-type chitinases. The other group of chitinases similar to hevein and Win proteins are given a separate family number, PR-4, which consists of class I and class II type chitinases (Van Loon, 1999).

PR-3 proteins consist of several classes of chitinases. Class I chitinases are basic proteins. These chitinases are synthesized as precursors with an N-terminal signal sequence, which directs them to the secretory pathway. At the N-terminal end of the enzyme a cysteine-rich domain is found. Most class I chitinases also have a C-terminal polypeptide, which is required for their targeting to the vacuole. Within the mature protein, the N-terminal CBD is linked to the catalytic domain by a spacer, variable in length and sequence, but usually rich in proline or glycine (Neuhaus, 1999). Tobacco basic chitinase belongs to this class (Shinshi et al., 1990; Eyal and Fluhr, 1991). A basic 32 kDa chitinase has been detected in pepper (Kim and Hwang, 1997).

The class II chitinases include the acidic chitinases. They do not contain the hevein-like region but otherwise have high homology to the other main structural features of class I. Chitin-binding domain is absent in these acidic chitinases (Shinshi et al., 1990). Tobacco PR-P and PR-Q are type members belonging to this class (Neuhaus, 1999). Class IV chitinases include bean chitinases originally called PR-4 (de Tapia et al., 1987). These chitinases are not serologically related to class I and class II chitinases. They also show low homology to the class I and II chitinases that prevented identification by DNA hybridization. A loop is missing in the chitin-binding domain in the class IV chitinases, without affecting its sugar-binding properties. Within the catalytic domain, there are three deletions (Neuhaus, 1999). A maize chitinase (Huynh et al., 1992), the basic sugar beet Ch4, and rapeseed ChB4 (Nielsen et al., 1994b) also belong to this class.

The class V chitinase includes a single protein. When the cDNA for the precursor of stinging nettle *(Urtica dioica)* lectin was cloned, it was found that the precursor was synthesized as a chitinase homolog with two chitin-binding domains (Lerner and Raikhel, 1992). The only class VI chitinase known so far was isolated from sugar beet (Berglund et al., 1995). Its heavily truncated chitin-binding domain lacks four out of eight cysteines and can at best form a single correct disulfide bond instead of four. This chitinase possesses by far the longest spacer sequence known, a stretch of more than 135 amino acids, of which 90 are prolines (Neuhaus, 1999). The class VII chitinases include a rice chitinase without a chitin-binding domain but with a catalytic domain highly homologous to the domain of class IV chitinases (Neuhaus, 1999).

Several chitinases have been detected in diseased plants. Ten chitinase isoforms have been shown to accumulate in roots of Norway spruce infected by a fungal pathogen (Sharma et al., 1993). Four chitinase isozymes have been detected in tomato genotypes (Lawrence et al., 1996). Three acidic and seven basic chitinase isoforms were detected in pepper plants (Lee and Hwang, 1996). These chitinases are generally detected in healthy plants at lower levels, and they accumulate in infected tissues (Voisey and Slusarenko, 1989; Lawrence et al., 1996). The genes encoding PR-3 proteins are designated *chia* genes or *ypr3* (Neuhaus, 1999; Van Loon, 1999).

PR-4 Proteins

There were four members of 13-14 kDa PR-4 proteins in tobacco (Stintzi et al., 1993). One of the tobacco PR-4 proteins, CBP 20, was found to have chitinase activity (Ponstein et al., 1994). Hence, the PR-4 family has been included in a chitinase nomenclature. Two classes of the PR-4 family have been recognized. Class I includes hevein and Win proteins. Hevein is a small protein found in high concentration in the latex fluid of the rubber tree *(Hevea brasiliensis)* (van Parijs et al., 1991). The three-dimensional structure of hevein is similar to each of the four domains of wheat-germ agglutinin and other cereal lectins (Neuhaus, 1999). Win (wound-induced) proteins from potato were found to include a chitin-binding domain related to hevein, the cereal chitin-binding lectins, and the chitin-binding domain of class I chitinases (Stanford et al., 1989).

The class II PR-4 proteins are considered defense-related proteins (Neuhaus, 1999). Tobacco PR-4 is the typical member of PR proteins of tobacco (Ponstein et al., 1994). The tomato protein P2 (15 kDa, pI 10.4) was serologically related to PR-4 from tobacco (Joosten et al., 1990). Characterization of cDNA clones encoding P2 not only revealed homology with tobacco PR-4, but also with the potato *win1* and *win2* gene products and with pre-pro-hevein from *Hevea brasiliensis* (Linthorst et al., 1991). In barley, a PR-4 protein with chitin-binding ability has been identified in grains and stressed leaf (Hejgaard et al., 1992). The genes encoding PR-4 proteins are designated *chid* genes (Neuhaus, 1999) or *ypr4* (Van Loon, 1999).

PR-5 Proteins

The PR-5 group of proteins occurs widely in the plant kingdom. Several PR-5 proteins have been detected in many parts of healthy plants (Neale et al., 1990; Hejgaard et al., 1991; Koiwa et al., 1994). They are induced in some plant tissues only after infection with pathogens (Neale et al., 1990; Ruiz-Medrano et al., 1992) and that is why they are called PR proteins. However, some of the PR-5 type proteins detected in floral parts in tobacco could not be detected in any other part of the plants even after infection (Richard et al., 1992). A flower-specific gene encoding an osmotin-like protein has been identified in tomato. No expression of the gene was detected in vegetative organs. The protein has 30 to 32 percent amino acid sequence identity to pathogenesis-related osmotins (Chen et al., 1996). However, it cannot be called a PR protein.

The PR-5 proteins have close resemblance to a sweet-tasting protein, thaumatin, that occurs in the fruit of the West African shrub *Thaumatococcus danielli* (Cusack and Pierpoint et al., 1988). Hence, these proteins are also called thaumatin-like proteins (TLPs). Three subclasses of PR-5 proteins have been recognized in tobacco based on their isoelectric points (Singh et al., 1987; Woloshuk et al., 1991): the basic forms (osmotins), neutral forms (osmotin-like proteins, OLPs), and acidic (PR-S) proteins (Table 4.1; Koiwa et al., 1994).

The basic PR-5 proteins are identical to osmotin, the salt stress protein detected in tobacco (Stintzi et al., 1991) and hence, they are called osmotins. The osmotin proteins seem to exert their antifungal function by altering the permeability of fungal membranes, and hence they are also called permatins (Roberts and Selitrennikoff, 1990; Vigers et al., 1991, 1992; Woloshuk

TABLE 4.1. Structure of Various PR-5 Proteins of Tobacco

PR-5 protein	**pI**	**Sequence**
PR-S	5.4	ATFDIVNQXTYTVW
Osmotin-like protein 1	7.0	SGVFEVHNNXPYTVWAXATPVG
Osmotin-like protein 2	7.5	SGVFEVHNNXPYTVWAAATPVGGRR
Osmotin I	> 10.0	ATIEVRNNXPYTVWAASTPIGGGRR

Source: Koiwa et al., 1994.

Note: A = Alanine, D = Aspartate, E = Glutamate, F = Phenylalanine, G = Glycine, H = Histidine, I = Isoleucine, N = Asparagine, P = Proline, Q = Glutamine, R = Arginine, S = Serine, T = Threonine, V = Valine, W = Tryptophan, X = Unknown amino acid, Y = Tyrosine

et al., 1991). They accumulate in vacuoles; acidic PR-5 proteins accumulate in the apoplast. The neutral osmotin-like proteins have been detected in roots and cultured cells of tobacco (Koiwa et al., 1994). The deduced amino acid sequence of the isolated cDNA of tobacco osmotin-like protein shares 76 percent identity with tobacco osmotin (Takeda et al., 1991). The neutral PR-5 proteins are not found in leaves of healthy tobacco plants, but induced by ethylene (Koiwa et al., 1994).

PR-5 proteins have been detected in many plants including rice (Reimann and Dudler, 1993; Velazhahan et al., 1998), wheat (Rebmann et al., 1991b), oats (Vigers et al., 1991; Lin et al., 1996), barley (Bryngelsson and Green, 1989; Hejgaard et al., 1991; Vale et al., 1994), maize (Roberts and Selitrennikoff, 1990; Malehorn et al., 1994; Batalia et al., 1996), sorghum (Vigers et al., 1991), potato (Pierpoint et al., 1990; Zhu et al., 1995a), tomato (King et al., 1988; Ruiz-Medrano et al., 1992), tobacco (Singh et al., 1987; Van Loon and Gerritsen, 1989; Van Loon et al., 1987), soybean (Graham et al., 1992), and bean (Sehgal et al., 1991). An acidic 21 kDa protein termed PR-4d is induced in the primary leaves of bean *(Phaseolus vulgaris)* cultivar Pinto inoculated with incompatible pathogens; it is absent from the comparable healthy leaves (Mohamed and Sehgal, 1987). The 'Pinto' bean PR-4d and 'Samsun NN' tobacco PR-5 proteins were reciprocally immunoreactive (Sehgal et al., 1991). It suggests that PR-4d is a PR-5 protein and the PR-4d type protein has been found to be induced in 20 plant species including cucumber *(Cucumis sativa),* pumpkin *(Cucurbita pepo),* okra *(Abelmoschus esculentus),* sugar beet *(Beta vulgaris),* pea *(Pisum sativum),* azuki bean *(Phaseolus angularis),* lima bean *(Phaseolus lunatus),* broad bean *(Vicia fabae),* and jack bean *(Canavalia ensiformis)* (Vidhyasekaran, 1997).

The function of PR-5 proteins is not yet known. PR-5 proteins alter permeability of fungal membranes (Vigers et al., 1992). The PR-5 protein of tobacco (PR-R = PR-S) has identity to a maize-trypsin/α-amylase inhibitor (Richardson et al., 1987). A PR-5 protein from bean cultivar Pinto, PR-4d, shows β-1,3-glucanase activity (Sehgal and Mohamed, 1990). Another PR protein from bean cultivar Saxa that shows resemblance to tobacco PR-5 protein, Saxa PR-4d, possesses chitinase activity (Awade et al., 1989).

PR-6 Proteins

Some of the proteinase (protease) inhibitor proteins are considered PR-6 proteins. Proteinase inhibitor proteins commonly occur in plants (Vidhyasekaran, 1997). Some of them are induced by pathogens (Ryan, 1990). Proteinase inhibitors are generally categorized according to the class of proteinases that they inhibit (Koiwa et al., 1997). Four types of proteinases have been identified as serine, cysteine, aspartic, or metallo-proteinases based on the active amino acid in the reaction center. Serine proteinase inhibitors which inhibit trypsin and chymotrypsin contain several families: the Kunitz family (Soybean trypsin inhibitor family), Bowman-Birk family, Barley trypsin inhibitor family, Potato inhibitor I family, Potato inhibitor II family, Squash inhibitor family, Ragi I-2/maize trypsin inhibitor family, and Serpin family. Cysteine proteinase inhibitors (phytocystatins) inhibit papain and cathepsin B, H, and L. Aspartic proteinase inhibitors inhibit cathepsin D and metallo-proteinase inhibitors inhibit papain, cathepsin B, H, and L (Koiwa et al., 1997). Tomato inhibitor I (belonging to the Potato inhibitor I family) is the type member of the PR-6 protein family (Green and Ryan, 1972). Tomato inhibitor II is also commonly induced in tomato due to stresses (Doares et al., 1995). Cordero et al. (1994) have isolated and characterized a cDNA clone encoding a *m*aize *p*roteinase *i*nhibitor, the MPI gene. Its mRNA accumulates in germinating maize embryos upon fungal infection. Local and systemic induction of the level of MPI mRNA occurs in leaves of maize plants (Cordero et al., 1994).

PR-7 Proteins

PR-7 proteins show endoproteinase activity. PR-7 protein P69 has been detected in tomato infected by citrus exocortis viroid (Vera and Conejero, 1988). Tornero et al. (1997) identified P69B, a second member of the family of plant proteinases induced during the response of tomato plants to pathogen attack. P69B represents a new plant subtilisin-like proteinase based on amino acid sequence conservation and structural organization which is highly related to the previously identified PR-69 proteinase (Tornero et al., 1997).

PR-8 Proteins

PR-8 proteins show class III chitinase activity possessing lysozyme activity. These chitinases strongly differ from other chitinases in sequence and/or substrate preference (Van Loon et al., 1994). These PR proteins show structural homology to a bifunctional lysozyme/chitinase from *Parthenocissus quinquifolia* (Metraux et al., 1989). They were first described as lysozymes (Bernasconi et al., 1987). In cucumber, a PR-8 protein is induced in response to a bacterial infection (Metraux and Boller, 1986). Both acidic and basic forms of PR-8 proteins have been reported. Acidic PR-8 proteins have been detected in cucumber, tobacco, and chickpea (Metraux et al., 1989; Samac et al., 1990; Vogelsang and Barz, 1993; Lawton et al., 1994a). Basic forms have been described in tobacco and *Arabidopsis* (Ward et al., 1991; Neuhaus, 1999). The genes encoding PR-8 proteins have been designated *chib* (Neuhaus, 1999) or *ypr8* (Van Loon, 1999).

PR-9 Proteins

PR-9 proteins show peroxidase activity. However, all peroxidases are not considered PR-9 proteins since most of them are constitutively expressed and are not induced during pathogenesis. Over 60 peroxidase genes from diverse plant species have been isolated and characterized, including multiple genes from individual species (Chittoor et al., 1999). Only some of the peroxidase genes are induced during pathogen stress. In wheat, six peroxidase genes have been detected and only two of them, *pox2* and *pox3,* were induced during pathogen stress (Rebmann et al., 1991a; Reimann et al., 1992; Baga et al., 1995). Five peroxidase genes have been characterized in rice, and two of them *(POX 8.1* and *POX 22.3)* were found to be induced during pathogenesis (Chittoor et al., 1997). A lignin-forming peroxidase from tobacco is the type member of the PR-9 protein family (Lagrimini et al., 1987). Two peroxidase genes, *sphx6a* and *sphx6b,* were shown to be induced by methyl jasmonate in *Stylosanthes humulis* (Curtis et al., 1997). Two tomato anionic peroxidase genes, *tap1* and *tap2,* are expressed in response to wounding, elicitor treatment, and fungal attack. They are not constitutively expressed (Robb et al., 1991; Mohan et al., 1993). In wheat roots, a peroxidase gene is induced during pathogenesis (Vale et al., 1994).

PR-10 Proteins

PR-10 proteins include intracellular defense-related proteins that have ribonuclease-like structure. The PR-10 family includes parsley PR1 (Somssich et al., 1986, 1988), STH-2 in potato (Matton and Brisson, 1989; Constabel and Brisson, 1992), asparagus AoPR1 (Warner et al., 1994), pea pI49 (Fristensky et al., 1988), and bean pvPR1 and pvPR2 (Walter et al., 1990; Warner et al., 1989). In rice a PR-10 protein, PBZ1, has been identified (Midoh and Iwata, 1996). This PR protein is induced by *Pyricularia oryzae* and probenazole. Significant homology at the amino acid level exists between the PBZ1 and several other PR-10 proteins such as pea pI49 protein, asparagus AoPR1 protein, potato STH-2 protein, bean pvPR1 protein, and parsley PoPR1-I protein (Midoh and Iwata, 1996). Soybean SAM22 protein and pea ABR17 protein were also identified as PR-10 proteins (Midoh and Iwata, 1996).

All PR-10 proteins are acidic and in all these proteins a signal peptide was absent, suggesting that this family of proteins is intracellular. These proteins were induced by pathogens (Constabel and Brisson, 1992; Midoh and Iwata, 1996). PR-10 genes form a multigene family (Crowell et al., 1992; Iturriaga et al., 1994). A high sequence similarity between a ribonuclease from ginseng and parsley *PR-2*-encoded protein was observed (Moiseyev et al., 1994). It suggests that the PR-10 gene family may encode ribonucleases.

Constitutive expression of genes homologous to PR-10 proteins has also been found. Potato *STH-2* homologous genes have been found to be expressed during pea embryogenesis (Barrat and Clark, 1991), in auxin-starved soybean cell cultures (Crowell et al., 1992), as well as in birch pollen (Breiteneder et al., 1989). The PR-10a is found in healthy, unstressed potato plants exclusively in the stigma. But this protein was not detected in style, ovary, sepal, petal, or anther extracts (Constabel and Brisson, 1995).

PR-11 Proteins

PR-11 proteins also show chitinase activity. A new type of chitinase from tobacco was purified using a zinc affinity column and subsequently cloned (Heitz et al., 1994; Melchers et al., 1994). This chitinase shows no sequence similarity to the previously identified chitinases (Melchers et al., 1994). The most related sequences belong to bacterial chitinases. However, unlike the bacterial chitinases, the plant enzymes were endochitinases lacking detectable exochitinase activity (Ohl et al., 1994). The PR-11 proteins in tobacco are encoded by a small multigene family and induced by various stresses such as wounding, ultraviolet irradiation, and a virus infection (Ohl et al., 1994).

PR-12 Proteins (Defensins)

Defensins are a family of small (about 5 kDa), usually basic, peptides which are rich in disulfide-linked cysteine residues (Broekaert et al., 1995, 1997; Conceicao and Broekaert, 1999). Defensins are also considered a novel group of thionins which are called γ-thionins (Colilla et al., 1990; Mendez et al., 1990). They have eight cysteine residues. The comparison of the amino acid sequences clearly shows that the γ-thionins do not belong to the classical thionin protein family (Bohlmann, 1994). Thionins and defensins are structurally unrelated. Terras et al. (1995) introduced the term "defensins" for this group of peptides, based on the structural and functional similarities with insect defensins. All defensins (γ-thionins) are less than 50 amino acids in length and contain eight cysteine residues. The preproproteins of the defensins have a precursor structure different from that of classical thionins. The predomain is located N-terminally to the mature defensin, between the signal sequence and the mature defensin (Dimarcq et al., 1990). The defensins have been classified as PR-12 proteins (Van Loon and Van Strien, 1999).

Defensins have been isolated from over 20 different plant species (Terras et al., 1992, 1993; Conceicao and Broekaert, 1999). They have been shown to be induced during pathogenesis in vegetative tissues of pea (Chiang and Hadwiger, 1991), radish (Terras et al., 1995), tobacco (Gu et al., 1992), and *Arabidopsis* (Penninckx et al., 1996). In *Arabidopsis,* five different defensin genes *(PDF1.1, PDF1.2, PDF2.2, PDF2.3)* have been identified. In healthy *Arabidopsis* plants, *PDF1.1* is expressed exclusively in seeds (Penninckx et al., 1996), *PDF2.1* in roots and seeds (Thomma and Broekaert, 1998), and *PDF2.3* in all organs except roots (Epple et al., 1997). The gene *PDF1.2,* on the other hand, is not expressed constitutively but is strongly induced in leaves upon attack by fungal pathogens (Penninckx et al., 1996). Hence, *PDF1.2* alone can be called a PR protein. *PDF1.2* is induced upon treatments with methyl jasmonate, silver nitrate, and by inoculation with plant pathogenic fungi (Epple et al., 1997). Ethylene also induced *PDF1.2* (Penninckx et al., 1998).

Radish seeds have been shown to contain two homologous, 5 kDa cysteine rich proteins designated *Raphanus sativus*-antifungal protein 1 (Rs-AFP1) and Rs-AFP2 (Terras et al., 1992). These proteins were barely detectable in healthy uninfected leaves but accumulated systemically at high levels after localized fungal infection. The induced leaf proteins (designated Rs-AFP3 and Rs-AFP4) were purified and shown to be homologous to seed Rs-AFPs (Terras et al., 1995b). These proteins are typical members of the defensins family and are considered PR proteins. Homologs of Rs-

AFPs have been detected in pea pods (Chiang and Hadwiger, 1991), tobacco flowers (Gu et al., 1992), potato flowers (Moreno et al., 1994), and *Petunia inflata* pistils (Karunanandaa et al., 1994). The γ-thionins isolated from barley and wheat have the same arrangement of their cysteine residues as the Rs-AFPs (Bohlmann, 1994). Three γ-thionins have been detected in *Sorghum bicolor*. All these proteins consist of 47 residues, containing eight cysteine residues similar to other defensins. They are α-amylase inhibitors (Nitti et al., 1995).

PR-13 Proteins (Thionins)

Thionins are small (5 kDa), basic, cysteine-rich proteins (Bohlmann and Apel, 1991; Bohlmann, 1994). They have been classified as PR-13 proteins (Van Loon and Van Strien, 1999). Thionins were first isolated from barley (Mendez et al., 1990) and wheat (Colilla et al., 1990). They have been isolated from seeds, roots, and leaves of oats, rye, and maize (Bunge et al., 1992; Castagnaro et al., 1992), and tomato, papaya, mango, and walnut (Daley and Theriot, 1987). Although thionins are constitutively expressed in seeds and roots, some leaf thionins were found to be induced during pathogenesis (Bohlmann et al., 1988). Hence, those thionins are considered PR proteins (Van Loon, 1999). An induction of leaf thionins was demonstrated at the mRNA and protein levels after infection with barley powdery mildew fungus (Bohlmann et al., 1988). Methyl jasmonate leads to an accumulation of leaf thionin transcripts and the mature proteins in barley (Andresen et al., 1992). Jasmonic acid and ethylene induced a thionin named THI 2.1 in *Arabidopsis* (Epple et al., 1995, 1997).

Leaf thionins are encoded by a large gene family containing as many as 50 genes that are differentially regulated (Bohlmann and Apel, 1987). Thionins are synthesized as much larger precursors with a molecular weight of about 15 kDa. N-terminal to the thionin domain is a typical sequence that directs the proprotein into the endoplasmic reticulum. The proprotein consists of the actual thionin and a C-terminal extension, the so-called acidic domain, which contains a large number of acidic amino acids. Although the amino acids at most positions are variable, some amino acids are highly conserved, including a tyrosine at position 61 and a glycine at position 65. The six cysteine residues are absolutely conserved (Bohlmann, 1994).

PR-14 (Lipid Transfer Proteins)

Lipid transfer proteins (LTPs) stimulate the transfer of a broad range of lipids between membranes in vitro (Yamada, 1992). They are generally secreted (Mundy and Rogers, 1986; Sterk et al., 1991) and externally associated with the cell wall (Molina and Garcia-Olmedo, 1993; Molina et al., 1993b; Segura et al., 1993; Thoma et al., 1993; Pyee et al., 1994). They may be involved in the secretion or deposition of extracellular lipophilic materials such as cutin or wax (Sterk et al., 1991; Thoma et al., 1993; Pyee et al., 1994; Hendriks et al., 1994). LTPs are distributed at high concentrations in the epidermis of exposed surfaces and in the vascular tissues (Fleming et al., 1992; Molina and Garcia-Olmedo, 1993; Thoma et al., 1993; Pyee et al., 1994). Expression of LTP genes has been shown to be induced well above basal levels in some plant-pathogen interactions (Molina and Garcia-Olmedo, 1993, 1994; Garcia-Olmedo et al., 1995). Some LTPs have been considered PR-14 proteins (Van Loon and Van Strien, 1999).

Chitosanases

Some chitosanases have been identified as PR proteins. Chitosanases act on chitosan (poly [1-4]-β-D-glucosamine, the deacetylated chitin) without any activity on chitin, and they differ from chitinases by their molecular weight (10-24 kDa) and substrate specificity. No chitosanase activity was detected in uninfected roots of Norway spruce *(Picea abies),* but three chitosanases are induced after a fungal infection (Sharma et al., 1993). Grenier and Asselin (1990) detected six chitosanases (both acidic and basic isoforms) in barley, four acidic chitosanases in cucumber, and one basic chitosanase in tomato following chemical stress.

Glycine-Rich Proteins

A new glycine-rich protein was induced in tobacco infected with tobacco mosaic virus (Hooft van Huijsduijnen et al., 1986). Salicylic acid treatment also induced this protein (Van Kan et al., 1988). Induction of two genes encoding glycine-rich proteins due to mercuric chloride treatment has been observed in maize leaves (Didierjean et al., 1992).

Other Inducible/PR Proteins

hsr and str Gene-Encoded Proteins

Some new genes have been detected in plants inoculated with pathogens. When tobacco plants were inoculated with *Ralstonia solanacearum,* two groups of genes were induced. The first group of genes, the *hsr* (hypersensitivity-related) genes, are specifically activated during the HR while the other group of genes, the *str* (sensitivity-related) genes, responds in a similar way in inoculations with *R. solanacearum* leading to resistance or disease development (Marco et al., 1990; Pontier et al., 1994). These two classes of genes were different from already described PR protein genes in tobacco. The function of *hsr* and *str* gene products remains undetermined since no homology could be detected with other previously characterized genes. Transcripts of one of the *str* cDNA clones, pNT246, accumulate during early steps of compatible and incompatible interactions with *R. solanacearum* or upon elicitor treatment (Godiard et al., 1991). The nucleotide sequence of the *str246C* gene was found to be identical to that of the tobacco *parA* gene (Froissard et al., 1994), previously characterized as a gene activated by auxin (Takahashi et al., 1990), and similar to that of *pLS216,* a *Nicotiana plumbaginifolia* gene responding to a cytokinin treatment (Dominov et al., 1992). The actual role of these genes in disease resistance is not known; but they are involved in resistant interactions suggesting some role in disease resistance.

Ribosome-Inactivating Proteins

Methyl jasmonate has been shown to induce accumulation of specific gene products in plants. It induced a 60 kDa protein in barley (Andresen et al., 1992; Becker and Apel, 1992). This protein was designated JIP60 (jasmonate-induced protein). N-terminal region of JIP60 is homologous to plant ribosome-inactivating proteins. Proteins with specific RNA-*N*-glycosidase activity responsible for eukaryotic ribosome-inactivation are denoted ribosome-inactivating proteins (RIPs) (Mundy et al., 1994). Plant RIPs inactivate ribosomes from phylogenetically distant species including animals and microorganisms. The barley JIP60 is synthesized as a precursor, which is processed in vivo. JIP60 is a novel ribosome-inactivating protein requiring at least two processing events for full activation (Chaudhry et al., 1994). It is induced in young and senescing leaves in response to pathogens or other stresses mediated by methyl jasmonate and abscisic acid (Reinbothe et al., 1993).

Phytoalexin Biosynthetic Enzymes

Van Loon et al. (1994) suggested that PR proteins may also include newly induced enzymes/isoforms of enzymes that are present in amounts too small for detection of gels by general stains, such as enzymes involved in the synthesis of isoflavonoid phytoalexins. Phenylalanine ammonia-lyase (PAL) is the key enzyme in biosynthesis of several phytoalexins.

When primary leaves of 'Red Mexican' bean plants were inoculated with an avirulent isolate of *P. savastanoi* pv. *phaseolicola,* PAL transcripts accumulated rapidly (Meier et al., 1993). No PAL transcripts could be detected in uninoculated healthy tissues, suggesting that this enzyme can be considered a PR protein. Chalcone synthase (CHS) is also a key enzyme in the biosynthesis of isoflavonoid phytoalexins. Seven CHS genes have been detected in the bean genome. Among them, the *CHS8* gene product can be considered a PR protein. *CHS8* transcripts were present in roots, stems, and floral buds whereas the transcript levels in leaves were too low to be detected (Schmid et al., 1990). The induction of *CHS8* transcripts was observed in elicitor-treated bean suspension cultures and in wounded hypocotyls (Ryder et al., 1987). An incompatible isolate of *P. syringae* induced bean *CHS8* gene in transgenic tobacco leaves (Schmid et al., 1990). Tobacco also has a CHS gene, but it was not induced by any stress. Hence, bean *CHS8* may be considered a PR protein.

Another bean chalcone synthase gene, *CHS15,* is also induced by elicitors (Dron et al., 1988; Stermer et al., 1990; Harrison et al., 1991a,b; Arias et al., 1993). Bean *CHS8* and *CHS15* were each fused upstream of the coding region of the β-glucuronidase (GUS) reporter gene and transformed into tobacco. Short wave ultraviolet (UV) light or mercuric chloride induced the expression of both gene fusions in transgenic tobacco leaves (Stermer et al., 1990). Thus CHS15 may also be a PR protein. There are many *CHS* genes in alfalfa plants, but five of them (*CHS1, CHS2, CHS4, CHS8,* and *CHS9*) were not expressed in healthy leaves and were induced in leaves in response to a fungal infection (Junghans et al., 1993). CHS transcripts accumulated in alfalfa leaves inoculated with *P. syringae* pv. *pisi,* an incompatible pathogen; and no such CHS transcripts could be detected in healthy leaves (Esnault et al., 1993). These *CHS* gene products can also be considered PR proteins. Accumulation of another important phytoalexin synthesizing enzyme isoflavone reductase (IFR) transcripts was observed in alfalfa leaves inoculated with the incompatible pathogen *P. syringae* pv. *pisi* while this enzyme could not be detected in uninoculated regions in the leaves (Esnault et al., 1993).

Stilbene synthase was almost absent in nonelicited leaves of *Vitis* (Fritzemeir and Kindl, 1981). The stilbene synthase gene from peanut has been cloned. The transgenic tobacco calli containing the stilbene synthase gene, when treated with the elicitors, produced resveratrol (Hain et al., 1990; Kishore and Somerville, 1993). Several stilbene synthase genes may be present in the peanut genome (Schroder et al., 1988). As stilbene synthase is newly induced by pathogen stress, it can also be considered a PR protein.

No casbene synthetase mRNA has been detected in uninfected castor bean seedlings. Casbene synthetase activity preceded the formation of cyclic diterpene phytoalexin casbene in castor bean seedlings infected with the pathogen (West et al., 1985). Messenger RNA for casbene synthetase increased in castor bean seedlings treated with the elicitor of the pathogen *(Rhizopus stolonifer)* within 6 h of treatment (Moesta and West, 1985). Farnesyl transferase involved in diterpene phytoalexin synthesis in castor bean (Dudley et al., 1986), and diterpene hydrocarbon synthase (cyclase) involved in synthesis of diterpene phytoalexins in rice (Ren and West, 1992) can also be considered as PR proteins.

BACTERIAL PATHOGENS INDUCE PR PROTEINS

Several PR proteins are induced when the plants were inoculated with incompatible bacterial pathogens. When rice leaves were inoculated with the incompatible pathogen *Pseudomonas syringae,* PR-1, PR-2, PR-3, PR-5, and PR-9 proteins were induced (Smith and Metraux, 1991; Schweizer et al., 1997). The nonhost pathogen *P. syringae* pv. *syringae* induced a peroxidase in rice (Reimmann et al., 1992). A peroxidase gene was induced in rice during resistant and susceptible interactions with *Xanthomonas oryzae* pv. *oryzae* (Chittoor et al., 1997). A PR-5 protein was induced in rice leaves inoculated with *P. syringae* pv. *syringae* (Reimann and Dudler, 1993). Ten new proteins appeared in rice leaves inoculated with *X. oryzae* pv. *oryzae* (Kim and Yoo, 1992). When tobacco was inoculated with *Erwinia carotovora* subsp. *carotovora,* a rapid induction of local and systemic accumulation of transcripts for PR proteins, including class I β-1,3-glucanase was observed (Vidal et al., 1997). When tobacco leaves were inoculated with the incompatible pathogen *P. syringae* pv. *phaseolicola,* methyl salicylate which induced PR-1 transcripts was induced (Seskar et al., 1998). Several PR proteins are induced in tobacco plants following infection with *P. syringae* (Ahl et al., 1981). *Pseudomonas tabaci* induces chitinase (PR-3) and β-1,3-glucanase (PR-2) in tobacco plants (Meins and Ahl, 1989). Several PR-2 and PR-3 proteins are synthesized in tobacco leaves infiltrated with different isolates of *Ralstonia solanacearum* (Godiard et al., 1990). *R. solanacearum* strain R296 induces large amounts of PR-1 proteins in tobacco leaves (White et al., 1996). Several PR genes were induced in tobacco leaves inoculated with various isolates of *R. solanacearum* (Marco et al., 1990; Godiard et al., 1991; Pontier et al., 1994; Froissard et al., 1994).

PR proteins, including peroxidase, β-1,3-glucanase, and chitinases, were induced in cucumber plants inoculated with *P. syringae* pv. *syringae* (Strobel et al., 1996). *P. syringae* pv. *tomato,* an incompatible pathogen, induced PR proteins in radish (Hoffland et al., 1996). Infection with *X. vesicatoria* induced synthesis of several PR proteins with molecular weights of 15, 18, 21, 23, 26, 29, and 54 kDa in tomato (Kim and Hwang, 1995). PR-6 proteins (proteinase inhibitor I and II) accumulated rapidly in tomato infected by *P. syringae* pv. *tomato* (Pautot et al., 1991). Activity of both β-1,3-glucanase and chitinase increased in pepper leaves after inoculation with *X. vesicatoria* (O'Garro and Charlemange, 1994). Infection of pepper (*Capsicum* sp.) leaves by *X. vesicatoria* induced synthesis of six β-1,3-glucanases and 10 chitinases (Lee and Hwang, 1996). *P. syringae,* an incompatible pathogen, induced acidic chitinase in potato leaves (Stromberg, 1996). In leaves of bean *(Phaseolus vulgaris)* cv. Red Mexican inoculated with avirulent race 1 isolate of *P. savastanoi* pv. *phaseolicola,* an increase in chitinase activity was observed (Voisey and Slusarenko, 1989). *X. campestris* pv. *campestris* (compatible pathogen) and *X. campestris* pv. *armoraciae* and *X. campestris* pv. *raphani* (incompatible pathogens) induced β-1,3-glucanase in turnip *(Brassica campestris)* (Newman et al., 1994, 1995).

Messenger RNA encoding PR-1 accumulated to high levels in *Arabidopsis thaliana* inoculated with *X. campestris* pv. *campestris* (Buell and Somerville, 1995). PR-1 protein mRNA levels increased in *Arabidopsis* tissues challenged with *P. syringae* pv. *tomato* (Uknes et al., 1992). Inoculation of *Arabidopsis* ecotypes Columbia and Pr-O with *P. syringae* pv. *syringae* D20, which elicits an HR in both ecotypes, resulted in the accumulation of PR-1 proteins (Buell and Somerville, 1995). β-1,3-Glucanase genes *bgl1, bgl2,* and *bgl3* were induced in *A. thaliana* inoculated with *P. syringae* pv. *maculicola* (Dong et al., 1991). β-1,3-glucanase mRNA levels were induced in *A. thaliana* inoculated with *P. syringae* virulent and avirulent strains (Dong et al., 1991). Chitinase mRNA levels increased in *Arabidopsis* inoculated with *X. campestris* pv. *campestris* (Lummerzheim et al., 1993). Thus, several studies have revealed that bacterial pathogens induce several types of PR proteins, paricularly in resistant interactions.

MOLECULAR MECHANISMS OF INDUCTION OF PR PROTEINS

Genes Encoding PR Proteins

It is now well established that PR proteins are induced by bacterial pathogens and other stresses. Since these proteins are involved in disease resistance in many host-pathogen interactions, several research efforts have been made to unravel the mechanism of induction of PR-proteins (Cornelissen et al.,

1986; Pfitzner and Goodman, 1987; Oshima et al., 1987, 1990). Several genes encoding PR proteins have been identified in different plants; they are almost silent in healthy plants (Cornelissen et al., 1987; Sharma et al., 1992; Gordon-Weeks et al., 1997). Generally most PR protein-encoding genes belong to a multigene family (Sharma et al., 1992). But not all genes are expressed in plants, suggesting that there may be pseudogenes (Van Kan et al., 1992). At least 16 genes are present in tobacco plants coding for various PR-1 proteins (Cornelissen et al., 1987; Pfitzner et al., 1988). A gene coding a basic type PR-1 protein *(prb-1b)* was cloned from tobacco and characterized (Eyal et al., 1992). Tobacco contains 13 to 14 genes encoding PR-2 proteins (Ohme-Takagi and Shinshi, 1990). In tobacco, a class I PR-2 gene, *Glb,* has been cloned (Hart et al., 1993; Livne et al., 1997). Van de Rhee et al. (1993) have reported another tobacco class I β-1,3-glucanase gene, *ggl50.* In tobacco, class II *PR-2b* and *PR-2d* genes have been identified (Henning et al., 1993; Van de Rhee et al., 1993; Shah and Klessig, 1996). 'Bright Yellow' tobacco contains another PR-2 gene, *Gln2* (Ohme-Takagi and Shinshi, 1995). A gene encoding β-1,3-glucanase, *gn1,* has been cloned from *Nicotiana plumbaginifolia* (Castresana et al., 1990). Acidic and basic chitinases are each encoded by two to four genes (Hooft van Huijsduijnen et al., 1987). A similar complexity was observed for the genes encoding tobacco PR-5 proteins. A small gene family encodes tobacco AP 24, and *ap24* is part of the gene family (Melchers et al., 1993). Four genes encoding PR-5 proteins, *tlp1, tlp2, tlp3,* and *tlp4,* have been cloned from oat (Lin et al., 1996). A gene encoding a PR-5 protein has been cloned from *Arabidopsis* (Hu and Reddy, 1995). In barley, a gene family encoding PR-5 has been identified (Vale et al., 1994). A PR-6 protein gene (MPI) has been characterized in maize (Cordeo et al., 1994). Genes encoding PR-9 have been identified in barley (Vale et al., 1994).

Bean PvPR3 is a PR-10 protein, and it is encoded by approximately 15 genes (Sharma et al., 1992). The other two bean PR-10 proteins, PvPR 1 and PvPR 2 families together consist of a minimum of 12 genes within the bean genome (Walter et al., 1990). Genes encoding PR-10 protein have been detected in pea, parsley, and bean (Fristensky et al., 1988; Somssich et al., 1988; Walter et al., 1990), asparagus (Warner et al., 1992), and soybean (Crowell et al., 1992). In rice a PR-10 gene, *PBZ1,* has been identified (Midoh and Iwata, 1996). In general, PR-10 genes form a multigene family (Chiang and Hadwiger, 1990; Crowell et al., 1992; Breiteneder et al., 1993; Iturriage et al., 1994). The PR-11 proteins in tobacco are encoded by a small multigene family (Ohl et al., 1994). Defensin gene *PDF1.2* has been cloned from *Arabidopsis* (Penninckx et al., 1996). A gene family encoding thionin has been reported in barley (Vale et al., 1994).

cis-Acting Elements and trans-Acting Factors in Gene Transcription

PR genes are quiescent in healthy plants. Regulation of PR protein biosynthesis occurs primarily at the level of transcription. The regulation of the PR gene expression depends on the binding of certain transcription factors to particular regions of the gene promoter. Expression of these PR genes depends on the presence of certain elements located in the flanking regions of the protein-coding genes. A gene is a sequence of DNA that codes for a diffusible product, which diffuses away from its site of synthesis to act elsewhere. An element is a sequence of DNA that is not converted into any other form, but that functions exclusively as a DNA sequence in situ. Because it affects only the DNA to which it is physically linked, it is described as a *cis*-acting element. Several *cis*-acting elements, which activate PR genes, have been identified; they are discussed later in this chapter. Genes that code for proteins, which in turn regulate the expression of other genes, are called regulator genes. Because the products of regulator genes are free to diffuse to their appropriate targets, they are described as *trans*-acting factors. A regulator gene codes for a regulator protein that controls transcription by binding to particular site(s) on DNA. Recognition of a *cis*-acting sequence by a *trans*-acting factor can regulate a target gene. The *cis*-acting sites are usually located just upstream of the target gene. Sequences prior to the starting point of transcription (a transcription unit is a sequence of DNA transcribed into a single RNA, starting at the promoter and ending at the terminator) are described as upstream of it; those after the start point (within the transcribed sequence) are downstream of it. Base positions are numbered in both directions away from the start point, which is assigned the value +1; numbers increase going downstream. The base before the start point is numbered –1, and the negative numbers increase going upstream. The sites that mark the beginning and end of the transcription unit, the promoter and terminator, are examples of *cis*-acting sites. A promoter serves to initiate transcription only of the gene or genes physically connected to it on the same stretch of DNA. Transcription is regulated by the interaction between *trans*-acting factors and *cis*-acting sites.

Several signals are required for regulation of the interaction between *cis*-acting elements and *trans*-acting factors resulting in transcription of PR genes.

Signals Involved in Inducing PR Gene Transcription

When bacterial pathogens invade host tissues, several PR proteins are induced. It suggests that signals for inducing PR proteins may be from bacteria themselves. Bacterial elicitors have been shown to induce PR proteins. PR proteins, including β-1,3-glucanase, chitinase, and peroxidase, were induced in cucumber plants inoculated with *Pseudomonas syringae* pv. *syringae* or treated with the elicitor produced by it, $HrpZ_{Pss}$. The induction patterns of PR proteins by $HrpZ_{Pss}$ and *P. syringae* were the same (Strobel et al., 1996). Infection of tobacco plants with *Erwinia carotovora* subsp. *carotovora* or treatment of plants with an *Erwinia*-derived elicitor preparation leads to the induction of genes encoding PR proteins (Vidal et al., 1997). Newman et al. (1995) have shown that purified lipopolysaccharides (LPS) from *X. campestris* pv. *campestris* can induce β-1,3-glucanase transcript accumulation in turnip. LPS from *Ralstonia solanacearum* induces synthesis of new polypeptides in tobacco (Leach et al., 1983).

Bacterial elicitors may not be the sole signal for induction of PR genes. Bacterial pathogens may also induce other signals, which may induce PR proteins. Strobel et al. (1996), who showed that the elicitor produced by *P. syringae* pv. *syringae* induced three PR proteins in tobacco, also reported that the *hrpH* mutant which lost the ability to produce the elicitor $HrpZ_{Pss}$ induced two of the three identified PR proteins. It suggests that at least some PR proteins can be induced by bacterial factors other than elicitors.

Even wounding induces PR proteins in plants (Godiard et al., 1990; Warner et al., 1992; Clarke et al., 1998). Hence, wound-induced signals (such as salicylic acid and jasmonic acid) may also be involved in PR gene activation (Sano et al., 1994). Bacterial infection or induced stress can lead to production of ethylene (Pennazio and Roggero, 1990; Chang et al., 1995; O'Donnell et al., 1996), cytokinins (Sano and Ohashi, 1995), abscisic acid (Zhu et al., 1993; Pena-Cortes et al., 1995), systemin (Bergey et al., 1996), jasmonic acid and methyl jasmonate (Farmer and Ryan, 1990; Schweizer et al., 1997), salicylic acid (Dann et al., 1996; Seskar et al., 1998), methyl salicylate (Shulaev et al., 1997), oligouronides (Pautot et al., 1991), and protein kinases (Subramaniam et al., 1997). All of them are known to act as signals for induction of PR proteins.

Molecular Mechanism of Induction of PR Genes by Elicitors

Several types of elicitors have been isolated from microbial pathogens and plants (see Chapter 1). These elicitors induce PR-genes. *Cis*-acting elicitor-responsive elements (EREs) have been identified in promoters of many PR-proteins. A *cis*-acting elicitor-responsive element (ERE) has been identified in the promoter of the class I chitinase gene *CHN50* of tobacco (Fukuda, 1997). Both expression and responsiveness to the elicitor disappeared when the region of the promoter that included the ERE sequence had been deleted. The nuclear factor(s) bound specifically to the sequence motif $^{-534}$GGTCANNNAGTC^{-523} (Fukuda, 1997).

Binding the elicitor message to the *cis*-acting elements in the promoter of the PR gene appears to be important for transcription of the gene. Several second messengers are needed for this process. Treatment with calcium channel blockers, calmodulin antagonists, phospholipase inhibitors, and proteinase inhibitors prior to an elicitor treatment suppressed PR gene activation in potato (Furuse et al., 1999). An elicitor (arachidonic acid from *Phytophthora infestans*) induced activation of the potato PR gene *PR-10a*. This activation is positively controlled by a protein kinase that affects the binding of the nuclear factors PBF-1 and PBF-2 to an elicitor response element in the promoter of the gene (Subramaniam et al., 1997). Treatment of potato tuber discs with specific inhibitors of protein kinase abolished elicitor-induced binding of PBF-2 to the ERE. This correlated with a reduction in the accumulation of the PR-10a protein. In contrast, treatment of potato tuber discs with 12-*O*-tetradecanoyl-phorbol-13-acetate, an activator of protein kinase, led to an increase in binding of PBF-2 to the ERE and a corresponding increase in the level of the PR-10a protein (Subramaniam et al., 1997).

Protein kinases appear to be important in binding proteins. Bacterial disease resistance genes have been cloned from rice *(Xa21)* and tomato *(Pto)* (Ronald, 1997; Tobias, 1996). Both the resistance genes code for protein kinases which are involved in protein binding with DNA (Vidhyasekaran, 1998). All resistance genes (except one in maize) cloned so far are involved in the signal transduction system and they contain leucine-rich repeats (LRR), leucine zippers, and nucleotide binding sites (NBS) (Ausubel et al., 1995; Grant et al., 1995; Song et al., 1995; Lawrence et al., 1995; Salmeron et al., 1996; Wang et al., 1999). Leucine zippers and LRRs serve as the sites of protein-protein binding; leucine zippers may directly interact with DNA. The NBS may direct transport of the signals to the plant nucleus (Yang and Gabriel, 1995). Generally, it is believed that a major function of resistance genes is binding elicitor signals with the promoter of the defense genes such as PR genes.

Analysis of the promoter region of the *PR-10a* gene in transgenic potato plants indicates that a 1015 bp 5' upstream region of the gene is sufficient to confer elicitor-inducibility of the β-glucuronidase (GUS) reporter gene in potato tuber disks (Matton et al., 1993). The *STH-2* gene family (PR-10 protein gene) in potato is induced by elicitor treatment (Constabel and Brisson, 1992). About 1 kb of 5' flanking DNA was sufficient to direct wound and elicitor induction of the *STH-2* gene (Matton et al., 1993). Deletion analysis of the *STH-2* gene 5' flanking region suggested that 135 bp of upstream region may be sufficient for full promoter activity (Matton et al., 1993). An important regulatory sequence is localized between nucleotides –135 and –105 (Matton et al., 1993). A 125 bp promoter fragment of the PCPR2 (PR-10 protein) gene was demonstrated to be sufficient for elicitor-mediated gene expression (Van de Locht et al., 1990).

Deletion analysis of the 5' region of the proteinase inhibitor II gene (PR-6 protein gene) revealed that the region from –136 to –165 and from –520 to –620 may contain a complex of *cis* elements that interact with elicitor-inducible *trans*-acting factors to regulate PR genes (Ryan, 1992). Promoter region of a rice peroxidase gene *(POX22.3)* (PR-9 protein gene) contained sequences identical to the G-box element GCACGTG and an upstream highly conserved region (Chittoor et al., 1997).

The tobacco osmotin (PR-5) promoter fragment –248 to –108 upstream of the transcription start site (fragment A) was sufficient to direct reporter gene expression when fused to a minimal CaMV35S promoter. Three conserved promoter elements in fragment A interact specifically with nuclear factors. These elements are: (1) a cluster of G-box-like sequences; (2) an AT-1 box-like sequence, 5'-AATTATTTTATG-3'; and (3) a sequence highly conserved in ethylene-induced PR gene promoters, 5'-TAAGACGCCGCC-3' (Liu et al., 1995).

Molecular Mechanism of Induction of PR Genes by Ethylene

Ethylene accumulates during pathogenesis in many host-pathogen interactions (Vidhyasekaran, 1997). The increased production of ethylene is one of the earliest chemically detectable events in pathogen-infected plants or in plants treated with elicitors (Toppan and Esquerre-Tugaye, 1984). Ethylene acts as a signal molecule inducing various PR protein genes (Xu et al., 1994). It was found to induce high accumulations of basic and neutral PR-5 proteins in tobacco (Koiwa et al., 1994). Ethylene induced a basic, not acidic-type, PR-1 transcript in tobacco (Eyal et al., 1992). Ethylene induced a basic class I β-1,3-glucanase, GLB, in tobacco (Vogeli-Lange et al., 1994b). Van de Rhee et al. (1993) reported that another tobacco class I β-1,3-glucanase gene, *gglb50,* is also induced by ethylene. The HRGP gene in carrot is induced by ethylene (Granell et al., 1992). After treatment with ethylene, seven basic PR proteins accumulated in bean (Del-Campillo and Lewis, 1992). Among them, two isoforms of β-1,3-glucanase, multiple isoforms of chitinase, a PR-5 protein, and a 15 kDa polypeptide serologically related to PRP1 (P14) from tomato were identified (Del-Campillo and Lewis, 1992). Ethylene induced a thionin named THI 2.1 in *Arabidopsis*

(Epple et al., 1997). Tobacco plants were transformed with the mutant *etr1-1* gene from *Arabidopsis,* conferring dominant ethylene insensitivity (Knoester et al., 1998). These transformants hardly expressed basic PR proteins PR-1g and PR-5c. Expression of acidic PR-1 genes was not affected. It suggests that in tobacco, basic PR genes are regulated by ethylene (Knoester et al., 1998). In *Arabidopsis* mutants insensitive to ethylene, *ein2* and *etr1* plants, ethylene did not induce the synthesis of defensin PDF1.2 while in wild plants it induced the defensin (Penninckx et al., 1996). It suggests that ethylene acts as a signal in induction of the PR protein defensin PDF1.2.

Cellulases have been reported to be protein-type elicitors (Vogeli and Chappell, 1990). A cellulase elicitor induced osmotin (PR-5) gene expression in tobacco seedlings (Chang et al., 1995). It activated the osmotin promoter fused to a β-glucuronidase (GUS) reporter gene in tobacco seedlings and induced the accumulation of osmotin RNA and protein (Chang et al., 1995). Increases in osmotin-promoter driven GUS activity (Figure 4.1; Chang et al., 1995) and accumulation of osmotin mRNA induced by the cellulase elicitor in tobacco leaves were reversed by norbornadiene, an ethylene action inhibitor, indicating that ethylene is involved in the induction of the osmotin gene by the elicitor (Figure 4.1; Chang et al., 1995).

Ethylene-responsive *cis*-elements have been detected in the promoters of several PR genes. AGCCGCC sequences are highly conserved in the promoters of PR genes with ethylene-inducible expressions (Hart et al., 1993) and are thought to constitute an ethylene-responsive *cis*-element. Sato et al. (1996) have reported the regulation of ethylene-responsive expression of an osmotin-like protein (neutral PR-5 protein) gene in tobacco by AGCCGCC sequences which are found at –46 to –52 and –161 to –167. The promoter sequence –248 to –108 bp is absolutely required for osmotin gene induction by ethylene in tobacco, and it contains the AGCCGCC motif (Eyal et al., 1993). More than 20 plant genes are reported to contain the GCC box (Zhou et al., 1997). The region from –1452 to –1193 of the β-1,3-glucanase gene *GLB* promoter, contains a 61 bp enhancer of transcription (Hart et al., 1993). This region is important for ethylene-regulated expression of the *GLB* in tobacco leaves (Vogeli-Lange et al., 1994b). This region contains two copies of the heptanucleotide AGCCGCC (Hart et al., 1993). Elements in the region –568 to – 402 are required for ethylene induction of *GLB* in tobacco leaves (Vogeli-Lange et al., 1994b). Van de Rhee et al. (1993) reported that in another tobacco class I β-1,3-glucanase gene, *ggl50,* which is also induced by ethylene, elements for responsiveness to ethylene are present in the –0.45 to –1.5 kb region of the 1.5 kb tobacco *ggl50* promoter. Two copies of the AGC box have been detected in a 1030 bp segment of *ggl50* (Van de Rhee et al., 1993). Enhancer activity and ethylene responsiveness depend on the integrity of two copies of the AGC box, AGCCGCC, present

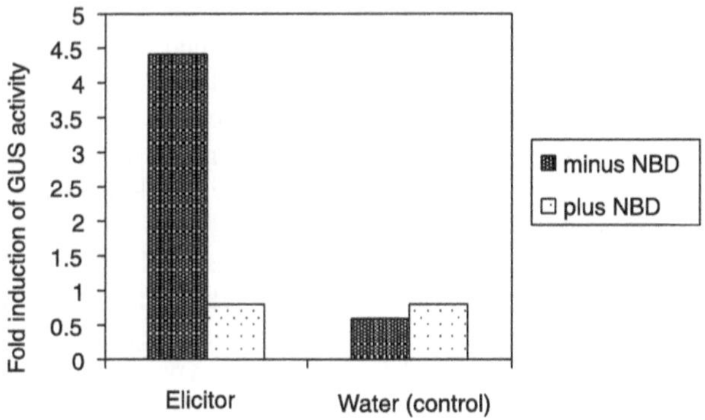

FIGURE 4.1. Effect of norbornadiene (NBD), an ethylene action inhibitor, on osmotin-promoter-driven GUS activity in tobacco seedlings by a cellulase elicitor (*Source:* Chang et al., 1995).

in the promoters of several ethylene-responsive genes (Hart et al., 1993; Sato et al., 1996; Buttner and Singh, 1997).

Promoter analysis of the basic *PRB-1b* gene of tobacco showed that the promoter sequence from position –213 to the transcription start site was sufficient to direct ethylene responsiveness (Eyal et al., 1993). Meller et al. (1993) analyzed the binding of proteins to the ethylene responsive PRB-1b promoter region (–213 and –142). Two binding activities were separated which show differential specificity toward the protein binding sites G (–200 to –178) and Y (–179 to –154). The first factor interacts with both sequences G and Y. The other factor specifically binds to the region G. Mutations in the G region, which alter the GCC box or the G-box motif, disrupt factor binding and abolish the ethylene inducibility of the minimum –213 bp basic PRB-1b promoter/GUS transgene. Besides the G-box motif, the G region also contains the 11 bp sequence, TAAGAGCCGCC (GCC box), which is highly conserved in promoters of ethylene-induced genes of *Nicotiana* species.

OLP (Neutral PR-5) protein promoter::GUS fusion genes bearing the wild or AGCCGCC-mutated promoter of the OLP gene (1099 bp) were introduced to tobacco by *Agrobacterium*-mediated transformation (Sato et al., 1996). GUS activity in normal healthy leaves was low when leaves were not treated with ethylene. When transgenic plants were treated with ethylene, transgenic tobacco bearing the wild-type OLP promoter showed a several-fold increase in GUS activity, whereas transgenic tobacco with the mutated AGCGGCC promoter had very low GUS activity in both untreated and ethylene-treated leaves. This clearly indicates that complete AGCCGCC sequences are necessary for ethylene-induced expression of the PR-5 gene (Sato et al., 1996).

The ethylene-induced expression of PR genes is regulated by a *trans*-acting factor(s), *e*thylene-*r*esponsive *e*lement *b*inding *p*rotein(s) (EREBPs). Four DNA-binding proteins called EREBPs, EREBP-1, EREBP-2, EREBP-3, and EREBP-4, which specifically bind the ethylene-responsive element (ERE) AGC box, have been detected in tobacco (Ohme-Takagi and Shinshi, 1995). These proteins share a central domain of 59 amino acids that is responsible for binding to the GCC box (Ohme-Takagi and Shinshi, 1995). Ethylene induces mRNAs for these EREBPs in tobacco leaves. The EREBP-1 gene was also found to be induced by *Pseudomonas* spp. (Zhou et al., 1997). The EREBPs may be transcription factors important for ethylene-dependent transcription of PR genes. EREBP-2 was found to bind to the AGCCGCC sequence(s) of the tobacco PR-5 gene (Sato et al., 1996).

A homolog of the tobacco EREBP in *Arabidopsis thaliana,* AtEBP, binds specifically to TAAGAGCCGCC, an AGC-box-containing sequence, and confers ethylene responsiveness to promoters of genes encoding PR proteins (Buttner and Singh, 1997). AtEBP interacts with *ocs* element binding factors (OBFs) belonging to a specific class of basic-region leucine zipper (bZIP) transcription factors. It suggests that cross-coupling between EREBP and bZIP transcription factors is important in regulating plant defense-related gene expression (Buttner and Singh, 1997). AtERB1 (also called AtERF) acts directly downstream of EIN3, a nuclear component of the PR gene expression. AtERF1 (=AtEBP) gene was found to be induced by ethylene in wild-type *Arabidopsis* but not in an *ein3* mutant. The EIN3 protein was capable of binding to the promoter sequence of AtERF1, suggesting that EIN3 functions are transcription regulators of AtERF1. The AtERF1 protein binds to the GCC box and activates a basic chitinase gene and a defensin (PDFf1.2) (Solano et al., 1998).

Pto is the resistance gene against *P. syringae* pv. *tomato* and it codes for protein kinase. Three other tomato genes, *Pti4, Pti5,* and *Pti6,* physically interact with *Pto* product (Zhou et al., 1997). Each of these three genes encodes a protein that is similar to the tobacco EREBP. These proteins bind a DNA sequence present in the promoter region of a larger number of PR genes. Thus, there is a relationship between EREBPs, a disease-resistance gene, and the specific activation of PR genes (Zhou et al., 1997).

Molecular Mechanism of Induction of PR Genes by Salicylic Acid

Bacterial infection leads to rapid accumulation of salicylic acid. When tobacco leaves are inoculated with *P. savastanoi* pv. *phaseolicola,* salicylic acid accumulated within 12 h after inoculation (Seskar et al., 1998). When ^{14}C-benzoic acid was administered to cucumber cotyledons infected with a fungal pathogen, ^{14}C-labeled salicylic acid was detected in upper uninoculated leaves (Molders et al., 1996). It suggests systemic induction of salicylic acid. When cucumber plants were inoculated with *P. syringae* pv. *lachrymans* on leaf 1, a significant increase in salicylic acid occurred 9 and 12 h postinoculation in leaf 1 and in leaf 2, respectively. Thus, local and systemic increases in endogenous salicylic acid were observed in cucumber (Meuwly et al., 1995).

It is well established that infection by different pathogens leads to synthesis of salicylic acid. Several signals for induction of salicylic acid have been identified, and it is described in detail in Chapter 1. H_2O_2 appears to be important for salicylic acid accumulation. High levels of H_2O_2 stimulate salicylic acid biosynthesis (Leon et al., 1995; Neuenschwander et al., 1995; Summermatter et al., 1995). When catalase expression leading to accumulation of H_2O_2 was suppressed in leaves of transgenic tobacco plants through sense cosuppression or antisense suppression, most plants failed to show constitutive *PR* gene expression (Chamnongpol et al., 1996; Takahashi et al., 1997). H_2O_2 and H_2O_2-inducing chemicals were unable to induce *PR* gene expression in transgenic plants expressing NahG (salicylate hydroxylase from *Pseudomonas putida,* which degrades salicylic acid) although they could activate *PR-1* genes in wild-type tobacco (Bi et al., 1995; Neuenschwander et al., 1995). These results suggest that salicylic acid may act downstream of H_2O_2 in inducing PR genes.

G-proteins, which modulate phosphorylation, also induce salicylic acid. Transgenic tobacco plants expressing the cholera toxin gene which modulates the signaling system induced by G-proteins constitutively accumulate high levels of salicylic acid and express PR genes (Beffa et al., 1995). Transgenic tobacco plants expressing the *Halobacterium* opsin gene with a proton pump function show elevated levels of salicylic acid and high expression of *PR* genes (Dangl et al., 1996).

Salicylic acid may also act as a signal molecule in inducing other signals for induction of PR genes. Salicylic acid binds to catalase, inhibits its activity, and thereby increases the intracellular concentration of H_2O_2, which might then serve as a second messenger for the induction of a defense response (Chen et al., 1993). Lipid peroxides are potent signaling molecules in animals (Durner et al., 1997). Salicylic acid induces lipid peroxidation in tobacco suspension cells, and the lipid peroxides activate *PR-1* genes in these cells. Salicylic acid binds with a protein (SABP2) (Du and Klessig, 1997) which triggers induction of PR proteins (Gorlach et al., 1996; Du and Klessig, 1997).

Salicylic acid has been shown to induce several PR genes (Malamy et al., 1990; Metraux et al., 1990; Gaffney et al., 1993). Increased levels of salicylic acid induced PR genes in various plants. Several *Arabidopsis* mutants such as *acd2, lsd1, lsd2, lsd3, lsd4, lsd5, lsd6, lsd7, cep1, cpr1,* and *cim3* constitutively express elevated levels of salicylic acid, and all of them show constitutively high *PR* gene expression (Dangl et al., 1996; Ryals et al., 1996). The *s*alicylic *a*cid *i*nsensitive *(sai1) Arabidopsis* mutant plants do not induce *PR* genes even after treatment with salicylic acid (Cao et al., 1997). When tobacco plants that were deficient in catalase were transformed with the tobacco catalase gene, elevated levels of salicylic acid were observed. These transgenic plants showed a constitutively high expression of PR genes (Chamnongpol et al., 1996; Takahashi et al., 1997).

Exogenous application of salicylic acid also induces PR genes in different plants (Uknes et al., 1992). Salicylic acid induced acidic PR-1a, PR-1b, and PR-1c protein genes in tobacco (Eyal et al., 1992). Treatment of tobacco plants with salicylic acid strongly induces accumulation of mRNAs of class II and class III β-1,3-glucanases and certain other PR proteins (Ward et al., 1991; Niki et al., 1998). Promoter activity of the class II PR-2b and PR-2d genes is induced in tobacco in response to salicylic acid (Eyal et al., 1992; Van de Rhee et al., 1993). TIMP genes encoding inhibitors of microbial proteinases (PR-6 proteins) accumulated in tobacco plants treated with salicylic acid (Heitz et al., 1999). The salicylic acid-dependent pathways activate expression of class II and III β-1,3-glucanase genes in *Arabidopsis* (Silverman et al., 1993; Lawton et al., 1994b; Ryals et al., 1996).

Exogenous application of salicylic acid normally induces acidic/intercellular PR proteins in many plants. Excised healthy tobacco leaves were fed salicylic acid for 72 h through the cut petiole and salicylic acid and PR-1 proteins were analyzed in opposite half-leaves (Yalpani et al., 1991). The level of salicylic acid in a leaf was proportional to the concentration of salicylic acid in the solution in which the petiole was immersed. Induction of acidic PR-1 proteins was positively correlated with leaf salicylic acid (Yalpani et al., 1991). Salicylic acid induced several acidic PR proteins in tobacco (White, 1979; Antoniw and White, 1980; Van Loon and Antoniw, 1982; Pennazio et al., 1987; Van de Rhee et al., 1993). Spraying of the leaves of young potato plants with salicylic acid induces the appearance of eight intercellular acidic PR proteins including β-1,3-glucanase, chitinase, and PR-5 (Pierpoint et al., 1990).

Many basic PR proteins such as class I PR-2 proteins Ggl50 and Glb are not induced by salicylic acid in tobacco (Castresana et al., 1990). Transcripts of tobacco class I β-1,3-glucanase and chitinases are not induced in response to salicylic acid (Linthorst et al., 1990; Ohme-Takagi and Shinshi, 1990; Beffa et al., 1995; Vidal et al., 1997; Niki et al., 1998). Salicylic acid induced only very small amounts of PR-S (a basic PR-5 protein) in tobacco (Koiwa et al., 1994). However, some basic PR proteins are also induced by salicylic acid (Eyal et al., 1992). The promoter of *Nicotiana plumbag-*

inifolia class I PR-2 protein Gn1 is strongly induced (about 14-fold) in transgenic tobacco plants treated with salicylic acid (Castresana et al., 1990). In sunflower, multiple PR-5 isomers of similar molecular weight but of different isoelectric points were excreted from the plant cells in response to the salicylic acid treatment (Jung et al., 1993). Salicylic acid also induced some intracellular PR proteins (PR-10 proteins) such as SAM22 (soybean PR protein) and AoPR1 (asparagus PR protein) (Crowell et al., 1992; Warner et al., 1994). But another PR-10 protein, PBZ1, is not induced by salicylic acid in rice (Midoh and Iwata, 1996).

Presence of *cis*-acting elements and presence or absence of certain critical *trans*-acting factors may determine the PR gene expression induced by salicylic acid. A deletion analysis of the promoter region of a gene encoding an acidic PR-1 protein revealed the presence of salicylic acid-responsive elements between nucleotides 625 and 902 upstream of the transcription site (Van de Rhee et al., 1990). The *cis*-acting elements for regulating gene expression of the acidic PR-1a protein gene were analyzed in transgenic tobacco plants (Oshima et al., 1990). To identify the *cis*-acting element more precisely, a series of 5'-deleted chimeric genes were constructed and transformed into tobacco plants. A 0.3 kb DNA sequence of fragment was sufficient to allow the regulated expression of the PR-1a gene. The existence of *cis*-acting element(s) in this 0.3 kb region just upstream from the coding region can also be inferred from the comparison of the sequence of the acidic PR-1 protein gene family members. Approximately 0.2 kb of the 5'-flanking region just upstream of the TATA box of active PR-1 protein genes (PR-1a, PR-1b, and PR-1c) are highly conserved, whereas sequence insertions further upstream of the conserved region generate divergent nucleotide sequences. The PR-1 protein genes are coordinatively expressed, and hence it was suggested that the *cis*-acting regulatory element(s) exist in this conserved 0.2 kb region (Oshima et al., 1990). As little as 300 bp of 5' flanking sequence was sufficient to appropriately express the *PR-1a* gene (Oshima et al., 1990). The first 300 bp of the *PR-1a* promoter are sufficient to drive salicylic acid-dependent reporter gene expression (Ohashi and Ohshima, 1992). But others could not reproduce this result; induction of *PR-1a* by salicylic acid could not be demonstrated using a 300 bp *PR-1a* promoter fragment fused to the GUS genes (Van de Rhee et al., 1990; Beilmann et al., 1991; Uknes et al., 1993a). A *PR-1a* promoter fragment containing nucleotides –902 to + 29 was able to confer high levels of salicylic acid-induced expression to the GUS reporter gene in transgenic tobacco (Van de Rhee et al., 1990). When this *PR-1a* promoter fragment was gradually truncated from the 5' end to position –689, the expression level was lowered several times (Van de Rhee and Bol, 1993). Van de Rhee et al. (1990), using the same *PR-1a* gene from tobacco, concluded that induction required at least 685 bp.

Uknes et al. (1993a) have reported that 661 bp of 5' flanking sequence is sufficient for fivefold induction of the *PR-1* gene by salicylic acid. The region between position –689 and –643 is important for *PR-1a* gene induction by salicylic acid treatment (Van de Rhee et al., 1990).

The *PR-1a* promoter has been reported to contain four regulatory elements, located between nucleotides –902 and –691 (element 1), –689 and –643 (element 2), –643 and –287 (element 3), and –287 and +29 (element 4) (Van de Rhee and Bol, 1993). All four elements are required for maximum induction of the reporter (GUS) gene by salicylic acid. Elements 1 to 3 positively regulate the *PR-1a* promoter while element 4 appears to be important for a correct spacing between the other three elements and the transcription site (Van de Rhee and Bol, 1993). The *PR-1a* promoter consists of at least two functional domains. One is located upstream of position –335 and contains a strong positive regulatory element. The other domain resides within the region between –71 and +28 of the *PR-1a* gene (Beilmann et al., 1991).

Several *trans*-acting factors binding nuclear sites in promoters of acidic *PR-1* genes in tobacco have been reported (Buchel et al., 1996). Formation of a slow migrating protein-DNA complex with nuclear proteins from healthy 'Samsun NN' tobacco has been demonstrated (Buchel et al., 1996). The same nuclear factor(s) bind(s) various promoter fragments. The factors involved in these interactions are probably GT-1-like factors (Buchel et al., 1996). The *PR-1a* promoter contains a number of putative GT-1 binding sites distributed over the entire 900 bp upstream region (Buchel et al., 1996).

The tobacco *PR-1a* gene promoter contains several binding sites for Myb protein in the region between positions –643 and –169. Some of these sites can be bound by recombinant tobacco Myb1 protein in vitro (Yang and Klessig, 1996). Expression of the tobacco *myb1* gene is rapidly (within 15 minutes) induced by salicylic acid. Hence, it is possible that Myb1 binding activity is involved in transduction of the salicylic acid signal to the PR promoters (Yang and Klessig, 1996). However, Myb1 by itself may not be sufficient for salicylic acid inducibility of the *PR-1a* gene, because in vivo analysis of this promoter has suggested that more than one region is involved in the salicylic acid-mediated activation (Yang and Klessig, 1996). The binding activity of GT-1-like proteins, which have also been shown to bind various fragments of the *PR-1a* promoter in vitro, is reduced in extracts from salicylic acid-treated leaf tissue. It suggests that the Myb1 and GT-1-like proteins may be involved in the salicylic acid-dependent expression of the *PR-1a* gene. Myb1 proteins are known to be involved in activation of gene expression in combination with other transcription fractions (Martin and Paz-Ares, 1997).

Hagiwara et al. (1993) demonstrated the binding of nuclear proteins from healthy tobacco to a region in the *PR-1a* promoter between positions –68 and –51 and a region between positions –184 and –172. TGA1 is a transcription factor from tobacco that specifically binds to the salicylic acid-responsive motifs *as-1* and *ocs* (Qin et al., 1994; Ulmasov et al., 1994). TGA1a belongs to the basic-leucine-zipper (bZIP) class of transcriptional factors. The N-terminus of TGA1a is important for transactivation. The DNA-binding domain resides in the bZIP region, present in the central part of the protein (Katagiri et al., 1992; Neuhaus et al., 1994a). TGA1a specifically binds to an *as-1*-like sequence in the *PR-1a* promoter present at position –593. Salicylic acid enhances the activity of promoters containing *as-1* type elements. Salicylic acid may induce the *PR-1a* gene, at least partly by mediating through binding TGA1a to the *as-1*-like element present in the promoter (Buchel and Linthorst, 1999).

For induction of *PR-2* genes by salicylic acid, *cis*-acting elements have been detected in tobacco. Multiple regions of the approximately 1.7 kb tobacco class II *PR-2b* and *PR-2d* promoters contain elements for inducibility by salicylic acid (Van de Rhee et al., 1993; Henning et al., 1993). For the *PR-2d* gene, this includes a major *cis*-acting element in the region –364 to –288, which confers high-level expression to a core CaMV 35S promoter in response to salicylic acid (Shah and Klessig, 1996).

A 10 bp TCA element that is common to the promoters of several tobacco PR genes was shown to bind a 40 kDa nuclear protein in a salicylic acid-dependent manner (Goldsbrough et al., 1993). However, this TCA element was not required for salicylic acid-mediated induction of the tobacco *PR-2d* promoter in vivo (Shah and Klessig, 1996). Salicylic acid may also interact with several other effector proteins besides those involved in redox regulation (Myb proteins). Induction of PR genes by salicylic acid is sensitive to inhibitors of protein synthesis (Durner et al., 1997). Cycloheximide prevented induction of the *PR-1a* gene in tobacco by salicylic acid (Uknes et al., 1993a). It suggests that a newly synthesized protein is required for the induction of the PR gene. A soluble high-affinity, salicylic acid-binding protein (SABP22) was identified (Gorlach et al., 1996; Du and Klessig, 1997). Salicylic acid may be bound by a receptor and the binding may trigger a signal transduction cascade that has an ultimate effect on transcription factors that regulate PR gene expression. Several proteins may act as *trans*-acting factors in inducing PR proteins by salicylic acid.

Molecular Mechanism of Induction of PR Genes by Methyl Salicylate

Besides salicylic acid, methyl salicylate is naturally produced by a number of plants (Wilson et al., 1987; Loughrin et al., 1993). Bacterial pathogens induced accumulation of methyl salicylate in plants. When tobacco leaves were inoculated with the incompatible pathogen *Pseudomonas savastanoi* pv. *phaseolicola*, methyl salicylate concentration increased severalfold even at 12 h after inoculation (Seskar et al., 1998). Salicylic acid accumulation seems to be required for methyl salicylate production. Transgenic tobacco plants expressing the salicylate hydroxylase gene *(nahg)* from *Pseudomonas putida*, which converts salicylic acid to catechol, were unable to produce salicylic acid as well as methyl salicylate (Seskar et al., 1998). Methyl salicylate induced PR-1 transcripts in tobacco. NahG tobacco plants were exposed to gaseous methyl salicylate. Methyl salicylate treatment did not induce PR-1 transcripts in NahG plants, in contrast to its dramatic effect in wild-type plants (Seskar et al., 1998). The results suggest that methyl salicylate is synthesized from salicylic acid and methyl salicylate induces PR proteins.

Methyl salicylate has been shown to be a translocatable form of salicylic acid (Shulaev et al., 1997). Methyl salicylate accumulation was detected in healthy tobacco leaves located above the inoculated leaf. It has been detected in phloem exudates of TMV-inoculated leaves (Shulaev et al., 1997). Besides phloem translocation of nongaseous methyl salicylate, the gaseous form of it may be the signal inducing *PR* genes in an adjacent plant (Seskar et al., 1998). There is controversy over whether salicylic acid is a systemic signal. Salicylic acid occurs in parallel to or even preceding *PR* gene activation, and salicylic acid is detected in the phloem of pathogen-infected tobacco or cucumber leaves (Hammond-Kosack and Jones, 1996; Ryals et al., 1996; Wobbe and Klessig, 1996). Salicylic acid was detected in upper uninoculated leaves when lower leaves were inoculated with TMV (Shulaev et al., 1995). However, it was argued that salicylic acid may be simply translocated in parallel with an unknown signal molecule. The signal for the development of systemic acquired resistance moved out of *P. syringae*-infected cucumber leaves before any increase in salicylic acid level could be detected in the phloem sap (Rasmussen et al., 1991). This unknown signal may be methyl salicylate.

Systemin-Abscisic Acid-Jasmonic Acid-Methyl Jasmonate: A Complex Signaling System in Inducing PR Genes

Systemin has been characterized as a powerful signal in induction of PR proteins in tomato plants (Bergey et al., 1996). Systemin is an 18-mer peptide:

AVQSKPPSKRDPPKMQTD

(Abbreviations: A-alanine; V-valine; Q-glutamine; S-serine; K-lysine; P-proline; R-arginine; D-aspartate; M-methionine; T-threonine)

A systemically induced protein of 200 amino acids, prosystemin, is processed to systemin (McGurl et al., 1994). The functionality of prosystemin was demonstrated with tomato plants transformed with an antisense prosystemin cDNA. These transformed plants expressing the antisense gene accumulated low levels of the PR protein (PR-6 protein) in leaves of wounded plants (McGurl et al., 1994). Overproduction of prosystemin in transgenic tomato plants resulted in constitutive accumulation of PR-6 proteins (Bergey et al., 1996).

Systemin acts as a systemic signal (Schaller and Ryan, 1995). Systemin is nearly 1 million times more powerful than the oligosaccharides, in inducing the synthesis of defensive genes in excised tomato leaves (Bergey et al., 1996). Radiolabeled systemin, when applied to a fresh wound on tomato leaf, is transported into the apoplast and xylem elements within 30 min, loaded into the phloem, and transported out into the upper, unwounded leaves of the plant within 60 to 90 min (Pearce et al., 1993). ρ-Chloromercuribenzene sulfonic acid (PCMBS) is a potent inhibitor of phloem loading of sucrose and it inhibited the translocation of systemin (Narvaez-Vasquez et al., 1994). It suggests that systemin is translocated through phloem.

A systemin-binding protein (SBP50) of 50 kDa has been isolated from the plasma membrane of tomato leaves and it resembles proteases in the Kex2p-like prohormone convertase family (Schaller and Ryan, 1995). The N-terminus of systemin binds to SBP50, but residues critical for induction of proteinase inhibitor I (PR-6 protein) accumulation are located at the C-terminus. It appears that systemin binds to SBP50, and the C-terminus, including the essential MQTD motif, is then released by proteolysis to mediate proteinase inhibitor gene expression (Schaller and Ryan, 1995).

Systemin activates a lipid-based signaling cascade (Heitz et al., 1997). After systemin perception by an uncharacterized reception system, probably SBP50 (Wasternack and Parthier, 1997), linolenic acid is released from plant membranes (Farmer and Ryan, 1992). Supplying systemin to young tomato plants resulted in a rapid and transient accumulation of α-linolenic acid (Conconi et al., 1996). α-Linolenic acid is released from cell membrane lipids by activated phospholipase A (Chandra et al., 1996). Linolenic acid is the substrate for the enzyme lipoxygenase (LOX) which is the key enzyme activating the octadecanoid pathway (Ohta et al., 1991). A group of jasmonates originates biosynthetically from linolenic acid via an inducible octadecanoid pathway consisting of at least seven enzymatic steps (Farmer, 1994; Schaller and Ryan, 1995). By means of 13-lipoxygenase located within the chloroplasts, hydroperoxide dehydratase, allene oxide synthase, and allene oxide cyclase, 13(S)-hydroperoxylinolenic acid followed by 12-oxo-phytodienoic acid are formed and finally metabolized by reduction and peroxisomal β-oxidation steps into (+)-7-isojasmonic acid, a physiologically active substance that is rapidly converted to its stereoisomer, stable (–)-jasmonic acid (Sembdner and Parthier, 1993; Wasternack and Parthier, 1997).

The importance of the octadecanoid pathway in synthesis of jasmonic acid has been demonstrated by developing a tomato mutant that is deficient in the octadecanoid pathway. The mutant produced only low levels of jasmonic acid in response to wounding or systemin (Bergey et al., 1996). Chemical inhibitors of the octadecanoid pathway substantially reduce the induction of PR-6 proteins by systemin and linolenic acid, but not by jasmonic acid (Koiwa et al., 1997). It suggests that the octadecanoid pathway is important in biosynthesis of jasmonic acid, which induces PR proteins. Supplementation of a tomato mutant called *def1* with intermediates of the octadecanoid pathway indicated that *def1* plants are affected in a biosynthetic step between hydroperoxylinolenic acid and 12-oxo-phytodienoic acid. These *def1* plants fail to accumulate PR-6 proteins, which are induced by jasmonic acid (Howe et al., 1996), suggesting the importance of octadecanoid pathway in synthesis of jasmonic acid and PR proteins. The octadecanoid-derived signaling pathway in plants has a parallel system in animals, in which prostaglandins, prostacyclins, leukotriene, and tromboxans are synthesized; they originate from arachidonic acid in a pathway that is very similar to that of jasmonate biosynthesis-lipid-based defense strategy (Bergey et al., 1996).

Jasmonic acid, its methyl ester, and certain (L)-amino conjugates, glucose esters, and hydroxylated forms can all be detected in plants (Sembdner and Parthier, 1993). Among them, methyl jasmonate and jasmonic acid are the important signal molecules in inducing PR genes. However, the intermediates of jasmonate biosynthesis 12-oxo-phytodienoic acid and 15,16-dihydro-12-oxo:: phytodienoic acid also act as inducers of genes (Blechert et al., 1995).

Methyl jasmonate was first isolated and chemically identified as a fragrant constituent of the essential oil of *Jasminum grandiflorum,* and the free acid (jasmonic acid) was isolated with high yield from the culture filtrate of the fungus *Lesiodiplodia theobromae* (Sembdner and Parthier, 1993). Methyl jasmonate is also present in essential oils of *Rosmarinus officinalis* (Creelman and Mullet, 1995). Jasmonic acid and methyl jasmonate are detected throughout the plant, with the highest concentrations in growing tissues such as the shoot apex, root tips, immature fruits, and young leaves (Creelman and Mullet, 1995). Jasmonic acid content increased in plants infected with pathogens (Penninckx et al., 1996). Leaves from the *Arabidopsis acd2* mutant, which spontaneously develops necrotic lesions in its leaves, have jasmonic acid contents much higher than those detected in wild-type plants (Penninckx et al., 1996).

Methyl jasmonate/jasmonic acid induced several PR proteins in various plants (Ryan, 1992; Farmer and Ryan, 1992; Andresen et al., 1992). Methyl jasmonate and jasmonic acid powerfully activated the synthesis of proteinase inhibitors I and II in tomato (Farmer and Ryan, 1990, 1992). Application of jasmonate or methyl jasmonate induced expression of PR-6 genes in a variety of plant species (Bergey et al., 1996; Wasternack and Partheir, 1997). Jasmonate induces PR-1 proteins in tobacco leaves following infection by a virulent pathogen (Green and Fluhr, 1995).

Exogenous application of methyl jasmonate strongly induced the expression of the defensin gene PDF1.2 in *Arabidopsis* (Penninckx et al., 1996). The *Arabidopsis acd2* mutant plants, which show increased jasmonic acid content, accumulate high levels of the PR-1 transcript (Penninckx et al., 1996). Jasmonic acid induces HRGP in tobacco (Rickauer et al., 1997), and a thionin named THI2.1 in *Arabidopsis* (Epple et al., 1997). Jasmonate induces JIP-6, which is a member of the thionin class in barley leaves (Andresen et al., 1992). Jasmonic acid induces PR-1, PR-3, PR-5, and PR-9 proteins/mRNAs in rice (Table 4.2; Schweizer et al., 1997).

Genes encoding chalcone synthase, phenylalanine ammonia-lyase, and caffeic acid methyl transferase have been induced by jasmonate in barley (Gundlach et al., 1992). Methyl jasmonate induces a ribosome-inactivating protein in barley (Farmer and Ryan, 1992). The application of methyl jasmonate to young barley leaves regulates gene expression whereby the mRNAs are preferentially transcribed and translated (Reinbothe et al., 1993). Jasmonate activates genes involved in cell wall proteins (HRGPs) in soybean (Creelman et al., 1992) and enzymes of phytoalexin synthesis in different plants (Blechert et al., 1995).

The importance of jasmonic acid/methyl jasmonate in inducing PR proteins has been demonstrated by several workers. Induction of PR proteins in plants was correlated with a preceding increase in the concentration of endogenous jasmonates (Creelman and Mullet, 1995; Lehmann et al., 1995). Inhibitors of the jasmonate biosynthesis pathway, such as tetracyclasis, urosolic acid, ibuprofen, or mefenamic acid prevented an increase of endogenous jasmonates and induction of PR proteins (Xu et al., 1994; Lehman et al., 1995). Tetracyclasis inhibits jasmonic acid synthesis, and it inhibits PR-1 accumulation in rice. PR-1 induction could be rescued by exogenous jasmonic acid (Schweizer et al., 1997). Jasmonic acid content is high in an *Arabidopsis acd2* mutant, and in this mutant mRNAs of the defensin PDF1.2 accumulated much more than in wild-type plants (Penninckx et al., 1996). When endogenous levels of jasmonic acid increased, systemic accumulation of defensins in *Arabidopsis* was observed (Penninckx et al., 1996). In *Arabidopsis* mutants insensitive to methyl jasmonate (*coi* plants), PDF1.2

TABLE 4.2. Induction of PR Proteins in Rice by Jasmonic Acid

Gene product	**Induction by jasmonic acid**
PR-1 protein	++
PR-2 protein	+ to ++
PR-3 protein	++
PR-5 mRNA	++
PR-9 mRNA	++

Source: Schweizer et al., 1997.

+ weak induction; ++ strong induction

is not induced (Feys et al., 1994). It suggests the importance of jasmonate acting as molecule in induction of PR proteins.

Artemisia tridentata (sagebrush) plants, when placed in chambers with tomato plants, activated synthesis of PR-6 proteins in tomato leaves. The volatile compound in sagebrush leaves was isolated and identified as methyl jasmonate (Farmer and Ryan, 1990). Systemic induction of a potato *pin2* promoter by methyl jasmonate in transgenic rice plants has been reported (Xu et al., 1993).

Exogenous precursors of jasmonic acid such as linolenic acid, 13(S)-hydroperoxylinolenic acid, and phytodienoic acid cause proteinase inhibitor II (PR-6 protein) induction in tomato leaves (Table 4.3; Farmer and Ryan, 1992). It suggests that all of the enzymes responsible for jasmonate synthesis from linolenic acid are present in healthy tomato leaves. A key event in the signaling of inducible PR proteins may be the release of linolenic acid by the activation, liberation, or synthesis of a lipase (Farmer and Ryan, 1992). A G-box sequence (CACGTGG) in the region of –574 nucleotides in the *cis*-acting promoter in the potato *pin2* (proteinase inhibitor, PR-6 protein) gene was found essential for jasmonic acid response (Kim et al., 1992). A palindrome TGACG element within the *LOX1* (lipoxygenase) promoter was found to be essential for jasmonate inducibility (Rouster et al., 1997). This element has been identified as a binding site for bZIP *trans*-acting factors (Rouster et al., 1997). Methyl jasmonate enhances the activity of promoters containing *as-1* type elements (e.g., CaMV 35S promoter and nopaline synthase promoter). These treatments have the same effect on induction of PR-1 genes. It is possible that the effect of methyl jasmonate on the expression of PR-1a gene may be at least partly mediated

TABLE 4.3. Induction of Proteinase Inhibitor II in Response to Octadecanoids and Jasmonic Acid in Tomato Leaves

Treatment	Proteinase inhibitor II (μg/g tissue)
Control	0
Linolenic acid	94
13(S)Hydroperoxylinolenic acid	60
Phytodienoic acid	186
Jasmonic acid	188

Source: Farmer and Ryan, 1992.

through binding of TGA1a to the *as-1*-like element present in the PR-1a promoter (Buchel and Linthorst, 1999).

The jasmonic acid and methyl jasmonate signal pathway may involve some other systemic signals such as abscisic acid (ABA). Systemin application induces accumulation of linolenic acid (Conconi et al., 1996). Systemin caused an increase in endogenous ABA (Herde et al., 1996) and jasmonic acid levels preceding PR-6 gene expression (Pena-Cortes et al., 1995). Induction of PR-6 genes by systemin does not occur in ABA-deficient tomato plants, but treatment with linolenic acid induces the PR-6 genes in these mutants, indicating that ABA may function at a step between systemin perception and the octadecanoid pathway (Pena-Cortes et al., 1996).

ABA itself has been shown as a systemic signal molecule in inducing PR protein genes. PR-5 (osmotin) gene expression and protein synthesis were upregulated in young tomato plants after a short exposure to 100 μm ABA (Grillo et al., 1995). ABA supplied to the petioles of detached leaves also induced proteinase inhibitor II (PR-6) mRNA accumulation in potato leaf blades, and spray treatment of potato leaves resulted in local and systemic induction of proteinase inhibitor II mRNA accumulation (Pena-Cortes et al., 1995). Wounding did not induce proteinase inhibitor mRNA accumulation in ABA-deficient tomato mutants, but induction occurred in these plants if ABA was exogenously supplied (Pena-Cortes et al., 1996). ABA induces the synthesis of osmotin (PR-5) in tobacco and several other species (Singh et al., 1987). ABA induces mRNA encoding osmotin in cultured tobacco (*N. tabacum* cv. Wisconsin 38) (Singh et al., 1989). ABA induces a PR-5 gene in *Solanum commersonii*. The gene was suppressed by fluridone, an ABA synthesis inhibitor, and the suppression was restored by exogenous ABA application (Zhu et al., 1993a). ABA may function at a point between the systemin and linolenic acid signaling pathway (Pena-Cortes et al., 1996). However, Dammann et al. (1997) have shown that the ABA signal may be transduced by a jasmonic acid-independent pathway that has protein kinase as an intermediate.

Ethylene may also be involved in the complex systemin-abscisic acid-jasmonate signal transduction pathway in inducing PR proteins. Norbornardiene and silver thiosulfate, both inhibitors of ethylene, could each inhibit PR-6 mRNA accumulation induced by jasmonate (O'Donnell et al., 1996). It suggests that ethylene functions downstream of or in parallel with jasmonate.

Jasmonic acid/methyl jasmonate may not induce certain PR proteins in some plants. Methyl jasmonate did not induce PR-1, PR-2, and PR-5 proteins in tobacco when this compound is administered alone (Heitz et al., 1999). It did not induce TIMP genes encoding proteinase inhibitors in tobacco. It stimulated a proteinase inhibitor (MPI) gene but not PR-1 mRNA in maize (Cordeo et al., 1994). Salicylic acid may act as an antagonist of jasmonic acid. Salicylic acid inhibited systemin- or jasmonate-induced PR protein synthesis in tomato (Wasternack and Parthier, 1997). It suggests that salicylic acid negatively regulates the octadecanoid pathway or jasmonate function.

Oligogalacturonides As Signal Molecules Inducing PR Genes

Oligogalacturonides derived from plant cell walls induce PR proteins (Bishop et al., 1981; Farmer et al., 1991). Oligouronides are produced at pathogen sites (Ryan, 1992). *P. syringae* pv. *tomato* induces these signals during pathogenesis in tomato (Pautot et al., 1991). The oligogalacturonides, homopolymers of α-1,4-linked D-galacturonic acid, are an important class of plant cell wall-derived signals. These molecules derive from a parent polysaccharide, homogalacturonan, which resides in the pectic matrix (Reymond et al., 1995).

Oligo- and polygalacturonic acids of DP (degree of polymerization) = 2 through DP = 20 are active inducers of PR-6 proteins in tomato leaves (Bishop et al., 1984). Reymond et al. (1995) reported that oligogalacturonides must have a degree of polymerization of 10 to 14 to be biologically active. Both di- and trigalacturonic acids have been shown to be active inducers of PR-6 proteins in tomato leaves (Ryan, 1992). Oligogalacturonides stimulate ion flux (Thain et al., 1990) and an oxidative burst in which G proteins may participate (Legendre et al., 1993). Linolenic acid hydroxylase has been shown as an intermediate in the oligogalacturonide-stimulated expression of PR-6 genes in tomato leaves (Farmer et al., 1994). Oligogalacturonides induce phosphorylation of proteins (Farmer et al., 1989, 1991). Many biological responses to oligogalacturonides are characterized by a dependence on the degree of polymerization of this ligand. Generally 10 or more residues long are necessary for activity. The ligands which activate the defense responses may be intermolecular "egg box" complexes of oligogalacturonides and calcium ions, known to form in solution with oligogalacturonates of DP > 9 (Reymond et al., 1995).

Auxins and Cytokinins

Auxins and cytokinins are known to induce synthesis of PR proteins in various plants (Kernan and Thornburg, 1989). Exogenous application of kinetin induced the synthesis of PR-1 and PR-3 mRNAs in tobacco (Memelink et al., 1987). A PR-2-like protein from *Nicotiana glutinosa* was induced by auxin treatment (de Loose et al., 1988). A group of four PR-2 proteins were found induced in IAA-treated bean leaves (Clarke et al., 1998). Cytokinins control the synthesis of jasmonic acid and salicylic acid in tobacco (Sano et al., 1996). Transgenic tobacco plants expressing a gene encoding a GTP-binding protein displayed high endogenous cytokinin concentrations (Sano and Ohashi, 1995). When these plants were wounded, they accumulated salicylic acid and jasmonic acid (Sano et al., 1994). Thus auxins and cytokinins may induce PR proteins.

Complexity of Signal Transduction Systems

In various host-pathogen systems, different signaling systems are induced. Induction of different PR proteins may require different signaling systems. Salicylic acid induced PR-1 and class II PR proteins in tobacco; but in contrast, the class I PR-2 proteins are not induced. Tissues of transgenic tobacco plants expressing cholera toxin accumulate high levels of salicylic acid and constitutively express PR protein genes including PR-1 and class II PR-2 protein genes, but not class I PR-2 protein genes (Beffa et al., 1995). Inoculation of *Erwinia carotovora* subsp. *carotovora* or treatment with its culture filtrate containing elicitor induced accumulation of transcripts for class I PR-2 proteins and not acidic PR-1a (Vidal et al., 1997). Induction of class I PR-2 transcripts in salicylic acid-deficient transgenic *NahG* tobacco plants and wild-type plants in response to the culture filtrate was comparable (Vidal et al., 1997). These two studies suggest that salicylic acid is not responsible for induction of class I PR-2 proteins in tobacco. Contrastingly, ethylene induces class I PR-2 proteins in tobacco (Ecker, 1995; Penninckx et al., 1996). Acidic PR-1a is induced by salicylic acid but not by ethylene in tobacco (Eyal et al., 1992).

Methyl jasmonate induced proteinase inhibitor (PR-6 protein), but not PR-1, PR-2, and PR-5 proteins in tobacco (Heitz et al., 1999). Methyl jasmonate stimulated a proteinase inhibitor (MPI) gene but not PR-1 mRNA in maize (Cordero et al., 1994). Methyl jasmonate or ethylene induced the expression of defensin gene *PDF1.2,* but salicylic acid does not induce the gene in *Arabidopsis* (Penninckx et al., 1996; Epple et al., 1997). Jasmonate-insensitive *coi1* and ethylene-insensitive *ein2 Arabidopsis* mutants are impaired in their ability to induce the defensin gene *PDF1.2* upon pathogen attack (Penninckx et al., 1996). But salicylic acid-deficient mutants are not impaired in their ability to induce the defensin gene (Thomma et al., 1998). Jasmonic acid and ethylene induced the *Arabidopsis* thionin *THI2.1;* but salicylic acid could not (Epple et al., 1995, 1997). Jasmonate and ethylene response pathways appear to act in parallel and not in sequence for the activation of the defensin PDF1.2 in *Arabidopsis* (Penninckx et al., 1998).

The salicylic acid-dependent pathways activate class II and III β-1,3-glucanases (PR-2 proteins). These pathways appear to be ethylene-independent in tobacco and *Arabidopsis* (Silverman et al., 1993; Lawton et al., 1994b; Ryals, 1996). Ethylene and not salicylic acid induces class I chitinase (Ecker, 1995; Penninckx et al., 1996). Ethylene induced basic, but not acidic type, PR-1 transcript in tobacco. But salicylic acid induced both basic-type *prb-1* and acidic-type *PR-1* transcript accumulation in tobacco (Eyal et al., 1992).

Salicylic acid induces accumulation of class II and III PR-2 proteins in tobacco (Ward et al., 1991; Van de Rhee et al., 1993; Niki et al., 1998). But transcripts of tobacco class I PR-2 and PR-3 proteins are not induced or are weakly induced in response to salicylic acid (Linthorst et al., 1990; Vidal et al., 1997; Niki et al., 1998). Salicylic acid induces several PR-10 proteins such as AoPR1 (asparagus PR protein) and SAM22 (soybean PR protein) (Growell et al., 1992; Warner et al., 1994). But another PR-10 protein of rice, PBZ1, was not induced by salicylic acid (Midoh and Iwata, 1996). Salicylic acid does not induce PR proteins in rice while jasmonic acid induces them (Schweizer et al., 1997).

Synergism between various signals has been reported. The combination of signal molecules of ethylene and methyl jasmonate is known to cause substantial accumulation of osmotin protein in the tobacco cultivar Wisconsin 38 (La Rosa et al., 1992; Xu et al., 1994). Antagonism between signals also has been reported. Salicylic acid inhibited synthesis of PR-6 protein in tomato leaves induced by systemin and jasmonic acid (Doares et al., 1995; Wasternack and Partheir, 1997). Salicylic acid is an inhibitor of the octadecanoid pathway (Bergey et al., 1996). *Erwinia* elicitor signal molecules antagonize the salicylic acid-mediated induction of PR genes. Similarly, salicylic acid appeared to inhibit the induction of PR genes elicited by *Erwinia carotovora* subsp. *carotovora* (Vidal et al., 1997).

Some signals may be dependent on other signals for their action, and they may be active upstream or downstream of other signals. Ethylene may function downstream of jasmonates in the signal transduction pathway (O'Donnell et al., 1996; Botella et al., 1996). Both oligosaccharide and systemin application increase jasmonate levels in tomato plants (Doares et al., 1995; Bergey et al., 1996). It suggests that they may be active at upstream of jasmonates. Abscisic acid may function at a step between systemin perception and the octadecanoid pathway inducing jasmonic acid (Pena-Cortes et al., 1996). Salicylic acid may act downstream of H_2O_2 (Bi et al., 1995; Neuenschwander et al., 1995; Summermatter et al., 1995; Leon et al., 1995). Methyl salicylate may act downstream of salicylic acid (Seskar et al., 1998). The salicylic acid-dependent pathway may act independently of ethylene (Silverman et al., 1993; Lawton et al., 1994b; Beffa et al., 1995; Ryals, 1996).

Some of the signals may downregulate the expression of some PR genes. Ethylene induces β-1,3-glucanase in tobacco (Vogeli-Lange et al., 1994a). But both auxin and cytokinin reduced the expression of β-1,3-glucanase. It was suggested that β-1,3-glucanase promoter may have elements for downregulation by auxin and cytokinin (Vogeli-Lange et al., 1994a). Accumulation of β-1,3-glucanase in tobacco cultured cells requires ethylene; but it is blocked by combinations of auxin and cytokinin added to the medium (Felix and Meins, 1987).

Some plants may induce PR proteins through some uncharacterized signals. Benzothiadiazole (BTH) treatment induces accumulation of mRNAs from genes encoding PR-1, PR-2, and PR-5 in *Arabidopsis thaliana* (Lawton et al., 1996). BTH treatment induced PR-1 mRNA in the ethylene-insensitive mutants *etr1* and *ein2,* in the methyl jasmonate-insensitive mutant *jar1,* and in transgenic plants expressing the *nahG* gene (suppressing salicylic acid accumulation). It suggests that BTH does not require ethylene or methyl jasmonate or salicylic acid to induce PR proteins. But BTH treatment failed to induce PR-1 mRNA accumulation in the noninducible immunity mutant *nim1* (Lawton et al., 1996). Hence, it appears that BTH activates the signal transduction pathway through mechanisms. Messenger RNA of *PBZ1* gene encoding PR-10 protein in rice was not induced after treatment with ethephon (ethylene releasing agent), salicylic acid and NAA (2-(1-naphthyl) acetic acid, which induces ethylene production). It suggests that these signal systems are not involved in *PBZ1* gene transcription.

COMPARTMENTALIZATION OF PR PROTEINS IN PLANT TISSUES

Bacterial pathogens multiply in specific tissues of infected plants. Most of them multiply in the extracellular space near the stomata, in the apoplast around mesophyll cells in epidermis, and in the vascular tissue. PR proteins expressed in these regions may have direct effects on bacterial pathogens. But PR proteins accumulate in various types of cells, and hence some of them may not have a direct role on bacterial disease resistance. Knowledge on location of PR proteins in plant tissues will help to develop technology to exploit PR proteins for bacterial disease management.

Several PR proteins accumulate in vacuoles of cells in the epidermis of plants treated with biotic and abiotic agents (Ohashi and Matsuoka, 1987; Mauch et al., 1992). Seven basic PR proteins were localized in vacuoles in tobacco leaves infected with TMV (Van Loon and Gerritsen, 1989). A basic chitinase was found localized in vacuoles in bean leaves (Boller and Vogeli, 1984; Broglie et al., 1986). Mauch et al. (1992) found significant accumulation of basic chitinases and β-1,3-glucanases in vacuoles of cells of the lower epidermis and in cells adjacent to the vascular tissue in tobacco. This accumulation was strictly correlated with the formation of protein aggregates in vacuoles of these cells. Wubben et al. (1992) observed accumulation of certain chitinases and β-1,3-glucanases in protein aggregates in the vacuoles in tomato leaves infected with pathogens. The osmotin (PR-5) protein accumulated in vacuoles in potato (Liu et al., 1996). Osmotins in tobacco are localized in vacuoles (Singh et al., 1987; Linthorst, 1991). The PR-5 protein AP24 was shown to be retained in vacuoles in tobacco (Melchers et al., 1993).

Acidic PR proteins normally accumulate in the apoplast while basic proteins are localized in the vacuole. PR proteins belonging to the acidic PR-1 family have been localized in the apoplast (Carr et al., 1987; Ohashi and Matsuoka, 1987; Hosokawa and Ohashi, 1988; Dumas et al., 1988). The extracellular acidic PR proteins were located in the extracellular space near the stomata and the mesophyll cells (Mauch et al., 1992).The acidic chitinases, β-1,3-glucanases, and thaumatin-like proteins were found to accumulate in extracellular pocketlike vesicles in 'Samsun NN' tobacco infected with TMV (Dore et al., 1991). The acidic PR protein PvPR1 was localized primarily in the extracellular space in bean leaves (De Tapia et al., 1986). An acidic chitinase was expressed in all cell types associated with the vascular bundles, especially in the phloem in potato leaves after inoculation with an incompatible bacterium, *Pseudomonas syringae* (Stromberg, 1996). PR protein genes such as *PR-1* and *win* also show expression in vascular bundles (Carr et al., 1987; Liang et al., 1989; Stanford et al., 1990). *PR-10a* expression in potato was associated with vascular bundles (Constabel and Brisson, 1995). In leaves, *PR-10a* expression was observed only in veins (Constabel and Brisson, 1995).

Some basic PR proteins have been shown to be secreted into the apoplast. Two basic PR proteins of barley, Hv PR-1a and Hv PR-1b, were detected in the intercellular fluid of barley leaves (Bryngelsson et al., 1994). Some of the PR proteins are found to accumulate in the cytoplasm. The acidic bean PR proteins PvPR2 and PvPR3 were localized to the cytoplasm (De Tapia et al., 1986; Sharma et al., 1992). Intracellular location of the acidic PR proteins STH-2 in potato, AoPR1 in asparagus, PoPR1 in parsley, and SAM22 in soybean has been reported (Somssich et al., 1986; Walter et al., 1990, 1993; Warner et al., 1994; Midoh and Iwata, 1996).

The mechanism of vacuolar targeting of basic PR proteins has been studied in detail with an aim of developing methods to make the basic PR proteins secrete into the apoplast so that the pathogens, which multiply in the apoplast, can be effectively controlled. A C-terminus 7 amino acid peptide was discovered to be required and sufficient for vacuolar targeting of a basic chitinase in tobacco (Neuhaus et al., 1991; Melchers et al., 1993). Further analysis of this C-terminal peptide indicated that deletion of the C-terminal methionine did not affect the intracellular location, but deletion of even a single internal amino acid caused predominantly extracellular secretion of chitinase (Neuhaus et al., 1994b). Substitution of amino acids within this peptide, including the C-terminal extension, with random sequences, has intermediate effects that cover the whole range from retention to secretion (Neuhaus et al., 1994b).

A vacuolar targeting signal may be present at the C-terminal end of basic PR-1, which contains a C-terminal extension compared to its acidic counterpart (Cornelissen et al., 1986). A 22 amino acid C-terminal extension was identified as the vacuolar targeting signal of β-1,3-glucanase in tobacco (Melchers et al., 1993). A PR-5 protein, AP24, was shown to contain a C-terminal sequence required for intracellular retention in tobacco (Melchers et al., 1993). The C-terminal peptide appeared to correspond to the amino acid sequence deduced from nucleotides 1102 to 1123, indicating that AP24 is processed C-terminally. Thus, the processing of tobacco AP24 protein includes the removal of both an N-terminal signal peptide of 25 amino acids and a C-terminal propeptide of 18 amino acids (Melchers et al., 1993).

Basic, vacuolar forms of PR-5 proteins have a C-terminal extension compared to the acidic forms, which are secreted extracellularly.

Transgenic tobacco and potato plants overexpressing osmotin with or without the 20 C-terminal amino acids were developed (Liu et al., 1996). Plants overexpressing an osmotin gene with a complete open reading frame accumulated osmotin mostly in an intracellular compartment, probably the vacuole. In contrast, in plants overexpressing a C-terminal 20 amino acid truncated gene, osmotin was totally secreted into the extracellular matrix.

Comparative sequence analysis between intracellular and extracellular isoforms of chitinases, β-1,3-glucanases, and PR-5 proteins have shown that, in general, intracellular proteins contain a C-terminal extension com-

Comparison of C-terminal amino acid sequences of basic and acidic forms of PR-5 proteins

Basic forms

NP24 (tomato)
AYSYPGDDPTSFTCPGGSTNYKLT.*FCPNGGAHPNFPLEMPGSDEVAK*
Osmotin (tobacco)
AYSYPGDDPTSTFTCPGGSTNYRVL.*FCPNGGAHPNFPLFPLEMPGSDEVAK*

Acidic forms

PR-5 *(Arabidopsis)* AYSYAYDDPTSTFTC. . . NVEIT. . . FCP
PR-5 (tobacco) AYSYPGDDPTSLFTCPSG. . . TNYRVVFCP
MAI (maize) AYSYPKDDPTSTFTCPAG. . . TNYKVVFCP

Abbreviations: A-alanine; Y-tyrosine; S-serine; P-proline; G-glycine; D-aspartate; T-threonine; F-phenylalanine; C-cysteine; N-asparagine; K-lysine; L-leucine; H-histidine; E-glutamate; M-methionine; R-arginine; I-isoleucine

Source: Richardson et al., 1987; Payne et al., 1988; King et al., 1988; Singh et al., 1989; Uknes et al., 1992.

pared to their extracellular homolog (Cornelissen et al., 1986; Van den Bulcke et al., 1989; Linthorst et al., 1990a,b). Deletion of C-terminal regions of 20, 7, and 22 amino acids resulted in the secretion of basic PR-5 protein AP24, basic chitinase, and basic β-1,3-glucanase, respectively, in tobacco (Melchers et al., 1993). It seems that the C-terminal propeptide of AP24, basic chitinase, and basic β-1,3-glucanase are necessary for efficient sorting of these proteins to vacuoles. The vacuolar proteins, barley lectin and β-1,3-glucanases of tobacco, and *Nicotiana plumbaginifolia* are processed to their mature form by the removal of a glycosylated C-terminal propeptide (CTPP) (Raikhel and Wilkins, 1987; Shinshi et al., 1988; Van den Bulcke et al., 1989; Wilkins et al., 1990). The propeptide of barley lectin (Bednarek et al., 1990) and tobacco β-1,3-glucanase (Melchers et al., 1993) have been shown to be required for vacuolar targeting. The CTPP from barley lectin and tobacco chitinase could redirect a secreted protein, cucumber chitinase, to vacuoles of the cells of the transgenic tobacco plants (Bednarek and Raikhel, 1991). These studies suggest that all vacuolar PR proteins may contain a C-terminal propeptide that is necessary for proper targeting of these proteins and is removed during or after transport to the plant vacuole. Some basic proteins such as basic proteins of barley, HvPR-1a and HvPR-1b, have been shown to be secreted into the apoplast, and they lack C-terminal sequences (Bryngelsson et al., 1994).

THE ROLE OF PR PROTEINS IN BACTERIAL DISEASE RESISTANCE

PR Proteins Are Associated with Disease Resistance

Although several PR proteins have been shown to be induced due to bacterial infection, the exact role of these proteins is not yet known. But much indirect evidence shows that PR proteins are involved in disease resistance. In the first instance, several studies have revealed that PR proteins are associated with incompatible interactions. When rice leaves are inoculated with a compatible pathogen *P. syringae,* PR-1, PR-2, PR-3, PR-5, and PR-9 proteins and their mRNAs were induced (Smith and Metraux, 1991; Schweizer et al., 1997). *P. syringae* pv. *tomato,* an incompatible pathogen, induces PR proteins in radish (Hoffland et al., 1996). PR proteins, including peroxidase, β-1,3-glucanase, and chitinases, were induced in cucumber plants inoculated with *P. syringae* pv. *syringae* strain 61 (Strobel et al., 1996). A thaumatin-like protein was induced in rice leaves inoculated with an incompatible pathogen, *P. syringae* pv. *syringae* (Reimann and Dudler, 1993). When potato leaves were infiltrated with the incompatible bacterial pathogen *P. syringae,* mRNA and the protein of acidic chitinase were expressed in all cell types of the vascular bundles, especially the phloem, in leaves within 72 h after inoculation (Stromberg, 1996). Infection of tobacco plants with *Erwinia carotovora* subsp. *carotovora* leads to the induction of genes encoding PR proteins (Vidal et al., 1997). *Ralstonia solanacearum* strain R296 induces large amounts of PR-1 proteins in infiltrated tobacco leaf tissues (White et al., 1996). Thus, in several incompatible interactions, accumulation of PR proteins is the distinct feature.

PR Proteins Are Induced Rapidly and to a Greater Degree in Resistant Interactions

Several studies have revealed that PR proteins are induced to a greater degree and more rapidly in resistant interactions. In leaves of the French bean *(Phaseolus vulgaris)* cultivar 'Red Mexican' inoculated with cells of the avirulent race 1 isolate of *P. savastanoi* pv. *phaseolicola* (incompatible combination), an increase in chitinase activity was detected between 6 and 9 h after inoculation. Chitinase activity was well established by 12 h, and by 48 h after inoculation was elevated almost 19-fold over basal levels. In contrast, in leaves of 'Red Mexican' inoculated with cells of the virulent race 3 isolate of *P. savastanoi* pv. *phaseolicola* (compatible combination), chitinase activity was still minimal by 24 h after inoculation, and even by 48 h the chitinase activity was only 75 percent of that in the resistant reaction (Voisey and Slusarenko,1989). Northern blot analysis of total DNA revealed an increase in hybridizable chitinase mRNA in the incompatible combination beginning between 3 and 6 h after inoculation. In the susceptible response, chitinase mRNA increased between 18 and 24 h after inoculation (Voisey and Slusarenko, 1989).

Pepper reproductive tissues (fruits, flowers) are less affected by *Xanthomonas vesicatoria* than vegetatative tissues (foliage) (O'Garro and Charlemange, 1994). Activity of β-1,3-glucanase (O'Garro and Charlemange, 1994) and chitinase increased over time in pepper leaves and flowers after inoculation. The sharpest increases in enzyme activity occurred between 1 and 4 days in flowers and 5 and 9 days in leaves. Maximal levels of enzyme activity in flowers and leaves were detected within 8 and 12 days and 16 and 20 days, respectively (O'Garro and Charlemange, 1994). The early induction of both β-1,3-glucanase and chitinase in resistant interactions suggests the involvement of these two PR proteins in bacterial disease resistance.

Differential accumulations of proteinase inhibitor II mRNA were observed in the bacterial speck disease *(P. syringae* pv. *tomato)*-resistant and -susceptible tomato lines (Pautot et al., 1991). Proteinase inhibitor II mRNAs were not detected in healthy leaves. Proteinase inhibitor II mRNA accumulated more rapidly in disease-resistant than in susceptible plants. In disease-resistant leaves, proteinase inhibitor II mRNA began to accumulate to higher levels between 12 and 24 h after infection. In contrast, there was a rapid increase in proteinase inhibitor II mRNA between 24 and 36 h after inoculation in disease-susceptible leaves (Pautot et al., 1991). When pepper (*Capsicum* sp.) leaves were inoculated with *X. vesicatoria,* several acidic and basic chitinases appeared (Lee and Hwang, 1996). When tomato leaves were inoculated with compatible and incompatible strains of *X. vesicatoria,* several PR proteins accumulated, especially in the incompatible interaction. Proteins with different molecular weight (21-29 kDa) and pI 8-9 were detected in the incompatible interaction (Kim and Hwang, 1995).

Various isolates of *Ralstonia solanacearum,* infiltrated into tobacco leaves, led to the isolation of 14 cDNA clones corresponding to different gene families whose mRNAs accumulated during the resistant interaction (Marco et al., 1990). The level of accumulation of mRNAs corresponding to many of the cDNA clones is 3- to 10-fold higher in incompatible interactions than that observed in compatible interactions (Marco et al., 1990). *R. solanacearum* GMI1000 and K60 isolates, incompatible and compatible in tobacco, were used to inoculate a cell suspension of *Nicotiana tabacum* cv. Wisconsin (Godiard et al., 1991). mRNAs encoding new proteins reached a maximum level which is 3- to 7-fold higher in tobacco cells undergoing an HR (inoculated with GMI1000) than in those inoculated with the K60 strain (Godiard et al., 1991). Activity of *Gn1* (β-1,3-glucanase) promoter of *N. plumbaginifolia* is strongly induced (about 21-fold) in tobacco leaves infected with the incompatible bacterium *P. syringae* pv. *syringae* (Alonso et al., 1995; Castresana et al., 1990). Induction of this promoter is much weaker in leaves infected with the compatible bacterium *Erwinia carotovora* subsp. *carotovora* (Castresana et al., 1990). In the compatible interaction between turnip (*Brassica campestris*) and *X. campestris* pv. *campestris,* the causal agent of black rot of crucifers, transcripts for β-1,3-glucanase accumulated up to a maximum level at 24 h after inoculation. This was a slower response than was seen in incompatible interactions with *X. campestris* pvs. *armoraceae* or *raphani* (Newman et al., 1994).

Early and high amounts of induction of PR proteins may retard the multiplication of the bacterial pathogens. However, in vivo inhibition of bacteria by PR proteins has not yet been demonstrated.

Induced Systemic Resistance and Systemic Acquired Resistance Against Bacterial Pathogens Are Associated with Accumulation of PR Proteins

When a plant is inoculated with an incompatible or sometimes even compatible pathogen or treated with abiotic inducers, systemic induction of resistance against pathogens has been seen. This type of resistance is called induced systemic resistance (ISR) or systemic acquired resistance (SAR). This resistance has been shown to be related to the accumulation of PR proteins in many plants (Metraux and Boller, 1986; Metraux et al., 1989; Schweizer et al., 1989; Tuzun et al., 1989; Irving and Kuc, 1990; Ye et al., 1990; Bredorde et al., 1991; Pan et al., 1991, 1992; Rebmann et al., 1991a; Ward et al., 1991; Enkerli et al., 1993; Cohen et al., 1994; Ji and Kuc, 1995). In *Arabidopsis thaliana,* infection of leaves with turnip crinkle virus was shown to induce resistance against *Pseudomonas syringae* DC3000. At the same time that resistance was established, mRNA from three PR protein genes, PR-1, PR-2, and PR-5, co-ordinately accumulated to high levels in uninfected leaves (Uknes et al., 1993). 2,6-Dichloroisonicotinic acid (INA) induced systemic resistance in tobacco against *P. syringae* pv. *tabaci* (Vernooij et al., 1995). In INA-treated tissues, PR-1 gene was induced between 8 and 16 h after application. Accumulation of PR-1 mRNA reached a plateau within 24 h and remained constant for another 8 h (Vernooij et al., 1995).

HrpZ$_{Pss}$ protein (= harpin$_{Pss}$), an elicitor isolated from *P. syringae pv. syringae,* induced SAR in cucumber to *P. syringae* pv. *lachrymans* (angular leaf spot bacterium). PR proteins, including peroxidase, β-1,3-glucanase, and chitinases, were induced in cucumber plants treated with the elicitor (Strobel et al., 1996). *P. syringae* (incompatible bacterium) induces SAR in potato against pathogens, and 72 h after SAR induction, mRNA and protein of acidic chitinase were expressed in leaves above the inducer leaves (Stromberg, 1996). Cucumber plants, when inoculated with *P. syringae* pv. *lachrymans* on leaf 1, showed SAR against pathogens on leaf 2. Chitinase was first detected after 12 and 18 h in leaves 1 and 2, respectively (Meuwly et al., 1994). Benzothiadiazole (BTH) induces systemic disease resistance in *Arabidopsis thaliana.* BTH-treated plants were resistant to infection by *P. syringae* pv. *tomato* DC3000 (Lawton et al., 1996). BTH treatment induced accumulation of mRNAs from PR genes *PR-1, PR-2,* and *PR-5* (Lawton et al., 1996). Three PR proteins were found to accumulate to high levels in intercellular washing fluid from leaves of *Arabidopsis* sp. in response to INA treatment, and resistance to pathogenic *P. syringae* pv. *tomato* was induced (Uknes et al., 1992). The same proteins were strongly induced by infection with *P. syringae* pv. *tomato* (Uknes et al., 1992).

Importance of PR Proteins in Bacterial Disease Resistance Is Demonstrated by Developing Transgenic Plants

Another way of demonstrating the importance of PR proteins in bacterial disease resistance is by developing transgenic plants. Tobacco plants were transformed with the mutant *etr-1* gene from *Arabidopsis,* conferring dominant ethylene insensitivity (Knoester et al., 1998). These transgenic plants hardly expressed basic *PR-1g* and *PR-5c* genes. When these basic PR proteins were suppressed, the plants became susceptible to *P. syringae* pv. *tomato* DC3000 (Knoester et al., 1998). Transgenic tobacco plants expressing the cholera toxin gene express PR genes and show enhanced resistance to the bacterial pathogen *P. syringae* pv. *tabaci* (Beffa et al., 1995).

Antimicrobial Action of PR Proteins

Several PR proteins such as PR-1, PR-2, PR-3, PR-4, and PR-5 have been shown to inhibit growth of fungi (Woloshuk et al., 1991; Vigers et al., 1992; Sela-Buurlage et al., 1993; Niderman et al., 1995; Abad et al., 1996; Batalia et al., 1996; Ji and Kuc, 1996; Lawrence et al., 1996; Maggio et al., 1996; Kim and Hwang, 1996). Antiviral activity of some PR proteins (tobacco gp35 (a β-1,3-glucanase) and gp22 (a PR-5 protein) has also been reported (Edelbaum et al., 1991).

Inhibitory action of some PR proteins on multiplication of bacterial pathogens has been reported. Some chitinases have inhibitory action against bacterial pathogens. Chitinase hydrolyzes the β-(1,4)-glycosidic bond in chitin, a polymer of N-acetyl-glucosamine. Chitin has features in common with the glycan portion of peptidoglycan, a major constituent of gram-positive and gram-negative bacterial cell walls. Peptidoglycan comprises alternate β-(1,4)-linked N-acetyl-muramic acid residues and is hydrolyzed by lysozyme. Some chitinases have lysozyme activity (Boller et al., 1983; Roberts and Selitrennikoff, 1988). PR-8 proteins (class III chitinases) possess lysozyme activity (Metraux et al., 1989). Chitinases have been shown to be associated with bacterial disease resistance in many plants (Voisey and Slusarenko, 1989; O'Garro and Charlemange, 1994).

Toxicity of thionins (PR-13 proteins) against bacteria has been reported (Fernandez de Calya et al., 1972; Cammue et al., 1992). The common denominator for these toxic effects seems to be a destruction of membranes. The amphipathic structure of the thionins indicates that the toxicity might be exerted by a direct, detergent-like interaction with the lipid bilayers of biological membranes (Bohlmann, 1994). Thionins may bind to a membrane receptor, probably a specific phospholipid (Bohlmann, 1994). Fernandez de Calya et al. (1972) found the growth of several phytopathogenic bacteria to be inhibited by purothionins. Cammue et al. (1992) showed that β-purothionin is toxic against some phytopathogenic bacteria. Purothionins from wheat are bactericidal (Stuart and Harris, 1942). Thionins may inhibit certain enzymes of bacteria and certain cysteine-containing proteins such as cytochrome *c* (Pineiro et al., 1995) and inhibit growth of bacteria. A 5 kDa polypeptide, pseudothionin *Solanum tuberosum* 1 (Pth-St1), was detected in flowers, tubers, stems, and leaves of potato. Magnesium chloride induced higher mRNA levels of the protein. It was active against potato pathogens *Clavibacter michiganensis* subsp. *sepedonicus* and *Ralstonia solanacearum* (Moreno et al., 1994). The Pth-st1 is structurally similar to thionins (Moreno et al., 1994). Some plant defensins have been reported to have antibacterial activity. A potato tuber defensin inhibits *Ralstonia solanacearum* and *Clavibacter michiganensis* (Moreno et al., 1994).

LTPS (PR-14 proteins) have been shown to inhibit bacterial plant pathogens in vitro (Terras et al., 1992a; Molina et al., 1993a,b; Segura et al., 1993). However, there were considerable differences in sensitivity to LTPs among different pathogenic bacterial species or even among strains or isolates of the same species (Molina and Garcia-Olmedo, 1991; Molina et al., 1993b; Segura et al., 1993). When barley LTP2 protein was simultaneously applied with *P. syringae* pv. *tabaci* to the leaf surface of tobacco, the disease symptom development was suppressed. The degree of disease suppression by LTP2 was in a concentration-dependent manner (Molina and Garcia-Olmedo, 1997). LTP-sensitive mutants of *Ralstonia solanacearum* have been shown to be avirulent (Molina and Garcia-Olmedo, 1997). The defense potential of LTPs was evaluated by developing transgenic tobacco plants that express the LTP2 from barley (Molina and Garcia-Olmedo, 1997). Cell-wall extracts from LTP2 transgenic tobacco plants inhibited growth of *P. syringae* pv. *tabaci* in vitro at concentrations at which extracts from nontransformed tobacco had no effect (Molino and Garcia-Olmedo, 1997). Growth of *P. syringae* pv. *tabaci* on the leaves of transgenic tobacco plants expressing LTP2 was retarded with respect to that in nontransformed leaves. The transgenic plants presented significantly lower percentages of necrotic lesions (Figure 4.2; Molina and Garcia-Olmedo, 1997). There was a reduction in total lesion area per inoculation to 38 percent of nontransformed controls (Molina and Garcia-Olmedo, 1997). Transgenic *Arabidopsis* plants expressing barley LTP2 were also developed, and these plants showed resistance to *P. syringae* pv. *tomato,* and the lesion area decreased significantly compared with nontransformed control plants (Molina and Garcia-Olmedo, 1997).

LTPs have been shown to act as defense barriers even at constitutive levels in some plants (Molina and Garcia-Olmedo, 1993; Molina et al., 1993b; Segura et al., 1993), but at induced high concentrations they appear to play a very important role in inducing resistance against bacterial diseases. Synergism between different inducible proteins has been reported (Zhu et al., 1994; Jach et al., 1995). Synergism of LTPs with thionins has been demonstrated in vitro (Molina et al., 1993a,b).

Some indirect evidence has been presented to show the possible role of PR proteins in suppression of bacterial growth. PR-6 proteins (proteinase inhibitors) may inhibit proteases produced by the bacterial pathogens and prevent bacterial spread. Several bacterial pathogens produce proteases, which are essential for pathogenesis. *Erwinia carotovora* and *E. chrysanthemi* secrete metallo-proteinases (Kyostio et al., 1991). A proteinase-deficient mutant of *X. campestris* pv. *campestris,* the black rot pathogen, which lacked a serine- and a metallo-proteinase activity, showed considerable loss of virulence in pathogenicity tests when the bacteria were introduced into mature turnip leaves (Dow et al., 1990). Bashan et al. (1986) demonstrated the protease activity in response to *P. syringae* pv. *tomato* infection in tomato. Proteolytic activities reached their maximal levels by 48 h after infection in disease-resistant plants and subsequently decreased. In disease-

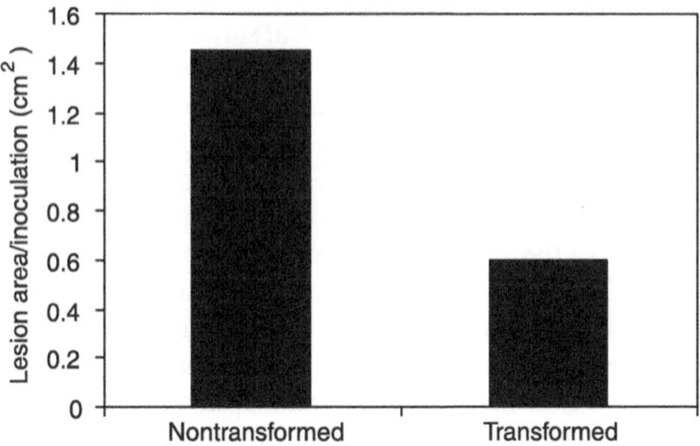

FIGURE 4.2. Total lesion area per inoculation caused in transformed (with LTP2 cDNA) and nontransformed tobacco plants (*Source:* Molina and Garcia-Olmedo, 1997).

sensitive plants, protease activity continued to increase until 120 h after inoculation (Bashan et al., 1986). The decline in protease activity correlates with the temporal induction of proteinase inhibitor I and II gene mRNAs (Pautot et al., 1991). It suggests that proteinase inhibitors may be an important mechanism to curtail *P. syringae* pv. *tomato* spread.

Toxic action of several other PR proteins against bacterial pathogens has not yet been demonstrated. However, it should be noted that although individual PR proteins may not have enough intrinsic toxicity, a combination of different PR proteins may have a highly toxic action. Coordinated induction of the different families of PR proteins may indeed confer a level of resistance considerably broader than that of any protein itself (Vidhyasekaran, 1997). Most of the phytopathogenic fungi that have been tested in vitro are resistant to β-1,3-glucanases and chitinases, but their growth is highly inhibited by a combination of these PR proteins (Mauch et al., 1988b; Ji and Kuc, 1996). Combination of several PR proteins may be necessary to reduce or prevent development of bacterial pathogens. Such a possibility has not yet been studied against bacterial pathogens. Further, the capacity to inhibit pathogens in vitro may not be sufficient to postulate an in planta defense role for a particular protein. The in planta concentrations at the particular site of infection are difficult to determine, and the distribution pattern of the protein after infection also has to be determined.

Some PR Proteins May Be Involved in Release of Elicitor Molecules in Planta

The function of some PR proteins may be to release elicitor molecules from the host and/or pathogen cell wall surface, and these elicitors may stimulate biosynthesis of phenolic compounds, phytoalexins, and PR proteins. The expression in tobacco of a β-1,3-glucanase from soybean increased the resistance of the transgenic plants against fungal pathogens (Yoshikawa et al., 1993). Phenylalanine ammonia-lyase transcripts increased earlier in the transgenic plants than in control plants, and the β-1,3-glucanase was not toxic to the fungal pathogens (Yoshikawa et al., 1993). This suggests that the increased resistance involved the release of active elicitor molecules and not the direct action of the PR-2 protein on the pathogen. Some chitinases may release signal molecules (specific oligosaccharides) from the plant cell walls (Van Loon, 1999).

Some PR Proteins May Be Involved in Reinforcement of Cell Wall Structures

Peroxidases (PR protein PR-9) are involved in lignification. They are also involved in cross-linking of extensin monomers, polysaccharide cross-linking, and suberization. The peroxidases may enhance resistance by the construction of a cell wall barrier that may impede bacterial pathogen ingress and spread (Venere et al., 1993).

Although the exact role of PR proteins in disease resistance is not known, some indirect evidence has indicated that they may play a very important role in bacterial disease resistance.

THE SECOND GROUP OF PATHOGEN-INDUCIBLE PROTEINS: CONSTITUTIVE, BUT INCREASINGLY INDUCED

Enzymes Involved in Phenylpropanoid Metabolism and the Octadecanoid Pathway

The second group of pathogen-inducible proteins is constitutively expressed ones; but they are induced severalfold in pathogen-infected plants. These inducible proteins are involved in phenylpropanoid metabolism and the octadecanoid pathway.

Phenylalanine Ammonia-Lyase

Phenylalanine ammonia-lyase (PAL) catalyzes the deamination of L-phenylalanine to *trans*-cinnamic acid, which is the first step of the phenylpropanoid pathway, which supplies the precursors for phenolics, lignin, and furanocoumarin phytoalexins. An increase in the amount of PAL mRNA has been shown to underlie the increase of PAL activity (Orr et al., 1993). Increased synthesis of PAL resulting from increased mRNA activity has been reported in pea infected by pathogens (Hahlbrock et al., 1981). PAL transcripts accumulated rapidly and massively in alfalfa cell cultures within 2 h after elicitor treatment, and PAL activity was induced 50-fold within 12 h after exposure to elicitor (Gowri et al., 1991). In all cases so far investigated, the changes of PAL activity are regulated at the transcription level (Lawton and Lamb, 1983; Lois et al., 1989). PAL is encoded by a small gene family in bean (Cramer et al., 1989), parsley (Lois et al., 1989), pea (Kawamata et al., 1992), *Arabidopsis* (Gowri et al., 1991), and poplar (Subramaniam et al., 1993). Two to four PAL genes have been detected in tobacco (Pellegrini et al., 1994). There are at least four PAL genes in rice (Zhu et al., 1995b). Potato contains approximately 40 to 50 PAL genes per haploid genome (Joos and Hahlbrock, 1992).

The transcripts of individual PAL genes show relatively different patterns of accumulation (Liang et al., 1989a; Lois et al., 1989). Three divergent classes of PAL genes, *gPAL1, gPAL2,* and *gPAL3,* have been cloned from bean (Cramer et al., 1989). *gPAL1* and *gPAL2* genes are induced by elicitor in cell cultures of bean cultivar Canadian Wonder while *gPAL3* is not activated; but all three genes are elicitor inducible in cell cultures of another bean cultivar, Immuna (Liang et al., 1989a; Ellis et al., 1989). Transgenic tobacco plants expressing the *gPAL2* gene have been developed (Liang et al., 1989b). In these plants, the gene was developmentally expressed in vascular tissue (Liang et al., 1989b). The activities of the bean *gPAL2* and *gPAL3* promoters in transgenic tobacco, potato, and *Arabidopsis* plants reflect the expression pattern of *gPAL2* and *gPAL3* genes in bean (Bevan et al., 1989; Liang et al., 1989a; Schufflebottom et al., 1993), indicating that the promoters of these genes control transcription activity. In potato, 40 to 50 genes have been detected, and out of them about 10 genes are potentially activated by elicitor treatment (Joos and Hahlbrock, 1992). In parsley, four PAL genes have been detected; only three of them are activated by elicitor (Lois et al., 1989a).

The upstream regulatory region of an efficiently expressed *PAL5* gene from tomato was characterized (Lee et al., 1994). Two transcription initiation sites giving rise to a long and a short transcript were identified. In general, basal levels of the short transcript were very low but preferentially and strongly stimulated by pathogen stress. In contrast, the longer transcript was less affected, essentially representing a constitutive expression of this gene. Thus, an actively expressed single tomato gene *(gPAL5)* is demonstrated to utilize alternate initiation sites in response to pathogen stress (Lee et al., 1994).

The expression of PAL genes is induced by elicitor or by pathogens (Edwards et al., 1985; Lamb et al., 1989). A PAL gene, *ZB8,* has been isolated from rice (Zhu et al., 1995). The *ZB8* gene is 4660 bp in length and consists of two exons and one intron. It encodes a polypeptide of 710 amino acids. The transcript start site was 137 bp upstream from the translation initiation site. In rice suspension-cultured cells, the *ZB8* PAL transcript was induced by a fungal elicitor (Zhu et al., 1995b). A dramatic increase in PAL activity was observed in cultured cells of peanut treated with an elicitor (Steffens et al., 1989). Increased PAL mRNA activity could be detected in soybean after a fungal infection (Lawton et al., 1983a). A pathogen-inducible PAL has been detected in soybean (Lawton et al., 1983b).

Importance of the inducible PAL in disease resistance has been demonstrated by developing transgenic plants. Transgenic tobacco plants with suppressed levels of PAL activity deposit less lignin and are more sensitive to pathogen attack than wild-type tobacco plants (Maher et al., 1994). Increased PAL activity induced the stilbenoid phytoalexin resveratrol in peanut (Steffens et al., 1989). PAL in parsley appears to be involved in biosynthesis of furanocoumarins (Collinge and Slusarenko, 1987). When PAL activity is inhibited in soybean roots by treatment with L-amino-oxy-3-phenylpropionic acid (AOPP) or R-(1-amino 2-phenyl-ethyl) -phosphonic acid (APEP), phytoalexin synthesis is prevented and the resistant plants become susceptible to pathogens (Moesta and Grisebach, 1982). The increase in phenolic synthesis was suppressed when a fungal pathogen-resistant *Eucalyptus calophylla* was treated with an inhibitor of PAL, aminooxyacetic acid (AOA), and the resistant plant became susceptible (Cahill and McComb, 1992). Thus, the inducible PAL appears to play a key role in disease resistance.

4-Coumarate:CoA Ligase

4-Coumarate:CoA ligase (4CL) is another important enzyme involved in biosynthesis of phenolics, lignin, and phytoalexins. Several 4CL isoenzymes have been reported in plants. Among several 4CL isoenzymes, isoenzyme 2 alone increased due to elicitor treatment in soybean, and this isoenzyme has been shown to be involved in flavonoid biosynthesis (Hahlbrock and Grisebach, 1979; Hille et al., 1982; Ebel et al., 1984). Increase in mRNA encoding 4CL was observed in soybean hypocotyls infected with a fungal pathogen, and it led to the accumulation of glyceollin (Schmelzer et al., 1984). 4CL occurs in two isoforms in parsley, each of which is encoded by a single-copy gene. Both the genes are activated in elicitor-treated cells (Douglas et al., 1987). The two genes encode 4CL isoenzymes that differ in three amino acid residues (Lozoya et al., 1988). 4CL mRNA rapidly accumulates in infected parsley leaves (Hahlbrock and Scheel, 1989). 4CL mRNA synthesis was observed even at 1.5 h after treatment with the elicitor in parsley cells (Chappel and Hahlbrock, 1984). 4CL activity increased in parsley cultured cells treated with elicitor, and it led to the accumulation of furanocoumarin phytoalexin (Kuhn et al., 1984). 4CL activity increased in cultured cells of bean treated with elicitor, and it led to synthesis of the phytoalexin phaseollin (Dixon and Bendall, 1978). mRNA of 4CL was induced before the increase in the enzyme activity (Hahlbrock et al., 1983). Increased 4CL activity leading to lignification in wheat inoculated with a pathogen also has been reported (Moerschbacher et al., 1988).

Cinnamic Acid 4-Hydroxylase

Cinnamic acid 4-hydroxylase (CA4H) is an important enzyme involved in biosynthesis of isoflavonoid phytoalexins. It converts *trans*-cinnamic acid (the product of conversion of phenylalanine by PAL) into *p*-coumaric acid, which is the important substrate for 4CL. CA4H activity increased fourfold in alfalfa cell suspension cultures after elicitor treatment (Kessmann et al., 1990). The enzyme activity is induced before accumulation of phytoalexin in bean cell cultures treated with elicitor (Bolwell et al., 1985).

Chalcone Isomerase

Chalcone isomerase (CHI) from bean catalyzes the stereospecific isomerization of both 2',4,4'-trihydroxy chalcone (isoliquiritigenin) and 2',4,4',6'-tetrahydroxychalcone to yield the corresponding (–) flavanones (Dixon et al., 1983). CHI in bean appears to be encoded by a single gene (Mehdy and Lamb, 1987). CHI activity is high in unelicited bean cell cultures. Its activity increases due to elicitor treatment (Robbins et al., 1985). In elicitor-treated cells, de novo synthesis of this enzyme has been shown (Cramer et al., 1985a,b). CHI in bean is involved in synthesis of phytoalexins. When alfalfa-cultured cells are treated with elicitor, CHI activity increased rapidly and extensively, resulting in phytoalexin accumulation (Dalkin et al., 1990a,b). CHI shows approximately fourfold increases within

10 to 14 h in alfalfa suspension-cultured cells due to elicitor treatment (Kessmann et al., 1990).

3-Hydroxy-3-Methylglutaryl-CoA Reductase (HMGR)

HMGR catalyzed the NADPH-dependent reduction of 3-hydroxy-3-methylglutaryl-CoA to mevalonate in sweet potato (Suzuki et al., 1975). HMGR activity increased rapidly and preceded the furanoterpenoid phytoalexin ipomeamarone formation (Suzuki et al., 1975). An elicitor from *Aspergillus parasitica* induced more HMGR than the level found in control tobacco cell cultures (Chappell et al., 1991). The importance of inducible HMGR activity for elicitor-stimulated sesquiterpenoid biosynthesis has been demonstrated. When HMGR activity was not induced, there was no induction of sesquiterpenoid phytoalexins (Chappell et al., 1991).

These inducible enzymes are involved in biosynthesis of phenolics and phytoalexins, and the role of phenolics and phytoalexins is discussed in Chapter 5.

Peroxidase

Peroxidase is the key enzyme in lignification. Many peroxidase isozymes have been reported in plants. Twelve peroxidase isozymes have been detected in tobacco. Not all peroxidase isozymes are pathogen-inducible. Two anionic isozymes, P37 and P35, alone increased due to infection in induced resistant plants (Ye et al., 1990). These two anionic peroxidases appear to be involved in increased lignification. There were four preformed peroxidases, and two of them alone were found to be pathogen-inducible in wheat (Flott et al., 1989). In barley, a peroxidase gene transcript was induced in incompatible interactions (Thordal Christensen et al., 1992). Thus, some specific peroxidases alone are induced in resistant interactions, and they appear to play an important role in lignification. Overexpression of anionic tobacco peroxidase in transgenic tomato plants resulted in a 20-fold increase in lignin content in fruit after wounding (Lagrimini et al., 1993). The importance of lignification in bacterial disease resistance has been discussed in Chapter 2.

Lipoxygenase (LOX)

A large number of LOX genes and isoenzymes has been reported in plants (Melan et al., 1993). In rice, three LOXs have been isolated from leaves (Ohta et al., 1991). LOX activity rapidly increases in cucumber infected with the incompatible pathogen *P. syringae* pv. *pisi* (Keppler and

Novacky, 1986) and bean infected with an incompatible race of *P. savastanoi* pv. *phaseolicola* (Croft et al., 1990). In rice leaves, among three LOXs, only activity of LOX-3 was induced to a greater degree after inoculation with the incompatible pathogen. Activities of the other two LOXs, leaf LOX-1 and leaf LOX-2, were only slightly affected (Ohta et al., 1991).

LOX mRNA was induced in the resistant bacterial interaction in bean leaves (Koch et al., 1992). When a cloned *avr* gene was transferred to a virulent isolate, the transformant rapidly induced LOX mRNA (Melan et al., 1993). LOXs are involved in octadecanoid pathway in inducing methyl jasmonate signals. LOX is involved in phytoalexin, cutin, and suberin synthesis; it is also capable of generating superoxide anions and singlet oxygen (Vidhyasekaran, 1997). The role of LOXs in bacterial disease resistance is discussed in Chapters 1 and 3.

HYDROXYPROLINE-RICH GLYCOPROTEINS

The third type of inducible plant proteins is hydroxyproline-rich glycoproteins (HRGPs). HRGPs are encoded by a multigene family whose constitutive level is generally low in healthy plants. When the plants are invaded by pathogens, a strong activation of HRGP genes occurs at the transcriptional level, which leads to the induction of several transcripts ranging from about 1.5 to 6.0 kb (Showalter et al., 1985; Rumeau et al., 1988). The HRGP is increased at the level of the vascular system and also in cortical and epidermal cells in pea epicotyls (Cassab et al., 1988). Systemic induction of HRGPs in plants has been reported (Niebel et al., 1993). Accumulation of HRGPs in plant cell walls has been shown in tomato inoculated with *Clavibacterium michiganense* (Benhamou, 1991), potato inoculated with *Ralstonia solanacearum* (Leach et al., 1982; Mazau and Esquerre-Tugaye, 1986) and *Erwinia carotovora* (Rumeau et al., 1990), in French bean inoculated with *P. savastanoi* pv. *phaseolicola* (Mazau and Esquerre-Tugaye, 1986), in tobacco inoculated with *R. solanacearum* (Mellon and Helgeson, 1982; Mazau and Esquerre-Tugaye, 1986), in melon inoculated with *P. savastanoi* pv. *phaseolicola* (Esquerre-Tugaye et al., 1979), and in turnip inoculated with *X. campestris* (Davies et al., 1997).

The HRGP gene is induced by several signals. Ethylene appears to be the major signal in induction of HRGPs (Esquerre-Tugaye et al., 1979). Nucleotide sequences in the promoter of an HRGP gene from carrot and nuclear proteins involved in regulation of the gene in response to ethylene have been identified (Granell et al., 1992). Wycoff et al. (1995) have shown that wound induction specificity lies in a region between –94 and –251 relative to the transcription start site and that activation by infection lies outside this region in the bean 4.1 kb transcript-encoding gene. Jasmonic acid has also been shown to induce HRGP in tobacco (Rickauer et al., 1997).

The role of HRGPs in bacterial disease resistance is discussed in Chapter 2. When HRGP monomers are secreted into the cell wall, they become tightly linked by covalent bonds to the wall network. During infection, intra- and intermolecular oxidative cross-linking of HRGP monomers occurs in the presence of H_2O_2 and peroxidase (Bradley et al., 1992; Brisson et al., 1994; Wojtaszek et al., 1995; Schnabelrauch et al., 1996). Cross-links occur between two tyrosine residues located on the same polypeptide backbone or on two adjacent backbones following the formation of isodityrosine (IDT) bridges (Brady et al., 1996). Such cross-links create a network of HRGPs interpenetrated by other wall polymer, contributing to cell wall reinforcement upon pathogen attack (Esquerre-Tugaye et al., 1999). The cross-linking occurs early during incompatible soybean-*P. savastanoi* pv. *glycinea* interaction and later in the compatible interaction, and during this process transcriptional activation of HRGP genes has been observed (Brisson et al., 1994).

HRGPs contain a high proportion of the basic amino acids lysine and histidine and, hence, they behave as polycations and display agglutinating activities against negatively charged surfaces. HRGPs from tobacco and potato agglutinate avirulent strains of *R. solanacearum* on these plants, but only weakly virulent ones (Mellon and Helgeson, 1982; Leach et al., 1982). Bacteria appear encased in an HRGP-containing material of host origin during incompatible interactions (O'Connell et al., 1990). The agglutinating properties of HRGPS may participate in the immobilization of bacteria observed during incompatible interactions (Esquerre-Tugaye et al., 1999).

LECTINS

Lectins are carbohydrate-binding proteins found in plants, and they are also referred to as agglutinins. Lectins are proteins that bind reversibly to specific mono- or oligosaccharides (Peumans and Van Damme, 1995). Although lectins cannot be considered inducible proteins, they are included in this chapter because they resemble some PR proteins and inducible proteins in their structure and function. Some lectins resemble ribosome-inactivating proteins (RIPs). The type II RIPs are synthesized as larger precursors and processed to a 30 kDa, RIP A-chain linked to a 30 kDa lectin B chain (Lord, 1985). Hevein is a lectin, and hevein-like structures are common among chitinases. Hevein contains a chitin-binding domain, and hence it has been placed under PR-4 proteins (Neuhaus, 1999). Lectins act on bacterial cell walls and in this respect they resemble PR-8 proteins with chitinase/

lysozyme activity. Lectins have higher affinity for oligosaccharides, which are commonly present in bacteria and fungi and uncommon or totally absent in plants. In this respect they resemble PR-3 proteins (chitinases). Lectins are resistant to proteases and hence they resemble some PR proteins.

Based on their overall structure, three major types of lectins are distinguished, namely merolectins, hololectins, and chimerolectins (Peumans and Van Damme, 1995). Merolectins are proteins that are built exclusively of a single carbohydrate-binding domain. They are small, single-polypeptide proteins, which, because of their monovalent nature, are incapable of precipitating glycoconjugates or agglutinating cells. Hevein belongs to this group. Hololectins also are built exclusively of carbohydrate-binding domains but contain two or more such domains that are either identical or very much homologous. This group comprises all lectins that have multiple binding sites, and hence are capable of agglutinating cells or precipitating glycoconjugates (Peumans and Van Damme, 1995). Chimerolectins are fusion proteins possessing a carbohydrate-binding domain tandemly arrayed with an unrelated domain, which has a well-defined catalytic activity that acts independently of the carbohydrate-binding domain. In this group, type 2 ribosome-inactivating proteins with two carbohydrate-binding sites on their B chain agglutinate bacterial cells (Peumans and Van Damme, 1995).

Lectins bind with bacterial cell wall peptidoglycans (Ayouba et al., 1994). Several legume seed lectins strongly interact with muramic acid, N-acetylmuramic acid, and muramyl dipeptide (Ayouba et al., 1994). Potato lectin immobilized avirulent strains of *Ralstonia solanacearum* in the plant cell wall (Sequeira and Graham, 1977). Virulent strains were not recognized by the lectin, escaped attachment to the cell wall, and therefore were able to multiply and spread over the plant (Sequeira and Graham, 1977). Lectins block the movement of normally motile bacteria (Broekaert and Peumans, 1986). In this process they may prevent invasion of the roots by soil-borne bacterial pathogens.

NOT ALL INDUCIBLE PROTEINS NEED BE INVOLVED IN INDUCING BACTERIAL DISEASE RESISTANCE

Although several inducible proteins are involved in bacterial disease resistance, it cannot be concluded that all inducible proteins will contribute for disease resistance. Buell and Somerville (1995) observed that mRNA encoding PR-1 protein accumulated to high levels in both *Arabidopsis thaliana* ecotypes Columbia (resistant) and Pr-O (susceptible) tissues when inoculated with *X. campestris* pv. *campestris*. PR-1 mRNA levels accumulated to higher levels in the susceptible ecotype in comparison to the resistant ecotype inoculated with the bacterial pathogen (Buell and Somerville, 1995). Probably induction of PR-1 may be only due to stress, and it may not be involved in disease resistance.

Several others have reported similar results. β-1,3-Glucanase mRNA levels were induced in both the compatible and incompatible interactions, with higher levels of β-1,3-glucanase mRNA accumulation occurring in the compatible interaction at 24 and 48 h postinoculation with *P. syringae* virulent and avirulent strains in *Arabidopsis* (Dong et al., 1991). Inoculation of *Arabidopsis* ecotypes Columbia and Pr-O with *P. syringae* pv. *syringae* PssD20, which elicits an HR in both ecotypes, resulted in the accumulation of several mRNA species including ASA1 (anthranilate synthase 1, the enzyme involved in indole synthesis and induction of the synthesis of the phytoalexin camalexin), CAD (cinnamyl alcohol dehydrogenase), C4H (cinnamate-4-hydroxylase), ELI3 (ELI3 plant defense gene), LOX, SOD, and PR-1. But all 10 defense-related mRNAs did not have preferential accumulation in Colombia tissues (*Arabidopsis* ecotype which is resistant to *X. campestris* pv. *campestris*) (Buell and Somerville, 1995).

Chitinase mRNA levels in *Arabidopsis* tissues, challenged with *X. campestris* pv. *campestris* strains, were similar in resistant and susceptible plants (Lummerzheim et al., 1993). Tobacco leaves were infiltrated with compatible (isolate K60), incompatible (GMI 1000), and avirulent (GMI 1178, a derivative of GMI 1000 which was shown to be deleted for some *hrp* genes involved in the HR and also has no apparent effect on the inoculated leaves) isolates of *R. solanacearum* (Godiard et al., 1990). A nonspecific accumulation of mRNAs of β-1,3-glucanase and chitinase, independent of the nature of the inoculum, was observed in the tobacco leaves (Godiard et al., 1990). Even mechanical stress such as leaf detachment and infiltration of water in vacuo induced two acidic and two basic chitinases in tobacco. It suggests that these PR proteins may be induced due to general stress and may not be involved in bacterial disease resistance (Godiard et al., 1990). Newman et al. (1994) suggested that accumulation of inducible proteins depends upon the growth of bacteria in plant tissues. These proteins accumulate in the later phases of compatible interactions associated with considerable bacterial growth (Dong et al., 1991).

In some cases, increased expession of certain PR proteins in a resistant variety may be characteristic of that variety rather than due to incompatible interaction. Increased expression of the PRb-1 gene was observed in resistant (to the powdery mildew fungus *Erysiphe graminis*) compared with near-isogenic susceptible barley plants following treatment with ethylene, salicylic acid, methyl jasmonate, and 2,6-dichloro-isonicotinic acid (Mouradov et al., 1993). A higher constitutive level of two chitinase isozymes (30 and 32 kDa proteins) was observed in all tomato cultivars that are resistant to *Alternaria solani* than in susceptible cultivars. These studies indicate that inducible proteins need not be wholly responsible for disease resistance. However, it has been shown that each inducible protein may have only a limited role, but in combination with other inducible proteins they may play a key role in disease resistance. Further, the action of individual inducible proteins against different pathogens may vary and their effectiveness may vary from host to host.

CONCLUSION

A number of genes coding for proteins are transcriptionally activated in plants when pathogens invade them. Most of the inducible proteins are involved in resistant interactions, suggesting their possible role in host defense mechanisms. Inducible plant proteins are classified into three major groups. Pathogenesis-related (PR) proteins are the ones that are newly induced by pathogens. The second group consists of proteins, which are constitutive, but increasingly induced by pathogens enhancing transcription of the genes encoding those proteins. The third group consists of hydroxyproline-rich glycoproteins (HRGPs).

Several PR proteins have been detected in all plants studied. The structure of these proteins varies widely, and it is difficult to classify them. At least 11 well-characterized PR protein families have been identified; several other groups of PR proteins have also been recognized. Most of the PR proteins are encoded by multigenes. A single PR protein (e.g., bean PvPR3) may be encoded by 15 genes. PR genes are quiescent in healthy plants. Regulation of PR protein biosynthesis occurs primarily at the level of transcription, and the PR gene expression depends on the binding of certain transcription factors to particular regions of the PR gene promoter.

Cis-acting elements, which activate PR genes, have been identified in promoter regions of different PR genes. *Cis*-acting elicitor-responsive elements, ethylene-responsive elements, salicylic acid-responsive elements, and jasmonate-responsive elements have been detected. Genes that code for proteins, which in turn regulate the expression of PR genes, have also been identified. The products of these regulator genes are the *trans*-acting factors. Recognition of the *cis*-acting sequence by the *trans*-acting factor regulates the PR gene expression. Elicitors, protein kinases, G-proteins, ethylene, salicylic acid, methyl salicylate, jasmonic acid/methyl jasmonate, abscisic acid, systemin, oligogalacturonides, auxins, and cytokinins are the important signals involved in the interaction of *cis*-acting elements and *trans*-acting factors to induce PR genes. The presence of *cis*-acting elements and the presence or absence of certain critical *trans*-acting factors may determine the PR gene expression induced by various signals.

Several *trans*-acting factors have been identified in plants. Four ethylene-responsive element-binding proteins (EREBPs) have been detected in tobacco. They are induced by ethylene, and the EREBPs were found to bind to the AGCCGCC sequence of the *cis* elements present in promoters of the PR genes. Similar *trans*-acting factors have been detected in tomato and *Arabidopsis*. Salicylic acid-responsive *cis* elements have *as-1* and *ocs* motifs. *Cis* elements, which contain the sequence TTCGACC (which is related to the W-boxes), are also involved in salicylic acid-inducible expression. Several *trans*-acting factors activated by salicylic acid have been recognized. Myb 1 proteins, GT-1-like proteins, SAB P22 protein, and TGA1a have been recognized as *trans*-acting factors induced by salicylic acid. A G-box sequence (CACGTGG) and a palindrome TGACG element, *as-1*-type elements in the *cis*-acting promoters have been found essential for induction of PR proteins by jasmonic acid. TGA1a and bZIP are the important *trans*-acting factors induced by jasmonic acid.

It appears that induction of a PR gene may require a specific signal or combination of signals, and/or a signal transduction system. Synergism and antagonism between signal molecules have also been reported. A single signal cannot induce all PR proteins in a plant, and each PR protein may require a different signaling system. However, in some plants a single signaling system may induce many PR proteins, if not all of them. Some PR proteins are localized in vacuoles, some in cytoplasm and others in apoplast. The vacuolar-targeting signal may be present at the C-terminal end of PR proteins, and deletion of the C-terminal region results in the secretion of these proteins into the apoplast. Thus the molecular biological studies have revealed that several genes and signals are involved in induction of PR proteins, and their interactions are complex.

Many PR proteins are induced in plants invaded by bacterial pathogens. Most of them are induced in incompatible interactions rather than compatible ones. PR proteins are induced more and/or more rapidly in resistant interactions. Several abiotic and biotic agents induce systemic resistance against bacterial pathogens, and in these induced-resistant plants, several PR proteins have been shown to accumulate. Transgenic plants expressing enhanced resistance to bacterial pathogens show enhanced expression of PR proteins. Inhibitory action of some PR proteins on multiplication of bacterial pathogens has been observed. Some PR proteins may release elicitor molecules from the host and/or pathogen and may induce biosynthesis of several defense-related compounds such as phenolics and phytoalexins. Some PR proteins may be involved in plant cell wall reinforcement and impede bacterial ingress and spread.

The second group of pathogen-inducible proteins is constitutively expressed but induced severalfold by the pathogens. These proteins include enzymes, which are involved in synthesis of defense chemicals. Some lipid-transfer proteins also belong to this group, and these proteins show antibacterial action. Hydroxyproline-rich glycoproteins are the third group of inducible proteins, and they are involved in cell wall reinforcement against bacterial pathogen ingress. Lectins show similarity to many PR proteins and are involved in agglutination of bacterial cells.

Thus, several studies have indicated the importance of inducible proteins in bacterial disease resistance. But when the literature on inducible proteins is critically reviewed, several questions arise whether these proteins are actually involved in disease resistance. The first puzzle is that these inducible proteins are induced not only by bacteria but also by many other biotic and abiotic factors. In fact, mere mechanical injury can induce many PR proteins. If it is so, how can we explain the relationship between PR proteins and bacterial disease resistance? The convincing answer is "not mere induction of proteins is important but the amount (overexpression) of induction is important." Even some of the PR proteins (so-called newly induced proteins) can be detected constitutively without giving any stress. But the basal level of them is always low. Toxicity of some of the well-known toxic inducible proteins depends upon their concentration gradient. Invariably, in the bacterial pathogen-plant interactions, greater PR protein expression was observed in resistant/incompatible interactions.

Another argument against the role of PR proteins can be that besides PR proteins, many other defense mechanisms are activated in the resistant/incompatible interactions and in the "induced resistant plants"; it may be the other factors rather than PR proteins that are involved in disease resistance. The best method to demonstrate the importance of a PR protein in disease resistance is to develop a transgenic plant expressing the particular PR gene and compare its disease reaction with the wild plant lacking the PR gene. It is true, no such transgenic plant has been developed and tested against a bacterial pathogen. But it is well known that not a single PR protein but a combination of PR proteins may contribute to disease resistance. It may be difficult, if not impossible, to develop transgenic plants overexpressing several PR proteins with effective concentrations. In the resistant interactions, no single PR protein but several PR proteins are expressed.

The other method is to develop transgenic plants by antisense transformation. In this method, expression of specific PR proteins can be suppressed in the resistant variety and thus the resistant variety should become susceptible to the bacterial pathogens. This type of demonstration has also not been done so far. But it has been shown that if one PR gene is suppressed, other related genes overexpress and compensate for the absence of the particular gene (McIntyre et al., 1996). Further, antisense transformations should aim to suppress all PR genes; that is, again, a very difficult task to perform.

Another major concern in assigning a role for PR proteins in bacterial disease resistance is that in vitro antibacterial activity has not been demonstrated with most of the well-characterized PR proteins such as PR-1, PR-2, PR-3, PR-4, PR-6, PR-7, PR-8, PR-9, PR-10, or PR-11. However, in many cases, no studies seem to have been made to test their toxicity against bacterial pathogens. In particular, the toxic effect of combinations of these proteins against bacterial pathogens has not yet been studied. PR proteins may induce disease resistance by several other mechanisms such as inducing plant cell wall reinforcement, suppressing enzymes produced by pathogens, by degrading proteins essential for pathogenesis, and inducing elicitor-inducible host-defense mechanisms, besides inhibiting bacterial growth. A few transgenic plants overexpressing PR proteins, which show resistance to bacterial diseases, have been developed. However, technologies should be developed to select effective PR proteins as well as a combination of PR proteins which show synergistic effects, and make them express in specific regions such as apoplast and vascular tissues to effectively manage bacterial diseases. Efficient induction of PR proteins by different signal systems should be studied, and methods should be developed to enhance the amount of endogenous signals, as higher amounts of signal molecules lead to more induction of PR proteins. Antagonism among various signaling systems should also be studied to develop an efficient combination of signaling systems to manage bacterial diseases exploiting PR proteins.

Another important aspect that needs the attention of molecular biologists is to find the possible molecular mechanism that would explain why inducible/PR proteins accumulate more rapidly and at higher amounts in resistant/incompatible interactions than they do in compatible bacteria-plant interactions. What are the signals involved in induction of these proteins in resistant interactions that are absent in the susceptible interactions? Intensive studies are indispensable to unravel this mechanism so that we can make these inducible proteins act in the susceptible varieties in a manner similar to their actions in the resistant varieties. It may be a very useful approach to manage bacterial diseases.

Chapter 5

Inducible Secondary Metabolites

WHAT ARE INDUCIBLE SECONDARY METABOLITES?

When bacterial pathogens invade host tissues, synthesis and accumulation of several secondary metabolites are observed in plants. Most of these secondary metabolites show antibacterial action, and these secondary metabolites accumulate rapidly in high amounts in resistant host-pathogen interactions, suggesting a possible role for them in disease resistance. The inducible antibacterial secondary metabolites can be classified into two major groups: (1) the secondary metabolites which are not detected in healthy host tissues but induced by pathogens, and (2) the secondary metabolites which are constitutively present in plants before infection, but their synthesis is increasingly induced by pathogens. The first group of newly induced metabolites are generally called phytoalexins. The second group of metabolites are called inhibitins (Ingham, 1973) or phytoanticipins (Van Etten et al., 1994).

Muller and Borger (1940) defined phytoalexins as plant antibiotics that are synthesized de novo after the plant tissue is exposed to microbial infection. Paxton (1980, 1981) further defined phytoalexins as low molecular weight, antimicrobial compounds that are both synthesized by and accumulated in plants after exposure to microorganisms. Two important criteria have been suggested to call a secondary metabolite a phytoalexin: (1) the secondary metabolite should be produced de novo in response to infection, and (2) the compound should accumulate to antimicrobial concentrations in the area of infection (Van Etten et al., 1995).

By definition, phytoalexins appear to be similar to pathogenesis-related (PR) proteins. Both are newly induced due to infection. But they differ in their mode of synthesis. PR proteins are produced by transcription of quiescent host genes (see Chapter 4), while phytoalexin production in plants requires a biosynthetic pathway involving coordinated action of several host enzymes (Ingham, 1993). Some of these biosynthetic enzymes may be called PR proteins (Van Loon, 1999), but several such enzymes are needed for biosynthesis of phytoalexins.

However, PR proteins and phytoalexins are similar in many of their characteristics. Similar to PR proteins, phytoalexins are induced by several pathogens, biotic and abiotic elicitors, various toxic chemicals, and environmental factors. Phytoalexins are induced by infection with fungal (Bennett et al., 1994), bacterial (Meier et al., 1993), and viral (Burden et al., 1972) pathogens. Elicitors isolated from fungi (Keen and Legrand, 1980; Sharp et al., 1984; Dalkin et al., 1990), bacteria (Davis et al., 1984), and plants (Nothnagel et al., 1983; Davis et al., 1986; Wingate et al., 1988) induce phytoalexins in plants. Several chemicals induce phytoalexins (Oba et al., 1976). Mercuric chloride (Hargreaves, 1979), silver nitrate (Yoshikawa, 1978), cupric chloride (Tamara et al., 1988), the herbicide glyphosate (Holliday and Keen, 1982), the fungicide fosetyl-Al (Nemestothy and Guest, 1990), and the detergent triton surfactant (Hargreaves, 1981) are all known to induce phytoalexins. Ultraviolet light (Chappell and Hahlbrock, 1984; Douglas et al., 1987; Kodama et al., 1988) and various other environmental stimuli (Wingender et al., 1989) induce phytoalexins in plants. Wounding itself induces phytoalexins in some plants (Kessmann and Barz, 1986).

Some compounds may be phytoalexins in one organ and constitutive in another organ of the same plant species. Momilactone A, which is induced in rice leaves as a phytoalexin (Cartwright et al., 1981), occurs constitutively in rice seeds (Grayer and Harborne, 1994). The same compound may be a phytoalexin in one plant species and a constitutive compound in another species. The flavanone sakuranetin is induced in rice leaves (Kodama et al., 1992) but is constitutive in blackcurrant *(Ribes nigrum)* leaves (Atkinson and Blakeman, 1982). The same phytoalexin can be detected in more than one plant species. Medicarpin is the phytoalexin detected in chickpea *(Cicer arietinum)* (Barz et al., 1989), alfalfa *(Medicago sativa)* (Dalkin et al., 1990), peanut *(Arachis hypogaea)* (Strange et al., 1985), and broad bean *(Vicia faba)* (Hargreaves et al., 1977). Luteolinidin has been detected as a phytoalexin in sugarcane *(Saccharum officinarum)* (Godshall and Lonergan, 1987) and sorghum *(Sorghum bicolor)* (Nicholson et al., 1987). Resveratrol has been detected in peanut (Bailey and Mansfield, 1982) and *Festuca versuta* (Powell et al., 1994).

It appears that induction of phytoalexins is universal in the plant kingdom (Grayer and Harborne, 1994). Several phytoalexins have been detected in each plant species and as many as 30 phytoalexins in carnation plants have been reported (Niemann, 1993). Structurally, phytoalexins vary greatly. They belong to four major groups: (1) phenylpropanoids, (2) terpenoids, (3) fatty acid derivatives, and (4) indole-based sulfur compounds.

Several kinds of phenylpropanoid phytoalexins have been reported in plants. The important isoflavonoid phytoalexins include phaseollin in bean, glyceollins in soybean, pisatin in pea, and medicarpin in alfalfa (Vidhyasekaran, 1997). In *Cassia obtusifolia,* a flavonoid phytoalexin was identified as 2-(*p*-hydroxy-phenoxy)-5,7-dihydroxy-chromone (Sharon and Gressel, 1991; Sharon et al., 1992). Resveratrol in peanut and grapevine (Langcake and Pryce, 1977; Calderon et al., 1993, 1994) and piceatannol in sugarcane are the important stilbene phytoalexins (Bailey and Mansfield, 1982; Brinker and Seigler, 1991). In *Dioscorea batatas,* the bibenzyl phytoalexin dihydropinosylvin has been detected (Takasugi et al., 1987). The coumarin phytoalexins include marmesin in parsley (Jahnen and Hahlbrock, 1988), psoralen and begapten in celery and parsley (Knogge et al., 1987), scopoletin in rubber and sunflower (Tal and Robeson, 1986), and xanthotoxin and angelicin in parsnip *(Pastinaca sativa)* (Desjardins et al., 1989). Luteolinidin in sugarcane and sorghum (Godshall and Lonergan, 1987; Nicholson et al., 1987) are the important anthocyanidin phytoalexins. The anthranilic acid phytoalexins include avenalumins in oat, HDIBOA glucoside in wheat, and dianthalexins and dianthramides in *Dianthus caryophyllus* (Niemann, 1993). The anthraquinone phytoalexins, which include purpurin 1-methyl ether, have been identified in *Cinchona ledgeirana* (Wijnsma et al., 1985). Two cyclic diones have been detected in onion (Tverskoy et al., 1991). A benzodioxin has been detected as a phytoalexin (yurinelide) in *Lilium maximowczii* (Monde et al., 1992). Benzofurans such as α-pyrufuran are formed in *Pyrus* as phytoalexins (Grayer and Harborne, 1994). An alkaloid phytoalexin sanguinarine has been reported in *Papaver bracteatum* cell cultures (Cline and Cosia, 1988).

Some terpenoids have been identified as phytoalexins. These include rishitin, lubimin, and solavetivone in potato (Coolbear and Threlfall, 1985), ipomeamarone and sesquiterpenes A1 and A2 in sweet potato (Brindle and Threlfall, 1983), hemigossypol in cotton (Bell et al., 1986), capsidiol, rishitin, phytuberin, and phytuberol in tobacco (Sato et al., 1985), rishitin in tomato (DeWit and Kodde, 1981), casbene in castor bean (Bruce and West, 1982), momilactones in rice (Kodama et al., 1988), cichoralexin in chicory (Monde et al., 1990), and lettuceninA in lettuce (Takasugi et al., 1985).

The fatty acid derivative phytoalexins include falcarindiol (polyacetylene) in tomato, safynol, and dehydrosafynol (acetylenic) in safflower, and wyerone acid in broad bean (furanoacetylene) (Tietjen and Matern, 1984). The indole-based sulfur phytoalexins have been reported in various plants belonging to the Cruciferae. These include spirobrassinin, cyclobrassinin, oxymethoxybrassinin, methoxybrassinin, brassinin, dioxybrassinin, and brassicanals A-C in *Brassica campestris* (Monde et al., 1990), brassilexin and cyclobrassinin in *Brassica juncea* (Devys et al., 1990), methoxybrassinin in *Brassica napus* (Dahiya and Rimmerl, 1988), and camalexin and methoxycamalexin in *Camelina sativa* (Browne et al., 1991). The phytoalexin 3-thiazol-2'-yl-indole has been detected in *Arabidopsis thaliana* (Tsuji et al., 1992).

Similar to PR proteins, phytoalexins also cannot be defined as compounds that are completely absent in healthy tissues. Trace amounts of phytoalexins have been detected in several healthy plants (Sallaud et al., 1997). Since induction of phytoalexin has been observed due to various environmental factors (Wingender et al., 1989), a small quantity of it may be present in healthy plants.

The second group of induced secondary metabolites also consists of antimicrobial substances. Their concentration in uninfected (healthy) plants may normally be low, but may increase enormously after infection (probably in order to combat attack by microorganisms). Ingham (1993) called this type of secondary metabolites "inhibitins." Van Etten et al. (1995) proposed another name, phytoanticipins, for this group of compounds. They defined phytoanticipins as low molecular weight antimicrobial compounds that are present in plants before challenge by microorganisms and/or produced after infection solely from preexisting constituents. Phytoanticipins also include preformed secondary metabolites, which are not induced after infection. The important second group of secondary metabolites is various forms of phenolics and terpenoids.

The distinction between the two groups of induced secondary metabolites is narrow. The same chemical may serve as both a phytoalexin and a phytoanticipin/inhibitin, even in the same plant. In the roots of red clover, the antimicrobial isoflavonoid derivative maackiain is present as the glycone of a preformed plant glucoside and is released from injured plant tissue by the action of a preformed plant glucosidase during tissue compartmentalization (McMurchy and Higgins, 1984). In this case, maackiain would be classified as a phytoanticipin. However, maackiain can also be synthesized de novo in this plant in response to microbial infection or other elicitors (Higgins and Smith, 1972; Dewick, 1975) making it, in this case, a phytoalexin.

BACTERIAL PATHOGENS INDUCE ACCUMULATION OF SECONDARY METABOLITES IN INFECTED TISSUES

When plants are inoculated with either incompatible or compatible bacterial pathogens, several secondary metabolites are induced. The phytoalexin glyceollin accumulated in various cultivars of soybean after infection with the pathogen *P. savastanoi* pv. *glycinea* (Long et al., 1985). The virulent pathogen *P. syringae* pv. *maculicola* elicited the synthesis and accumulation of the phytoalexin camalexin in *Arabidopsis* (Glazebrook and Ausubel, 1994). Phytoalexins are induced in potato tubers infected by *Erwinia* spp.

(Lyon, 1989). When stored potato tubers were inoculated with *Erwinia carotovora* subsp. *atroseptica,* the phytoalexins rishitin, solavetivone, and phytuberin accumulated (Hildenbrand and Ninnemann, 1994). Rober (1989) reported lubimin accumulation at 23 h and rishitin and other phytoalexins about 40 h after inoculation of tubers with *E. carotovora* subsp. *atroseptica.* Phytoalexins could be detected in potato roots after inoculation with the pathogen *E. carotovora* subsp. *atroseptica* (Abenthum et al., 1995).

Several flavonoids accumulated during an incompatible interaction in alfalfa leaves with the bacterium *Pseudomonas syringae* pv. *pisi* (Sallaud et al., 1997). The major accumulated compounds were 4',7-dihydroxyflavone and to a lesser extent 4',7-dihydroxyflavanone (liquiritigenin) and 2',4,4'-trihydroxychalcone (isoliquiritigenin) (Sallaud et al., 1997). When lettuce leaves were inoculated with an incompatible bacterium, *Pseudomonas savastanoi* pv. *phaseolicola,* the phytoalexin lettuceninA accumulated (Bestwick et al., 1995). Camalexin accumulated to high levels after infection with an avirulent pathogen, *Pseudomonas syringae* pv. *syringae,* in *A. thaliana* (Tsuji et al., 1992; Tsuji and Somerville, 1992). Phytoalexins accumulated in cotton cotyledons inoculated with an incompatible strain of *X. axonopodis* pv. *malvacearum* (Gorski et al., 1995). Isoflavonoid phytoalexins accumulate in the hypersensitive reaction of soybeans to *P. savastanoi* pv. *glycinea* (Keen and Kennedy, 1974). Antibacterial phytoalexins accumulated in leaves of bean plants inoculated with avirulent isolate of *P. savastanoi* pv. *phaseolicola* (Meier et al., 1993).

PHYTOALEXINS ACCUMULATE IN PLANTS AFTER IRREVERSIBLE CELL MEMBRANE DAMAGE

Phytoalexins accumulate rapidly in resistant/incompatible interactions, which are characterized by the hypersensitive reaction (HR). Cell death precedes major accumulation of phytoalexins within the infected tissues (Hargreaves and Bailey, 1978; Bailey, 1982; Mansfield, 1982; Bennett et al., 1994). Phytoalexins are usually not found to accumulate in plant tissues in the absence of host cell necrosis (Lyon and Wood, 1975; Meier et al., 1993). Phytoalexins are detected in plants after the appearance of cells undergoing irreversible membrane damage (Bennett et al., 1994). In other words, phytoalexins are produced in dying cells (cells undergoing apoptosis).

The HR developed rapidly in lettuce leaves inoculated with the bean halo blight bacterium, *P. savastanoi* pv. *phaseolicola* (a nonpathogen of lettuce). The bacteria-infiltrated tissue collapsed within 12 h. At the cellular level, failure to plasmolyse was first observed in mesophyll cells 3 h after inoculation. The phytoalexin lettuceninA was detected after the appearance of cells undergoing irreversible membrane damage, and concentrations subsequently increased rapidly (Bennett et al., 1994).

Bestwick et al. (1995) provided more evidence to show that dying cells are important in induction of phytoalexins. They inoculated lettuce leaves with a wild isolate of *P. savastanoi* pv. *phaseolicola* which induced collapse of tissue (characteristic of HR lesion) within 8 h after inoculation. When the leaves were inoculated with an *hrp* mutant (*hrpD*$^-$) of this bacterium, no tissue collapse was observed. Cell collapse was measured by (1) assessing membrane integrity by counting the number of cells failing to plasmolyze, and (2) assessing increase in electrolyte leakage (Bestwick et al., 1995).

The results showed that only the wild-type strain of the incompatible bacterium induced loss in membrane integrity; the *hrp* mutant did not induce any significant loss in membrane integrity. The wild-type strain, which induced collapse of cells, induced significant accumulation of the phytoalexin lettuceninA, while the *hrp* mutant, which did not induce cell collapse, did not induce an appreciable amount of the phytoalexin (Bestwick et al., 1995). It indicates that the phytoalexin is induced in apoptosis cells undergoing membrane integration.

Accumulation of phytoalexins in necrotic cells has been demonstrated in cotton leaves inoculated with an incompatible strain of *X. axonopodis* pv. *malvacearum* (Pierce and Essenberg, 1987; Gorski et al., 1995). In leaves of resistant cotton lines infiltrated with the bacterial pathogen, mesophyll cells closest to the bacterial colony collapse and turn brown and fluorescent yellow-green (Pierce and Essenberg, 1987). Mesophyll cells isolated from cotton cotyledons responding hypersensitively to *X. axonopodis* pv. *malvacearum* were subjected to fluorescence-active cell sorting. An average brightly fluorescent cell contained 40 times as much of the most potently antibacterial phytoalexin, 2,7-dihydroxycadalene (DHC), 10 to 25 times as much of the yellow-green fluorescent phytoalexins lacinilene C (LC) and lacinilene C7-methyl ether (LCME), and 10 times as much as 2-hydroxy-7-methoxycadalene (HMC) as an average less fluorescent cell (Pierce and Essenberg, 1987). The phytoalexins were predominantly localized in the necrotic cells at sites of bacterial infection (Essenberg et al., 1992). The yellow-green fluorescent cells were the predominant sites of phytoalexin localization (Essenberg et al., 1992).

Phytoalexin accumulates only in host tissues undergoing hypersensitive necrosis. *P. syringae* pv. *syringae*, a wheat pathogen, is nonpathogenic on *Arabidopsis thaliana* and elicits an HR. Phytoalexins accumulated in leaves of *A. thaliana* when inoculated with the wheat pathogen which induced rapid cell necrosis. No phytoalexin accumulation was detected after infiltration of leaves with a mutant of *P. syringae* pv. *syringae* deficient in the ability to elicit cell necrosis (Tsuji et al., 1992). The crucifer pathogen *X.*

campestris pv. *campestris,* which did not induce hypersensitive cell necrosis, did not induce an accumulation of phytoalexin (Tsuji et al., 1992). Coumestrol accumulates in soybean leaves that respond hypersensitively to incompatible races of *P. savastanoi* pv. *glycinea* (Martin and Dewick, 1980).

Jakobek and Lindgren (1993) provided evidence that cell necrosis is not needed for accumulation of phytoalexins in bean. They reported an increase in phytoalexins in macroscopically symptomless bean tissues infiltrated with Hrp$^-$ *P. syringae* pv. *tabaci.* However, even *Hrp*$^-$ bacteria may induce microscopically visible cell death. A very low number of dead cells have been observed in response to the *hrpD*$^-$ strain of *P. savastanoi* pv. *phaseolicola* in lettuce leaves (Bestwick et al., 1995). Hence, microscopic-level cell death in bean inoculated with *hrp*$^-$ bacteria cannot be ruled out. Meier et al. (1993) found no phytoalexins in bean in the absence of necrosis characteristic of the HR. Thus there was a close link between the occurrence of irreversible membrane damage and induction of phytoalexin synthesis. A threshold level of membrane dysfunction may be essential for the release of endogenous elicitors.

PHYTOALEXINS ACCUMULATE ONLY LOCALLY AND NOT SYSTEMICALLY

Phytoalexins appear to accumulate in localized areas around the attempted infection zone. This is in contrast to many PR proteins, which accumulate systemically. Meier et al. (1993) have shown that when bean leaves were infiltrated with an avirulent race of *P. savastanoi* pv. *phaseolicola,* flavonoid phytoalexins accumulated only in the infiltrated zone, and no phytoalexins could be detected in the remainder of the leaf. *P. syringae* pv. *pisi,* a nonpathogen of alfalfa, was infiltrated into alfalfa leaflets (Sallaud et al., 1997). The flavonoid phytoalexins accumulated only in the infiltrated zone, and in the noninfiltrated zone no phytoalexins could be detected (Sallaud et al., 1997).

Essenberg et al. (1992) have shown that phytoalexins are localized in the necrotic cells at sites of infection by avirulent isolates of *X. axonopodis* pv. *malvacearum* in cotton leaves. Elicitors are known to induce phytoalexins in plants, and the elicitor-induced phytoalexins accumulate locally. A fungal elicitor elicited the phytoalexin glyceollin to levels as high as 1800 nmoles/g of tissue, but only in the uppermost cell layers of treated soybean cotyledons. In underlying cell populations, 5 to 20 cells away from the point of elicitor application no phytoalexin could be detected (Graham and Graham, 1991).

MODE OF SYNTHESIS OF PHYTOALEXINS

It is now well established that cell membrane dysfunction precedes synthesis of phytoalexins in plants. Several bacterial elicitors, the products of various *hrp* and *avr* genes, are known to induce electrolyte leakage and hypersensitive cell death (see Chapter 1). These elicitors induce phytoalexin synthesis. Lipopolysaccharide preparations from *P. syringae* pv. *syringae* elicit phytoalexin synthesis in soybean hypocotyls (Barton-Willis et al., 1984). Oligogalacturonates released from plant cell walls by pectic enzymes produced by bacterial pathogens induce phytoalexins (Nothnagel et al., 1983; Davis et al., 1984; Forrest and Lyon, 1990). An oligogalacturonide elicitor could be isolated from soybean cell walls by treating them with polygalacturonate lyase obtained from *E. carotovora* culture filtrates. This elicitor induces soybean cotyledons to accumulate phytoalexins (Davis et al., 1986).

The breakdown of plant cell membranes induced by bacterial pathogens may release various elicitors of host origin. Various fatty acid esters consisting mainly of *cis*-linoleic acid and α-linolenic acid could be detected in potato tubers inoculated with *E. carotovora* subsp. *atroseptica* (Hildenbrand and Ninnemann, 1994). Linoleic and linolenic acids are efficient elicitors of defense mechanisms in plants (Cohen et al., 1991). Glutathione (γ-L-glutamyl-L-cysteinyl-glycine) is a low molecular weight thiol found in plants. Glutathione causes a massive and selective induction of the transcription of defense genes encoding enzymes of phytoalexin biosynthesis (Wingate et al., 1988).

Several second messengers are involved in the signal transduction systems and have been described in detail in Chapter 1. Ca^{2+} and cyclic AMP stimulate the phytoalexin 6-methoxymellein in carrot cells (Kurosaki et al., 1987a). These substances may participate in the elicitation process as second messengers and transduce primary stimuli from various types of elicitors into common signals for phytoalexin production in carrot cells. Breakdown of plant phospholipids may be involved in the elicitation of phytoalexin production in carrot cells (Kurosaki et al., 1987a).

Certain processes mediated by calmodulin and protein kinase C-like enzyme(s) are involved in elicitation of 6-methoxymellein production, and they cooperatively affect phytoalexin production. PMA (phorbol 12-myristate 13-acetate), a potent activator of protein kinase C, induced 6-methoxymellein production even in the absence of an elicitor (Kurosaki et al., 1987a). PA (phorbol 13-monoacetate), another derivative of phorbol that does not activate protein kinase C, did not show any significant effect on phytoalexin elicitation. A synthetic diacylglycerol OAG (1,-oleoyl-2-acetyl-*rac*-glycerol), a different type of protein kinase C activator, also induced 6-

methoxymellein production without the addition of an elicitor (Kurosaki et al., 1987a). Phytoalexin production in carrot was inhibited by H-7 [1-(5-isoquinolinesulfonyl)-2-methylpiperazine], a well-known inhibitor of protein kinase. W-7, a specific inhibitor of calmodulin-mediated processes, appreciably inhibited the elicitor-induced methoxymellein production in cultured carrot cells. Another calmodulin inhibitor, trifluoperazine, also inhibited the production of the phytoalexin (Kurosaki et al., 1987a).

Phosphatidylinositol-degrading phospholipase activity increased rapidly without a notable lag upon the addition of elicitor (pectin fragment of carrot cells released by the bacterial pectic enzymes). Maximal activity was observed after 3 to 5 minutes; then it decreased gradually (Kurosaki et al., 1987a). In elicitor-treated carrot cells, IP3 (inositol triphosphate) formation was observed at 3 to 5 min after addition of elicitor by which time phospholipase activity also increased. Phosphatidylinositol (PI) response of elicitor-treated carrot cells was found to occur in a very early stage of elicitation (Kurosaki et al., 1987b). By contrast, increase in Ca^{2+} flux and cyclic AMP content was shown to occur in a later stage (Kurosaki et al., 1987b). It is possible that PI response of elicitor-treated carrot cells initiates certain trigger responses which precede reactions for 6-methoxymellein production including Ca^{2+} flux and cyclic AMP generation (Kurosaki et al., 1987a).

Some proteins may be involved in signal transduction. Two cDNA clones of rice, EL2 and EL3, have been isolated as genes responsive within 6 min to N-acetylchitoheptaose, a potent biotic elicitor for phytoalexin biosynthesis (Minami et al., 1996). N-acetylchitoheptaose is the elicitor easily released from chitin polymer of bacterial cell wall, by plant chitinase (Yamada et al., 1993). The mRNA levels of EL2 and EL3 increased even 3 and 6 min after the treatment, respectively (Minami et al., 1996). The role of these proteins in signal transduction is not yet known. Since phytoalexins do not accumulate systemically, the role of systemic signals such as salicylic acid and jasmonic acid in induction of phytoalexins is doubtful.

The signaling system activates specific genes leading to phytoalexin accumulation (Chappell and Hahlbrock, 1984; Cramer et al., 1985). Phytoalexin accumulation precedes accumulation and translation of mRNAs encoding the enzymes responsible for phytoalexin synthesis (Ebel et al., 1984; Ryder et al., 1984). Accumulation of phytoalexins is regulated, at least in part, at the transcriptional level (Collinge and Slusarenko, 1987; Templeton and Lamb, 1988; Dixon and Lamb, 1990). 3-Hydroxy-3-methylgutaryl coenzyme A reductase (HMGR) is an important enzyme in biosynthesis of isoprenoid phytoalexins. HMGR catalyses the first committed step in a pathway leading to isoprenoids and plays a key role in controlling the flux of intermediates through the pathway (Garg and Douglas, 1983; Bach, 1986). Two isoforms of

HMGR have been identified in plants, one bound in the microsomal fraction and another associated with the organelle fraction (Bach et al., 1986).

An elicitor induces HMGR genes. Elicitor induction of HMGR gene transcription has been reported in potato (Stermer et al., 1991; Yang et al., 1991; Choi et al., 1992), tomato (Park et al., 1992), and rice (Nelson et al., 1994). Induction of the HMGR gene was shown to play a role in the complex regulation of phytoalexin accumulation (Stermer et al., 1991). Induction of HMGR leads to production of terpenoid phytoalexins in potato tubers (Stermer and Bostock, 1987; Chappell et al., 1991), sweet potato roots (Suzuki et al., 1975), and tomato cell cultures (Park et al., 1989). The accumulation of sesquiterpenoid phytoalexins in potato is preceded by a rapid increase in HMGR activity (Stermer et al., 1991). Specific inhibitors of HMGR block the rise in the activity and also markedly reduce the accumulation of the phytoalexins in potato (Stermer and Bostock, 1987).

Chalcone synthase (CHS) and isoflavone reductase (IFR) are involved in the synthesis of the phytoalexin medicarpin in alfalfa (Esnault et al., 1993). When alfalfa leaves were infiltrated with incompatible pathogen *P. syringae* pv. *pisi,* the accumulation of the CHS and IFR transcripts was observed 6 h after inoculation (Esnault et al., 1993). The phytoalexin medicarpin accumulation in alfalfa results from rapid increases in the activities of phenylalanine ammonia-lyase (PAL), 4-coumarate ligase (4CL), CHS, chalcone isomerase (CHI), and isoflavone-*O*-methyl transferase (Dalkin et al., 1990). When alfalfa leaves were infiltrated with the incompatible bacterium *P. syringae* pv. *pisi,* accumulation of CHS, CHI, chalcone reductase (CHR), and IFR transcripts was observed within 3 h after inoculation (Sallaud et al., 1997). Various isoflavonoid phytoalexins (other than medicarpin) accumulated at 48 h after inoculation (Sallaud et al., 1997).

Induction of CHS in soybean cultivars treated with *P. savastanoi* pv. *glycinea* has been reported (Dhawale et al., 1989). Induction of CHS results in accumulation of glyceollins in soybean (Habereder et al., 1989). When bean plants were inoculated with avirulent isolates of *P. savastanoi* pv. *phaseolicola,* PAL and CHS transcripts accumulated and it led to rapid synthesis of antibacterial phytoalexins in bean leaves (Meier et al., 1993). Transcripts for PAL, CHS, and CHI accumulated followed by the accumulation of phytoalexin in bean after infiltration with the incompatible bacterium *P. syringae* pv. *tabaci* (Jakobek and Lindgren, 1993).

Activation of the stilbene synthase gene has been shown to induce the phytoalexin resveratrol in tobacco, which does not produce resveratrol without this gene expression. The stilbene synthase gene was transferred from *Vitis vinifera,* which normally produces resveratrol, to tobacco, which does not produce the phytoalexin. The resulting transformed plant makes resveratrol (Hain et al., 1993). It suggests that stilbene synthase gene expression is essential for induction of the phytoalexin resveratrol. When tobacco cell suspension cultures were treated with elicitor, sesquiterpene cyclase activity was induced, resulting in accumulation of sesquiterpenoid phytoalexins (Zook and Kuc, 1991). When PAL, *O*-methyl transferases, and bibenzyl synthase were induced, the phytoalexins hircinol and orchinol accumulated in orchid (Reinecke and Kindl, 1994).

Transcription of the genes involved in the phytoalexin synthetic pathway alone may not be sufficient to lead to the accumulation of phytoalexins. Many reports indicate that, irrespective of the transcription of genes involved in synthesis of all enzymes involved in phytoalexin synthesis, the final product, phytoalexin, is not induced. Total RNA from the infiltrated and noninfiltrated zones was extracted from alfalfa leaflets at different times after infiltration with the incompatible bacterium *P. syringae* pv. *pisi,* and accumulation of CHS, CHR, CHI, and IFR transcripts was analyzed by Northern blots (Table 5.1; Sallaud et al., 1997). None of the four transcripts was detectable in the control leaflets. Both in the infiltrated zone and noninfiltrated tissue, accumulation of all four messenger RNAs was observed (Sallaud et al., 1997). However, the flavonoid phytoalexins accumulated only in the infiltrated zone, and in the noninfiltrated zone no phytoalexins could be detected (Sallaud et al., 1997). It suggests that the accumulation of CHS, CHR, CHI, and IFR transcripts in the cells of the noninfiltrated zone did not lead to the synthesis of phytoalexins in this part of the leaf. Moreover, no phytoalexins were detected during the compatible interaction (Sallaud et al., 1997), while CHS, CHI, and IFR transcripts significantly accumulated during this interaction (Esnault et al., 1997). These results suggest that induction of the genes involved in the phytoalexin synthetic pathway is not sufficient to lead to the accumulation of phytoalexins in alfalfa.

Similar results have been obtained in many other host-pathogen interactions. HMGR, which is the key enzyme in synthesis of sesquiterpenoid phytoalexins, was detected in control or nonelicited tissue of potato tuber tissue (Stermer and Bostock, 1987) or in tobacco cell suspensions (Chappell and Nable, 1987); but no phytoalexin accumulated in those tissues. PAL and CHS transcripts accumulated in the *P. savastanoi* pv. *phaseolicola*-infiltrated zones in bean (compatible interaction) primary leaves, but no phytoalexins accumulated (Meier et al., 1993). When cultured parsley cells were treated with an extracellular polysaccharide from *X. campestris,* a large increase of PAL and 4CL activity was observed, but the phytoalexins coumarins did not accumulate (Kombrink and Hahlbrock, 1986). All these studies suggest that besides transcription of genes encoding biosynthetic enzymes, enzymatic regulation and posttranslational modifications of the polypeptides may be involved in the control of the phytoalexin production.

TABLE 5.1. Flavonoid Accumulation and Peroxidase Activity in Cotton Cotyledons Infected with *Xanthomonas axonopodis* pv. *malvacearum*

		Reaction according to the time course of infection (h) after infiltration of the bacteria								
	Cultivar	**0**	**2**	**4**	**5**	**6**	**9**	**10**	**24**	**48**
Flavonoids	Resistant	–	–	–	–	–	+	+	+	+
	Susceptible	–	–	–	–	–	–	–	–	–
Peroxidase	Resistant	–	–	+	+	+	+	+	+	+
	Susceptible	–	–	–	–	–	–	–	–	–

Source: Dai et al., 1996.

– = no response; + = strong response

EVIDENCE THAT INDUCED SECONDARY METABOLITES ARE INVOLVED IN BACTERIAL DISEASE RESISTANCE

Secondary Metabolites Are Induced Rapidly and to a Greater Degree in Resistant Interactions

Conclusive proof for direct involvement of induced secondary metabolites in bacterial disease resistance is still lacking. But much indirect evidence suggests their role in the resistance of plants against bacterial pathogens. Newly induced secondary metabolites (phytoalexins) are induced specifically in resistant host-pathogen interactions. When leaves of *Arabidopsis* were inoculated with the nonhost pathogen *P. syringae* pv. *syringae,* the phytoalexin camalexin accumulated within 12 h, and maximum accumulation was observed at 36 h (Tsuji et al., 1992; Tsuji and Somerville, 1992). No phytoalexin could be detected during this period in *Arabidopsis* leaves inoculated with the pathogen *X. campestris* pv. *campestris* (Tsuji et al., 1992).

Flavonoid compounds have been shown to accumulate within 9 h after inoculation in resistant cotton-*X. axonopodis* pv. *malvacearum* interactions. In susceptible interactions flavonoids did not accumulate even at 48 h after inoculation (Table 5.1; Dai et al., 1996). The oxidized flavonoids show antimicrobial activity (Andary, 1993). Plant peroxidases have been reported to convert the flavonoids into oxidized or polymerized compounds which show high antimicrobial activity (Patzlaff and Barz, 1978; Takahama, 1988; Morales et al., 1993). Further, peroxidase may be associated in isoflavonoid incorporation into cell wall polysaccharides, making the cell wall resistant to bacterial multiplication (Hrubcova et al., 1994; Dai et al., 1996). Peroxidase activity is also induced in resistant cotton-*X. axonopodis* pv. *malvacearum* interactions and not in susceptible interactions (Table 5.1; Dai et al., 1996).

Flavonoids accumulated at 9 h after inoculation with *X. axonopodis* pv. *malvacearum* while the sesquiterpenoids were observed later, 48 h after inoulation. The sesquiterpene cyclase activity, the first enzyme in the biosynthetic pathway of sesquiterpenoid defense compounds, increases only after 20 h postinoculation (Davis et al., 1996). This indicates that the flavonoids are first synthesized in response to *X. axonopodis* pv. *malvacearum* and the sequiterpenoid phytoalexins are synthesized later. Both these induced compounds are synthesized in the resistant interactions. Besides these phytoalexins, two new phytoalexins have been detected: the *cis* and *trans*-diastereomers of 7-hydroxycalamenen-2-one and *trans*-7'-hydroxycalamenene in cotton leaves inoculated with *X. axonopodis* pv. *malvacearum*. Both of them were induced only in resistant interactions during the period of sesquiterpenoid phytoalexin synthesis (Davila-Huerta et al., 1995).

Stored potato tubers rotted slowly while immature tubers that were inoculated with *E. carotovora* subsp. *atroseptica* rotted very fast (Hildenbrand and Ninnemann, 1994). When the stored potato tubers were inoculated with the bacterial pathogen, rishitin, solavetivone, and phytuberin accumulated rapidly after 24 h, and maximum accumulation was observed at three to five days after inoculation. No phytoalexins could be detected in immature freshly harvested tubers inoculated with the pathogen. No phytoalexins could be detected in healthy potato tissues (Hildenbrand and Ninnemann, 1994). Tubers incubated anaerobically rotted rapidly and extensively and no phytoalexins were detected. When tubers were incubated aerobically, stored mature tubers accumulated rishitin, solavetivone, and phytuberin and exhibited less extensive rotting (Hildenbrand and Ninnemann, 1994). It indicates that phytoalexins accumulated only in resistant potato-*E. carotovora* subsp. *atroseptica* interactions. The sesquiterpenoid phytoalexins 2,7-dihydroxycadalene (DHC), lacinilene C (LC), and lacinilene C7-methyl ether (LCME) accumulated in leaves of upland cotton *(Gossypium hirsutum)* after inoculation with incompatible races of *X. axonopodis* pv. *malvacearum* (Essenberg et al., 1982). Induction of phytoalexins in incompatible host-bacterial pathogen interactions has been reported in several other plants (Keen and Kennedy, 1974; Martin and Dewick, 1980; Tsuji et al., 1992).

Rapid induction of phytoalexins in resistant interactions may occur even at the stage of transcription of genes encoding biosynthetic enzymes. When alfalfa leaves were infiltrated with the incompatible pathogen *P. syringae* pv. *pisi,* maximum accumulation of the CHS and IFR transcripts was observed 6 h after inoculation. When the alfalfa leaves were inoculated with the compatible pathogen *X. campestris* pv. *alfalfae,* the induction of these transcripts was delayed until 25 to 30 h postinoculation, and the level of their accumulation was considerably lower (Esnault et al., 1993). A strong correlation of a rapid accumulation of CHS and IFR mRNAs with resistance was thus observed in alfalfa. CHS and IFR are involved in the synthesis of the phytoalexin medicarpin.

Early and rapid induction of the CHS gene has been observed in soybean cultivars resistant to *P. savastanoi* pv. *glycinea* during pathogenesis (Dhawale et al., 1989). Accumulation of PAL and CHS transcripts was more rapid and generally of greater magnitude in bean leaves inoculated with the incompatible isolate of *P. savastanoi* pv. *phaseolicola* than in the compatible interaction (Meier et al., 1993). Both PAL and CHS mRNAs accumulated in the incompatible interaction by 3 h postinoculation, while in the compatible interaction PAL transcripts began to accumulate about 15 h postinoculation, and CHS transcripts accumulated very weakly around 18 h postinoculation. Phytoalexins were detected in the incompatible interaction by 24 h postinoculation, while no phytoalexins accumulated in the compatible interactions, probably due to a lack of enough induction of the biosynthetic enzymes in the compatible interaction (Meier et al., 1993). Transcripts for PAL, CHS, and CHI accumulated, and phytoalexins were produced in bean after infiltration with the incompatible pathogen *P. syringae* pv. *tabaci.* In contrast, transcript accumulation and phytoalexin production did not occur in bean after infiltration with the compatible pathogen *P. savastanoi* pv. *phaseolicola* (Jakobek and Lindgren, 1993).

Accumulation of phytoalexins specifically in resistant interactions suggests a role for the phytoalexins in bacterial disease resistance. However, the possibility of fortuitous accumulation of phytoalexins in the resistant interactions cannot be ruled out.

Phytoalexin Accumulation Restricts Growth and Spread of Bacterial Pathogens in Infected Host Tissues

The importance of the phytoalexins in bacterial disease resistance has been demonstrated by assessing bacterial growth in host tissues in which phytoalexins start to accumulate during pathogenesis. When lettuce leaves were inoculated with an incompatible bacterium, *P. savastanoi* pv. *phaseolicola,* the phytoalexin lettuceninA accumulated in the inoculated tissues. But when these leaves were inoculated with an *hrp* mutant *(hrpD$^-$)* of this bacterium, only trace amounts of the phytoalexin were noticed (Bestwick et al., 1995). The wild-type bacterium was quickly eliminated (within 48 h) from the leaf tissues where phytoalexin accumulated, while the *hrpD* mutant of the bacterium survived in the leaf tissues where no (or trace amount) phytoalexin accumulated (Bestwick et al., 1995).

When *Arabidopsis* leaves were inoculated with an incompatible pathogen, *P. syringae* pv. *syringae*, the phytoalexin 3-thiazol-2'-yl-indole accumulated within 12 h, and maximum accumulation was observed at 36 h (Figure 5.1; Tsuji et al., 1992). Populations of *P. syringae* pv. *syringae* exhibited a decline in the number of viable bacteria after 24 h in host tissues in which the phytoalexin was observed (Tsuji et al., 1992).

The spread of bacteria in infected tissues may be suppressed by the phytoalexins. Infiltration of leaves with the compatible pathogen *X. axonopodis* pv. *alfalfae* resulted in spreading water-soaked lesions in alfalfa. The incompatible pathogen *P. syringae* pv. *pisi* induced a hypersensitive reaction (HR). Both the compatible and incompatible pathogens grew within the infiltrated area almost similarly. But *X. axonopodis* pv. *alfalfae* began to spread in the leaf blade from the fourth day onward while the *P. syringae* pv. *pisi* population was strictly limited to the infiltrated area (Esnault et al., 1993).

The phytoalexin medicarpin and its biosynthetic enzymes accumulated only in the *P. syringae* pv. *pisi*-inoculated zone and not in the compatible pathogen-inoculated tissues (Esnault et al., 1993). The results suggest that the accumulated phytoalexins may be bacteriostatic and would have pre-

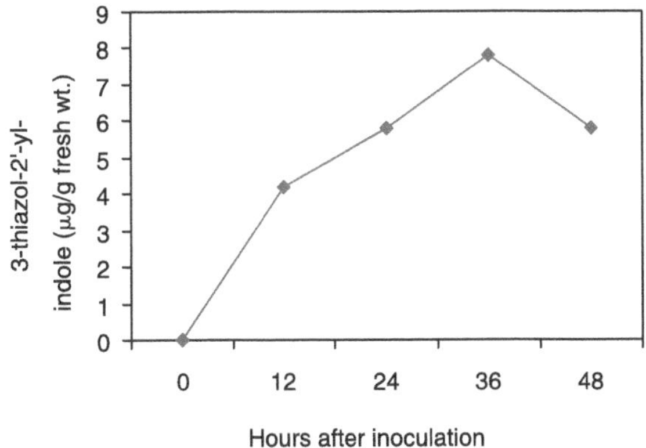

FIGURE 5.1. Time course of accumulation of 3-thiazol-2'-yl-indole in *Arabidopsis thaliana* following inoculation with *Pseudomonas syringae* pv. *syringae* (*Source:* Tsuji et al., 1992).

vented further spread of the incompatible bacteria. In contrast, the compatible pathogen, which did not induce the phytoalexins, spread to the uninjected zone.

The HR developed rapidly in lettuce leaves infiltrated with the incompatible pathogen *P. savastanoi* pv. *phaseolicola*. Observation of infiltrated tissue under the electron microscope revealed no increase in bacterial numbers between 12 and 24 h after inoculation. The phytoalexin lettuceninA accumulated within the tissue undergoing HR within 24 h after inoculation (Bennett et al., 1994). The highest concentration, 4.40 $\mu g\ g^{-1}$ fresh weight, was found in collapsed brown tissue within the zone of infiltration. Tissue dissected from the lesion edge including mostly green, apparently unaffected cells contained 0.39 $\mu g\ g^{-1}$ fresh weight and in the surrounding completely green leaf no lettuceninA was detected. There was close correlation between the accumulation of lettuceninA and the restriction of bacterial growth. The phytoalexin is found in or around the dead plant cells where bacterial growth is restricted (Bennett et al., 1994). The levels of the phytoalexin glyceollin accumulation in various cultivars of soybean after infection with *P. syringae* pv. *glycinea* were inversely correlated with bacterial multiplication (Long et al., 1985). Although these studies indicate the possible role of phytoalexins in suppression of bacterial growth, possible involvement of other induced defense chemicals such as PR proteins in inhibition of bacterial growth cannot be ruled out.

Phytoalexins Are Inhibitory to Bacterial Pathogens in Vitro

Several phytoalexins have been shown to inhibit bacterial pathogens in vitro. The lettuce phytoalexin lettuceninA was found to be highly inhibitory to *P. savastanoi* pv. *phaseolicola*. Clear zones of inhibition of bacterial growth were produced by 0.5 μg samples of lettuceninA applied to lawns of *P. savastanoi* pv. *phaseolicola* (Bennett et al., 1994). In the *P. savastanoi* pv. *phaseolicola*-infected lettuce leaves, the phytoalexin lettuceninA reached concentrations in vivo that are antibacterial in vitro (Bennett et al., 1994). The phytoalexin coumestrol has been reported to be toxic to various bacterial pathogens (Keen and Kennedy, 1974; Fett and Osman, 1982). The bean phytoalexin phaseollin shows antibacterial activity (Wyman and Van Etten, 1978). The soybean phytoalexin glyceollin inhibits several pathogens (Fett and Osman, 1982). Flavonoid phytoalexins have been demonstrated to have bactericidal effects against several *Xanthomonas* species (Venere, 1980). Camalexin, the phytoalexin of *Arabidopsis thaliana,* was found to be toxic to several bacterial pathogens (Tsuji et al., 1992; Rogers et al., 1996). There is a critical threshold concentration of camalexin for killing bacteria. Below the threshold concentration, camalexin had only a modest effect on viability of the bacterial pathogens; above the threshold, viability dropped by several orders of magnitude. The threshold concentration for killing *P. syringae* pv. *maculicola, P. savastanoi* pv. *phaseolicola,* and *X. campestris* pv. *campestris* was between 250 and 500 μg/ml (Rogers et al., 1996). Camalexin inhibits growth of *P. syringae* pv. *syringae* at 1 μg/ml concentration (Tsuji et al., 1992).

Camalexin disrupts the integrity of *P. syringae* membranes (Rogers et al., 1996). The effect of camalexin on the membrane integrity of *P. syringae* pv. *maculicola* was assessed by estimating the rate of proline uptake by the bacteria (Rogers et al., 1996). Failure of bacteria to take up labeled proline is good evidence for membrane disruption (Galvez et al., 1991). Addition of 500 μg/ml camalexin immediately stopped proline uptake by *P. syringae* pv. *maculicola* and also caused the efflux of most of the labeled proline taken up prior to camalexin treatment (Rogers et al., 1996). Camalexin caused increases in ion leakage, protein leakage, and supernatant serine deaminase activity (Rogers et al., 1996).

Rishitin and phytuberin at concentrations detected in potato tubers infected with *E. carotovora* subsp. *atroseptica* showed antimicrobial activity in biotests (Lyon, 1978). Targets of the lipophilic sesquiterpenoid phytoalexins may be membranes, as structural modifications of the outer membranes of the bacteria were observed both in vitro and in planta (Robertson et al., 1985; Lyon et al., 1989). Several phytoalexins have been shown to disrupt microbial membrane function (Smith, 1982). Phaseollin (Smith, 1982), kievitone (Smith, 1982), maackiain (Smith, 1982), and rishitin (Robertson et al., 1985) are known to disrupt membrane.

The cotton sesquiterpenoid phytoalexins 2,7-dihydroxycadalene (DHC) and lacinilene C (LC) show antibacterial activity (Sun et al., 1989). In the dark, 0.1 mM DHC only partially inhibited multiplication of *X. axonopodis* pv. *malvacearum,* but when irradiated, this concentration of DHC was bactericidal. DHC induces single-strand breaks in bacterial DNA. LC also is photoactivated to nick DNA. Catalytic activities of deoxyribonuclease I and malate dehydrogenase were greatly reduced after incubation with DHC plus radiation (Sun et al., 1989).

Lectin, a hydroxyproline-rich glycoprotein (HRGP), plays an important role in bacterial disease resistance. It acts as an agglutinin and immobilizes the bacteria. The role of lectins in disease resistance is discussed in Chapter 2. Some of the phytoalexins facilitate the action of lectins in limiting growth of bacteria in plant tissues. Robertson et al. (1985) showed that the outer surface of cells of *E. carotovora* subsp. *atroseptica* was modified by rishitin (a phytoalexin of potato) in vitro. The potato lectin did not bind to the surface of normal (untreated) cells of *E. carotovora* subsp. *atroseptica*, but did bind to the surface of rishitin-treated cells, suggesting that a surface layer had been removed to expose N-acetylglucosamine oligomers.

Although toxicity of several phytoalexins against bacterial pathogens has been reported, there are also reports that some of the phytoalexins do not inhibit bacterial pathogens. The effect of the alfalfa flavonoid phytoalexins 4',7-dihydroxyflavone and 2',4,4'-trihydroxychalcone was tested separately at 1, 5, and 10 μM concentrations, on the growth of *P. syringae* pv. *pisi* and *X. axonopodis* pv. *alfalfae*. Even at 10 μM, none of the treatments significantly affected bacterial growth (Sallaud et al., 1997). It should also be noted that the concentration of phytoalexins accumulating at the cellular level in the infection zone is difficult to assess. It is still not possible to readily measure the phytoalexin concentrations to which the bacterial pathogens are exposed in planta, and hence in vitro tests may not reflect the in planta events.

PHYTOALEXINS MAY BE SUPPRESSED, DEGRADED, OR INACTIVATED IN SUSCEPTIBLE INTERACTIONS

In many host-bacterial pathogen susceptible interactions, induction of phytoalexins is suppressed. When bean plants were inoculated with the incompatible pathogen *P. syringae* pv. *tabaci*, phytoalexins were induced. Prior infiltration with the compatible pathogen *P. savastanoi* pv. *phaseolicola* suppressed the accumulation of phytoalexins induced by the incompatible pathogen (Jakobek et al., 1993). Prior infiltration with *P. savastanoi* pv. *phaseolicola* suppressed accumulation of the phytoalexin synthetic enzymes PAL, CHI, and CHS that occurs in bean after infiltration with *P. syringae* pv. *tabaci* (Table 5.2; Jakobek et al., 1993) or the elicitor glutathione.

The suppressor activity was lost when *P. savastanoi* pv. *phaseolicola* cells were heat killed or treated with protein synthesis inhibitors (kanamycin or neomycin), indicating that active metabolism is a prerequisite for suppressor activity. The experiments with glutathione also suggest that suppressor activity is not due to some form of interaction between bacterial strains, such as competition for plant recognition or binding sites (Jakobek et al., 1993). The compatible pathogen *P. savastanoi* pv. *phaseolicola* may produce suppressor(s) of the phytoalexin synthesis.

Transposon insertion mutants of *X. campestris* pathovars that induce hypersensitive reactions on normally susceptible plants have been isolated (Daniels et al., 1984; Kamoun et al., 1992). It suggests that these bacterial pathogens might produce suppressors. Phaseolotoxin, commonly produced by *P. savastanoi* pv. *phaseolicola,* has been shown to inhibit the production of the hypersensitive response and phytoalexin production in resistant bean

TABLE 5.2. Suppression of PAL, CHS, and CHI Transcript Accumulation by *Pseudomonas savastanoi* pv. *phaseolicola* in Bean

		Intensity of transcript accumulation		
8 hours	**8 hours**	**PAL**	**CHS**	**CHI**
Water	Water	–	–	–
Water	*P. s.* pv. *tabaci*	++	+++	++
P. s. pv. *phaseolicola*	*P. s.* pv. *tabaci*	–	+	–
P. s. pv. *phaseolicola*	Water	–	–	–

Source: Jakobek et al., 1993.

Note: P. syringae pv. *tabaci* = nonpathogen of bean; *P. savastanoi* pv. *phaseolicola* = pathogen of bean; – : no transcript detected; +, ++, +++ : increasing concentration of transcript accumulation

cultivars (Gnanamanickam and Patil, 1977). However, phaseolotoxin could not be shown to be responsible for the suppressor activity. Toxin-minus mutants of the same strain of *P. savastanoi* pv. *phaseolicola* suppressed PAL, CHS, and CHI transcript accumulation in a manner identical to the wild-type strain (Jakobek et al., 1993).

The produced phytoalexin may be inactivated in the susceptible interactions. Rogers et al. (1996) showed that while King's medium B containing 500 μg of camalexin was toxic to *P. syringae* pv. *maculicola,* intracellular fluid extracted from *Arabidopsis* leaves with camalexin added to a concentration of 500 μg/ml was not toxic to *P. syringae* pv. *maculicola.* When *P. syringae* pv. *maculicola* cells were grown in intercellular fluid and then transferred to King's medium B containing camalexin, killing was indistinguishable from that observed when cells were grown in King's medium B, suggesting that intercellular fluid does not induce resistance to camalexin. Camalexin was not degraded in the intracellular fluid. It suggests that a component of intracellular fluid sequesters camalexin in a nontoxic form (Rogers et al., 1996).

Phytoalexins are degraded by fungal pathogens (Vidhyasekaran, 1997). Degradation of phytoalexins by bacterial pathogens has not yet been demonstrated. There is a report that a bacterial pathogen does not degrade phytoalexin (Rogers et al., 1996). Camalexin (sublethal dose of 50 μg/ml) was added to a culture of *P. syringae* pv. *maculicola.* Over five days, the total amount of camalexin in media with and without *P. syringae* pv. *maculicola* remained unchanged, and the cells grew to a high density (Rogers et al.,

1996). It suggests that *P. syringae* pv. *maculicola* does not seem to have the capacity to degrade camalexin. However, there is indirect evidence indicating that bacterial pathogens also may degrade phytoalexins. When potato roots were inoculated with *E. carotovora* subsp. *atroseptica,* the phytoalexin rishitin could be detected from day 2 and the maximum phytoalexin could be detected at four days postinoculation. However, on day 6 no phytoalexin could be detected in the disintegrated root tissues (Figure 5.2; Abenthum et al., 1995). It suggests that the phytoalexin is quickly degraded.

SOME PHYTOALEXINS MAY NOT HAVE ANY ROLE IN DISEASE RESISTANCE

Phytoalexins also have been shown to accumulate in many susceptible interactions. Several phytoalexins, including rishitin, solavetivone, phytuberin, phytuberol, and hydroxysolavetivone, accumulated in potato stem tissues inoculated with *E. carotovora* subsp. *atroseptica.* There was no evident correlation between phytoalexin content and resistance against *E. carotovora* subsp. *atroseptica* (Abenthum et al., 1995). In fact, rishitin content was greater in stem tissues of susceptible potato cultivars than in the resistant cultivar infected with *E. carotovora* subsp. *atroseptica* (Abenthum et al., 1995). The virulent pathogen *P. syringae* pv. *maculicola* elicits the synthesis of camalexin to high levels in *Arabidopsis* ecotype Columbia (Glazebrook and Ausubel, 1994). Three *p*hyto*a*lexin-*d*eficient *(pad) Arabidopsis* mutants that synthesize decreased levels of the phytoalexin camalexin have been isolated, and their reaction to

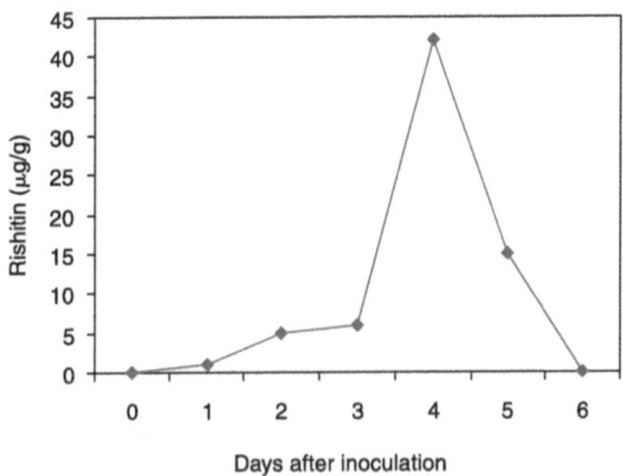

FIGURE 5.2. Time course of accumulation of the phytoalexin rishitin in potato roots after inoculation with *Erwinia carotovora* subsp. *atroseptica* (*Source:* Abenthum et al., 1995).

avirulent strains of *P. syringae* was not different from the wild ecotype (Glazebrook and Ausubel, 1994; Ausubel et al., 1995). It suggests that phytoalexins may not have any role in disease resistance in this host-pathogen interaction. However, it is known that phytoalexins accumulate locally in necrotic regions, and in susceptible interactions necrotic regions are larger than they are in resistant interactions. When local concentrations in the infected tissues are measured, phytoalexin concentrations in the hypersensitive cell death regions may be higher. Further, the phytoalexins normally accumulate earlier in the resistant interactions.

CONSTITUTIVE, BUT INDUCED SECONDARY METABOLITES DURING PATHOGENESIS

Some of the induced secondary metabolites are also detected constitutively. But these secondary metabolites accumulate severalfold during pathogenesis. Plant polyphenols are the important secondary metabolites belonging to this group. Resistance to some bacterial diseases has been shown to be due to induced phenolics. When bulbs of purple onion cv. 'Red Creole' (a resistant cultivar) were inoculated with *Burkholderia cepacia,* phenolic concentration increased severalfold (Figure 5.3; Omidiji and Ehimidu, 1990). Another group of phenolic compounds, termed tsibullins, also has been shown to increase in concentration due to infection (Dmitriev et al., 1986, 1988).

PAL is the key enzyme involved in increasing concentrations of phenolics in infected host tissues. The levels of PAL mRNA increased in *Arabidopsis* leaves inoculated with an incompatible strain of *X. campestris* pv. *campestris*. The strongest PAL mRNA accumulation was noticed 48 h after the incompatible interaction was established. In contrast, PAL was very weakly induced by the compatible bacterial strain (Lummerzheim et al., 1993).

Phenolics are shown to inhibit several bacterial pathogens. Two phenolic compounds, isohamnetin 3-glucoside and quercetin 3'-glucoside, were identified in onion bulb tissues, and the two glycosides were most abundant in the onion cv. 'Red Creole.' Isohamnetin 3-glucoside was highly toxic to *B. cepacia* (Omidiji and Ehimidu, 1990). It was 50 times more toxic to *B. cepacia* than quercetin 3'-glucoside.

Polyphenols (tannins) inhibit the growth of several mutants of *Erwinia chrysanthemi,* altered in their siderophore-mediated iron transport pathway. Mutant growth was restored by the addition of iron (III) to the medium (Mila et al., 1996). Polyfunctional polyphenols, with their several chelating *O*-dihydroxyphenyl groups per molecule, also removed iron (III) from other iron/ligand complexes more efficiently than monofunctional low molecular

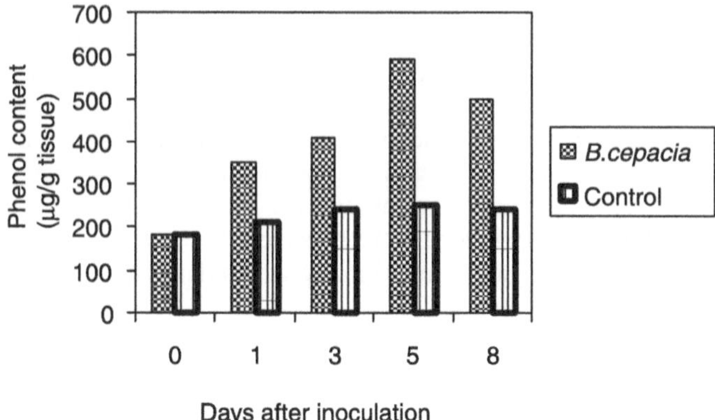

FIGURE 5.3. Accumulation of phenolics in onion bulbs after inoculation with *Burkholderia cepacia* (*Source:* Omidiji and Ehimidu, 1990).

weight phenols. Growth inhibition of the mutants is thus explained by their inability to assimilate iron in the presence of polyphenols. Polyphenols mimic animal iron-binding proteins such as transferrin and may protect plants by withholding iron from pathogens (Mila et al., 1996).

Jalali et al. (1976) reported that anthocyanin compounds play a vital role in cotton resistance to *X. axonopodis* pv. *malvacearum* by inactivating both polygalacturonase and pectin methyl esterase activities of the pathogen. Flavonoid compounds have been found to be inhibitory to several *Xanthomonas* spp. (Venere, 1980). α-Pinene, a monoterpenoid, was found induced in *Pinus sylvestris* after wounding and was highly inhibitory to the growth of *Agrobacterium tumefaciens* (Aronen, 1997). Conifers contain the monoterpene α-pinene and a phenolic constituent *trans*-stilbene (Aronen, 1997). These compounds did not inhibit virulence induction in *A. tumefaciens* but caused a reduction in the frequency of gall formation in susceptible birch *(Betula pendula)* seedlings when applied after the *vir*-gene induction. The inefficiency of *Agrobacterium* spp. in infecting many conifers may not be the result of insufficient *vir*-gene induction, but rather may be the result of interference by chemical defense compounds during the later stages of T-DNA transformation.

Although phenolics and terpenoids may play an important role in disease resistance, it is also known that the bacterial pathogens may degrade or adapt to the induced chemicals. Microbial degradation of phenolics (Wolfgang and Wolfgang, 1975) and bacterial adaptation to these toxic chemicals (Omidiji and Ehimidu, 1990) have been reported. Probably early and high accumulation of phenolics may contribute for disease resistance. Transgenic potato plants constitutively expressing the pectate lyase gene from *Erwina carotovora* induced the transcription of the PAL gene to a high level and showed resistance to *E. carotovora* (Wegener et al., 1996).

CONCLUSION

Several secondary metabolites are induced during incompatible plant-bacterial interactions. Most of the secondary metabolites are produced de novo during pathogenesis, and they are called phytoalexins. Phytoalexins usually are not found to accumulate in host tissues in the absence of cell necrosis. Irreversible cell membrane damage appears to be important in the induction of phytoalexins. It suggests that some endogenous elicitors may be released during the membrane damage. *hrp* and *avr* gene products of the bacteria may also be involved in induction of phytoalexins, as these genes are known to induce host cell membrane damage. Bacterial enzymes may release oligogalacturonides from host cell walls, and these oligomers induce phytoalexins in plants.

Phytoalexins accumulate only around infection sites, and systemic accumulation of phytoalexins has not been observed in plant-bacterial interactions. PR proteins normally accumulate systemically, and hence signaling systems for the induction of phytoalexins may be different from those involved in the induction of PR proteins. Transcription of several genes encoding the biosynthetic enzymes precedes phytoalexin synthesis in plants. PAL, CHS, CHI, IFR, and HMGR are the common enzymes involved in biosynthesis of various phytoalexins. Each one of them exists in multiple forms encoded by several genes. Not all the genes are induced by pathogens or elicitors; only a few of them are inducible. Even when all the enzymes needed for a phytoalexin synthesis are induced, the specific phytoalexin is not induced. A coordinated induction of these enzymes and posttranslational modifications of the transcribed gene products may be necessary for phytoalexin synthesis.

Phytoalexins may accumulate in both compatible and incompatible interactions, as cell membrane dysfunction (evidenced by increased electrolyte leakage) and cell necrosis (hypersensitive and normosensitive necrosis) are commonly observed in both types of interactions. Phytoalexins are quickly degraded, and pathogens can adapt to the phytoalexins. Early induction of electrolyte leakage and rapid cell necrosis (hypersensitive reaction) are characteristic of resistant interactions. Hence, early induction of phytoalexins in resistant interactions is possible. Rapid and large accumulations of phytoalexins may lead to disease resistance. Signals required for early induction should be studied in detail to manipulate them for inducing resistance against bacterial pathogens. Some constitutive secondary metabolites are also induced after infection. Overexpression of these metabolites may also contribute to disease resistance. Molecular manipulation to overexpress these metabolites may be another ideal tool to manage bacterial diseases. The next chapter deals with various options to manipulate disease resistance at the molecular level.

Chapter 6

Biotechnological Applications: Molecular Manipulation of Bacterial Disease Resistance

INTRODUCTION

Molecular biological studies have unraveled several new options to manage bacterial diseases in plants. Several strategies are available to engineer plants against bacterial pathogens, and they are based on the manipulation of (1) the host's regulatory mechanisms involving signal perception and transduction, (2) single-gene defense mechanisms (such as PR proteins), and (3) multigene defense mechanisms (such as biosynthesis of phytoalexins). Genes from bacteria, fungi, viruses, insects, and mammals can also be inserted into plants to confer resistance against plant bacterial pathogens. This chapter reviews the attempts made so far exploiting this new technology for effective management of bacterial diseases and suggests future biotechnological strategies to control bacterial diseases effectively.

MANIPULATION OF THE SIGNAL TRANSDUCTION SYSTEM FOR INDUCTION OF DISEASE RESISTANCE

Molecular biological studies have revealed that bacterial disease resistance is governed by various disease resistance genes and defense genes. Several disease resistance genes have been cloned during this decade, and interestingly all of them have been shown not to code for any toxic defense chemicals, which can contribute to disease resistance. All of them act as regulatory genes, which regulate action of defense genes. In other words, only the defense genes are functional genes, which encode several defense chemicals. Resistance genes appear to be involved in recognition of signals from invading pathogens and subsequent activation of defense genes.

Molecular biological studies have also shown that both disease resistance genes and defense genes are present in both resistant and susceptible varieties. Disease resistance genes found in one plant species have also been detected in other plants. Functional homologs of the resistance genes *RPM1* and *RPS2* detected in *Arabidopsis thaliana* could be detected in pea *(Pisum sativum),* bean *(Phaseolus vulgaris),* and soybean *(Glycine max)* (Whalen et al., 1991; Dangl et al., 1992; Innes et al., 1993b). Nine disease resistance genes cloned from soybean are highly homologous to the *N* gene of flax, and the other two are similar to *RPS2* in *Arabidopsis* (Yu et al., 1996). Analogs of already cloned resistance genes have been detected in rice, wheat, barley, cotton, soybean, pepper, chickpea, flax, tomato, tobacco, and *Brassica napus* (Lyon et al., 1998; Graham et al., 1998; Vallad et al., 1998). Homologs of the disease resistance genes cloned from resistant varieties have been detected in susceptible varieties at the corresponding locus (Mindrinos et al., 1994; Jones et al., 1994; Whitham et al., 1994). However, most of these resistance genes are highly specific in their function, being effective against particular strains of the pathogen. It suggests that specific signals from the pathogen are required to activate these resistant genes. Even single amino acid changes in the resistance gene products result in nonrecognition of signals from pathogens (Bent et al., 1994; Grant et al., 1995).

The structure of the already cloned resistance genes reveals that all of them are involved in the signal transduction system (Table 6.1). A perusal of Table 6.1 reveals that all resistance gene products except tomato Pto contain leucine-rich repeat (LRR) domains. Even the *Pto* locus contains two genes, *Pto* and *Prf,* and both of them appear to be required for conferring resistance to the bacterial pathogen *P. syringae* pv. *tomato;* Prf contains the LRR domain. LRRs have been implicated in protein-protein interactions and ligand binding in signal-transducing proteins (Baker et al., 1997). Besides serving as receptors of pathogen signals, LRRs may act downstream of the signal transduction pathway because of its function in protein-protein interactions (Baker et al., 1997). The presence of nucleotide binding sites (NBS) and receptor kinases (serine/threonine kinases) in resistance gene products indicates that all resistance genes encode proteins involved in signal recognition and transduction (Zhou et al., 1995).

Defense genes are the functional genes encoding PR proteins, phytoalexins, phenolics, lignin, callose, hydroxyproline-rich glycoproteins, and other defense chemicals. Defense genes are also present in both resistant and susceptible plants. In normal healthy plants, they are almost undetectable; they are all induced by specific signals. In other words, defense genes are sleeping genes in healthy plants, and signals are required to activate them. This theory is supported by the well-known phenomenon called systemic induced resistance (SAR).

Plants which are normally susceptible to a particular pathogen can be made resistant by a predisposing infection with a pathogen (compatible or

TABLE 6.1. Structure of Cloned Resistance Genes

Plant	R gene	Structure
Arabidopsis	*RPS2*	LZ-NBS-LRR
	RPM1	LZ-NBS-LRR
	RPP5	TIR-NBS-LRR
Rice	*Xa21*	Serine/threonine kinase-LRR
	Pib	NBS-LRR
Tomato	*Pto*	Serine/threonine kinase
	Pti	Serine/threonine kinase
	Prf	LZ-NBS-LRR
	Cf9	LRR
	Cf2	LRR
	Cf4	LRR
	Cf5	LRR
Tobacco	*N*	TIR-NBS-LRR
Flax	*L6*	TIR-NBS-LRR
	M	TIR-NBS-LRR
Sugar beet	*HS1*$^{pro-1}$	LRR
Wheat	*LR10*	Protein kinase-LRR

Source: Martin et al., 1993; Bent et al., 1994; Jones et al., 1994; Mindrinos et al., 1994; Whitham et al., 1994, 1996; Grant et al., 1995; Ori et al., 1995; Lawrence et al., 1995; Song et al., 1995; Dixon et al., 1996; Salmeron et al., 1996; Tang et al., 1996; Cai et al., 1997; Feuillet et al., 1997; Hammond-Kosack and Jones, 1997; Wang et al., 1999

Note: LZ, Leucine zipper; NBS, Nucleotide binding site; LRR, Leucine-rich repeats; TIR, TOLL and interleukin-I receptor

incompatible) which causes a necrosis (normosensitive or hypersensitive) (Ross, 1961; Uknes et al., 1993b; Cameron et al., 1994; Mauch-Mani and Slusarenko, 1994). This phenomenon has been termed systemic acquired resistance (SAR) (Chester, 1933; Ross, 1961; Kuc, 1982). Resistance so induced is typically effective against a wide range of pathogens, bacteria, fungi, and viruses (Kuc, 1983, 1987; Madamanchi and Kuc, 1991). The induced resistance lasts for several weeks to months after the initial inducing infection (Kuc, 1982). Certain saprophytes (rhizobacteria) and chemicals [salicylic acid, 2,6-dichloroisonicotinic acid (INA), and benzo-(1,2,3)-thiodiazole-7-carbothioic acid S-methyl ester (BTH)] also induce SAR in plants (Ward et al., 1991; Palva et al., 1994; Lawton et al., 1995; Friedrich et al., 1996; Leeman et al., 1996; May et al., 1996; Pieterse et al., 1996; Wei et al., 1996; Van Wees et al., 1997; Vidhyasekaran, 1998, 1999). SAR has been reported in many plants (Kuc, 1982, 1983, 1987; Metraux et al., 1990; Malamy et al., 1990, 1996; Smith et al., 1991; Smith and Metraux, 1991; Rasmussen et al., 1991; Uknes et al., 1992, 1993; Silverman et al., 1993; Mauch-Mani and Slusarenko, 1994; Cameron et al., 1994; Lawton et al., 1995; Vernooij et al., 1995; Pieterse et al., 1996; Schaffrath et al., 1997).

Since SAR is induced by different types of biotic and abiotic inducers, it is also called induced systemic resistance (ISR). ISR has been defined as active resistance dependent on the host plant's physical or chemical barriers, activated by biotic or abiotic agents (Kloepper et al., 1992). ISR can be better defined as an enhancement of a plant's defensive capacity against a broad spectrum of pathogens that is acquired after stimulation (Hammerschmidt and Kuc, 1995). Induction of ISR has been shown to be due to induction of various defense genes (Schweizer et al., 1989; Brederode et al., 1991; Rebmann et al., 1991a,b; Yalpani et al., 1991; Ward et al., 1991; Bull et al., 1992; Meir et al., 1993; Reimann and Dudler, 1993; van de Rhee and Bol, 1993; Uknes et al., 1993b; Mauch-Mani and Slusarenko, 1994; Delaney et al., 1995; Kmecl et al., 1995; Gorlach et al., 1996; Keller et al., 1996a,b; Malamy et al., 1996; Pieterse et al., 1996; Schaffrath et al., 1997). Salicylic acid, jasmonic acid, and other systemic signals have been shown to induce ISR (White, 1979; Mills and Wood, 1984; Ye et al., 1989; Rasmussen et al., 1991; Malamy et al., 1996). These studies suggest that signal transduction systems can be manipulated to activate defense genes and make susceptible plants resistant.

MANIPULATION OF RESISTANCE GENES INVOLVED IN THE SIGNAL TRANSDUCTION SYSTEM

Resistance genes have now been cloned from several plant species, and they have been shown to be initiators or components of signal transduction chains leading to expression of various defense genes. Resistance genes cloned from resistant varieties can be transferred to susceptible cultivars, making them resistant to bacterial pathogens. It is also possible to transfer R genes from one plant species to another species (Thilmony et al., 1995; Rommens et al., 1995; Michelmore et al., 1996; Whitham et al., 1996). Already several resistance genes against bacterial pathogens have been cloned, and the available resistance genes can be exploited to develop transgenic plants expressing disease resistance. However, proper resistance genes should be selected to develop useful transgenic plants.

Several resistance genes have been shown to act against specific races of a pathogen. The specific Avr product may act as the elicitor molecule and the resistance gene product may be the receptor of the signal. However, many *avr* genes have been shown to be conserved in many races of bacterial pathogens. For example, *avrBs3* from *X. vesicatoria* could be detected in *X. axonopodis* pv. *citri, X. axonopodis* pv. *malvacearum, X. oryzae* pv. *oryzae, X. oryzae* pv. *oryzicola,* and many other bacteria. *P. syringae* pv. *tomato* contains an avirulence gene homologous to *avrA,* the avirulence gene cloned from *P. savastanoi* pv. *glycinea* (see Chapter 1). It is not known whether the resistance gene expressing resistance against the particular *avr* gene from one bacterial pathogen can express its action against another bacterium containing an analog of the same *avr* gene, and this possibility should be studied. It has been demonstrated that the *Pto* gene expressing resistance against *P. syringae* pv. *tomato* with *avrPto* in tomato confers resistance against *P. syringae* pv. *tabaci* containing the *avrPto* gene in tobacco (Thilmony et al., 1995). In such cases the cloned resistance genes can be used in various crops against different bacterial pathogens containing the specific *avr* gene analogs.

However, the resistance genes acting against specific races of bacterial pathogens may not be of much use, as several races exist in most of the bacterial pathogens. But there are reports that resistance genes may act against several races of the pathogen. Although it has been suggested that R gene products interact with *avr* gene products of the pathogen, interaction between R gene and *avr* gene products has not yet been demonstrated. There is only one report indicating interaction between an R gene product and an *avr* gene product. The tomato *Pto* gene product, Pto kinase, has been shown to physically interact with the *P. syringae* pv. *tomato* gene product AvrPto (Tang et al., 1996). But it is known that the *Pto* gene itself is a member of a tightly clustered family of five genes (Martin et al., 1993). Another gene, *Prf,* found in the *Pto* locus is involved in the functioning of *Pto,* and both are required for the induction of resistance against *P. syringae* pv. *tomato* in tomato (Staskawicz et al., 1995; Salmeron et al., 1996). It is not known whether the *Prf* product interacts with the R gene product. Besides *avr* gene products, there may be many other signal molecules from pathogens (see Chapter 1) which activate resistance genes. It has been hypothesized that the resistance genes may be capable of recognizing many conserved or diverse pathogen determinants (Kearney and Staskawicz, 1990; Ronald et al., 1992; Wang et al., 1996).

The *Xa21* gene from rice has been shown to confer resistance against many isolates of *X. oryzae* pv. *oryzae* found in the Philippines, India, Nepal, Indonesia, China, Thailand, Colombia, and Korea. These isolates include well-characterized races of the Philippines, race 1, race 2, race 3, race 4, race 5, race 6, and race 7, probably with different *avr* genes (Wang et al., 1996). Studies were undertaken to determine whether the multi-isolate resistance observed for rice line IRBB21 with the *Xa21* gene was due to a single gene or multiple genes at the *Xa21* locus (the *Xa21* locus contains at least seven genes). It is possible that the locus may encode a single gene product, *Xa21,* that specifies resistance to multiple pathogen isolates, or the locus may be composed of a cluster of tightly linked genes, each of which recognizes a unique isolate-specific determinant (Wang et al., 1996). Transgenic rice plants expressing the cloned *Xa21* gene confer multi-isolate resistance to 29 diverse isolates from eight countries. The resistance spectrum of the engineered line was identical to that of the donor line, indicating that the single cloned gene is sufficient to confer multi-isolate resistance (Wang et al., 1996). It suggests that the *Xa21* gene may be active against many diverse pathogen products. It is also possible that the *Xa21* gene may receive some signal from pathogens or plant cells and several defense genes may be induced. Ronald (1997) has suggested that the Xa21 LRR domain is extracellular, and its function is to bind a polypeptide produced by the pathogen or plant cell and that this specific interaction is mediated by a finite subset of amino acids in the LRR. Specific binding may lead to activation of the Xa21 kinase with subsequent phosphorylation on specific serine or threonine residues. Phosphorylated residues may then serve as binding sites of proteins that can initiate down-stream responses. Another resistance gene *N* cloned from tobacco confers resistance against tobacco mosaic virus (Whitham et al., 1996), in which existence of any *avr* gene is not yet known. Such resistance genes without much specificity will be useful to develop transgenic plants expressing disease resistance against many pathogens.

Resistance genes mostly belong to multigene families. The selection of a useful resistance gene from the clustered family of genes is a prerequisite for development of effective transgenic plants with enhanced resistance to pathogens. The *Pto* gene confers resistance against *P. syringae* pv. *tomato* in tomato (Martin et al., 1993). *Pto* itself is a member of a tightly clustered family of five genes encoding serine/threonine kinase (Martin et al., 1993). Another serine/threonine kinase, *Pti,* is also involved in *Pto*-mediated disease resistance (Zhou et al., 1995). Another gene, *Prf,* found in the *Pto* locus is also essential for activation of the *Pto* gene (Loh and Martin, 1995; Rommens et al., 1995; Salmeron et al., 1996). Zhou et al. (1997) identified three genes, *Pti4, Pti5,* and *Pti6,* encoding proteins showing direct interaction with Pto kinase.

Seven *Xa21* gene family members from the resistant rice line IRBB21 have been cloned. They have been designated *A1, A2, B (Xa21), C, D, E,* and *F* (Song et al., 1995; Wang et al., 1995). These genes have been grouped into two classes based on DNA similarity. One class (designated the *Xa21* class) contains *Xa21, D,* and *F* (Song et al., 1997). Another class (designated the *A2* class) contains *A1, A2, C,* and *E.* Within each class, family members share striking nucleotide sequence identity (98.0 percent average identity for the members of the *Xa21* class; 95.2 percent average identity for the members of the *A2* class). In contrast, a low level of DNA sequence identity was observed between members of the two classes (about 63.5 percent identity between *Xa21* and *A2*) (Song et al., 1997). Both *Xa21* and *A1* encode serine/threonine kinase. But transgenic rice plants containing *Xa21* expressed disease resistance against *X. oryzae* pv. *oryzae,* while the transgenic plants containing the *A1* sequence were susceptible to all *X. oryzae* pv. *oryzae* isolates tested (Song et al., 1997).

In tomato, *Cf2* and *Cf9* loci are composed of five or more related genes (Jones et al., 1994; Dixon et al., 1996). Two nearly identical *Cf2* genes have also been detected in tomato (Dixon et al., 1996). At least 23 tightly linked resistance genes or alleles at the *Mla* locus in barley have been reported (De Scenzo et al., 1994). Hence, careful selection of resistance genes from the clusters may be needed to obtain most successful results. Several failures have been reported in this type of attempt when the selection of proper resistance genes was not made. Jia et al. (1997) isolated alleles of the *Pto* gene from a cultivated tomato *(Lycopersicon esculentum)* and *Lycopersicon pimpenellifolium,* and a transgenic tomato expressing these alleles did not show any enhanced resistance to *P. syringae* pv. *tomato.* Probably other needed genes such as *Pti* and *Prf* would not have expressed in these transgenic plants. Single amino acid changes in the LRR domain of the resistance genes *RPS2* and *RPM1* (of *Arabidopsis*) and *N* (of tobacco) result in failure to develop resistance against pathogens (Bent et al., 1994; Grant et al., 1995).

Resistance genes have been exploited to develop bacterial disease-resistant plants in rice, tomato, and tobacco. The resistance gene *Xa21* has been cloned from a wild rice, *Oryza longistaminata* (Khush et al., 1990). The cloned gene could be transferred to different rice cultivars varying widely in their phenotypic characters, such as *O. sativa* ssp. *japonica* and *O. sativa* ssp., *indica* types (Wang et al., 1996; Tu et al., 1998). The japonica rice cultivar TP309 was transformed with the *Xa21* gene by a particle bombardment technique (Song et al., 1995). The transgenic plants expressing the *Xa21* gene showed resistance to several races of the pathogen (Table 6.2; Wang et al., 1996).

TABLE 6.2. Comparative Reaction of Nontransformed Rice Cultivar TP309 and the Transgenic Plants Expressing the *Xa21* Gene to Different Races of *Xanthomonas oryzae* pv. *oryzae*

	Lesion length in cm	
Race	**TP309**	**Transgenic plants (T1 progeny)**
Race 1	13.0	0.3
Race 2	7.3	0.2
Race 3	8.3	0.1
Race 4	16.5	0.4
Race 5	7.3	0.3
Race 6	16.6	1.0
Race 7	13.5	0.2

Source: Wang et al., 1996.

The indica rice cultivar IR 72 was transformed with the *Xa21* gene, and the transgenic plants showed resistance to both the tested races (race 4 and race 6) of *X. oryzae* pv. *oryzae* (Tu et al., 1998). The transgenic *Xa21* plants showed greater resistance to various isolates of the pathogen compared to the donor cultivar IRBB21 (Song et al., 1995; Wang et al., 1996). An *Xa21*-specific hybridization about 5 to 10 times stronger was observed in the transgenic *Xa21* plants compared to the donor line IRBB21, indicating multiple insertions in a single locus in the transformed lines (Wang et al., 1996). Multiple copies of transgenes are often inherited as a single locus in transgenic lines generated by particle bombardment (Cooley et al., 1995).

Transgenic tomato plants expressing the cloned *Pto* gene showed high resistance to *P. syringae* pv. *tomato* (Martin et al., 1993). Resistance genes from one plant species could be successfully transferred to another plant species. The cloned *Pti* gene from tomato was transferred to tobacco and the transgenic plants showed resistance to *P. syringae* pv. *tabaci* (Zhou et al., 1995). The transgenic tobacco plants expressing the cloned tomato *Pto* gene showed resistance to *P. syringae* pv. *syringae* strains. Similar resistance was observed in *Nicotiana benthamiana* against *P. syringae* strains (Rommens et al., 1995; Thilmony et al., 1995). Transgenic tomato plants carrying the cloned *N* gene of tobacco show resistance to tobacco mosaic virus (Whitham et al., 1996).

There is a lot of potential to exploit disease resistance genes for bacterial disease management and more studies are needed in this aspect. However, the silencing or inactivation of the transgene in the disease-resistant transgenic plants poses a problem. It is known that homologs of resistance genes exist in susceptible plants, and the interactions between an inserted gene and its DNA sequence homologs may lead to silencing of the inserted gene (Jorgensen, 1990; Flavell, 1994). The consequence of the interactions between the loci with DNA sequence homology may result in chromatin restructuring or DNA sequence modification by methylation of different cytosine residues or inhibition of mRNA processing, transport, export, or translation (Flavell, 1994). The expression of the transgene in the transgenic plant may be finally cosuppressed and *trans*-activated (Tu et al., 1998). However, inactivation of the transgene may not occur in all plants. Hence, a large number of progeny should be obtained for selecting a homozygous line with a consistently high level of resistance to the pathogens.

MANIPULATION OF THE SIGNAL TRANSDUCTION SYSTEM BY ELICITORS

Elicitors of both pathogen and host origin are known to be primary signal molecules inducing disease resistance. Pectic enzymes of bacterial pathogens have been shown to be key factors in synthesis of endogenous elicitors in plants. The enzymes can form unsaturated oligogalacturonides from plant cell wall pectin, and these oligogalacturonates activate genes encoding phytoalexins, PR proteins, phenolics, and cell wall proteins (see Chapter 2). Early induction of defense genes will lead to disease resistance. The pathogen should invade host tissues and produce pectic enzymes, which in turn will lead to elicitor formation. This delay may result in pathogen invasion before induction of host defense mechanisms. Hence, it may be ideal that the elicitor is formed before pathogen invasion. Several genes encoding pectic enzymes have been cloned. These genes can be transferred to plants so that the transgenic plants may synthesize an elicitor before pathogen attack. Wegener et al. (1996) have successfully developed transgenic plants expressing the pectate lyase gene (PL3) of *Erwinia carotovora* subsp. *atroseptica,* the potato soft rot pathogen. The transgenic potato plants endogenously expressed the PL3 enzyme, and the defense mechanisms could be activated independently from an attack by *Erwinia*. However, two important factors have to be taken care of before developing transgenic plants: (1) the enzyme should be expressed only after wounding (wounds are important for *Erwinia* infection) and not constitutively (high constitutive expression of pectic enzymes will affect tuber development), and (2) the enzyme should be expressed in tubers, which are the bacterial infection site. In order to achieve a localization of the enzyme in the cytoplasm, the PL3 coding sequence was used without its signal sequence for extracellular secretion (Wegener et al., 1996). The idea was that the PL3 produced intracellularly would be liberated by mechanical cell destruction as a result of tissue wounding. Subsequently, the enzyme would release the oligogalacturonides from the cell wall pectins on the wound surface. Two types of promoters were tested for induction of PL3 genes in potato tuber tissues. Transgenic plants were developed using chimeric genes encoding PL3 of *E. carotovora* subsp. *atroseptica* under control of the patatin B33 gene promoter and the cauliflower mosaic virus (CaMV) 35S promoter. Enzyme production in plant lines transformed with the plasmid containing the patatin B33 promoter was confined to tuber tissue while plants harboring the CaMV 35S promoter exhibited constitutive expression of PL3 more in leaf tissues and less in tuber tissues. Hence, the patatin B33 promoter was found useful in expressing PL3 in potato tuber tissues (Wegener et al., 1996).

Compared with nontransformed plants, the transgenic lines expressing PL3 (under control of the patatin B33 gene promoter) were more resistant to tissue maceration by *E. carotovora* or its enzyme (Wegener et al., 1996). Growth of *E. carotovora* subsp. *atroseptica* on tuber tissue of transgenic plants was less than that in nontransformed plants (Table 6.3; Wegener et al., 1996).

Wounding induced the transcription of the plant defense-related gene encoding phenylalanine ammonia-lyase (PAL) to a high level in tubers of PL3-expressing transgenic line. There was no detectable PAL induction in the susceptible transformant expressing only small amounts of PL3 in tubers (the transformant was developed using chimeric gene CaMV 35S-PL3, which does not express PL3 in tubers) (Wegener et al., 1996). These studies

TABLE 6.3. Growth of *Erwinia carotovora* subsp. *atroseptica* on Potato Tissue of PL-Active Transformants and on Nontransgenic, PL-Inactive Tubers

	Colony forming units (cfu ml^{-1})			
Inoculum density	**PL-inactive control**	**PL-active line D29**	**PL-inactive control**	**PL-active line D19**
1×10^4	6.5×10^5	0	1×10^8	0
1×10^5	7×10^{11}	2×10^5	7×10^{11}	1×10^6

Source: Wegener et al., 1996.

Note: D29 and D19 are the transformants with patatin B33-PL3 fusion; the chimeric gene was expressed only in tubers and not in leaves.

suggest the importance of selection of proper promoter in the gene construction so as to obtain useful disease-resistant transgenic plants.

Elicitors of pathogen origin can also be used to induce resistance. The structure of these elicitors varies greatly. Among them proteinaceous elicitors can be exploited to induce resistance, as a single gene can be manipulated to produce these elicitors constitutively in plants. Harpins are the proteinaceous elicitors produced by *Erwinia amylovora, E. chrysanthemi, E. carotovora* subsp. *carotovora, Pantoea (= Erwinia) stewartii, P. syringae* pv. *syringae, P. savastanoi* pv. *glycinea,* and *P. syringae* pv. *tomato*. Harpins induce resistance in incompatible/nonhosts and incite disease symptoms in susceptible plants. Harpins induce leakage of nutrients. A moderate leakage in the host plant could provide the pathogens with various metabolites originating from the plant cell and thus permit growth of the invading organism. If improperly balanced, such leakage would result in the rapid death of the target plant cells and allow the development of a necrotic response characteristic of HR resistance. Thus, by overproducing harpin at early stages of infection, the susceptible reaction can be converted to a resistant one.

Harpin encoded by *hrpN* of *E. amylovora* induces resistance in seven different plants against various diseases caused by fungi, viruses, and bacteria (Wei and Beer, 1996). HarpinZ_{Pss} protein (=harpin$_{Pss}$), an elicitor isolated from *P. syringae* pv. *syringae,* induced systemic acquired resistance in cucumber to the angular leaf spot bacterial pathogen *P. syringae* pv. *lachrymans*. PR proteins including peroxidase, β-1,3-glucanase, and chitinases were induced in cucumber plants treated with the elicitor (Strobel et al., 1996). These results suggest that harpins can be exploited to manage bacterial diseases. Generally, harpins induce defense genes in resistant hosts. But if harpins are induced early and in high amounts, they may induce resistance, even in susceptible plants. Several harpin genes have been cloned (see Chapter 1) and they can be used to develop transgenic plants expressing the harpin gene in a precise manner.

Harpins induce disease symptoms in susceptible hosts when they induce normosensitive cell death. Inactivation of harpin by reaction of *E. amylovora* cells with an antiserum specific for harpin or a protease that degrades harpin resulted in a reduction in disease of pear caused by *E. amylovora* (Beer et al., 1994). It suggests that techniques should be developed that will inactivate, destroy, or bind with harpins. This will neutralize the roles of harpins in disease development and reduce disease in plants. Research efforts are needed to develop antibodies against harpins, sequence the antibodies produced, and construct nucleic acid sequences which when inserted properly into the genome of a plant would cause the plant to express the antibody and thus prevent bacteria from causing disease in plants. Thus disruption of harpin or of the improper balance of its production would be a novel approach to control bacterial diseases of plants.

The fungal pathogens, *Phytophthora* spp., produce proteinaceous elicitors called elicitins. Elicitins are known to induce systemic resistance against pathogens (Keller et al., 1996a,b). Elicitin induced expression of genes encoding PR-1a, PR-2, PR-5, basic class III chitinase, and basic PR-1 in tobacco. The elicitin genes can also be used to develop transgenic plants expressing ISR against bacterial pathogens, as elicitor-induced resistance is not specific against particular pathogens (Wei and Beer, 1996).

MANIPULATION OF THE SIGNAL TRANSDUCTION SYSTEM USING CHEMICALS

INA

In recent years, chemicals that can induce the signal transduction system have been identified. Among them, 2,6-dichloroisonicotinic acid (INA) has been reported to induce systemic resistance against many diseases. Two synthetic compounds structurally related to INA were also reported to induce resistance in rice (Yoshida et al., 1990; Seguchi et al., 1992). INA was found to protect rice (Metraux et al., 1991), bean (Dann and Deverall, 1995), barley (Kogel et al., 1994), cucumber (Metraux et al., 1991), sugar beet (Nielsen et al., 1994a), rose (*Rosa* sp.) (Hijwegen et al., 1996), and *Arabidopsis* (Uknes et al., 1992) from a number of pathogens.

INA was effective in inducing resistance against *P. syringae* pv. *tabaci* in tobacco (Press et al., 1997). INA induces resistance to *P. syringae* pv. *tomato* (Uknes et al., 1992, 1993) and *P. syringae* pv. *syringae* (Summermatter et al., 1994) in *Arabidopsis.* INA (10 μM) applied as a soil drench induced resistance to *P. syringae* pv. *tomato* in *Arabidopsis*. INA treatment caused a 100-fold reduction in the recovery of viable bacteria from inoculated leaves (Figure 6.1; May et al., 1996).

INA has been shown to act via activation of plant defense mechanisms. *A. thaliana* mutant plants that are no longer protected by INA, in contrast to the corresponding wild-type plants, have been obtained (Cao et al., 1994; Delaney et al., 1995), indicating that defense genes are essential for action of INA. INA shows a dual mode of action: induction of defense responses prior to infection, as well as potentiation of defense responses postinoculation. In sugar beet, INA does not induce β-1,3-glucanase or chitinase, but conditions the plant to induce these PR proteins faster upon attack by *Cercospora beticola* (Nielsen et al., 1994a). Similar results were found in INA-treated

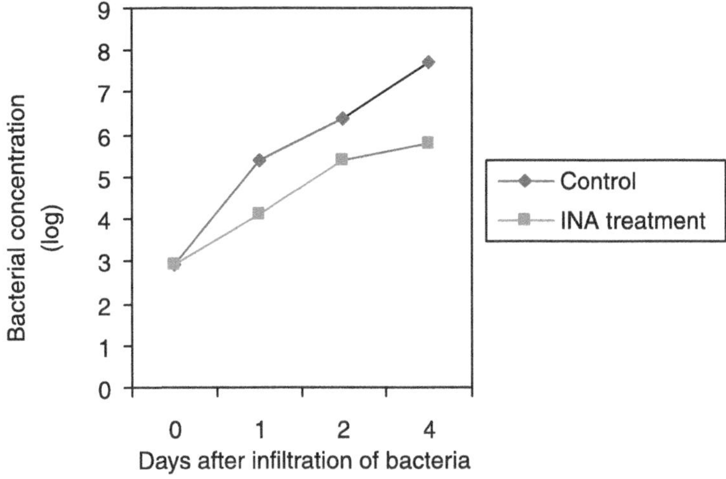

FIGURE 6.1. Effect of INA treatment on growth of *Pseudomonas syringae* pv. *tomato* in *Arabidopsis thaliana* ecotype Col-O (*Source:* May et al., 1996).

cucumber hypocotyls upon challenge by *Colletotrichum lagenarium* (Siegrist et al., 1994).

INA may induce salicylic acid, which may signal the induction of defense genes. SAR, which has been reported to be induced by compatible and incompatible pathogens in various plants, is associated with an early increase in endogenously synthesized salicylic acid (Malamy et al., 1990; Metraux et al., 1990). Accumulation of salicylic acid was critical in the signaling pathway that controls SAR, since plants that do not accumulate salicylic acid were incapable of expressing induced resistance (Delaney et al., 1994; Gaffney et al., 1993). SAR was characterized by activation of so-called SAR genes (Ward et al., 1991). Application of salicylic acid induces SAR and the accumulation of defense genes (White, 1979; Malamy et al., 1990; Cokmus and Sayar, 1991; Uknes et al., 1992, 1993; Summermatter et al., 1994). This type of SAR gene induced by salicylic acid was also found to be induced by INA. In tobacco, coordinated induction of SAR genes by INA has been reported. These genes include those encoding PR-1 to PR-9 proteins plus a gene called SAR 8.2 (Ward et al., 1991). In *Arabidopsis,* the same set of SAR genes and the spectrum of resistance specificity were activated by salicylic acid and INA (Uknes et al., 1992, 1993).

Genetically engineered tobacco and *Arabidopsis* plants unable to accumulate salicylic acid failed to exhibit SAR after induction with pathogens (Gaffney et al., 1993; Vernooij et al., 1994). However, application of INA still induced SAR and SAR gene activity in such plants (Delaney et al., 1995; Vernooij et al., 1994). INA did not induce salicylic acid glucoside accumulation in tobacco (Vernooij et al., 1995; Malamy et al., 1996). INA was effective in inducing resistance against *P. syringae* pv. *tabaci* in both wild-type Xanthi-nc and transgenic tobacco expressing salicylate hydroxylase (NahG) (Table 6.4; Press et al., 1997). It suggests that INA action may not be mediated by salicylic acid. INA may enter a common pathway downstream of salicylic acid synthesis (Malamy et al., 1996). This model is supported by the isolation and characterization of mutants in *Arabidopsis* that are insensitive to INA as well as salicylic acid (Cao et al., 1994; Delaney et al., 1995). INA, as well as salicylic acid, inhibits catalase and ascorbate peroxidase, two key enzymes for H_2O_2 degradation (Chen et al., 1993; Conrath et al., 1995; Durner and Klessig, 1995).

INA induces at least some of the SAR genes induced by salicylic acid in different plants (Uknes et al., 1992; Kogel et al., 1994). INA activates additional regulatory genes besides SAR genes (Nielsen et al., 1994a; Siegrist et al., 1994). INA was found to be a strong inducer of LOX enzyme activity (Schaffrath et al., 1994, 1997; Schweizer et al., 1997). Levels of jasmonic acid were enhanced in leaves of plants treated with INA (Schweizer et al., 1997). Hence, the octadecanoid pathway with (–)- jasmonic acid as a central component may be the INA-activated signaling pathway (Sembdner and Partier, 1993; Creelman and Mullet, 1995; Doares et al., 1995). However, Schweizer et al. (1997a) have presented evidence that INA may induce the genes through another, uncharacterized signal transduction pathway. In

TABLE 6.4. Induction of Systemic Resistance in Wild-Type Xanthi-nc and NahG-10 Tobacco to *Pseudomonas syringae* pv. *tabaci* by INA

	Disease severity*	
Treatment	**Xanthi-nc wild type**	**NahG-10**
INA	0.080	0.221
Control	0.179	0.417

Source: Press et al., 1997.

*Disease severity was calculated by arcsine transformations of percent disease as determined with a 0 to 5 disease rating scale in which 0 = no disease and 5 = a totally necrotic leaf.

rice, INA induced massive accumulation of an acid-soluble protein of 31 kDa, ASP31, that was not induced by salicylic acid, jasmonic acid, or ethylene or by auxin, kinetin, abscisic acid, or gibberellic acid and by *P. syringae* pv. *syringae,* a biological inducer of resistance (Schweizer et al., 1997a). ASP 31 appears to be specifically INA-induced protein. Moreover, a number of in vitro translation products (ITPs), ITP24, ITP25, ITP46, and ITP66 accumulated only in INA-treated rice plants (Schweizer et al., 1997a) and not in plants inoculated with *P. syringae* pv. *syringae,* an SAR inducer (Smith and Metraux, 1991; Reimann et al., 1992). Jasmonic acid-induced products did not accumulate in rice plants treated with INA (Schweizer et al., 1997a). Exogenously applied jasmonic acid enhanced INA-induced resistance and accumulation of PR-1 and ASP-31 in rice leaves. The fact that ASP-31 accumulation is enhanced by jasmonic acid, although jasmonic acid by itself could not induce accumulation of this protein, suggests the existence of a signaling network between the octadecanoid pathway and another INA-induced pathway (Schweizer et al., 1997b).

BTH

Benzo-(1,2,3)-thiadiazole-7-carbothioic acid S-methyl ester (BTH) is another chemical which is known to induce systemic resistance against different pathogens. BTH has been shown to induce resistance in rice (Schaffrath et al., 1997), wheat (Gorlach et al., 1996), tobacco (Friedrich et al., 1996), and *Arabidopsis* (Lawton et al., 1996). The signal transduction pathway induced by BTH appears to be different from the pathway involved in SAR induced by incompatible pathogens. Genetically engineered tobacco and *Arabidopsis* plants unable to accumulate salicylic acid failed to exhibit SAR after induction with pathogens. But application of BTH still induced SAR and SAR gene activity in such plants (Friedrich et al., 1996; Lawton et al., 1996).

Schaffrath et al. (1997) inoculated wheat plants with an incompatible pathogen or sprayed them with a 1 mM BTH solution. Inoculation with the nonhost pathogen resulted in accumulation of a set of WIR (wheat induced resistance) transcripts. WIR-1 and its homolog WIR-4 encode a putative cell wall protein (Bull et al., 1992); WIR-2, a thaumatin-like protein (PR-5 protein) (Rebmann et al., 1991b); and WIR-3, a peroxidase (Rebmann et al., 1991a). BTH induced another set of transcripts called WCI (wheat chemical induction) transcripts (Schaffrath et al., 1997). WCI-4 encodes a cysteine proteinase; WCI-2 encodes lipoxygenase; and WCI-1, WCI-3, and WCI-5 encode proteins of unknown function (Schaffrath et al., 1997). Nonhost pathogens did not induce WCI transcripts while BTH did not induce WIR transcripts. The fact that in wheat the biological and chemical (BTH) SAR inducers induce the accumulation of nonidentical sets of transcripts indicates the existence of different signal transduction pathways.

Similar results have been obtained in rice. SAR induced by *P. syringae* pv. *syringae* in rice results in the accumulation of transcripts homologous to the wheat WIR-1, WIR-2, and WIR-3 clones PIR-1 (Schaffrath et al., 1997), PIR-2 (Reimann and Dudler, 1993), and PIR-3 (Reimann et al., 1992). BTH induced SAR in rice. However, BTH treatment does not lead to the accumulation of PIR transcripts in rice (Schaffrath et al., 1997). The types of signal transduction systems activated by BTH appear to be similar to those of INA (Friedrich et al., 1996; Lawton et al., 1996; Schaffrath et al., 1997).

Other Chemicals

Several other chemicals are also known to induce the signaling system in plants. Polyacrylic acid and thiamine induced SAR in tobacco. A burst of salicylic acid production occurred within 10 and 6 h of injection of polyacrylic acid and thiamine, respectively (Malamy et al., 1996). Thiamine induced PR-1 protein accumulation in wild-type tobacco plants, but this induction was dramatically reduced in transgenic plants expressing the bacterial salicylate hydroxylase (NahG) gene (which degrades salicylic acid). In contrast, PR-1 accumulation in response to polyacrylic acid was similar in wild-type and NahG plants. But salicylic acid was reduced to undetectable levels in NahG plants following both polyacrylic acid and thiamine treatment. Therefore, the induction of the PR-1 gene by thiamine may be through salicylic acid signal transduction system while the induction of the PR-1 gene by polyarylic acid appears to be independent of salicylic acid accumulation, even though polyacrylic acid is a strong inducer of salicylic acid synthesis (Malamy et al., 1996).

DL-3-aminobutyric acid (β-aminobutyric acid, BABA) induces systemic resistance in tobacco (Cohen et al., 1994). Salicylic acid and INA-treated tobacco plants showed accumulation of PR-1, β-1,3-glucanase, and chitinases. BABA did not induce these PR proteins. But all three of them induced systemic resistance against *Peronospora tabacina* in tobacco (Cohen et al., 1994). It suggests that BABA induces ISR in a pathway different from that induced by salicylic acid or INA.

Probenazole (3-allyloxy-1,2-benzisothiazole-1,1-dioxide) induces resistance in rice. It induces several defense genes (Watanabe et al., 1979). A gene, *RPR1,* was identified as a probenazole-responsive gene (Sakamoto et al., 1999). RPR1 contains a nucleotide binding site and leucine-rich repeats, thus sharing structural similarity with known disease resistance genes. The expression of *RPR1* in rice could be upregulated by treatment with chemical inducers of SAR and by inoculation with pathogens. *RPR1* was induced during the systemic induced resistance (Sakamoto et al., 1999).

Practical Utility of Chemicals Inducing the Signal Transduction System in Management of Bacterial Diseases

It is almost three decades now since the effect of salicylic acid and other chemicals inducing disease resistance was reported. But large-scale field use of these chemicals has not yet been reported. The major drawback of the chemical induction of defense genes is that their effect is only transient; the resistance induced by chemicals lasts only for a few days. Methods of application such as infiltration by injection or root feeding are impractical under field conditions. When they are sprayed on foliage, they are easily washed away by rain or dew. These chemicals should be applied before infection as they sensitize the plants for induction of defense genes during the infection process. They are not curative; they cannot eliminate already established infection. Most of the chemicals are phytotoxic. However, recently identified chemicals such as INA and BTH persist for a long time, even 30 to 45 days. They can be sprayed or applied through soil. It appears that these two chemicals have the potential for large-scale field use in management of bacterial diseases. The time of application, dosage required, and mode application have to be standardized against each disease.

INA and BTH induce resistance by activating genes that are not normally activated during SAR induced by biological agents. It suggests that many other defense genes are not activated by INA or BTH. Hence, these chemicals alone cannot contribute to high resistance. In other words, they can only reduce the disease intensity and cannot completely control the disease (Press et al., 1997). These chemicals should be one of the treatments and not the only treatment to control diseases. For example, the efficacy of INA was enhanced with additional application of jasmonic acid in rice (Schweizer et al., 1997a). These chemicals may be useful in moderately disease-resistant cultivars rather than in susceptible cultivars and/or when they are combined with the rhizobacteria with the ability to induce ISR (see next section).

MANIPULATION OF THE SIGNAL TRANSDUCTION SYSTEM USING RHIZOBACTERIAL STRAINS

Several rhizobacterial strains have been shown to elevate plant resistance against several pathogens by inducing various signal transduction systems (Albert and Anderson, 1987; Alstrom, 1991; Frommel et al., 1991; Van Peer et al., 1991; Wei et al., 1991, 1996; Van Peer and Schippers, 1992; Zdor and Anderson, 1992; Meyer et al., 1992; Ohno et al., 1992; Meier et al., 1993; Kloepper et al., 1993; Maurhofer et al., 1994; Sayler et al., 1994; Schneider and Ulrich, 1994; Yao et al., 1994; Zhou and Paulitz, 1994; Chen et al., 1995; Liu et al., 1995a,b,c; Raaijmakers et al., 1995; Hoffland et al., 1995, 1996; Leeman et al., 1995a,b, 1996; Benhamou et al., 1996a,b,c; Duijff et al., 1996, 1997; Press et al., 1996; Raupach et al., 1996; Rabindran and Vidhyasekaran., 1996; Van Wees et al., 1997; De Meyer and Hofte, 1997; M'Piga et al., 1997; Vidhyasekaran et al., 1997; Vidhyasekaran and Muthamilan, 1998; Vidhyasekaran et al., 2000; Meena et al., 2000). Strains of *Pseudomonas fluorescens, P. putida, P. aeruginosa, Serratia marcescens,* and *Bacillus* spp. are the common rhizobacteria which induced ISR in rice (Ohno et al., 1992; Vidhyasekaran et al., 1997, 2000), cucumber (Liu et al., 1995b; Meyer et al., 1992), radish (Leeman et al., 1995b), tobacco (Maurhofer et al., 1994; Press et al., 1996), tomato (Van Wees et al., 1997; M'Piga et al., 1997), bean (Meier et al., 1993; De Myer and Hofte, 1997), pea (Benhamou et al., 1996a,b), carnation (Van Peer and Schippers, 1992), and *Arabidopsis* (Van Wees et al., 1997).

The rhizobacterial strains have been shown to induce different defense genes in plants. Maurhofer et al. (1994) showed that ISR induced by *P. fluorescens* strain CHAO in tobacco was associated with accumulation of various PR proteins (PR-1, PR-2, and PR-3 proteins). Inoculation of bean leaves with cells of *P. fluorescens* induced the accumulation of transcripts for chalcone synthase (CHS), chitinase, and lipoxygenase (Meier et al., 1993). Increase in peroxidase activity as well as an increase in the level of mRNAs encoding for phenylalanine ammonia-lyase (PAL) and CHS could be recorded in the early stages of the interaction between roots and various bacterial endophytes (Zdor and Anderson, 1992). Treatment with *P. fluorescens* causes increases in activities of peroxidase, lysozyme, and PAL in tobacco (Schneider and Ulrich, 1994). Van Peer et al. (1991) showed massive accumulation of phytoalexins in carnation roots colonized by the rhizobacterial strains after pathogen challenge. M'Piga et al. (1997) reported that *P. fluorescens* strain 63-28 induced accumulation of an electron-dense material in epidermal and outer cortical cells and coating of most intercellular spaces with similar substances in tomato. This aggregated material appeared to be mainly composed of phenolic compounds, especially phenols containing *o*-hydroxy groups. The deposition of β-1,3-glucans (callose) was also observed in host cell walls. Chitinases were also induced (M'Piga et al., 1997). Benhamou et al. (1996a,b,c) demonstrated that pea root bacterization with *P. fluorescens* or *Bacillus pumilus* triggered a set of defense reactions that resulted in the elaboration of permeability barriers. Increases in plant lignification, phytoalexins, various lytic enzymes, and other PR proteins have been observed upon treatment of plants with different specific strains of rhizobacteria (Albert and Anderson, 1987; Frommel et al., 1991; Kloepper et al., 1993; Sayler et al., 1994; M'Piga et al., 1997; Vidhyasekaran et al., 2000).

Studies on mode of induction of defense genes by the rhizobacterial strains reveal that different bacterial cell wall components may act as elicitors of induction of the signal transduction system. Lipopolysaccharides (LPS) of *P. fluorescens* strain WCS417r act as elicitors and induce resistance against different diseases (Duijff et al., 1997). LPS and LPS-containing cell wall preparations of *P. fluorescens* WCS417r are as effective as living WCS417r bacteria in inducing ISR in radish (Leeman et al., 1995b). The O-antigenic side chain of the outer membrane LPS of the strain WCS417r appears to be the main determinant for induction of ISR in radish and carnation (Van Peer and Schippers, 1992; Leeman et al., 1995b). A bacterial mutant lacking the O-antigenic side chain did not induce resistance. Duijff et al. (1997) showed involvement of the O-antigenic side chain in LPS in endophytic root colonization of *P. fluorescens* in tomato. Cell wall preparations of WCS417r and its mutant WCS417r OA^-, which lacks the O-antigenic side chain of the LPS, were tested for their action against *P. syringae* pv. *tomato* in tomato (Van Wees et al., 1997). Treatment of the tomato roots with cell walls of WCS417r reduced symptoms by 20 percent, whereas the cell walls of WCS417r OA^- were ineffective. The reduction was significantly less than the level of protection obtained with living bacteria, suggesting that the O-antigenic side chain of the LPS of WCS417r contributes to elicitation of ISR but is probably not sufficient for full induction (Van Wees et al., 1997). Comparison of the resistance-inducing ability of living cells of WCS417 and its OA^- mutant in *Arabidopsis* plants revealed that wild-type and mutant bacteria induced similar levels of protection against *P. syringae* pv. *tomato*. These results demonstrate that in *Arabidopsis,* elicitation of ISR by WCS417r is not dependent upon the O-antigenic side chain of the LPS (Van Wees et al., 1997).

The variable results obtained with radish, carnation, tomato, and *Arabidopsis* suggest that the induction of ISR depends not only on the bacterial determinant but also on some host factors. However, it is commonly found that cell walls of *P. fluorescens* strains can induce some resistance against pathogens. The cell walls of *P. fluorescens* WCS417r were able to elicit a full resistance response in radish (Leeman et al., 1995b) and carnation (Van Peer and Schippers, 1992) and partial resistance in *Arabidopsis* (Van Wees et al., 1997). It suggests that some cell wall components of the bacteria may contribute to the induced resistance, at least partial resistance, and there may be other bacterial products inducing ISR.

Leeman et al. (1996) demonstrated that the siderophore of WCS374r can act as an elicitor of ISR in radish. *P. fluorescens* strain CHAO induced systemic resistance in tobacco. A siderophore (pyoverdine)-deficient derivative of this strain no longer induced ISR (Maurhofer et al., 1994). Leeman et al. (1996) reported that the purified siderophore, pseudobactin, from *P. fluorescens* strain WCS374 induced ISR in radish. However, a pseudobactin-deficient *P. fluorescens* 374PSB retained ISR-inducing activity. It suggests that siderophore production by this strain was only partially responsible for induction of systemic resistance in radish.

Pyochelin, a siderophore, is produced by several rhizobacteria. Salicylic acid is a precursor of pyochelin synthesis (Ankenbauer and Cox, 1988; Leeman et al., 1996). Several genera of bacteria including fluorescent pseudomonads are known to synthesize salicylic acid (Hudson and Bentley, 1970; Ankenbauer and Cox, 1988; Lebuhn and Hartmann, 1994; Dowling and O'Garra, 1994; Buysens et al., 1996). *P. fluorescens* strain CHO produces salicylic acid (Meyer et al., 1992; Visca et al., 1993). Leeman et al. (1996) reported that *P. fluorescens* strain WCS374 produced salicylic acid in quantities that were iron dose-dependent. *P. fluorescens* strain WCS417r has the capacity to produce salicylic acid (Leeman et al., 1996). *P. aeruginosa* 7NSK2, which induces ISR in bean, produces salicylic acid (De Meyer and Hofte, 1997). The rhizobacterial strain *Serratia marcescens* 90-166, which induces ISR, produces salicylic acid in vitro (Press et al., 1997). Salicylic acid has been shown to be responsible for induction of resistance in radish (Leeman et al., 1996) and bean (De Meyer and Hofte, 1997).

Meyer et al. (1992) reported that salicylic acid itself might function as an endogenous siderophore. Leeman et al. (1996) and De Meyer and Hofte (1996) reported that rhizobacteria-mediated ISR is affected by iron concentration. Salicylic acid production is iron (Fe^{3+}) regulated (Leeman et al., 1996). Salicylic acid production is promoted by low iron concentrations. Increasing ferric iron concentrations in vitro reduced salicylic acid production below detectable limits by bacteria (Meyer et al., 1992). When ferric iron was applied as a soil drench, ferric iron concentration increased in planta, but it significantly reduced the level of ISR observed in cucumber. It suggests that salicylic acid may not be involved in ISR, but some other siderophores mediated by iron may be involved in induction of ISR in cucumber.

The signaling system involved in plant-microbe interactions may vary from plant to plant. This may be true in case of rhizobacterial strains-plant interactions. Specific interactions between the rhizobacterial strains and the plant ecotypes appear to determine induction of systemic resistance. Ecotypes of *Arabidopsis thaliana* were differentially responsive to *P. fluorescens* WCS417r treatment. In contrast to ecotypes Col and Ler, ecotype RLD did not develop ISR upon treatment of the roots with *P. fluorescens* WCS417r (Van Wees et al., 1997). *P. fluorescens* strain WCS358 triggered an ISR response in *Arabidopsis,* whereas *P. fluorescens* strain WCS374r did not (Van Wees et al., 1997). In contrast, in radish ISR is induced by strain WCS374 but not by WCS 358r (Leeman et al., 1995a). Cell wall preparations of WCS374r were ineffective in inducing ISR in *Arabidopsis* (Van Wees et al., 1997), but they induced ISR in *Arabidopsis* (Leeman et al., 1995b). It suggests that lack of response of the plant to the cell wall preparation may reflect the lack of response of the plant for ISR to the particular strain of the rhizobacteria. The ecotype-specific induction of resistance in *Arabidopsis* by WCW417r indicates that protection against *P. syringae* pv. *tomato* is dependent upon specific interaction between the bacteria and the plant.

Massive accumulation of phytoalexins could be detected in roots of carnation plants treated with *P. fluorescens* only after pathogen challenge (Van Peer et al., 1991). Induction of phenolics and callose in tomato by *P. fluorescens* strain 63-28 was substantially amplified upon infection with a pathogen (M'Piga et al., 1997). These results suggest that the rhizobacterial strains may be capable of evoking transcriptional activation of plant defense genes, the expression of which may be subsequently latent until perception of signals originating from contact with the pathogen. It is also possible that besides the rhizobacterial signal molecules, the pathogen's signal molecules may also be involved in the induction of ISR.

Some bacterial strains induced the same set of PR protein genes that are induced in SAR, which are induced by incompatible/compatible pathogens mediated by salicylic acid (Maurhofer et al., 1994). It suggests the possible involvement of salicylic acid in the signal transduction pathway induced by rhizobacterial strains. However, PR proteins did not accumulate in radish plants expressing ISR elicited by *P. fluorescens* strain WCS417r (Hoffland et al., 1995, 1996). Pieterse et al. (1996) demonstrated that in *Arabidopsis,* ISR induced by WCS417r was not associated with PR gene activation and was elicited in transgenic *Arabidopsis* plants unable to accumulate salicylic acid. This indicates that in contrast to pathogen-induced SAR, WCS417r-mediated ISR is controlled by a salicylic acid-independent signaling pathway.

In *A. thaliana* ecotypes Columbia and Landsberg *erecta,* colonization of the rhizosphere by *P. fluorescens* strain WCS417r induced ISR against

P. syringae pv. *tomato*. In contrast, ecotype RLD did not respond to WCS417r treatment, whereas all three ecotypes expressed ISR upon treatment with salicylic acid (Van Wees et al., 1997). It suggests that the mechanism of ISR induced by *P. fluorescens* may be different from that induced by salicylic acid.

Transgenic *Arabidopsis* NahG plants do not accumulate salicylic acid. *P. fluorescens* WCS417r induced protection against *P. syringae* pv. *tomato* in both wild-type and NahG plants (Figure 6.2; Van Wees et al., 1997). It suggests that WCS417r-mediated ISR is independent of endogenous salicylic acid accumulation. The level of protection induced by *P. fluorescens* WCS417r is somewhat lower in NahG plants, suggesting a modulating role for salicylic acid in the level of expression of ISR. *P. fluorescens* did not induce PR-1, PR-2, or PR-5 mRNAs (Van Wees et al., 1997). But it is possible that *P. fluorescens* would have simply sensitized plants to receive the signal from pathogens, and the PR proteins would have accumulated in the *P. fluorescens*-treated plants when challenge-infected by pathogens. Van Peer et al. (1991) have demonstrated massive accumulation of phytoalexins in carnation plants only after pathogen challenge. However, these results suggest that a salicylic acid-independent signaling pathway leading to ISR would have been induced by *P. fluorescens* in addition to a salicylic-dependent pathway.

The rhizobacterial strain *Serratia marcescens* strain 90-166 has been observed to induce systemic resistance in cucumber to *P. syringae* pv. *lachrymans*

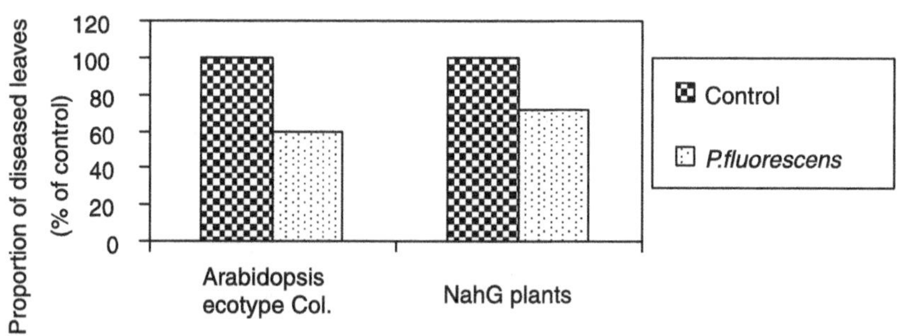

FIGURE 6.2. Quantification of induced systemic resistance against *Pseudomonas syringae* pv. *tomato* infection in *Arabidopsis thaliana* Col. or NahG plants treated with *P. fluorescens* WCS417r (*Source:* Van Wees et al., 1997).

(Liu et al., 1995b; Wei et al., 1996) and *Erwinia tracheiphila* (Yao et al., 1994). It induced disease resistance to *P. syringae* pv. *tabaci* in wild-type Xanthi-nc and transgenic NahG-10 tobacco expressing salicylate hydroxylase (Press et al., 1997). It produced salicylic acid in vitro; but an ISR mutant of the bacterium still produced salicylic acid (Press et al., 1997). It suggests that salicylic acid may not be the primary bacterial determinant of ISR. But the data presented in Table 6.5 suggest that NahG transgenic plants which degrade salicylic acid showed more disease intensity than wild plants (both control and *S. marcescens*-treated plants) when inoculated with *P. syringae* pv. *tabaci* (Press et al., 1997). It suggests that salicylic acid is also involved in signal transduction in tobacco, and *S. marcescens*-induced resistance may also partially be operated through the salicylic acid signal transduction system. Induction of H_2O_2 production due to *P. fluorescens* treatment in plants has also been reported (Jakobek and Lindgren, 1993).

Several signal transduction systems may be involved in rhizobacteria-induced systemic resistance in plants. Different strains of bacteria may operate through different signal transduction systems. The interactions may also vary from plant to plant. It is not yet known which and how many signaling proteins are required in the plant defense induced by rhizobacteria.

Practical utility of rhizobacterial strains in control of various bacterial diseases has been demonstrated in many greenhouse and field trials. Fluorescent pseudomonad strains induced systemic resistance against the angular leaf spot pathogen *P. syringae* pv. *lachrymans* in cucumber. Their efficacy in control of this disease has been demonstrated in three field trials (Wei et al., 1996). *P. fluorescens* WCS417r protected radish plants against *P. syringae* pv. *tomato* (Hoffland et al., 1996). The two rhizobacterial strains *P. putida* 89B-27 and *Serratia marcescens* strain 90-166 protected cucum-

TABLE 6.5. Induction of Systemic Resistance in Wild-Type Xanthi-nc and Transgenic NahG-10 Tobacco to *Pseudomonas syringae* pv. *tabaci* by *Serratia marcescens* Strain 90-166

	Disease severity*	
Treatment	**Xanthi-nc**	**NahG-10**
Serratia marcescens	0.094	0.156
Control	0.179	0.417

Source: Press et al., 1997.

*Disease severity was calculated by arcsine transformations of percent disease as determined with a 0 to 5 rating scale.

ber against bacterial angular leaf spot caused by *P. syringae* pv. *lachrymans* (Liu et al., 1995b). *S. marcescens* strain 90-166 protected cucumber plants against *Erwinia tracheiphila* in both greenhouse and field trials (Yao et al., 1994). This strain was able to protect tobacco plants against *P. syringae* pv. *tabaci* (Press et al., 1997).

Fluorescent pseudomonad strains protected rice plants against bacterial blight pathogen *X. oryzae* pv. *oryzae* (Ohno et al., 1992; Vidhyasekaran et al., 2000). These rhizobacterial strains induced resistance not only against bacterial pathogens, but also against fungal and viral pathogens. Vidhyasekaran et al. (1996, 1997a,b, 2000) and Vidhyasekaran and Muthamilan (1998) have shown that *P. fluorescens* Pf1 strain protected rice plants against the bacterial pathogen *X. oryzae* pv. *oryzae,* the fungal pathogens *Rhizoctonia solani, Pyricularia oryzae, Helminthosporium oryzae,* and *Sarocladium oryzae* and against the viral pathogen rice tungro virus. *P. fluorescens* WCS417 treatment protected radish plants against P. *syringae* pv. *tomato, Fusarium oxysporum* f. sp. *raphani,* and *Alternaria brassicae* (Hoffland et al., 1996). *S. marcescens* 90-166 induced resistance against *P. syringae* pv. *lachrymans* (Liu et al., 1995b; Wei et al., 1996), *Erwinia tracheiphila* (Yao et al., 1994), *Colletotrichum orbiculare* (Liu et al., 1995c), *Fusarium oxysporum* f. sp. *cucumerinum* (Liu et al., 1995a), and cucumber mosaic virus (Raupach et al., 1996) in cucumber.

P. fluorescens strains have been developed as powder formulations for easy handling and storage (Vidhyasekaran and Muthamilan, 1995, 1998; Rabindran and Vidhyasekaran, 1996; Vidhyasekaran et al., 1997a,b, 1998, 2000). Method of application, time of application, and effective dosage of formulations have been standardized against some bacterial pathogens (reviewed by Vidhyasekaran, 1999). Commercial formulations of these rhizobacterial strains are available in different countries. These formulations have to be applied before infection by pathogens as they induce resistance and are not curative. Their effect lasts only for 10 to 15 days and in some cases up to 30 days; hence their application has to be repeated at the appropriate time. These bacterial strains only enhance resistance, and hence their effectiveness is very much less in highly susceptible varieties under high disease pressure. This type of ISR can only be a component of integrated disease management (for review see Vidhyasekaran, 1998).

MANIPULATION OF THE SIGNAL TRANSDUCTION SYSTEM BY ENHANCED BIOSYNTHESIS OF SALICYLIC ACID

Salicylic acid has been shown to be an important systemic signal in inducing resistance against pathogens. Enhanced synthesis of salicylic acid enhances disease resistance. Addition of salicylic acid to the growth medium of axenically growing tobacco seedlings made them almost fully resistant to subsequent infection by the soft rot pathogen *E. carotovora* subsp. *carotovora* (Palva et al., 1994). Pectic enzymes of *E. carotovora* subsp. *carotovora* macerate the leaf tissue of the control plant but are not effective on salicylic acid-treated plants (Palva et al., 1994). It suggests that the ISR to *E. carotovora* may function by neutralizing the primary weapons of this pathogen, the plant cell wall-degrading enzymes.

Transgenic tobacco plants expressing a *Halobacterium halobium* gene encoding a light-driven bacterio-opsin proton pump showed higher systemic levels of salicylic acid. These plants showed enhanced resistance to the wildfire pathogen *P. syringae* pv. *tabaci* (Mittler et al., 1995). But the same gene, when expressed in potato, did not offer protection against tuber infection of *E. carotovora* (Abad et al., 1997). Probably this gene would not have expressed in tubers, although it would have expressed in leaves (the leaves of transgenic potato plants showed resistance to the fungal pathogen *Phytophthora infestans*). Abad et al. (1997) have used the figwort mosaic virus promoter in gene construction. Wegener et al. (1996) reported that when they used the CaMV 35S promoter for pectate lyase gene PL3 expression, the expression was greater in potato leaves and very much less in tubers, while when patatin B33 was used as a promoter, the gene expression was limited to tubers only. Probably different types of promoters should be tried to get maximum expression of the desired gene.

Transgenic tobacco plants expressing a gene encoding the A1 subunit of cholera toxin showed increased resistance to *P. syringae* pv. *tabaci* (Beffa et al., 1995). The cholera toxin is known to activate a signal transduction system dependent on G-proteins in animals. The transgenic plants showed enhanced accumulation of salicylic acid, and several PR proteins were also induced (Beffa et al., 1995).

H_2O_2 is an important component in the signal transduction system. A glucose oxidase gene from the fungus *Aspergillus niger* was transferred to potato. The transgenic potato plants expressing the glucose oxidase gene showed an increased level of resistance to *E. carotovora* (Wu et al., 1995). Even though glucose oxidase was produced constitutively and extracellularly in the transgenic plants, a significant increase in H_2O_2 was detected only following bacterial infection. Probably intracellular glucose is released only after bacterial infection, and the released glucose in the apoplast may serve as a substrate for the extracellular glucose oxidase. The induced H_2O_2 may induce salicylic acid and induce ISR.

H_2O_2 is degraded by catalase, and catalase may suppress signal transduction. When potato roots were infected with *E. carotovora* or *Corynebacterium sepedonicum,* a catalase gene *(Cat2St)* was induced. *Cat2St* was found to be systemically induced after compatible interaction (Niebel et al., 1995). Hence, suppression of the catalase gene may result in H_2O_2 signal transduction and induce resistance. Transgenic tobacco plants with severely depressed levels of catalase have been developed with antisense suppression (Takahashi et al., 1997). These plants showed enhanced accumulation of salicylic acid and various PR proteins. These transgenic plants showed enhanced disease resistance (Takahashi et al., 1997).

Rice cultivars contain high amounts of salicylic acid. The basal levels of salicylic acid in the rice cultivar M-201 are 50-fold higher than those observed in tobacco (Enyedi et al., 1992; Silverman et al., 1995). Salicylic acid content varies among various cultivars (Figure 6.3; Silverman et al., 1995). Rice cultivars with high resistance to a fungal pathogen, *Magnaporthe grisea* contain high amounts of salicylic acid (Silverman et al., 1995). Several *Arabidopsis* mutants such as *acd1, acd2, lsd1, lsd2, lsd3, lsd4, lsd5, lsd6, lsd7, cep1, cpr1,* and *cim3* constitutively express high levels of salicylic acid, and all of them constitutively show high *PR* gene expression (Dangl et al., 1996; Ryals et al., 1996). It may be possible to select cultivars with high salicylic acid content, which may show resistance to pathogens.

MANIPULATION OF THE SIGNAL TRANSDUCTION SYSTEM BY INDUCING ACCELERATED CELL DEATH

It is known that rapid and early cell death (hypersensitive reaction) leads to induction of various signals conferring disease resistance in plants. Cell

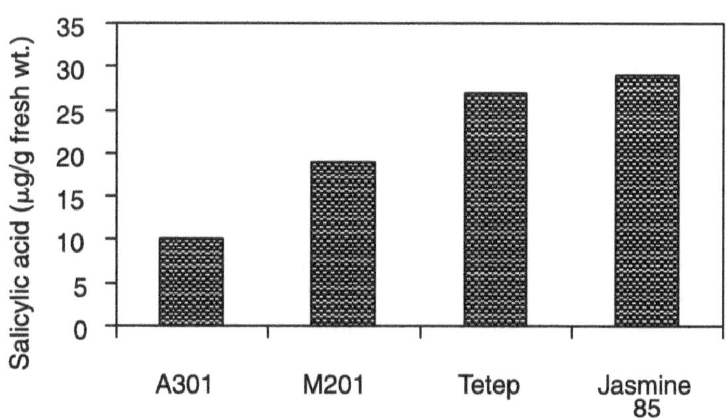

FIGURE 6.3. Variation in salicylic acid content among rice varieties (*Source:* Silverman et al., 1995).

death can be accelerated in susceptible plants to make them resistant. Programmed cell death may be a useful approach to develop disease resistant plants. Different mutants have been selected to exploit this phenomenon. *Arabidopsis* lesions simulating disease *(lsd)* and accelerated cell death *(acd2)* mutants that form spontaneous lesions show elevated SAR expression and exhibit increased disease resistance (Dietrich et al., 1994; Greenberg et al., 1994; Weymann et al., 1995; Hunt et al., 1997). Similar lesioned phenotypes exist in tomato (Langford, 1948) and barley (Wolter et al., 1993) displaying enhanced disease resistance.

Lesion-mimic phenotypes have also been engineered in planta via expression of a variety of transgenes. Transgenic tobacco plants expressing a bacterial proton pump (Mittler et al., 1995), the A1 subunit of cholera toxin (Beffa et al., 1995), or a ubiquitin variant (Becker et al., 1993) have been developed, and all of them show spontaneous cell death and enhanced disease resistance. Although accelerated cell death genes may be useful to develop disease-resistant plants, many of these transgenic plants show excessive chlorosis and necrosis beginning on the lower leaves and progressively appearing on the upper leaves as well. These plants have an irregular morphology and less vigor than control plants (Abad et al., 1997). Transgenic plants with less stunting and fewer detrimental phenotypic characteristics should be selected. In order to avoid the deleterious effects of a generalized lesion development, the transgenic induction should be restricted strictly to the sites of infections. The success of this type of strategy may depend upon the choice of a specific promoter. Some promoters of the gene such as the tobacco *hsr203* gene, which are unresponsive to abiotic signals but are specifically responsive to bacterial infection (Pontier et al., 1994), will be useful in developing useful disease-resistant plants.

MANIPULATION OF THE SIGNAL TRANSDUCTION SYSTEM BY ENHANCED BIOSYNTHESIS OF CYTOKININS

Cytokinins are known to control the synthesis of jasmonic acid and salicylic acid in plants. By precise manipulation of cytokinin synthesis, both these systemic signal molecules can be induced, resulting in enhanced disease resistance. Transgenic tobacco plants expressing a gene encoding a GTP-binding protein show high endogenous cytokinin concentrations (Sano and Ohashi, 1995). When these plants were wounded, accumulation of jasmonic acid and salicylic acid was seen (Sano et al., 1994). The disease reaction of these transgenic plants was not assessed, but similar plants may be useful in developing disease-resistant plants.

The *rolC* gene is one of four genes located on the TL-DNA of the Ri plasmid of *Agrobacterium rhizogenes*. Potato plants expressing the *rolC* gene have been developed and the *rolC* transformants display an up to fourfold increase in cytokinins (Fladung and Gieffers, 1993). The transgenic potato plants expressing the *rolC* gene behaved differently depending upon the level of *rolC* gene expression to *Erwinia carotovora* subsp. *atroseptica* (Fladung and Gieffers, 1993). The strongest *rolC*-expressing clone, T346, showed an 80 percent infection rate, while the low *rolC*-expressing transgenic clones T341 and T351 had an infection rate of about 20 percent. The transformed control plants showed about 69 percent infection (Fladung and Gieffers, 1993).

Similar results were obtained when the transgenic potato plants were inoculated with *E. carotovora* subsp. *carotovora* (Fladung and Gieffers, 1993). Any disturbance in growth regulators may affect the phenotype of plants. Growth regulators can promote plant growth and they can also act as an herbicide, depending upon their concentrations. Potato plants overexpressing *rolC* have reduced apical dominance, have a large number of side shoots, are pale green in color with decreased leaf area, and show significant changes in yield parameters (more but smaller tubers) (Fladung and Ballvora, 1992). But it may be possible to select the best transgenic plants with minimum disturbance in their yield characters and develop those plants as useful disease-resistant cultivars, because low *rol-C* expressing plants show high resistance to bacterial pathogens.

MANIPULATION OF INDUCIBLE PROTEINS FOR INDUCTION OF BACTERIAL DISEASE RESISTANCE

Several inducible proteins have been shown to induce disease resistance. Early and high induction of these inducible proteins contributes to resistance. In susceptible plants, these proteins are induced slowly and in smaller amounts. If these proteins are constitutively overexpressed, the susceptible plants can be made resistant. Hence, several attempts have been made to develop transgenic plants overexpressing specific inducible proteins constitutively.

Although several transgenic plants constitutively expressing inducible proteins have been developed (Table 6.6), the reaction of most of those plants to bacterial pathogens has not been studied. Some of the transgenic plants constitutively expressing the inducible proteins show enhanced resistance to bacterial pathogens. Tobacco plants expressing barley thionin genes show resistance to *P. syringae* pv. *tabaci* (Carmona et al., 1993). Transgenic tobacco plants expressing barley lipid-transfer proteins 2 (LTP2)

TABLE 6.6. List of Developed Transgenic Plants Expressing Inducible Proteins

Crop	Inducible proteins	References
Rice	Chitinase	Lin et al.,1995
	PR-5	Datta et al., 1999
Tomato	Chitinase	Logemann et al., 1994
	β-1,3-glucanase	Jongediik et al., 1995
	Tobacco anionic peroxidase	Lagrimini et al., 1993
Potato	Tobacco osmotin	Liu et al., 1994
	Tobacco AP20	Woloshuk et al., 1991
	PR-5	Liu et al., 1995; Zhu et al., 1996
	PR-10a	Matton et al., 1993
	Bean PAL	Bevan et al., 1989
Cucumber	Rice chitinase	Tabei et al., 1998
Alfalfa	β-1,3-glucanase	Masoud et al., 1996
Barley	β-1,3, β-1,4-glucanase	Jensen et al., 1996
Rose	Chitinase	Marchant et al., 1998
Soybean	Tobacco β-1,3-glucanase	Yoshikawa et al., 1993
Tobacco	AP24	Melchers et al., 1993
	Chitinase	Linthorst et al., 1990; Broglie et al., 1991
	β-1,3-glucanase	Yoshikawa et al., 1993
	Rice chitinase	Zhu et al., 1993
	Barley chitinase	Jach et al., 1995
	Barley β-1,3-glucanase	Jach et al., 1995
	Sugar beet chitinase	Nielsen et al., 1993
	Ribosome-inactivating protein	Logemann et al., 1992
	PR-1a	Alexander et al., 1993
	PR-1b	Cutt et al., 1989
	PR-5	Oshima et al., 1990
	Bean chitinase	Roby et al., 1990
	Peanut chitinase	Kellmann et al., 1996
	Maize RIP	Maddaloni et al., 1997
	Potato wun1	Siebertz et al., 1989
	PAL	Elkind et al., 1990
	Radish Rs-AFP2	Terras et al., 1995
	AP20	Woloshuk et al., 1991
	Osmotin	Liu et al., 1994
	Barley thionin	Carmona et al., 1993; Florack et al., 1994
	LTP2	Molina and Garcia-Olmedo,1997
	Bean PAL2	Liang et al.,1989
	PAL	Maher et al., 1994
Nicotiana sylvestris	Tobacco chitinase	Neuhaus et al., 1991; Vierheilig et al., 1992; Hart et al., 1992
Arabidopsis	LTP2	Molina and Garcia-Olmedo, 1997
	Bean PAL	Schufflebottom et al., 1993

show high resistance to *P. syringae* pv. *tabaci* (Molina and Garcia-Olmedo, 1997). Similarly, transgenic *Arabidopsis* plants expressing barley LTP2 showed enhanced resistance to *P. syringae* pv. *tomato* (Molina and Garcia-Olmedo, 1997) (Table 6.6).

The usefulness of several other inducible proteins against bacterial pathogens has not yet been studied. Chitinases with lysozyme activity (PR-8 proteins) will be highly useful as they show antibacterial activity (Chapter 4). Lysozymes with a lytic activity can act on bacterial cell walls, which contain peptidoglycans. The bifunctional chitinases with lysozyme activity are generally localized in vacuoles (Jolles and Jolles, 1984; Audy et al., 1988; Bernasconi et al., 1987). But the bacterial pathogens multiply in the apoplast. For effective control of the bacterial pathogen, the bifunctional chitinases should be secreted into the apoplast. It is possible to make the protein secrete into the apoplast by fusing the proper signal sequence to the chitinase gene. Lund et al. (1989) reported that when *Serratia marcescens* chitinase ChiA was fused to the bacterial signal sequence, it was inefficiently secreted by plant cells. But Lund and Dunsmuir (1992) have shown that when *S. marcescens* chitinase ChiA was fused to the signal sequence derived from tobacco PR1b, the ChiA protein was fully secreted. The portion of the bacterial *chia* gene encoding the signal sequence of ChiA was replaced with that encoding the signal sequence from tobacco PR-1b so that the resulting fusion protein contained the PR-1b signal sequence plus the first two amino acids of the PR-1b mature protein (Gln-Asn) in place of the first two amino acids of the mature ChiA protein (Ala-Ala). Possession of this signal sequence improved the ability of the ChiA protein to enter the secretory pathway of the plant cells in which it was expressed.

Another potential inducible protein for management of bacterial diseases is peroxidase, which is involved in the formation of lignin. Lignin confers resistance against various bacterial diseases (see Chapter 2). Tobacco plants transformed with a chimeric tobacco anionic peroxidase gene synthesized high levels of peroxidase in all tissues throughout the plant (Lagrimini et al., 1987). The percentage of lignin and lignin-related polymers in cell walls was nearly twofold greater in peroxidase-overproducer plants compared to control plants. Wound-induced lignification occurred 24 to 48 h sooner in plants overexpressing the anionic peroxidase (Lagrimini, 1991). Most bacterial pathogens enter through wounds in plants, and hence it is possible that early induction of lignin in infected transgenic plants would result in bacterial disease resistance.

Inducible phenylalanine ammonia-lyase (PAL) is another potential protein that can be exploited for bacterial disease management. Enhanced PAL activity can lead to enhanced lignification and production of phytoalexins and phenolics, which are all involved in bacterial disease resistance. Several transgenic plants expressing PAL genes have been developed (Liang et al., 1989; Bevan et al., 1989; Schufflebottom et al., 1993), and these plants should be screened for their resistance reaction to bacterial pathogens.

Another important factor is the synergistic action of inducible/PR proteins. A single PR protein may not contribute for disease resistance, but a combination of PR proteins may confer higher resistance (see Chapter 4). Synergism of LTPs with thionins has been reported (Molina et al., 1993b). Hence, transgenic plants expressing more than one PR protein should be developed. Transfer of more genes with suitable promoters may be difficult to achieve; scientists prefer to transfer one or two genes rather than multiple genes. One good method to transfer multiple genes has been suggested by Zhu et al. (1994). They developed transgenic tobacco plants expressing a gene encoding rice basic chitinase and transgenic tobacco plants expressing alfalfa acidic glucanase. Hybrid plants were generated by crossing transgenic parental lines. Homozygous selfed progeny that expressed the two transgenes could be selected from the hybrids. Logeman et al. (1994) could develop transgenic tomato plants simultaneously expressing class I chitinase and β-1,3-glucanase. This type of approach will be useful in developing bacterial disease-resistant plants.

Transport of the inducible proteins to the intercellular spaces is another important strategy to prevent colonization of plant tissue by bacterial pathogens. Florack et al. (1994) reported that transgenic tobacco plants expressing barley thionin did not show any improved resistance to *P. syringae* pv. *tabaci*. They attributed it to the absence of thionin secretion into the intercellular spaces, where the bacteria are generally found. Hence, the selected gene for conferring resistance should be fused to the proper signal sequence for secretion of the induced protein.

SUPPRESSION OF VIRULENCE FACTORS OF BACTERIAL PATHOGENS TO MANAGE BACTERIAL DISEASES

Bacterial pathogens produce toxins, which have been shown to be responsible for disease symptom development (Vidhyasekaran et al., 1989; Vidhyasekaran, 1993). Mutants that do not produce toxins have reduced virulence (Bender et al., 1987). Toxin-insensitive plants show enhanced resistance (Vidhyasekaran et al., 1986, 1990). The resistance gene cloned from maize has been shown to be involved in detoxification of the toxin produced by a fungal pathogen (Meeley and Walton, 1991; Johal and Briggs, 1992). Hence, it is possible to develop bacterial disease-resistant plants by inactivating toxins produced by bacterial pathogens. *P. savastanoi* pv. *phaseolicola,* the bean halo blight pathogen, produces a toxin which induces typical symptoms of the disease in bean. The pathogen is tolerant to its own toxin and it has been shown to be due to the bacterial enzyme ornithine carbomyl transferase. The corresponding gene *(argK)* has been cloned. Transgenic bean plants expressing *argK* showed high resistance to *P. savastanoi* pv. *phaseolicola* (Herrera-Estrella and Simpson, 1995). *P. syringae* pv. *tabaci,* the wildfire pathogen of tobacco, produces tabtoxin, which is responsible for symptom development. A gene, *ttr,* which encodes an enzyme that inactivates the tabtoxin, has been cloned from *P. syringae* pv. *tabaci.* The transgenic tobacco plants expressing the *ttr* gene show high resistance to the pathogen (Anzai et al., 1989).

Coronatine is a phytotoxin produced by several members of the *P. syringae* group of pathovars. The toxin acts as a virulence factor in *P. syringae* pv. *tomato.* The toxin induces spreading chlorosis in leaves. Feys et al. (1994) attempted to develop toxin-insensitive plants by a mutation using *Arabidopsis* as a type plant. They screened more than 200,000 seedlings from M2 populations of ethyl methane sulfonate-mutagenized seeds of *Arabidopsis* germinating on Murashige and Skoog plates containing 1 μm coronatine and isolated 14 *coi* (*co*ronatine *i*nsensitive) mutants. A coronatine-producing strain of *P. syringae* pv. *atropurea* grew in leaves of wild-type *Arabidopsis* to a population more than 100 times greater than that reached in the *coi1* mutant (Figure 6.4; Feys et al., 1994).

The wild-type plants showed typical chlorotic symptoms at six days after inoculation with *P. syringae* pv. *atropurpurea,* but this symptom was absent from inoculated leaves of the *coi1* mutant even at 14 days after inoculation

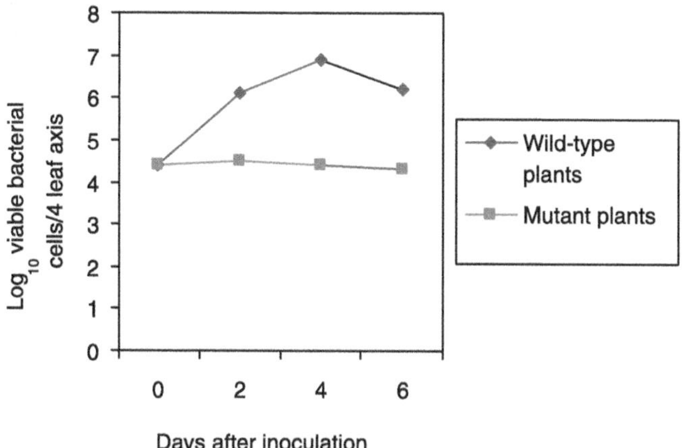

FIGURE 6.4. Growth of *Pseudomonas syringae* pv. *atropurpurea* in leaves of *Arabidopsis* wild-type and *coi* mutant plants (*Source:* Feys et al., 1994).

(Feys et al., 1994). Although the present work has been done in a weed plant, similar approach may be useful in important crops such as tomato which are also sensitive to coronatine, a toxin produced by various *P. syringae* pathovars infecting a number of crop plants.

Albicidins are the toxins produced by *X. albilineans,* the sugarcane leaf scald pathogen. The toxins have been shown to be important in disease development. A bacterium *Pantoea dispersa (= Erwinia herbicola)* was found to detoxify albicidin. The *P. dispersa* gene *(albD)* for enzymatic detoxification of albicidin has been cloned (Zhang and Birch, 1997). When *albD* was inactivated in *P. dispersa* by site-directed mutagenesis, the mutants were unable to detoxify albicidin. The bacterium was developed as a biocontrol agent. *P. dispersa* effectively controlled the leaf scald disease caused by *X. albilineans,* probably by detoxifying the toxin produced by the pathogen. This type of biocontrol agent will be highly useful in management of bacterial diseases.

EXPLOITATION OF INSECT GENES ENCODING ANTIBACTERIAL PROTEINS FOR BACTERIAL DISEASE MANAGEMENT

Antibacterial proteins have been detected in insects. These proteins also can be exploited to develop bacterial disease-resistant plants. Cecropins are a family of homologous antibacterial peptides of 35 to 37 residues derived from the giant silk moth, *Hyalophora cecropia* (Boman et al., 1991). The antibacterial activity of cecropins appears to be linked to a molecular architecture consisting of a cationic *N*-terminal amphipathic region and a hydrophobic *C*-terminal amphipathic half separated by a hinge (Gururaj Rao, 1995). The cecropins have no effect on eukaryotic cells (Steiner et al., 1981) but affect prokaryotic cells through the formation of voltage-dependent ion channels (Christensen et al., 1988). The lytic peptides form pores in bacterial membranes (Destefano-Beltran et al., 1993). Cecropin B shows antimicrobial activity against pathogenic bacteria at concentrations that do not affect the plant cells (Mills and Hammerschlag, 1993).

The cecropin gene was introduced into potato. The protein extracted from the transgenic plants showed antibacterial activity (Montanelli and Nascari, 1991). However, Hightower et al. (1994) reported that transgenic tobacco plants expressing cecropin mRNA and protein are just as susceptible to infection with *P. syringae* as the control plants. Probably the cecropin is found inside the cells and not secreted. Secretion of cecropin is important to reduce bacterial growth in the apoplast. No resistance to *Ralstonia solanacearum* or *P. syringae* pv. *tabaci* was found in transgenic tobacco plants expressing the cecropin B gene, as a result of the degradation of cecropins by plant proteases (Florack et al., 1995). Several stable analogs of cecropin B, such as SB-37, Shiva-1, and MB39 have been synthesized (Fink et al., 1989; Norelli et al., 1994a; Huang et al., 1997). Jaynes (1990) has reported on the superior biological activity of Shiva-1, a cecropin-like synthetic peptide exhibiting 46 percent homology to the natural molecule. Transgenic tobacco plants expressing Shiva-1 have been developed, and they showed enhanced resistance to *R. solanacearum*.

The stable analog of cecropin, MB39, was used to develop transgenic tobacco plants (Huang et al., 1997). A chimeric gene fusion cassette consisting of a secretory sequence from barley α-amylase joined to the MB39 coding sequence, and placed under control of the promoter and terminator from the potato proteinase inhibitor (PiII) gene, was introduced into tobacco by *Agrobacterium*-mediated transformation. No disease symptom development was observed in leaf tissues infiltrated with *P. syringae* pv. *tabaci* in MB39 transgenic plants while disease developed rapidly in control plants. Bacterial multiplication in leaves of the transgenic plants was suppressed more than 10-fold compared with control plants (Huang et al., 1997). The pathogen-induced promoter and the secretory sequence used may be the cause for successfully transforming a cecropin gene into an effective disease-control gene in plants.

The attacins are another class of lytic peptides isolated from the giant silk moth. The *attacinE* gene has been introduced into apple plants. The apple rootstock M.26, which is very susceptible to fire blight *(E. amylovora),* was transformed with the gene for attacinE to yield transgenic line T1 (Norelli et al., 1994a). T1 was significantly more resistant to fire blight than M.26. Twenty-one days after inoculation, 30 percent of the shoot length of T1 was necrotic, compared with 57 percent of M.26 and 23 percent of Liberty (resistant control cultivar). After inoculation of in vitro grown plants, 50 percent of the plants were killed by $10^4 \times 14$ cells of *E. amylovora* while 50 percent of the M.26 plants were killed by $10^4 \times 4$ cells of the bacterial pathogen (Norelli et al., 1994a). Several transgenic clones showed resistance to *E. amylovora* (Norelli et al., 1994b). One of these lines showed 50 percent symptom reduction in the field when inoculated with the pathogen (Norelli et al., 1994b).

EXPLOITATION OF BACTERIOPHAGE GENES FOR BACTERIAL DISEASE MANAGEMENT

Lysozymes are the enzymes with a specific lytic activity against bacterial cell walls. Bacteriophages lyse the bacterial cell walls, producing lysozymes

(Tsugita, 1971). A lysozyme has been purified from bacteriophage T4, and it was active against both gram-negative and gram-positive bacteria. Phage T4 shows lytic action against *E. carotovora,* and a receptor for phage T4 has been reported in the bacterium (Pirhonen and Palva, 1988).

Attempts were made to induce disease resistance by transferring the bacteriophage T4 lysozyme gene into plants and making it secrete into the intercellular spaces where the bacterial pathogens multiply. A plant expression vector was constructed bearing a chimeric gene constructed from the α-amylase signal peptide and the bacteriophage T4 lysozyme gene under control of the CaMV 35S promoter and the NPT-II selectable marker gene between spaces. In the developed transgenic tobacco plants the T4 lysozyme was found to be secreted into the intercellular spaces. The same chimeric gene cassette was transferred to potato through *Agrobacterium tumefaciens*-mediated transformation. The T4 lysozyme could be detected in intercellular washing fluids of transgenic potato plants (During et al., 1993). The transgenic potato plants expressing the bacteriophage T4 lysozyme showed high resistance to *E. carotovora* subsp. *atroseptica* as shown by reduced tissue maceration in transgenic plants (Figure 6.5).

No black leg symptoms developed in transgenic plants till harvest, while no control plants survived under greenhouse conditions (During et al., 1993). Field evaluation of these transgenic plants is needed; however, this technology appears to be a potential tool to manage plant bacterial diseases.

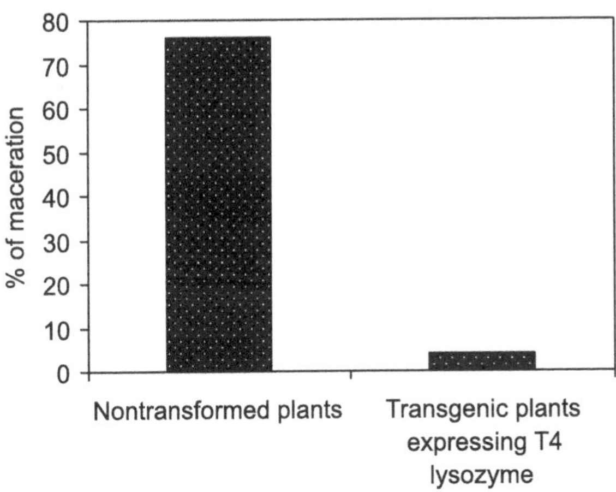

FIGURE 6.5. Maceration of transgenic or nontransformed potato tissue by *Erwinia carotovora* subsp. *atroseptica* (*Source:* Dong et al., 1993).

EXPLOITATION OF GENES FROM HUMAN BEINGS, HENS, AND CRABS FOR MANAGEMENT OF PLANT BACTERIAL DISEASES

Human lysozyme is a powerful lytic enzyme degrading bacterial cell walls. Tobacco plants were transformed with the human lysozyme gene (Nakajima et al., 1997). These transgenic plants showed reduced disease symptoms induced by *P. syringae* pv. *tabaci* in tobacco. Lactoferrin is an iron-binding glycoprotein known to have antibacterial properties. The expression of a human lactoferrin gene in tobacco delayed the onset of symptoms caused by *Ralstonia solanacearum* from 5 to 25 days (Zhang et al., 1996).

A gene encoding a lytic peptide (tachyplesin) has been isolated from the horseshoe crab. Transgenic potato plants expressing this gene were developed, and these plants showed reduction in tuber rot caused by *E. carotovora* (Allefs et al., 1996). A hen-egg-lysozyme gene has been cloned, and transgenic tobacco plants expressing this gene have been developed (Trudel et al., 1995). Extracts from transgenic tobacco plants producing hen-egg-lysozyme inhibited the growth of several species of bacteria (Trudel et al., 1995). However, reaction of these transgenic plants to bacterial diseases has not yet been studied.

E. chrysanthemi synthesizes a siderophore called chrysobactin, which provides the bacteria with iron under the iron-limited conditions that prevail in planta and is required for the bacteria to provoke symptoms in the plant (see Chapter 1). Expert et al. (1996) have suggested that engineering plants with genes encoding transferrins (which are powerful iron scavengers from vertebrates) in order to deprive the bacteria of nutritional iron during infection may enhance host resistance to the bacterial pathogen.

Thus several studies have indicated that genes from heterologous sources can be transferred to plants by using proper plant promoters. These transgenic plants show enhanced disease resistance, although spectacular control of bacterial diseases is not seen. Probably site and amount of accumulation may be important, and proper gene constructs may enhance utility of these biotechnological techniques.

CONCLUSION

Bacterial pathogens affect almost all important crops of the world, causing heavy losses. Some diseases, such as rice bacterial blight, citrus canker, moko wilt of banana, wilt diseases of vegetables, cotton black arm, apple fire blight, and soft rot of potato affect the national economy of many countries. The bacterial diseases are difficult to control. Many efficient fungicides are commercially available to manage fungal diseases. But no effective chemicals are available to control bacterial diseases. Use of antibiotics is impractical under field conditions. Several races in each bacterial pathogen exist, which makes it difficult to develop durable resistant varieties.

Molecular biological studies have revealed several new options for management of bacterial diseases. Activation of defense genes, which are quiescent in susceptible plants, by suitable signals may be one of the best approaches for management of bacterial diseases, because (1) disease resistance is induced by a multitude of defense genes, not by one or two; and (2) coordinated induction of defense genes is needed for induction of significant disease resistance. Several signaling systems exist in plants, and signaling systems may vary from plant to plant and among different plant-pathogen systems. Transcription factors that bind to *cis*-elements present in the promoter region of the defense genes also may vary. Which and how many signaling proteins are required in the defense of plants are still not known. However, different chemicals and biotic inducers activate different signaling pathways and different sets of defense genes. Probably an integration of chemical and biotic inducers may induce a broad spectrum of defense genes and enhance disease resistance. However, caution is needed as antagonism between different signaling systems is also known.

Resistance genes are also involved in signal transduction systems. Successful transfer of resistance genes to heterologous plant species gives another new option to develop disease-resistant plants. These resistance genes activate several defense genes and induce resistance. Homologs of resistance genes are detected in susceptible varieties. It raises a problem; gene silencing or inactivation of introduced resistance genes may be common because of the interaction between the loci with DNA sequence homology. Hence, scrupulous selection of homozygous transgenic plants expressing the transferred resistance gene is important.

Unlike the resistance genes, which are race-specific, defense genes can confer general resistance to various pathogens, bacteria, fungi, and viruses. Hence, manipulation of defense genes may be more ideal. Several transgenic plants expressing defense genes have been developed. But these defense gene products should be secreted into intercellular spaces where the bacterial pathogens multiply in the plant tissues. Efforts are needed to standardize technology to increase transcription, translation, and protein stability in transformed tissue in order to enhance the resistance effect. Transgenic plants expressing a greater number of defense genes should be developed. Enhanced resistance is seen when more copies of the candidate gene are inserted in a single locus in the transformed lines. The defense proteins must be synthesized in sufficient quantities, exported from the cell, and transported to their desired location quickly and without undergoing major modification during the process, and must also be stable at their destination, avoiding degradation by plant proteases. Most of the defense proteins are stable and are not degraded by proteases.

Antibacterial proteins from bacteriophages, humans, insects, birds, and crabs also can be transferred to plants to confer resistance to bacterial pathogens. Several reports of success have been published recently. The important criteria for exploiting this technology are the durability of expression of these foreign genes in the transgenic plants, absence of toxicity, and low environmental impact. Field evaluation of all these technologies is important. In the near future, we can expect this type of biotechnological approach to result in the development of practical strategies for efficient management of bacterial diseases in crop plants.

References

Abad, L. R., D'Urzo, M. P., Liu, D., Narasimhan, M. L., Reuveni, M., Zhu, J. K., Niu, X., Singh, N. K., Hasegawa, P. M., and Bressan, R. A. (1996). Antifungal activity of tobacco osmotin has specificity and involves plasma membrane permeabilization, *Plant Sci.,* 118:11-23.

Abad, M. S., Hakimi, S. M., Kaniewski, W. K., Rommens, C. M. T., Shulaev, V., Lam, E., and Shah, D. M. (1997). Characterization of acquired resistance in lesion-mimic transgenic potato expressing bacterio-opsin, *Mol. Plant-Microbe Interact.,* 10:635-645.

Abeles, F. B. (1986). Plant chemiluminescence, *Annu. Rev. Plant Physiol.,* 37:49-72.

Abeles, F. B., and Biles, C. L. (1991). Characterization of peroxidases in lignifying peach fruit endocarp, *Plant Physiol.,* 95:269-273.

Abeles, F. B., and Forrence, L. E. (1970). Temporal and hormonal control of β-1,3-glucanase in *Phaseolus vulgaris* L., *Plant Physiol.,* 45:395-400.

Abenthum, K., Hildenbrand, S., and Ninnemann, H. (1995). Elicitation and accumulation of phytoalexins in stems, stolons and roots of *Erwinia*-infected potato plants, *Physiol. Mol. Plant Pathol.,* 46:349-359.

Adam, A., Farkas, T., Somlyai, G., Hevesi, M., and Kiraly, Z. (1989). Consequence of O_2^- generation during a bacterially induced hypersensitive reaction in tobacco: deterioration of membrane lipids, *Physiol. Mol. Plant Pathol.,* 34:13-26.

Adam, A. L., Bestwick, C. S., Barna, B., and Mansfield, J. W. (1995). Enzymes regulating the accumulation of active oxygen species during the hypersensitive reaction of bean to *Pseudomonas syringae* pv. *phaseolicola, Planta,* 197:240-249.

Adams, C. A., Nelson, W. S., Nunberg, A. N., and Thomas, T. L. (1992). A wound-inducible member of the hydroxyproline-rich glycoprotein gene family in sunflower, *Plant Physiol.,* 99:775-776.

Addy, S. K. (1976). Leakage of electrolytes and phenols from apple leaves caused by virulent and avirulent strains of *Erwinia amylovora, Phytopathology,* 66:1403-1405.

Ahl, P., Antoniw, J. F., White, R. F., and Gianinazzi, S. (1985). Biochemical and serological characterization of b-proteins from *Nicotiana* species, *Plant Mol. Biol.,* 4:31-37.

Ahl, P., Benjama, A., Sanson, R., Gianinazzi, S. (1981). Induction chez le tabac par *Pseudomonas syringae* de nouvelles proteines (proteines b) associees au developement d'une resistance non specifique a une dexieme infection, *Phytopathol. Z.,* 102:201-202.

Ahmad, M., Majerczak, D. R., and Coplin, D. L. (1996). Harpin is not necessary for the pathogenicity of *Erwinia stewartii* on maize, *Phytopathology,* 86:S77.

Albert, F., and Anderson, A. J. (1987). The effect of *Pseudomonas putida* colonization on root surface peroxidase, *Plant Physiol.,* 85:537-541.

Albright, L. M., Huala, E., and Ausubel, F. M. (1989). Prokaryotic signal transduction mediated by sensor and regulator protein pairs, *Annu. Rev. Genet.,* 23:311-336.

Albright, L. M., Yanofsky, M. F., Leroux, B., Ma, D., and Nester, E. W. (1987). Processing of the T-DNA of *A. tumefaciens* generates border nicks and linear, single stranded T-DNA, *J. Bacteriol.,* 169:1046-1055.

Alexander, D., Goodman, R. M., Gut-Rella, M., Glascock, C., Weymann, K., Friedrich, L., Maddox, D., Ahl Goy, P., Luntz, T., Ward, E., and Ryals, J. (1993). Increased tolerance to two oomycete pathogens in transgenic tobacco expressing pathogenesis-related protein 1a, *Proc. Natl. Acad. Sci. U.S.A.,* 90:7327-7331.

Alexandre, J., Lassalles, J. P., and Kado, R. T. (1990). Opening of Ca^{2+} channels in isolated red beet root vacuole membrane by inositol 1,4,5-triphosphate, *Nature,* 343:567-570.

Alfano, J. R., Bauer, D. W., Milos, T. M., and Collmer, A. (1996). Analysis of the role of the *Pseudomonas syringae* pv. *syringae* HrpZ harpin in elicitation of the hypersensitive response in tobacco using functionally nonpolar deletion mutations, truncated HrpZ fragments, and *hrmA* mutations, *Mol. Microbiol.,* 19:715-728.

Alfano, J. R., and Collmer, A. (1996). Bacterial pathogens in plants: Life up against the wall, *Plant Cell,* 8:1683-1698.

Alfano, J. R., Ham, J. H., and Collmer, A. (1995). Use of Tn5tac1 to clone a *pel* gene encoding a highly alkaline, asparagine-rich pectate lyase isozyme from an *Erwinia chrysanthemi* mutant with deletions affecting the major pectate lyase isozymes, *J. Bacteriol.,* 177:4553-4556.

Alfano, J. R., Kim, H.-S., Delaney, T. P., and Collmer, A. (1997). Evidence that the *Pseudomonas syringae* pv. *syringae hrp*-linked *hrmA* gene encodes an Avr-like protein that acts in an *hrp*-dependent manner within tobacco cells, *Mol. Plant-Microbe Interact.,* 10:580-588.

Allefs, S. J. H. M., de Jong, E. R., Florack, D. E. A., Hoogendoom, C., and Stiekema, W. J. (1996). *Erwinia* soft rot resistance of potato cultivars expressing antimicrobial peptide tachyplesin I, *Mol. Breed.,* 2:97-105.

Allen, C., Huang, Y., and Sequeira, L. (1991). Cloning of genes affecting polygalacturonase production in *Pseudomonas solanacearum, Mol. Plant-Microbe Interact.,* 4:147-154.

Allen, C., Reverchon, S., and Robert-Baudouy, J. (1989). Nucleotide sequence of the *Erwinia chrysanthemi* gene encoding 2-keto-3-deoxygluconate permease, *Gene,* 38:233-241.

Allen, C., Stromberg, V. K., Smith, F. D., Lacy, G. H., and Mount, M. S. (1986). Complementation of an *Erwinia carotovora* protease mutant with a protease-encoding cosmid, *Mol. Gen. Genet.,* 202:276-279.

Allen, D., Gupta, A. S., Webb, R. P., and Holaday, A. S. (1994). Coordinate expression of oxidative stress over-express chloroplastic Cu/Zn superoxide dismutase, *J. Cellular Biochem. Suppl.,* 18A:96.

Alonso, E., de Carvalho Niebel, F., Obregon, P., Gheysen, G., Inze, D., Van Montagu, M., and Castresana, C. (1995). Differential in vitro DNA binding activity to a promoter element of the *gn1* β-1,3-glucanase gene in hypersensitively reacting tobacco plants, *Plant J.,* 7:309-320.

Alstrom, S. (1991). Induction of disease resistance in common bean susceptible to halo blight bacterial pathogen after seed bacterization with rhizosphere pseudomonads, *J. Gen. Appl. Microbiol.,* 37:495-501.

Altabe, S., de Lannino, N. I., de Mendoza, D., and Ugalde, R. A. (1990). Expression of the *Agrobacterium tumefaciens chvB* virulence region in *Azospirillum* spp. *J. Bacteriol.,* 172:2563-2567.

Alt-Morbe, J., Kuhlmann, H., and Schroder, J. (1989). Differences in induction of Ti plasmid virulence genes *virG* and *virD* and continued control of *virD* expression by four external factors, *Mol. Plant-Microbe Interact.,* 2:301-308.

Alt-Morbe, J., Neddermann, P., Von Lintig, J., Weiler, E. W., and Schroder, J. (1988). Temperature-sensitive step in Ti plasmid *vir*-region induction and correlation with cytokinin secretion by *Agrobacterium, Mol. Gen. Genet.,* 213:1-8.

Andary, C. (1993). Caffeic acid glycoside esters and pharmacology. In A. Scolbert (Ed.), *Polyphenolic Phenomena,* INRA editions, Paris, pp. 237-245.

Anderson, D. M., and Mills, D. (1985). The use of transposon mutagenesis in the isolation of nutritional and virulence mutants in two pathovars of *Pseudomonas syringae, Phytopathology,* 75:104-108.

Anderson, J. M. (1985). Evidence for phloem transport of jasmonic acid, *Plant Physiol.,* 77:S-75.

Andresen, I., Becker, W., Schluter, K., Burges, J., Parthier, B., and Apel, K. (1992). The identification of leaf thionin as one of the main jasmonate-induced proteins of barley *(Hordeum vulgare), Plant Mol. Biol.,* 19:193-204.

Andro, T., Chambost, J. P., Kotoujansky, A., Cattaneo, J., Bertheu, Y., Barras, F., Van Gijsegem, F., and Coleno, A. (1984). Mutants of *Erwinia chrysanthemi* defective in secretion of pectinase and cellulase. *J. Bacteriol.,* 160:1199-1203.

Ankenbauer, R. G., Best, E. A., Palanca, C. A., and Nester, E. W. (1991). Mutants of the *Agrobacterium tumefaciens virA* gene exhibiting acetosyringone-independent expression of the *vir* regulon, *Mol. Plant-Microbe Interact.,* 4:400-406.

Ankenbauer, R. G., and Nester, E. W. (1990). Sugar-mediated induction of *Agrobacterium tumefaciens* virulence genes: Structural specificity and activities of monosaccharides, *J. Bacteriol.,* 172:6442-6446.

Antoniw, J. F., Ritter, C. E., Pierpoint, W. S., and Van Loon, L. C. (1980). Comparison of three pathogenesis-related proteins from plants of two cultivars of tobacco infected with TMV, *J. Gen. Virol.,* 47:79-87.

Antoniw, J. F., and White, R. F. (1980). The effects of aspirin and polyacrylic acid on soluble leaf proteins and resistance to virus infection in five cultivars of tobacco, *Phytopathol. Z.,* 98:331-341.

Anzai, H., Yoneyama, K., and Yamaguchi, I. (1989). Transgenic tobacco resistant to a bacterial disease by the detoxification of a pathogenic toxin, *Mol. Gen. Genet.,* 219:492-494.

Apostol, I., Heinstein, P. F., and Low, P. S. (1989). Rapid stimulation of an oxidative burst during elicitation of cultured plant cells, *Plant Physiol.,* 90:109-116.

Arias, J. A., Dixon, R. A., and Lamb, C. J. (1993). Dissection of the functional architecture of a plant defense gene promoter using a homologous in vitro transcription initiation system, *Plant Cell,* 5:485-496.

Arlat, M., Barberis, P., Trigalet, A., and Boucher, C. (1990). Organization and expression in *hrp* genes in *Pseudomonas solanacearum. Proceedings of the 7th International Conference on Plant Pathogenic Bacteria,* Budapest, Hungary, pp. 419-424.

Arlat, M., and Boucher, C. (1991). Identification of a *dsp* DNA region controlling aggressiveness of *Pseudomonas solanacearum, Mol. Plant-Microbe Interact.,* 4:211-213.

Arlat, M., Gough, C. L., Barber, C. E., Boucher, C., and Daniels, M. J. (1991). *Xanthomonas campestris* contains a cluster of *hrp* genes related to the large *hrp* cluster of *Pseudomonas solanacearum, Mol. Plant-Microbe interact.,* 4:593-601.

Arlat, M., Gough, C. L., Zischek, C., Barberis, P. A., Trigalet, A., and Boucher, C. A. (1992). Transcriptional organization and expression of the large *hrp* gene cluster of *Pseudomonas solanacearum, Mol. Plant-Microbe Interact.,* 5:187-193.

Arlat, M., Van Gijsegem, F., Huet, J. C., Pernollet, J. C., and Boucher, C. A. (1994). POPA1, a protein, which induces a hypersensitive-like response on specific *Petunia* genotypes, is secreted via the Hrp pathway of *Pseudomonas solanacearum, EMBO J.,* 13:543-553.

Aronen, T. S. (1997). Interaction between *Agrobacterium tumefaciens* and coniferous defense compounds α-pinene and *trans*-stilbene. *Eur. J. Forest Pathol.,* 27:55-67.

Asita, M., Cui, Y., Liu, Y., Dumenyo, C. K., and Chatterjee, A. K. (1996). Global regulation in *Erwinia carotovora rsmA,* a homologue of *Escherichia coli esrA:* Repression of secondary metabolites, pathogenicity and hypersensitive reaction, *Microbiology (Reading)* 142:427-434.

Aspinall, G. O. (1980). Chemistry of cell wall polysaccharides. In J. Preiss (Ed.), *The Biochemistry of Plants,* Vol. 3, Academic Press, New York, pp. 473-500.

Asselin, A., Grenier, J., and Cote, F. (1985). Light-influenced extracellular accumulation of b (pathogenesis-related) proteins in *Nicotiana* green tissue induced by various chemicals or prolonged floating on water, *Can. J. Bot.,* 63:1276-1283.

Atkinson, M. M., and Baker, C. J. (1987a). Alteration of plasmalemma sucrose transport in *Phaseolus vulgaris* by *Pseudomonas syringae* pv. *syringae* and its association with K^+/H^+ exchange, *Phytopathology,* 77:1573-1578.

Atkinson, M. M. and Baker, C. J. (1987b). Association of host plasma membrane K^+/H^+ exchange with multiplication of *Pseudomonas syringae* pv. *syringae* in *Phaseolus vulgaris, Phytopathology,* 77:1273-1279.

Atkinson, M. M., Bina, J., and Sequeira, L. (1993). Phosphoinositide breakdown during the K^+/H^+ exchange response of tobacco to *Pseudomonas syringae* pv. *syringae, Mol. Plant-Microbe Interact.,* 6:253-260.

Atkinson, M. M., Huang, J. S., and Knopp, J. A. (1985a). Hypersensitivity of suspension-cultured tobacco cells to pathogenic bacteria, *Phytopathology,* 75:1270-1274.

Atkinson, M. M., Huang, J. S., and Knopp, J. A. (1985b). The hypersensitive reaction of tobacco to *Pseudomonas syringae* pv. *pisi:* Activation of a plasmalemma K^+/H^+ exchange mechanism, *Plant Physiol.,* 79:843-847.

Atkinson, M. M., Keppler, L. D., Orlandi, E. W., Baker, C. J., and Mischke, C. F. (1990). Involvement of plasma membrane calcium influx in bacterial induction of the K^+/H^+ and hypersensitive responses in tobacco, *Plant Physiol.,* 92:215-221.

Atkinson, P., and Blakeman, J. P. (1982). Seasonal occurrence of an antimicrobial flavanone, sakuranetin associated with glands on leaves of *Ribes nigrum, New Phytol.,* 92:63-74.

Audy, P., Benhamou, N., Trudel, J., and Asselin, A. (1988). Immunocytochemical localization of a wheat germ lysozyme in wheat embryo and coleoptile cells and cytochemical study of its interaction with the cell wall, *Plant Physiol.,* 88:317-1322.

Auh, C. K., and Murphy, T. M. (1995). Plasma membrane redox enzyme is involved in the synthesis of O_2^- and H_2O_2 by *Phytophthora* elicitor-stimulated rose cells, *Plant Physiol.,* 107:1241-1247.

Ausubel, F. M., Katagiri, F., Mindrinos, M., and Glazebrook, J. (1995). Use of *Arabidopsis thaliana* defence-related mutants to dissect the plant response to pathogens, *Proc. Natl. Acad. Sci. U.S.A.,* 92:4189-4196.

Awade, A., de Tapia, M., Didierjean, L., and Burkard, G. (1989). Biological function of bean pathogenesis-related (PR3 and PR4) proteins, *Plant Sci.,* 63:121-130.

Ayouba, A., Causse, H., Van Damme, E. J. M., Peumans, W. J., Cambillau, C., and Rouge, P. (1994). Interactions of plant lectins with the components of the bacterial cell wall peptidoglycan, *Biochem. Syst. Ecol.,* 22:153-159.

Bach, T. J. (1986). Hydroxymethylglutaryl-CoA reductase, a key enzyme in phytosterol synthesis?, *Lipids,* 21:82-88.

Baer, D., and Gudmestad, N. C. (1995). In vitro cellulolytic activity of the plant pathogen *Clavibacter michiganensis* subsp. *sepedonicus, Can. J. Microbiol.,* 41:877-888.

Baga, M., Chibbar, R. N., and Kartha, K. K. (1995). Molecular cloning and expression analysis of peroxidase genes from wheat, *Plant Mol. Biol.,* 29:647-662.

Bailey, J. A. (1982). Mechanisms of phytoalexin accumulation. In A. Bailey and J.W. Mansfield, (Eds.), *Phytoalexins,* Blackie Press, Glascow, pp. 289-318.

Baker, B., Zambryski, P., Staskawicz, B., and Dinesh-Kumar, S. P. (1997). Signaling in plant-microbe interactions, *Science,* 276:726-733.

Baker, C. J., Atkinson, M. M., and Collmer, A. (1987). Concurrent loss in Tn5 mutants of *Pseudomonas syringae* pv. *syringae* of the ability to induce the hypersensitive response and host plasma membrane K^+/H^+ exchange in tobacco, *Phytopathology,* 77:1268-1272.

Baker, C. J., Atkinson, M. M., Roy, M. A., and Collmer, A. (1986). Inhibition of the hypersensitive response in tobacco by pectate lyase, *Physiol. Mol. Plant Pathol.,* 29:217-225.

Baker, C. J., Mock, N., Atkinson, M. M., and Hutcheson, S. (1990). Inhibition of the hypersensitive response in tobacco by pectate lyase digests of cell wall and of polygalacturonic acid, *Physiol. Mol. Plant Pathol.,* 37:133-167.

Baker, C. J., O'Neill, N., Keppler, L. D., and Orlandi, E. W. (1991). Early responses during plant-bacteria interactions in tobacco cell suspensions. *Phytopathology,* 81:1504-1507.

Baker, C. J., and Orlandi, E. W. (1995). Active oxygen in plant pathogenesis. *Annu. Rev. Phytopathol.,* 33:299-321.

Baker, C. J., Orlandi, E. W., and Mock, N. M. (1993). Harpin, an elicitor of the hypersensitive response in tobacco caused by *Erwinia amylovora,* elicits active oxygen production in suspension cells, *Plant Physiol.,* 102:1341-1344.

Banta, L. M., Joerger, R. D., Howitz, V. R., Campbell, A. M., and Binns, A. N. (1994). Glu-225 outside of the predicted ChvE binding site in VirA is crucial for sugar enhancement of acetosyringone perception by *Agrobacterium tumefaciens, J. Bacteriol.,* 176:3242-3249.

Barny, M. A., Guinebretiere, M. H., Marcais, B., Coissac, A., Paulin, J. P., and Laurent, J. (1990). Cloning of a large gene cluster involved in *Erwinia amylovora* CFBP1430 virulence, *Mol. Microbiol.,* 4:777-787.

Barras, F., Van Gijsegem, F., and Chatterjee, A. K. (1994). Extracellular enzymes and pathogenesis of soft-rot Erwinia, *Annu. Rev. Phytopathol.,* 32:201-234.

Barrat, D. H. P., and Clark, J. A. (1991). Proteins arising during the late stages of embryogenesis in *Pisum sativum* L., *Planta,* 184:14-23.

Barron, D., Varin, L., Ibrahim, R. K., Harborne, J. B., and Williams, C. A. (1988). Sulphated flavonoids: An update, *Phytochemistry,* 27:2375-2395.

Bartling, S., Derkx, P., Wegener, C., and Olsen, O. (1995a). *Erwinia* pectate lyase differences revealed by action pattern analyses. In J. Visser and R. G. J. Voragen (Eds.), *Pectins and Pectinases.* Progress in Biotechnology, Vol. 14, Elsevier Science, Amsterdam, pp. 282-293.

Bartling, S., Wegener, C., and Olsen, O. (1995b). Synergism between *Erwinia* pectate lyase isoenzymes that depolymerize both pectate and pectin, *Microbiology,* 141:873-881.

Barton-Willis, P. A., Roberts, P. D., Guo, A., and Leach, J. E. (1989). Growth dynamics of *Xanthomonas campestris* pv. *oryzae* in leaves of rice differential cultivars, *Phytopathology,* 79:573-578.

Barton-Willis, P. A., Wang, M. C., Holliday, M. J., Long, M. R., and Keen, N. T. (1984). Purification and composition of lipopolysaccharides from *Pseudomonas syringae* pv. *glycinea, Physiol. Plant Pathol.,* 25:387-398.

Barz, W., Bless, W., Daniel, S., Gunia, W., Hinderer, W., Jaques, U., Kessmann, H., Meier, D., Tiemann, K., and Wittkampf, U. (1989). Elicitation and suppression of isoflavones and pterocarpan phytoalexin in chickpea (*Cicer arietinum* L.) cell cultures. In W. G. K. Kurz (Ed.), *Secondary Metabolism of Plant Cell Cultures II,* Springer-Verlag, Berlin, pp. 208-218.

Bashan, Y., Okon, Y., and Henis, Y. (1986). A possible role for proteases and deaminases in the development of the symptoms of bacterial speck disease in tomato caused by *Pseudomonas syringae* pv. *tomato, Physiol. Mol. Plant Pathol.,* 28:15-31.

Batalia, M. A., Monzingo, A. F., Emst, S., Roberts, W., and Robertus, J. D. (1996). The crystal structure of the antifungal protein zeamatin, a member of the thaumatin-like, PR-5 protein family, *Nature Structural Biology,* 3:19-23.

Bauer, D. W., and Beer, S. V. (1987). Cloning of *Erwinia amylovora* DNA responsible for pathogenicity and the induction of the hypersensitive reaction. *Acta Hortic.,* 217:169-170.

Bauer, D. W., and Beer, S. V. (1991). Further characterization of an *hrp* gene cluster of *Erwinia amylovora, Mol. Plant-Microbe Interact.,* 4:493-499.

Bauer, D. W., Bogdanove, A. J., Beer, S. V., and Collmer, A. (1994). *Erwinia chrysanthemi hrp* genes and their involvement in soft rot pathogenesis and elicitation of the hypersensitive response, *Mol. Plant-Microbe Interact.,* 7:573-581.

Bauer, D. W., and Collmer, A. (1997). Molecular cloning, characterization, and mutagenesis of a *pel* gene from *Pseudomonas syringae* pv. *lachrymans* encoding a member of the *Erwinia chrysanthemi* PelADE family of pectate lyase, *Mol. Plant-Microbe Interact.,* 10:369-379.

Bauer, D. W., Wei, Z.-M., Beer, S. V., and Collmer, A. (1995). *Erwinia chrysanthemi* $harpin_{Ech}$: An elicitor of the hypersensitive response that contributes to soft-rot pathogenesis, *Mol. Plant-Microbe Interact.,* 8:484-491.

Bavage, A., Vivian, A., Atherton, G., Taylor, J., and Malik, A. (1991). Molecular genetics of *Pseudomonas syringae* pathovar *pisi:* Plasmid involvement in cultivar-specific incompatibility, *J. Gen. Microbiol.,* 137:2231-2239.

Bayles, C. J., Ghemawat, M. S., and Aist, J. R. (1990). Inhibition by 2-deoxy-D-glucose of callose formation, papilla deposition, and resistance to powdery mildew in an ml-O barley mutant, *Physiol. Mol. Plant Pathol.,* 36:63-72.

Beaulieu, C., Minsavage, G. V., Canteros, B. I., and Stall, R. E. (1991). Biochemical and genetic analysis of a pectate lyase gene from *Xanthomonas campestris* pv. *vesicatoria, Mol. Plant-Microbe Interact.,* 4:446-451.

Beaulieu, C., and Van Gijsegem, F. (1990). Identification of plant-inducible genes in *Erwinia amylovora, Mol. Plant-Microbe Interact.,* 4:493-499.

Becker, F., Buschfeld, E., Schell, J., and Bachmair, A. (1993). Altered response to viral infection by tobacco plants perturbed in ubiquitin system, *Plant J.,* 3:875-881.

Becker, W., and Apel, K. (1992). Isolation and characterization of a cDNA clone encoding a novel jasmonate-induced protein of barley (*Hordeum vulgare* L.), *Plant Mol. Biol.,* 19:1065-1067.

Bednarek, S., and Raikhel, N. V. (1991). The barley lectin carboxy-terminal propetide is a vacuolar protein-sorting determinant in plants, *Plant Cell,* 3:1195-1206.

Bednarek, S., Wilkins, T. A., Dombrowski, J. E. and Raikhel, N. V. (1990). A carboxyl-terminal propeptide is necessary for proper sorting of barley lectin to vacuoles of tobacco, *Plant Cell,* 2:1145-1155.

Beer, S. V., Bauer, D. W., Jiang, X. H., Laby, R. J., Sneath, B. J., Wei, Z. M., Wilcox, D. A., and Zumoff, C. H. (1991). The *hrp* gene cluster of *Erwinia amylovora*. In H. Hennecke and D. P. S. Verma (Eds.), *Advances in Molecular Genetics of Plant-Microbe Interactions. Vol. 1.* Kluwer Academic Publishers, Dordrecht, The Netherlands, pp. 53-60.

Beer, S. V., Laby, R. J., and Coplin, D. L. (1990). Complementation of *hrp* mutants of *Erwinia amylovora* with DNA of *Erwinia stewartii, Phytopathology,* 80:985 (Abstr.).

Beer, S. V., Wei, Z. M., Bauer, D. W., Collmer, A., He, S. Y., and Laby, R. (1994). Elicitor of the hypersensitive response in plants, International Application Published under the Patent Cooperation Treaty, International Publication Number: W094/01546.

Beerhues, L., and Kombrink, E. (1994). Primary structure and expression of mRNAs encoding basic chitinase and 1,3-β-glucanase in potato, *Plant Mol. Biol.,* 24:353-367.

Beffa, R., Szell, M., Meuwly, P., Pay, A., Vogeli-Lange, R., Metraux, J. P., Neuhaus, G., Meins, F., and Ferenc, N. (1995). Cholera toxin elevates pathogen resistance and induces pathogenesis-related gene expression in tobacco, *EMBO J.,* 14:5753-5761.

Beilmann, A., Pfitzner, A. J. P., Goodman, H. M., and Pfitzner, U. M. (1991). Functional analysis of the pathogenesis-related 1a protein gene minimal promoter region: comparison of reporter gene expression in transient and in stable transfections, *Eur. J. Biochem.,* 196:415-421.

Beimen, A., Bermpohl, A., Meletzus, D., Eichenlaub, R., and Barz, W. (1992). Accumulation of phenolic compounds in leaves of tomato plants after infection with *Clavibacter michiganense* strains differing in virulence, *Z. Naturforsch.,* 47C:898-909.

Belanger, C., Loubens, I., Nester, E. W., and Dion, P. (1997). Variable efficiency of a Ti plasmid-encoded VirA protein in different agrobacterial hosts, *J. Bacteriol.,* 179:2305-2313.

Bell, A. A., Mace, M. E., and Stipanovic, R. D. (1986). Biochemistry of cotton *(Gossypium)* resistance to pathogens, *ACS Symp. Ser.,* 296:36-56.

Bender, C. L., Stone, H. E., Sims, J. J., and Cooksey, D. A. (1987). Reduced pathogen fitness of *Pseudomonas syringae* pv. *tomato* Tn5 mutants defective in coronatine production, *Physiol. Mol. Plant Pathol.,* 30:273-283.

Benhamou, N. (1991). Cell surface interactions between tomato and *Clavibacter michiganense* subsp. *michiganense:* Localization of some polysaccharides and hydroxyproline-rich glycoproteins in infected host leaf tissues, *Physiol. Mol. Plant Pathol.,* 38:15-38.

Benhamou, N. (1995). Ultrastructural and cytochemical aspects of the response of eggplant parenchyma cells in direct contact with *Verticillium*-infected xylem vessels, *Physiol. Mol. Plant Pathol.,* 46:321-338.

Benhamou, N., Belanger, R. R., and Paulity, T. (1996a). Induction of differential host responses by *Pseudomonas fluorescens* in Ri-TDNA transformed pea roots upon challenge with *Fusarium oxysporum* f.sp. *pisi* and *Pythium ultimum, Phytopathology,* 86:1174-1185.

Benhamou, N., Belanger, R. R., and Paulity, T. (1996b). Ultrastructural and cytochemical aspects of the interaction between *Pseudomonas fluorescens* and Ri T-DNA transformed pea roots: host response to colonization by *Pythium ultimum Trow., Planta,* 199:105-117.

Benhamou, N., Grenier, J., and Asselin, A. (1991a). Immunogold localization of pathogenesis-related protein P14 in tomato root cells infected by *Fusarium oxysporum* f.sp. *radicis-lycopersici, Physiol. Mol. Plant Pathol.,* 38:237-253.

Benhamou, N., Lafitte, C., Barthe, J. P., and Esquerre-Tugaye, M. T. (1991b). Cell surface interactions between bean leaf cells and *Colletotrichum lindemuthianum, Plant Physiol.,* 97:234-244.

Benhamou, N., Kloepper, J. W., Quadt-Hallmann, A., and Tuzun, S. (1996c). Induction of defense-related ultrastructural modifications in pea root tissues inoculated with endophytic bacteria, *Plant Physiol.,* 112:919-929.

Bennett, M. H., Gallagher, M. D. S., Bestwick, C. S., Rossiter, J. T., and Mansfield, J. W. (1994). The phytoalexin response to lettuce to challenge by *Botrytis cinerea, Bremia lactucae* and *Pseudomonas syringae* pv. *phaseolicola, Physiol. Mol. Plant Pathol.,* 44:321-333.

Benov, L. T., and Fridovich, I. (1994). *Escherichia coli* expresses a copper- and zinc-containing superoxide dismutase, *J. Biol. Chem.,* 269:25310-25314.

Bent, A. F. (1996). Plant disease resistance genes: Function meets structure, *Plant Cell,* 8:1757-1771.

Bent, A., Kunkel, B., Dahlbeck, D., Brown, K., Schmidt, R., Giraudat, J., Leung, J., and Staskawicz, B. J. (1994). RPS2 of *Arabidopsis thaliana:* A leucine-rich repeat class of plant disease resistance genes, *Science,* 265:1856-1860.

Bequin, P. (1990). Molecular biology of cellulose degradation, *Annu. Rev. Microbiol,* 44:219-248.

Beraha, L., and Garber, E. D. (1971). Avirulence and extracellular enzymes of *Erwinia carotovora, Phytopathol. Z.,* 70:335-344.

Bergey, D. R., Howe, G. A., and Ryan, C. A. (1996). Polypeptide signaling for plant defensive genes exhibits analogies to defense signaling in animals, *Proc. Natl. Acad. Sci. U.S.A.,* 93:12053-12058.

Berglund, L., Brunstedt, J., Nielsen, K. K., Chen, Z., Mikkelsen, J. D., and Marcker, K. A. (1995). A proline-rich chitinase from *Beta vulgaris, Plant Mol. Biol.,* 27:211-216.

Bergmann, C., Ito, Y., Singer, D., Albersheim, P., Darvill, A. G., Benhamou, N., Nuss, L., Salvi, G., Cervone, F., and De Lorenzo, G. (1994). Polygalacturonase-inhibiting protein accumulates in *Phaseolus vulgaris* L. in response to wounding, elicitors, and fungal infection, *Plant J.,* 5:625-634.

Berhane, K., Widersten, M., Engstrom, A., Kozarich, J. W., and Mannervik, B. (1994). Detoxification of base propenals and other α, β-unsaturated aldehyde products of radical reactions and lipid peroxidation by human glutathione transferases, *Proc. Natl. Acad. Sci. U.S.A.,* 91:1480-1484.

Bernasconi, P., Locher, R., Pilet, P. E., Jolles, J., and Jolles, P. (1987). Purification and N-terminal amino-acid sequence of a basic lysozyme from *Parthenocissus quinquifolia* cultured *in vitro, Biochim. Biophys. Acta,* 915:254-260.

Berridge, M. J., and Irvine, R. F. (1989). Inositol phosphates and cell signalling, *Nature,* 341:197-205.

Bertoni, G., and Mills, D. (1987). A simple method to monitor growth of bacterial populations in leaf tissue, *Phytopathology,* 77:832-835.

Bestwick, C. S., Bennett, M. H., and Mansfield, J. W. (1995). Hrp mutant of *Pseudomonas syringae* pv. *phaseolicola* induces cell wall alterations but not membrane damage leading to the hypersensitive reaction in lettuce *Lactuca sativa, Plant Physiol.,* 108:503-516.

Bevan, M., Schufflebottom, D., Edwardas, K., Jefferson, R., and Schuch, W. (1989). Tissue- and cell-specific activity of a phenylalanine ammonia-lyase promoter in transgenic plants, *EMBO J.,* 8:1899-1906.

Bi, Y.-M., Kenton, P., Mur, L., Darby, R., and Draper, J. (1995). Hydrogen peroxide does not function downstream of salicylic acid in the induction of PR protein expression, *Plant J.,* 8:235-245.

Biles, C. L., and Abeles, F. B. (1991). Xylem sap proteins, *Plant Physiol.,* 96:597-601.

Billah, M. M. (1993). Phospholipase D and cell signaling, *Curr. Opin. Immunol.,* 5:114-123.

Binns, A. N., Beaupre, C. E., and Dale, E. M. (1995). Inhibition of VirB-mediated transfer of diverse substates from *Agrobacterium tumefaciens* by the IncQ plasmid RSF 1010. *J. Bacteriol.,* 177:4890-4899.

Bishai, W. R., Smith, H. O., and Barcak, G. J. (1994). A peroxide/ascorbate-inducible catalase from *Haemophilus influenzae* is homologous to *Escherichia coli katE* gene product, *J. Bacteriol.,* 176:2914-2921.

Bishop, C. D., and Cooper, R. M. (1984). Ultrastructure of vascular colonization by fungal wilt pathogens. II. Invasion of resistant cultivars, *Physiol. Plant Pathol.,* 24:277-289.

Bishop, P. D., Pearce, G., Bryant, J. E., and Ryan, C. A. (1984). Isolation and characterization of the proteinase inhibitor-inducing factor from tomato leaves, *J. Biol. Chem.,* 259:13172-13177.

Bizarri, M., Fiore, N., Ranalli, P., and Stefani, E. (1996). Transmission of induced resistance to *Pseudomonas syringae* pv. *tabaci* in tobacco plants regenerated in vitro, *Phytopathologia Medit.,* 35:152-156.

Blechert, S., Brodschelm, W., Holder, S., Kammerer, L., Kutchan, T. M., Mueller, M. J., Xia, Z. Q., and Zenk, M. H. (1995). The octadecanoid pathway: Signal molecules for the regulation of secondary pathways, *Proc. Natl. Acad. Sci. U.S.A,* 92:4099-4105.

Boccara, M., Diolez, A., Rouve, M., and Kotoujansky, A. (1988). The role of individual pectate lyases of *Erwinia chrysanthemi* strain 3937 in pathogenicity on saintpaulia plants, *Physiol. Mol. Plant Pathol.,* 33:95-104.

Bogdanove, A. J., Beer, S. V., Bonas, U., Boucher, C. A., Collmer, A., Coplin, D. L., Cornelis, G. R., Huang, H. C., Hutcheson, S. W., Panopoulos, N. J., and Van Gijsegem, F. (1996a). Unified nomenclature for broadly conserved *hrp* genes of phytopathogenic bacteria, *Mol. Microbiol.,* 20:681-683.

Bogdanove, A. J., Wei, Z. M., Zhao, L., and Beer, S. V. (1996b). *Erwinia amylovora* secretes harpin via a type III pathway and contains a homolog of *yopN* of *Yersinia, J. Bacteriol.,* 178:1720-1730.

Bogre, L., Olah, Z., and Dudits, D. (1988). Ca^{2+}-dependent protein kinase from alfalfa *(Medicago varia):* Partial purification and autophosphorylation, *Plant Sci.,* 85:135-144.

Boher, B., Kpemoua, K., Nicole, M., Luisetti, J., and Geiger, J. P. (1995). Ultrastructure of interactions between cassava and *Xanthomonas campestris* pv. *manihotis*. Cytochemistry of cellulose and pectin degradation in a susceptible cultivar, *Phytopathology,* 85:777-788.

Bohlmann, H. (1994). The role of thionins in plant protection, *Crit. Rev. Plant Sci.,* 13:1-16.

Bohlmann, H., and Apel, K. (1987). Isolation and characterization of cDNAs coding for leaf-specific thionins closely related to the endosperm-specific hordothionin of barley (*Hordeum vulgare* L.), *Mol. Gen. Genet.,* 207:446-454.

Bohlmann, H., and Apel, K. (1991). Thionins, *Annu Rev. Plant Physiol. Plant Mol. Biol.,* 42:227-240.

Bohlmann, H., Clausen, S., Behnke, S., Giese, H., Hiller, C., Reimann-Philipp, U., Schrader, G. Barkholt, V., and Apel, K. (1988). Leaf-specific thionins of barley—a novel class of cell wall proteins toxic to plant pathogenic fungi and possibly involved in the defense mechanism of plants, *EMBO J.,* 7:1559-1565.

Bol, D. K., and Yasbin, R. E. (1990). Characterization of an inducible oxidative stress system in *Bacillus subtilis, J. Bacteriol.,* 172:3503-3506.

Boller, T. (1988). Ethylene and the regulation of antifungal hydrolases in plants, *Oxford Surv. Plant Mol. Cell Biol.,* 5:145-174.

Boller, T., Gehri, A., Mauch, F., and Vogeli, U. (1983). Chitinase in bean leaves: Induction by ethylene, purification, properties, and possible function, *Planta,* 157: 22-31.

Boller, T., and Vogeli, U. (1984). Vacuolar localization of chitinase in bean leaves, *Plant Physiol.,* 74:442-444.

Bolton, G. W., Nester, E. W., and Gordon, M. P. (1986). Plant phenolic compounds induce expression of the *Agrobacterium tumefaciens* loci needed for virulence, *Science,* 232:983-985.

Bolwell, G. P. (1993). Dynamic aspects of the plant extracellular matrix, *International Review of Cytology,* 146:261-324.

Bolwell, G. P., Robbins, M. P., and Dixon, R. A. (1985). Metabolic changes in elicitor-treated bean cells. Enzymic responses associated with rapid changes in cell wall components, *Eur. J. Biochem.,* 148:571-578.

Boman, H. G., Faye, I., Gudmundson, G. H., Lee, J. Y., and Lidhom, D. A. (1991). Cell-free immunity in cecropia, *Eur. J. Biochem.,* 201:23-31.

Bonas, U. (1994). *hrp* genes of phytopathogenic bacteria, *Curr. Top. Microbiol. Immunol.,* 192:79-98.

Bonas, U., Schulte, R., Fenselau, S., Minsavage, G. V., Staskawicz, B. J., and Stall, R. E. (1991). Isolation of a gene cluster from *Xanthomonas campestris* pv. *vesicatoria* that determines pathogenicity and the hypersensitive response on pepper and tomato, *Mol. Plant-Microbe Interact.,* 4:81-88.

Bonas, U., Stall, R. E., and Staskawicz, B. (1989). Genetic and structural characterization of the avirulence gene *avrBs3* from *Xanthomonas campestris* pv. *vesicatoria, Mol. Gen. Genet.,* 218:127-136.

Boss, W. F., and Massel, M. O. (1985). Phosphoinositides are present in plant tissue culture cells. *Biochem. Biophys. Res. Commun.,* 132:1018-1023.

Bostock, R. M. (1989). Metabolism of lipids containing arachidonic and eicosapentaenoic acids in race-specific interactions between *Phytophthora infestans* and potato, *Phytopathology,* 79:989-992.

Bostock, R. M., Yamamoto, H., Choi, D., Ricker, K. E., and Ward, B. L. (1992). Rapid stimulation of 5-lipoxygenase activity in potato by the fungal elicitor arachidonic acid, *Plant Physiol.,* 100:1448-1456.

Boucher, C. A., Barberis, P. A., and Arlat, M. (1988). Acridine orange selects for deletion of *hrp* genes in all races of *Pseudomonas solanacearum, Mol. Plant-Microbe Interact.,* 1:282-288.

Boucher, C. A., Barberis, P. A., Trigalet, A. P., and Demery, D. A. (1985). Transposon mutagenesis of *Pseudomonas solanacearum:* Isolation of Tn5-induced avirulent mutants. *J. Gen. Microbiol.,* 131:2449-2457.

Boucher, C., Martinel, A., Barberis, P., Alloing, G., and Zischek, C. (1986). Virulence genes are carried by a megaplasmid of the plant pathogen *Pseudomonas solanacearum, Mol. Gen. Genet.,* 205:270-275.

Boucher, C. A., Van Gijsegem, F., Barberis, P. A., Arlat, M., and Zischek, C. (1987). *Pseudomonas solanacearum* genes controlling both pathogenicity on tomato and hypersensitivity on tobacco are clustered, *J. Bacteriol.,* 169:5626-5632.

Bowles, D. J. (1990). Defense-related proteins in higher plants, *Annu. Rev. Biochem.,* 59:873-907.

Bowling, S. A., Guo, A., Cao, A., Gordon, A. S., Klessig, D. F., and Dong, X. (1994). A mutation in *Arabidopsis* that leads to constitutive expression of systemic acquired resistance, *Plant Cell,* 6:1845-1857.

Bradley, D. J., Kjellbom, P., and Lamb, C. J. (1992). Elicitor- and wound-induced oxidative cross-linking of a proline-rich plant cell wall protein: A novel, rapid defense response, *Cell,* 70:21-30.

Brady, J. D., Sadler, I. H., and Fry, S. C. (1996). Di-isodityrosine, a novel tetrameric derivative of tyrosine in plant cell wall proteins: a new potential cross-link, *Biochem. J.,* 315:323-327.

Brederode, F. T., Linthorst, H. J. M., and Bol, J. F. (1991). Differential induction of acquired resistance and PR gene expression in tobacco by virus infection, ethephon treatment, UV light and wounding, *Plant Mol. Biol.,* 17:1117-1125.

Breiteneder, H., Ferreira, F., Hoffmann-Sommergruber, K., Ebner, C., Breitenbach, M., Rumpold, H., Kraft, D., and Scheiner, O. (1993). Four recombinant isoforms of Coral, the major allergen of hazel pollen, show different IgE-binding properties, *Eur. J. Biochem.,* 212:355-362.

Breiteneder, H., Pettenburger, K., Bito, A., Valenta, R., Kraft, D., Rumpold, H., Scheiner, O., and Breitenbach, M. (1989). The gene coding for the major birch pollen allergen BetvI is highly homologous to a pea disease resistance response gene, *EMBO J.,* 8:1935-1938.

Bretschneider, K. E., Gonella, M. P., and Robeson, D. J. (1989). A comparative light and electron microscopical study of compatible and incompatible interactions between *Xanthomonas campestris* pv. *campestris* and cabbage *(Brassica oleracea), Physiol. Mol. Plant Pathol.,* 34:285-297.

Breviario, D., Morello, L., and Gianai, S. (1995). Molecular cloning of two novel rice cDNA sequences encoding putative calcium-dependent protein kinases, *Plant. Mol.Biol.,* 27:953-967.

Brightwell, G., Hussain, H., Tiburtius, A., Yeoman, K. H., and Johnston, A. W. B. (1995). Pleiotropic effects of regulatory *ros* mutants of *Agrobacterium radiobacter* and their interaction with Fe and glucose, *Mol. Plant-Microbe Interact.,* 8:747-754.

Brindle, P. A., and Threlfall, D. R. (1983). The metabolism of phytoalexins, *Biochem. Soc.Transactions,* 11:516-522.

Brinker, A. M., and Seigler, D. S. (1991). Isolation and identification of piceatannol as a phytoalexin from sugarcane, *Phytochemistry,* 30:3229-3232.

Brisset, M. N., and Paulin, J. P. (1991). Relationships between electrolyte leakage from *Pyrus communis* and virulence of *Erwinia amylovora, Physiol. Mol. Plant Pathol.,* 39:443-453.

Brisson, L. F., Tenhaken, R., and Lamb, C. J. (1994). Functions of oxidative cross-linking of cell wall structural proteins in plant disease resistance, *Plant Cell,* 6:1703-1712.

Broekaert, W. F., Cammue, B. P. A., De Bolle, M. F. C., Thevissen, K., De Samblanx, G. W., and Osborn, R. W. (1997). Antimicrobial peptides from plants, *Crit. Rev. Plant Sci.,* 16:297-323.

Broekaert, W. F., and Peumans, W. J. (1986). Lectin release from seeds of *Datura stramonium* and interference of the *Datura stramonium* lectin with bacterial motility. In T. C. Bog-Hansen and E. Van Dreiessche (Eds.), *Lectins, Biology, Biochemistry, Clinical Biochemistry,* Vol. 5, Walter de Gruyter, Berlin, pp. 57-65.

Broekaert, W. F., Terras, F. R. G., Cammue, B. P. A., and Osborn, R. W. (1995). Plant Defensins: novel antimicrobial peptides as components of the host defense system, *Plant Physiol.,* 108:1353-1358.

Broglie, K., Chet, I., Holliday, M., Cressman, R., Biddle, P., Knowlton, S., Mauvais, J. C., and Broglie, R. (1991). Transgenic plants with enhanced resistance to the fungal pathogen *Rhizoctonia solani, Science,* 254:1194-1197.

Broglie, K. E., Gaynor, J. J., and Broglie, R. M. (1986). Ethylene-regulated gene expression: Molecular cloning of the genes encoding an endochitinase from *Phaseolus vulgaris, Proc. Natl. Acad. Sci. USA,* 83:6820-6824.

Brown, I. R., and Mansfield, J. W. (1988). An ultrastructural study, including cytochemistry and quantitative analyses of the interactions between pseudomonads and leaves of *Phaseolus vulgaris* L., *Physiol. Mol. Plant Pathol.,* 33:351-376.

Brown, I., and Mansfield, J. (1991). Interactions between pseudomonads and *Phaseolus vulgaris.* In K. Mengen and D. E. Lesemann (Eds.), *Electron Microscopy of Plant Pathogens,* Springer-Verlag, Berlin, pp.185-196.

Brown, I., Mansfield, J., and Bonas, U. (1995). *hrp* genes in *Xanthomonas campestris* pv. *vesicatoria* determine ability to suppress papilla deposition in pepper mesophyll cells, *Mol. Plant-Microbe Interact.,* 8:825-836.

Brown, I., Mansfield, J., Irlam, I., Conrads-Strauch, J., and Bonas, U. (1993). Ultrastructure of interactions between *Xanthomonas campestris* pv. *vesicatoria* and pepper, including immunocytochemical localization of extracellular polysaccharides and AvrBs3 protein, *Mol. Plant-Microbe Interact.,* 6:376-386.

Brown, J. H., Paliyath, G., and Thompson, J. E. (1990). Influence of acyl chain composition on the degradation of phosphatidylcholine by phospholipase D in carnation microsomal membranes, *J. Exp. Bot.,* 41:979-986.

Browne, L. M., Conn, K. L., Ayer, W. A., and Tewari, I. P. (1991). The camalexins: New phytoalexins produced in the leaves of *Camelina sativa* (Cruciferae), *Tetrahedron,* 47:3909-3914.

Brownleader, M., Golden, K. D., and Dey, P. M. (1993). An inhibitor of extensin peroxidase in cultured tomato cells, *Phytochemistry,* 33:755-758.

Bruce, R. J., and West, C. A. (1982). Elicitation of casbene synthetase activity in castor bean: The role of pectic fragments of the plant cell wall in elicitation by a fungal endopolygalacturonase, *Plant Physiol.,* 69:1181-1188.

Brumbley, S. M., Carney, B. F., and Denny, T. P. (1993). Phenotype conversion in *Pseudomonas solanacearum* due to spontaneous inactivation of PhcA, a putative LysR transcriptional activator, *J. Bacteriol.,* 175:5477-5487.

Brumbley, S. M., and Denny, T. P. (1990). Cloning of wild-type *Pseudomonas solanacearum phcA,* a gene that when mutated alters expression of multiple traits that contribute to virulence, *J. Bacteriol.,* 172:5677-5685.

Brun, E., Gans, P., Marion, D., and Barras, F. (1995). Overproduction, purification and characterization of cellulose-binding domain of the *Erwinia chrysanthemi* secreted endoglucanase EGZ, *Eur. J. Biochem.,* 231:142-148.

Bryngelsson, T., and Green, B. (1989). Characterization of a pathogenesis-related, thaumatin-like protein isolated from barley challenged with an incompatible race of mildew, *Physiol. Mol. Plant Pathol.,* 35:45-52.

Bryngelsson, T., Sommer-Knudsen, J., Gregersen, P. L., Collinge, D. B., Ek, B., and Thordel-Christensen, H. (1994). Purification, characterization, and molecular cloning of basic PR-1 type pathogenesis-related proteins from barley. *Mol. Plant-Microbe Interact,* 7:267-275.

Buchel, A. S., and Linthorst, H. J. M. (1999). PR-1: A group of plant proteins induced upon pathogen infection. In S. K. Datta and S. Muthukrishnan (Eds.), *Pathogenesis-related Proteins in Plants,* CRC Press, Boca Raton, pp. 21-47.

Buchel, A. S., Molenkamp, R., Bol, J. F., and Linthorst, H. J. M. (1996). The *PR-1a* promoter contains a number of regulatory elements that bind GT-1 like factors with different affinity, *Plant Mol. Biol.,* 30:493-504.

Buell, C. R., and Somerville, S. C. (1995). Expression of defense-related and putative signaling genes during tolerant and susceptible interactions of *Arabidopsis* with *Xanthomonas campestris* pv. *campestris, Mol. Plant-Microbe Interact.,* 8:435-443.

Bull, J., Mauch, F., Hertig, C., Rebmann, G., and Dudler, R. (1992). Sequence and expression of a wheat gene that encodes a novel protein associated with pathogen defense, *Mol. Plant-Microbe Interact.,* 5:516-519.

Bunge, S., Wolters, J., and Apel, K. (1992). A comparison of leaf thionin sequences of barley cultivars and wild barley species, *Mol. Gen. Genet.,* 231:460-468.

Buonaurio, R., Torre, G. D., and Montalbini, P. (1987). Soluble superoxide dismutase (SOD) in susceptible and resistant host-parasite complexes of *Phaseolus vulgaris* and *Uromyces phaseoli, Physiol. Mol. Plant Pathol.,* 31:173-184.

Burden, R. S., Bailey, J. A., and Dawson, G. W. (1972). Structures of three new isoflavnoids from *Phaseolus vulgaris* infected with tobacco necrosis virus, *Tetrahedron Lett.,* 41:4175-4178.

Burkowicz, A., and Goodman, R. N. (1969). Permeability alterations induced in apple leaves by virulent and avirulent strains of *Erwinia amylovora, Phytopathology,* 59:314-318.

Bussink, H. J. D., Buxton, F. P., and Visser, J. (1991). Expression and sequence comparison of the *Aspergillus niger* and *Aspergillus tubigensis* genes encoding polygalcturonase II, *Curr. Genet.,* 19:467-474.

Buttner, M., and Singh, K. B. (1997). *Arabidopsis thaliana* ethylene-responsive element binding protein (AtEBP), ethylene-inducible, GCC box DNA-binding protein interacts with an *ocs* element binding protein, *Proc. Natl. Acad. Sci. U.S.A.,* 94:5961-5966.

Buvat, R. (1989). *Ontogeny, Cell Differentiation, and Structure of Vascular Plants,* Springer Verlag, Berlin, pp. 481.

Caelles, C., Delseney, M., and Puidomenech, P. (1992). The hydroxyproline-rich glycoprotein gene from Oryza sativa, *Plant Mol. Biol.,* 18:617-619.

Cahill, D. M., and McComb, J. A. (1992). A comparison of changes in phenylalanine ammonia-lyase activity, lignin and phenolic synthesis in the roots of *Eucalyptus calophylla* (field resistant) and *E. marginata* (susceptible) when infected with *Phytophthora cinnamomi, Physiol. Mol. Plant Pathol.,* 40:315-332.

Cai, D., Kleine, M., Kifle, S., Harloff, H. J., Sandal, N. N., Marcker, K. A., Klein-Lankhorst, R. M., Salentijn, E. M. J., Lange, W., and Stiekema, W. J. (1997). Positional cloning of a gene for nematode resistance in sugar beet, *Science,* 275:832-834.

Calderon, A. A., Zapata, J. M., and Barcelo, A. R. (1994). Peroxidase-mediated formation of resveratrol oxidation products during the hypersensitive-like reaction of grapevine cells to an elicitor from *Trichoderma viride, Physiol. Mol. Plant Pathol.,* 44:289-299.

Calderon, A. A., Zapata, J. M., Munoz, R., Pedreno, M. A., Ros Barcelo, A. (1993). Resveratrol production as a part of the hypersensitive-like response of grapevine cells to an elicitor from *Trichoderma viride, New Phytol.,* 124:455-463.

Camacho Henriquez, A., and Sanger, H. L. (1982). Analysis of acid extractable tomato leaf proteins after infection with a viroid, two viruses and a fungus and partial purification of the "pathogenesis-related" protein P14, *Arch. Virol.,* 74: 181-195.

Cameron, R. K., Dixon, R. A., and Lamb, C. J. (1994). Biologically induced systemic acquired resistance in *Arabidopsis thaliana, Plant J.,* 5:715-725.

Cammue, B. P. A., De Bolle, M. F. C., Terras, F. R. G., Proost, P., Van Dammme, J., Rees, S. B., Vanderleyden, J., and Broekaert, W. F. (1992). Isolation and characterization of a novel class of plant antimicrobial peptides from *Mirabilis jalapa* L. seeds, *J. Biol. Chem.,* 267:2228-2233.

Campa, A. (1991). Biological roles of plant peroxidases: known and potential function. In J. Everse and M. B. Grisham (Eds.), *Peroxidases in Chemistry and Biology, Vol. II,* CRC Press, Boca Raton, FL., pp. 25-50.

Cangelosi, G., Ankenbauer, R., and Nester, E. (1990). Sugars induce the *Agrobacterium* virulence genes through periplasmic binding and a transmembrane signal protein, *Proc. Natl. Acad. Sci. U.S.A,* 87:6708-6712.

Cangelosi, G. A., Hung, L., Puvanesarajah, V., Stacey, G., Ozga, D. A., Leigh, J. A., and Nester, E. W. (1987). Common loci of *Agrobacterium tumefaciens* and *Rhizobium meliloti* extracellular polysaccharide synthesis and their roles in plant interactions, *J. Bacteriol.,* 169:2086-2091.

Canteros, B., Minsavage, G., Bonas, U., Pring, D., and Stall, R. (1991). A gene from *Xanthomonas campestris* pv. *vesicatoria* that determines avirulence in tomato is related to *avrBs3, Mol. Plant-Microbe Interact.*, 4:628-632.

Cao, H., Bowling, S. A., Gordon, A. S., and Dong, X. (1994). Characterization of an *Arabidopsis* mutant that is nonresponsive to inducers of systemic acquired resistance, *Plant Cell,* 6:1583-1592.

Caprari, C., Mattei, B., Basile, M. L., Salvi, G., Crescenzi, V., De Lorenzo, G., and Cervone, F. (1996). Mutagenesis of endopolygalacturonase from *Fusarium moniliforme:* Histidine residue 234 is critical for enzymatic and macerating activities and not for binding to polygalacturonase inhibiting protein (PGIP), *Mol. Plant-Microbe Interact.,* 9:617-624.

Carmona, M. J., Molina, A., Fernandez, J. A., Lopez-Fando, J., and Garcia-Olmedo, F. (1993). Expression of the alpha-thionin gene from barley in tobacco confers enhanced resistance to bacterial pathogens, *Plant J.,* 3:457-462.

Carney, B. F., and Denny, T. P. (1990). A cloned avirulence gene from *Pseudomonas solanacearum* determines incompatibility on *Nicotiana tabacum* at the host species level, *J. Bacteriol.,* 172:4836-4843.

Carpita, N. C., and Gibeaut, D. M. (1993). Structural models of primary cell walls in flowering plants: Consistency of molecular structure with the physical properties of the walls during growth, *Plant J.,* 3:1-30.

Carr, J. P., Dixon, D. C., and Klessig, D. F. (1985). Synthesis of pathogenesis-related proteins in tobacco is regulated at the level of mRNA accumulation and occurs on membrane bound polysomes, *Proc. Natl. Acad. Sci. U.S.A.,* 82:7999-8003.

Carr, J. P., Dixon, D. C., Nikolau, B. J., Voelkerding, K. V., and Klessig, D. F. (1987). Synthesis and localization of pathogenesis-related proteins in tobacco, *Mol. Cell. Biol.,* 7:1580-1583.

Cartwright, D. W., Langcake, P., Pryce, R. J., Leworthy, D. P., and Ride, J. P. (1981). Isolation and characterization of phytoalexins from rice as momilactones A and B, *Phytochemistry,* 20:535-537.

Casacuberta, J. M., Puigdomenech, P., San Segundo, B. (1991). A gene coding for a basic pathogenesis-related (PR-like) protein from *Zea mays.* Molecular cloning and induction by a fungus *(Fusarium moniliforme)* in germinating maize seeds, *Plant Mol. Biol.,* 16:527-536.

Cassab, G. I., Lin, J. J., Lin, L. S., and Varner, J. E. (1988). Ethylene effect on extensin and peroxidase distribution in the sub-apical region of pea epicotyls, *Plant Physiol.,* 88:522-524.

Castagnaro, A., Marana, C., Carbonero, P., and Garcia-Olmedo, F. (1992). Extreme divergence of a novel wheat thionin generated by a mutational burst specifically affecting the mature protein domain of the precursor, *J. Mol. Biol.,* 224:1003-1009.

Castresana, C., de Carvalho, F., Gheysen, G., Habets, M., Inze, D., and Van Montague, M. (1990). Tissue-specific and pathogen-induced regulation of a *Nicotiana plumbaginifolia* β-1,3-glucanase gene, *Plant Cell,* 2:1131-1143.

Cervone, F., Castoria, R., Leckie, F., and De Lorenzo, G. (1997). Perception of fungal elicitors and signal transduction. In P. Aducci (Ed.), *Signal Transduction in Plants,* Birkauser Verlag, Basel, Switzerland, pp.153-177.

Chamnongpol, S., Mongkolsuk, S., Vattanaviboon, P., Fuangthong, M. (1995a). Unusual growth phase and oxygen tension regulation of oxidative stress protection enzymes, catalase and superoxide dismutase, in the phytopathogen *Xanthomonas oryzae* pv. *oryzae*, *Appl. Environ. Microbiol.,* 61:393-396.

Chamnongpol, S., Vattanaviboon, P., Loprasert, S., and Mongkolsuk, S. (1995b). Atypical oxidative stress regulation of a *Xanthomonas oryzae* pv. *oryzae* monofunctional catalase, *Can. J. Microbiol.,* 41:541-547.

Chamnongpol, S., Willekens, H., Langebartels, C., Van Montagu, M., Inze, D., and Van Camp, W. (1996). Transgenic tobacco with a reduced catalase activity develops necrotic lesions and induces pathogenesis-related expression under high light, *Plant J.,* 10:491-503.

Chandra, S., Heinstein, P. F., and Low, P. S. (1996). Activation of phospholipase A by plant defense elicitors, *Plant Physiol.,* 110:979-986.

Chang, C.-H., and Winans, S. C. (1992). Functional roles assigned to the periplasmic, linker, and receiver domains of the *Agrobacterium tumefaciens* VirA protein, *J. Bacteriol.,* 174:7033-7039.

Chang, C.-H., Zhu, J., and Winans, S. C. (1996). Pleiotropic phenotypes caused by genetic ablation of the receiver module of the *Agrobacterium tumefaciens* VirA protein, *J. Bacteriol.,* 178:4710-4716.

Chang, P.-F. L., Cheah, K. T., Narasimhan, M. L., Hasegawa, P. M., and Bressan, R. A. (1995). Osmotin gene expression is controlled by elicitor synergism, *Physiologia Plant.,* 95:620-626.

Chanock, S. J., Elbenna, J., Smith, R. M., and Babior, B. M. (1994). The respiratory burst oxidase, *J. Biol. Chem.,* 269:24519-24522.

Chappell, J., and Hahlbrock, K. (1984). Transcription of plant defense genes in response to UV light or fungal elicitor, *Nature,* 311:76-78.

Chappell, J., and Nable, R. (1987). Induction of sesquiterpenoid biosynthesis in tobacco cell suspension cultures by fungal elicitor, *Plant Physiol.,* 85:469-473.

Chappell, J., Von Lanken, C., and Vogeli, U. (1991). Elicitor-inducible 3-hydroxy-3-methylglutaryl coenzyme A reductase activity is required for sesquiterpene accumulation in tobacco cell suspension cultures, *Plant Physiol.,* 97:693-698.

Charles, T. C., Jin, S. G., and Nester, E. W. (1992). Two-component sensory transduction systems in phytobacteria, *Annu. Rev. Phytopathol.,* 30:463-484.

Chatterjee, A., Cui, Y., Liu, Y., Dumenyo, C. K., and Chatterjee, A. K. (1995). Inactivation of *rsmA* leads to overproduction of extracellular pectinases, cellulases, and proteases in *Erwinia carotovora* subsp. *carotovora* in the absence of the starvation/cell density sensing signal, N-(3-oxohexanoyl)-L-homoserine lactone, *Appl. Environ. Microbiol.,* 61:1959-1967.

Chatterjee, A. K., Thurn, K. K., and Tyrell, D. J. (1985). Isolation and characterization of *Tn5* insertion mutants of *Erwinia chrysanthemi* that are deficient in polygalacturonate catabolic enzymes oligogalacturonate lyase and 3-deoxy-D-glycero-2,5-hexodiulosonate dehydrogenase, *J. Bacteriol.,* 162:708-714.

Chatterjee, A. K., and Vidaver, A. K. (1986). Genetics of pathogenicity factors: Application to phytopathogenic bacteria, *Advances in Plant Pathology,* 4:1-224.

Chaudhry, B., Muller-Uri, F., Cameron-Mills, V., Gough, S., Simpson, D., Skriver, K., and Mundy, J. (1994). The barley 60 kDa jasmonate-induced protein (JIP 60) is a novel ribosome-inactivating protein, *Plant J.,* 6:815-824.

Chelsky, D., Ralph, R., and Jonak, G. (1989). Sequence requirements for synthetic peptide-mediated translocation to the nucleus, *Mol. Cell Biol.,* 9:2487-2492.

Chen, C., Bauske, E. M., Musson, G., Rodriquez-kabana, R., and Kloepper, J. W. (1995). Biological control of *Fusrium* wilt in cotton by use of endophytic bacteria, *Biological Control,* 5:83-91.

Chen, C.-Y., and Winans, S. C. (1991). Controlled expression of the transcriptional activator gene *virG* in *Agrobacterium tumefaciens* by using the *Escherichia coli lac* promoter, *J. Bacteriol.,* 173:1139-1144.

Chen, J., and Varner, J. E. (1985). Isolation and characterization of cDNA clones for carrot extensin and a proline-rich 33-kDa protein, *Proc. Natl. Acad. Sci. U.S.A.,* 82:4399-4403.

Chen, R., Wang, F., and Smith, A. G. (1996). A flower-specific gene encoding an osmotin-like protein from *Lycopersicon esculentum, Gene,* 179:301-302.

Chen, Z., Ricgliano, J. W., and Klessig, D. F. (1993a). Purification and characterization of a soluble salicylic acid-binding protein from tobacco, *Proc. Natl. Acad. Sci. U.S.A.,* 90:9533-9537.

Chen, Z., Silva, H., and Klessig, D. F. (1993b). Active oxygen species in the induction of plant systemic acquired resistance by salicylic acid, *Science,* 262:1883-1886.

Chester, K. S. (1933). The problem of acquired physiological immunity in plants, *Q. Rev. Biol.,* 8:275-324.

Chiang, C. C., and Hadwiger, L. A. (1991). The *Fusarium solani*-induced expression of a pea gene family encoding high cysteine content proteins, *Mol. Plant-Microbe Interact.,* 4:324-331.

Chittoor, J. M., Leach, J. E., and White, F. F. (1997). Differential induction of a peroxidase gene family during infection of rice by *Xanthomonas oryzae* pv. *oryzae, Mol. Plant-Microbe Interact.,* 10:861-871.

Chittoor, J. M., Leach, J. E., and White, F. F. (1999). Induction of peroxidase during defense against pathogens. In S. K. Datta and S. Muthukrishnan (Eds.), *Pathogenesis-Related Proteins in Plants,* CRC Press, Boca Raton, FL, pp. 171-193.

Choi, D., Ward, B. L., and Bostock, R. M. (1992). Differential induction and suppression of potato 3-hydroxy 3-methylglutaryl coenzyme A reductase genes in response to *Phytophthora infestans* and to its elicitor arachidonic acid, *Plant Cell,* 4:1333-1344.

Choi, J., and Han, K. (1996). Plant cell wall, an inducer of pectate lyase of *Erwinia rhapontici, Korean J. Plant Pathol.,* 12:129-131.

Christensen, B., Fink, J., Merrifield, R. B., and Mauzerall, D. (1988). Channel-forming properties of cecropins and related model compounds incorporated into plant lipid membranes, *Proc. Natl. Acad. Sci. U.S.A.,* 85:5072-5076.

Citovsky, V., De Vos, G., and Zambryski, P. (1988). A novel single stranded DNA binding protein encoded by the *virE* locus is produced following activation of the *A. tumefaciens* T-DNA transfer process, *Science,* 240:501-504.

Citovsky, V., and Zambryski, P. (1993). Transport of nucleic acids through membrane channels: Snaking through small holes, *Annu. Rev. Microbiol.,* 47:167-197.

Clarke, S. F., Burritt, D. J., Jameson, P. E., and Guy, P. L. (1998). Influence of plant hormones on virus replication and pathogenesis-related proteins in *Phaseolus vulgaris* L. infected with white clover mosaic potex virus, *Physiol. Mol. Plant Pathol.,* 53:195-207.

Cline, S. D., and Coscia, C. J. (1988). Stimulation of sanguinarine production by combined fungal elicitation and hormonal deprivation in cell suspension cultures of *Papaver bracteatum, Plant Physiol.,* 86:161-165.

Close, T. J., Tait, R. C., and Kado, C. I. (1985). Regulation of Ti plasmid virulence genes by a chromosomal locus of *Agrobacterium tumefaciens, J. Bacteriol.,* 164:774-781.

Clough, S. J., Schell, M. A., Denny, T. P. (1994). Evidence for involvement of a volatile extracellular factor in *Pseudomonas solanacearum* virulence gene expression, *Mol. Plant-Microbe Interact.,* 7:621-630.

Cohen, Y., Gisi, U., and Mosinger, E. (1991). Systemic resistance of potato plants against *Phytophthora infestans* induced by unsaturated fatty acids, *Physiol. Mol. Plant Pathol.,* 38:255-263.

Cohen, Y., Niderman, T., Mosinger, E., and Fluhr, R. (1994). β-aminobutyric acid induces the accumulation of pathogenesis-related proteins in tomato plants and resistance to late blight caused by *Phytophthora infestans, Plant Physiol.,* 104:59-66.

Cokmus, C., and Sayar, A. H. (1991). Effect of salicylic acid on the control of bacterial speck of tomato caused by *Pseudomonas syringae* pv. *tomato, J. Turk. Phytopathol.,* 20:27-32.

Colilla, F. J., Rocher, A., and Mendez, E. (1990). Purothionins: Amino acid sequence of two polypeptides of a new family of thionins from wheat endosperm, *FEBS Lett.,* 270:191-194.

Collinge, D. B., and Slusarenko, A. J. (1987). Plant gene expression in response to pathogens, *Plant Mol. Biol.,* 9:389-410.

Collmer, A., and Bauer, D. W. (1994). *Erwinia chrysanthemi* and *Pseudomonas syringae:* Plant pathogens trafficking in virulence proteins. In J. L. Dangl (Ed.), *Current Topics in Microbiology and Immunology,* Vol.192, *Bacterial Pathogenesis of Plants and Animals; Molecular and Cellular Mechanisms,* Springer Verlag, Berlin, pp. 43-48.

Collmer, A., and Keen, N. T. (1986). The role of pectic enzymes in plant pathogenesis, *Annu. Rev. Phytopathol.,* 24:383-409.

Conceicao, A. de Silva, and Broekaert, W. F. (1999). Plant defensins. In S. K. Datta and S. Muthukrishnan (Eds.), *Pathogenesis-Related Proteins in Plants,* CRC Press, Boca Raton, FL, pp. 247-260.

Conconi, A., Miquel, M., Browse, J. A., and Ryan, C. A. (1996). Intracellular levels of free linolenic and linoleic acids increase in tomato leaves in response to wounding, *Plant Physiol.,* 111:797-803.

Condemine, G., Dorel, C., Hugouvieux-Cotte-Pattat, N., and Robert-Baudouy, J. (1992). Some of the *out* genes involved in the secretion of pectate lyases in *Erwinia chrysanthemi* are regulated by *kdgR, Mol. Microbiol.,* 6:3199-3211.

Condemine, G., Hugouvieux-Cotte-Pattat, N., and Robert-Baudouy, J. (1986). Isolation of *Erwinia chrysanthemi kduD* mutants altered in pectin degradation, *J. Bacteriol.,* 165:937-941.

Condemine, G., and Robert-Baudouy, J. (1987a). Tn5 insertion in *kdgR,* a regulatory gene of the polygalacturonate pathway in *Erwinia chrysanthemi, FEMS Microbiol. Lett.,* 42:39-46.

Condemine, G., and Robert-Baudouy, J. (1987b). 2-Keto-deoxygluconate transport system in *Erwinia chrysanthemi, J. Bacteriol.,* 169:1972-1978.

Condemine, G., and Robert-Baudouy, J. (1991). Analysis of an *Erwinia chrysanthemi* gene cluster involved in pectin degradation, *Mol. Microbiol.,* 5:2191-2202.

Condemine, G., and Robert-Baudouy, J. (1995). Synthesis and secretion of *Erwinia chrysanthemi* virulence factors are coregulated, *Plant-Microbe Interact.,* 8:632-636.

Conrath, U., Chen, Z., Ricigliano, J. R., and Klessig, D. F. (1995). Two inducers of plant defense responses, 2,6-dichloroisonicotinic acid and salicylic acid, inhibit catalase activity in tobacco, *Proc. Natl. Acad. Sci. U.S.A.,* 92:7143-7147.

Conrath, U., Jeblick, W., and Kauss, H. (1991). The protein kinase inhibitor, K-252a, decreases elicitor-induced Ca^{2+} uptake and K^+ release, and increase coumarin synthesis in parsley cells, *FEBS Lett.,* 279:141-144.

Constabel, C. P. and Brisson, N. (1992). The defense-related STH-2 gene product of potato shows race-specific accumulation after inoculation with low concentrations of *Phytophthora infestans* zoospores, *Planta,* 188:289-295.

Constabel, C. P., and Brisson, N. (1995). Stigma- and vascular-specific expression of the pathogenesis-related gene, *Mol. Plant-Microbe Interact.,* 8:104-113.

Cook, A. A., and Stall, R. E. (1968). Effect of *Xanthomonas vesicatoria* on loss of electrolytes from leaves of *Capsicum annuum, Phytopathology,* 58:617-619.

Coolbear, T., and Threlfall, D. R. (1985). The biosynthesis of lubimin from (I-^{14}C) isopentenyl pyrophosphate by cell-free extracts of potato tuber tissue inoculated with an elicitor preparation from *Phytophthora infestans, Phytochemistry,* 24:1963-1971.

Cooley, M. B., D'Souza, M. R., and Kado, C. I. (1991). The *virC* and *virD* operons of the *Agrobacterium* Ti plasmid are regulated by the *rosAT* chromosomal gene: Analysis of the cloned *rosAT* gene, *J. Bacteriol.,* 173:2608-2616.

Cooper, J. C., and Salmond, G. P. C. (1993). Molecular analysis of the major cellulase (cel IV) of *Erwinia carotovora:* Evidence for an evolutionary "mix-and-match" of enzyme domains, *Mol. Gen. Genet.,* 241:341-350.

Cooper, R. M. (1983). The mechanisms and significance of enzyme degradation of host cell walls by parasites. In J. A. Callow (Ed.), *Biochemical Plant Pathology,* John Wiley and Sons Ltd, New York, pp. 101-135.

Coplin, D. L., Frederick, R. D., Majerczak, D. R., and Tuttle, L. D. (1992). Characterization of a gene cluster that specifies pathogenicity in *Erwinia stewartii, Mol. Plant-Microbe Interact.,* 5:81-88.

Cordero, M. J., Raventos, D., and Segundo, B. S. (1994). Expression of a maize proteinase inhibitor gene is induced in response to wounding and fungal infection: Systemic wound-response to a monocot gene. *Plant J.,* 6:141-150.

Cornelissen, B. J. C., Hoof Van Huijsduijnen, R. A. M., Van Loon, L. C., and Bol, J. F. (1986). Molecular characterization of messenger RNAs for pathogenesis-related proteins 1a, 1b and 1c induced by TMV infection of tobacco, *EMBO J.,* 5:37-40.

Cornelissen, B. J. C., Horowitz, J., Van Kan, J. A., Goldberg, R. B., and Bol, J. F. (1987). Structure of tobacco genes encoding pathogenesis-related proteins from the PR-1 group, *Nucl. Acids Res.,* 15:6799-6811.

Coughlan, M. P., and Mayer, F. (1992). The cellulose-decomposing bacteria and their enzyme systems. In A. Balows, H. G. Truper, M. Dworkin, W. Harder, and K. H. Schleifer (Eds.), *The Prokaryotes,* Vol. I. Springer-Verlag, New York, pp. 460-516.

Cournoyer, B., Sharp, J. D., Astuto, A., Gibbon, M. J., Taylor, J. D., and Vivian, A. (1995). Molecular characterization of the *Pseudomonas syringae* pv. *pisi* plasmid-borne avirulence gene *avrPpiB* which matches the R3 resistance locus in pea, *Mol. Plant-Microbe Interact.,* 8:700-708.

Crafts, A. S., and Crisp, C. E. (1971). *Phloem Transport in Plants,* W. H. Freeman and Co., San Francisco, pp. 481.

Cramer, C. L., Bell, J. N., Ryder, T. B., Bailey, J. A., Schuch, W., Bolwell, G. P., Robbins, M. P., Dixon, R. A., and Lamb, C. J. (1985a). Co-ordinated synthesis of phytoalexin biosynthetic enzymes in biologically-stressed cells of bean (*Phaseolus vulgaris* L.), *EMBO J.,* 4:285-289.

Cramer, C. L., Edwards, K., Dron, M., Liang, X., Dildine, S. L., Bolwell, G. P., Dixon, R. A., Lamb, C. J., and Schuch, W. (1989). Phenylalanine ammonia-lyase gene organization structure, *Plant Mol. Biol.,* 12:367-383.

Cramer, C. L., Ryder, T. B., Bell, J. N., and Lamb, C. J. (1985b). Rapid switching of plant gene expression induced by fungal elicitor, *Science,* 227:1240-1243.

Creelman, R. A., and Mullet, J. E. (1995). Jasmonic acid distribution and action in plants: regulation during development and response to biotic and abiotic stress, *Proc. Natl. Acad. Sci. U.S.A.,* 92:4114-4119.

Creelman, R. A., Tierney, M. L., and Mullet, J. E. (1992). Jasmonic acid/methyl jasmonate accumulate in a wounded soybean hypocotyls and modulate wound gene expression, *Proc. Natl. Acad. Sci. U.S.A.,* 89:4938-4941.

Croft, K. P. C., Juttner, F., and Slusarenko, A. J. (1993). Volatile products of the lipoxygenase pathway evolved from *Phaseolus vulgaris* (L.) leaves inoculated with *Pseudomonas syringae* pv. *phaseolicola, Plant Physiol.,* 101:13-24.

Croft, K. P. C., Voisey, C. R., and Slusarenko, A. J. (1990). Mechanism of hypersensitive cell collapse: Correlation of increased lipoxygenase activity with membrane damage in leaves of *Phaseolus vulgaris* (L.) inoculated with an avirulent race of *Pseudomonas syringae* pv. *phaseolicola, Physiol. Mol. Plant Pathol.,* 36:49-62.

Crowell, D. N., John, M. J., Russell, D., and Amasino, R. M. (1992). Characterization of a stress-induced developmentally regulated gene family from soybean, *Plant Mol. Biol.,* 18:459-466.

Cui, Y., Chatterjee, A., Liu, Y., Dumenyo, C. K., and Chatterjee, A. K. (1995). Identification of a global repressor gene, *rsmA,* of *Erwinia carotovora* subsp. *carotovora* that controls extracellular enzymes, N-(3-oxohexanoyl)-L-homoserine lactone, and pathogenicity in soft-rotting *Erwinia* spp. *J. Bacteriol.,* 177:5108-5115.

Cui, Y., Madi, L., Mukherjee, A., Dumenyo, C. K., and Chatterjee, A. K. (1996). The $RsmA^-$ mutants of *Erwinia carotovora* subsp. *carotovora* strain Ecc 71 overexpress $hrpN_{Ecc}$ and elicit a hypersensitive reaction-like response in tobacco leaves, *Mol. Plant-Microbe Interact.,* 9:565-573.

Cuppels, D. A. (1986). Generation and characterization of Tn5 insertion mutations in *Pseudomonas syringae* pv. *tomato, Appl. Environ. Microbiol.,* 511:323-327.

Curtis, M. D., Rae, A. L., Rusu, A. G., Harrison, S. J., and Manners, J. M. (1997). A peroxidase gene promoter induced by phytopathogens and methyl jasmonate in transgenic plants, *Mol. Plant-Microbe Interact.,* 10:326-338.

Cusack, M., and Pierpoint, W. (1988). Similarities between sweet protein thaumatin and a pathogenesis-related protein of tobacco, *Phytochemistry,* 27:3817-3821.

Cutt, J. R., Dixon, D. C., Carr, J. P., and Klessig, D. F. (1988). Isolation and nucleotide sequence of cDNA clones for the pathogenesis-related proteins PR1a, PR1b and PR1c of *Nicotiana tabacum* cv. Xanthi nc induced by TMV infection, *Nucl. Acids. Res.,* 16:9861.

Cutt, J. R., Harpster, M. H., Dixon, D. C., and Carr, J. P. D. (1989). Disease response to tobacco mosaic virus in transgenic tobacco plants that constitutively express the pathogenesis related PR-1b gene, *Virology,* 173:89-97.

Czernic, P., Huang, H. C., and Marco, Y. (1996). Characterization of *hsr201* and *hsr15,* two tobacco genes preferentially expressed during the hypersensitive reaction provoked by phytopathogenic bacteria, *Plant Mol. Biol.,* 31:255-265.

Dahiya, J. S., and Rimmer, S. R. (1988). Phytoalexin accumulation in tissues of *Brassica napus* inoculated with *Leptospheria maculans, Phytochemistry,* 27:3105-3107.

Dahler, G. S., Barras, F., and Keen, N. T. (1990). Cloning of genes encoding extracellular metalloproteases from *Erwinia chrysanthemi* EC16, *J. Bacteriol.,* 172:5803-5815.

Dai, G. H., Nicole, M., Andary, C., Martinez, C., Bresson, E., Boher, B., Daniel, J. F., and Geiger, J. P. (1996). Flavonoids accumulate in cell walls, middle lamellae and callose-rich papillae during an incompatible interaction between *Xanthomonas campestris* pv. *malvacearum* and cotton, *Physiol. Mol. Plant Pathol.,* 49:285-306.

Daley, L. S., and Theriot, L. J. (1987). Electrophoretic analysis, redox activity, and other characteristics of proteins similar to purothionins from tomato *(Lycoperscicum esculenta),* mango *(Mangifera indica),* papaya *(Carica papaya),* and walnut *(Juglans regia), J. Agric. Food Chem.,* 35:680-687.

Dalkin, K., Edwards, R., Edington, B., and Dixon, R. A. (1990a). Stress responses in alfalfa (*Medicago sativa* L.). I. Induction of phenylpropanoid biosynthesis and hydrolytic enzymes in elicitor-treated cell suspension cultures, *Plant Physiol.,* 92:440-446.

Dalkin, K., Jorrin, J., and Dixon, R. A. (1990b). Stress responses in alfalfa (*Medicago sativa* L.). VII. Induction of defence related mRNAs in elicitor-treated cell suspension cultures, *Physiol. Mol. Plant Pathol.,* 37:293-307.

Dammann, C., Rojo, E., and Sanchez-Serrano, J. J. (1997). Abscisic acid and jasmonic acid activate wound-inducible genes in potato through separate, organ-specific signal transduction pathways, *Plant J.,* 11:773-782.

Dane, F., and Shaw, J. J. (1993). Growth of bioluminescent *Xanthomonas campestris* pv. *campestris* in susceptible and resistant host plants, *Mol. Plant-Microbe Interact.,* 6:786-789.

Dangl, J. L. (1994). The enigmatic avirulence genes of phytopathogenic bacteria. In J. L. Dangl (Ed.), *Bacterial Pathogenesis of Plants and Animals: Molecular and Cellular Mechanisms,* Vol. 192, Springer-Verlag, Berlin, pp. 99-108.

Dangl, J. L., Dietrich, R. A., and Richberg, M. H. (1996). Death don't have mercy: Cell death programs in plant-microbe interactions, *Plant Cell,* 8:1793-1807.

Dangl, J. L., Ritter, C., Gibbon, M. J., Mur, L. A. J., Wood, J. R., Goss, S., Mansfield, J., Taylor, J. D., and Vivian, A. (1992). Functional homologs of the *Arabidopsis Rpm1* disease resistance gene in bean and pea, *Plant Cell,* 4:1359-1369.

Daniels, M. J., Barber, C. E., Turner, P. C., Cleary, W. G., and Sawczye, M. K. (1984). Isolation of mutants of *Xanthomonas campestris* pv. *campestris* showing altered pathogenicity, *J. Gen. Microbiol.,* 130:2447-2455.

Dann, E. K., and Deverall, B. J. (1995). Effectiveness of systemic resistance in bean against foliar and soil-borne pathogens as induced by biological and chemical means, *Plant Pathol,* 44:458-466.

Dann, E. K., Meuwly, P., Metraux, J. P., and Deverall, B. J. (1996). The effect of pathogen inoculation or chemical treatment on activities of chitinase and β-1,3-glucanase and accumulation of salicylic acid in leaves of green bean, *Phaseolus vulgaris* L., *Physiol. Mol. Plant Pathol.,* 49:307-319.

Darvill, A. G., and Albersheim, P. (1984). Phytoalexins and their elicitors—A defense against microbial infection in plants, *Annu. Rev. Plant Physiol.,* 35:243-275.

Darvill, A., Augur, C., Bergmann, C., Carlson, R. W., Cheong, J. J., Eberhard, S., Hahn, M. G., Lo, V. M., Marfa, V., Meyer, B., Mohnen, D., O'Neill, M. A., Spiro, M. D., van Halbeek, H., York, W. S., and Albersheim, P. (1992). Oligosaccharins—oligosaccharides that regulate growth, development and defence in plants, *Glycobiology,* 2:181-198.

Darvill, A., McNeil, M., Albersheim, P., and Delmer, D. P. (1980). The primary cell walls of flowering plants. In N. E. Tolbar (Ed.), *The Biochemistry of Plants,* Vol. I., Academic Press, New York, pp. 91-92.

Darvis, K. R., and Hahlbrock, K. (1987). Induction of defense responses in cultured plant cells by plant cell wall fragments, *Plant Physiol.,* 85:1286-1290.

Datta, K., Velazhahan, R., Oliva, N., Ona, I., Mew, T., Khush, G., Muthukrishnan, S., and Datta, S. K. (1999). Overexpression of the cloned rice thaumatin-like protein (PR-5) gene in transgenic rice plants enhances environmental friendly resistance to *Rhizoctonia solani* causing sheath blight disease, *Theor. Appl. Genet.,* 98:1138-1145.

Davies, H. A., Daniels, M. J., and Dow, J. M. (1997a). A novel proline-rich glycoprotein associated with the extracellular matrix of vascular bundles of *Brassica* petioles, *Planta,* 202:28-35.

Davies, H. A., Daniels, M. J., and Dow, J. M. (1997b). Induction of extracellular matrix glycoproteins in *Brassica* petioles by wounding and in response to *Xanthomonas campestris, Mol. Plant-Microbe Interact.,* 10:812-820.

Davila-Huerta, G., Hamada, G., Davis, G. D., Stipanovic, R. D., Adams, C. M., and Essenberg, M. (1995). Cadiane-type sesquiterpenes induced in *Gossypium* cotyledons by bacterial inoculation, *Phytochemistry,* 39:531-536.

Davis, D., Merida, J., Legendra, L., Low, P. S., and Heinstein, P. (1993). Independent elicitation of the oxidative burst and phytoalexin formation in cultured plant cells, *Phytochemistry,* 32:607-611.

Davis, E. M., Tsuji, J., David, G. D., Pierce, M. L., and Essenberg, M. (1996). Purification of (+)-delta-cadinene synthase, a sesquiterpene cyclase from bacteria-inoculated cotton foliar tissue, *Phytochemistry,* 41:1047-1055.

Davis, K. R., and Ausubel, F. M. (1989). Characterization of elicitor-induced defense responses in suspension-cultured cells of *Arabidopsis, Mol. Plant-Microbe Interact.,* 2:363-368.

Davis, K. R., Darvill, A. G., and Albersheim, P. (1984). Host pathogen interactions. XXV. Endopolygalacturonic acid lyase from *Erwinia carotovora* elicits phytoalexin accumulation by releasing plant cell wall fragments, *Plant Physiol.,* 74:52-60.

Davis, K. R., Darvill, A. G., Albersheim, P., and Dell, A. (1986a). Host-pathogen interactions. XXX. Characterization of elicitors of phytoalexin accumulation in soybean released from cell walls by endopolygalacturonic acid lyase, *Z. Naturforsch.,* 41c:39-48.

Davis, K. R., Darvill, A. G., Albersheim, P., and Dell, A. (1986b). Host-pathogen interactions. XXIX. Oligogalacturonides released from sodium polypectate by endopolygalacturonic acid lyase are elicitors of phytoalexins in soybean, *Plant Physiol.,* 80:568-577.

Davis, K. R., Lyon, G. D., Darvill, A. G., and Albersheim, P. (1984). Host-pathogen interactions. XXV. Endopolygalacturonic acid lyase from *Erwinia carotovora* elicits phytoalexin accumulation by releasing plant cell wall fragments, *Plant Physiol.,* 74:52-60.

Davis, T., Lehnackers, H., Arnold, M., and Dangl, J. L. (1991). Identification and molecular mapping of a single *Arabidopsis thaliana* locus determining resistance to a phytopathogenic *Pseudomonas syringae* isolate, *Plant J.,* 1:289-302.

De Feyter, R., and Gabriel, D. W. (1991). At least six avirulence genes are clustered on a 90-kilobase plasmid in *Xanthomonas campestris* pv. *malvacearum, Mol. Plant-Microbe Interact.,* 4:423-432.

De Feyter, R., Yang, Y., and Gabriel, D. W. (1993). Gene-for-genes interactions between cotton R genes and *Xanthomonas campestris* pv. *malvacearum avr* genes, *Mol. Plant-Microbe Interact.,* 6:225-237.

De Loose, M., Alliotte, T., Gheysen, G., Genetello, C., Gielen, J., Soetaert, P., Van Montagu, M., and Inze, D. (1988). Primary structure of a hormonally regulated β-glucanase of *Nicotiana plumbaginifolia, Gene,* 70:13-23.

De Lorenzo, G., and Cervone, F. (1997). Polygalacturonase-inhibiting proteins (PGIPs): Their role in specificity and defense against pathogenic fungi. In G. Stacey and N. T. Keen, (Eds.), *Plant-Microbe Interactions,* Vol. 3, Chapman and Hall, New York, pp. 76-93.

De Scenzo, R. A., Wise, R. P., and Mahadevappa, M. (1994). High-resolution mapping of the *Hor1/Mla/Hor2* region on chromosome 5S in barley, *Mol. Plant-Microbe Interact.,* 7:657-666.

de Tapia, M., Bergman, P., Awade, A., and Burkard, G. (1986). Analysis of acid extractable bean leaf proteins induced by mercuric chloride treatment and alfalfa mosaic virus infection. Partial purification and characterization, *Plant Sci.,* 45: 167-177.

de Tapia, M., Dietrich, A., and Burkard, G. (1987). In vitro synthesis and processing of a bean pathogenesis-related (PR4) protein, *Eur. J. Biochem.,* 166:559-563.

De Vos, G., and Zambryski, P. (1989). Expression of *Agrobacterium* nopaline-specific VirD1, VirD2, and VirC1 proteins and their requirement for T-strand production in *E. coli, Mol. Plant-Microbe Interact.,* 2:43-52.

Debener, T., Lehnackers, H., Arnold, M., and Dangl, J. L. (1991). Identification and molecular mapping of a single *Arabidopsis thaliana* locus determining resistance to a phytopathogenic *Pseudomonas syringae* isolate, *Plant J.,* 1:289-302.

Delaney, T. P., Friedrich, L., and Ryals, J. (1995). *Arabidopsis* signal transduction mutant defective in chemically and biologically induced disease resistance, *Proc. Natl. Acad. Sci. U.S.A.,* 92:6602-6606.

Delaney, T. P., Uknes, S., Vernooij, B., Friedrich, L., Weymann, K., Negrotto, D., Gaffney, T., Gut-Rella, M., Kessmann, H., Ward, E., and Ryals, J. (1994). A central role of salicylic acid in plant disease resistance, *Science,* 266:1247-1250.

Del-Campillo, E., and Lewis, L. N. (1992). Identification and kinetics of accumulation of proteins induced by ethylene in bean abscission zones, *Plant Physiol.,* 98:955-961.

Delepelaire, P., and Wandersman, C. (1989). Protease secretion by *Erwinia chrysanthemi.* Proteases B and C are synthesized and secreted as zymogens without a signal peptide, *J. Biol. Chem.,* 264:9083-9089.

Delmer, D. P., Solomon, M., and Read, S. M. (1991). Direct photolabeling with (^{32}P) UDP-glucose for identification of a subunit of cotton fiber callose synthase, *Plant Physiol.,* 95:556-563.

Demarty, M., Morvan, C., and Thellier, M. (1984). Calcium and the cell wall, *Plant Cell and Environ.,* 7:119-139.

Demple, B. (1991). Regulation of bacterial oxidative stress genes, *Annu. Rev. Genet.,* 25:315-337.

Dempsey, D. A., and Klessig, D. F. (1995). Signals in plant disease resistance, *Bull. Inst. Pasteur,* 93:167-186.

Dennis, E. A., Rhee, S. G., Billah, M. M., and Hannun, Y. A. (1991). Role of phospholipases in generating second messengers in signal transduction, *FASEB J.,* 5:2068-2077.

Denny, T. P., Carney, B. F., and Schell, M. A. (1990). Inactivation of multiple virulence genes reduces the ability of *Pseudomonas solanacearum* to cause wilt symptoms. *Mol. Plant-Microbe Interact.,* 3:293-300.

Desiderio, A., Aracri, B., Leckie, F., Mattei, B., Salvi, G., Tigelaar, H., Van Roekel, J. S., Baulcombe, D. C., Melchers, L. S., De Lorenzo, G., and Cervone, F. (1997). Polygalacturonase-inhibiting proteins (PGIPs) with different specificities are expressed in *Phaseolus vulgaris, Mol. Plant-Microbe Interact.*, 10:825-860.

Desikan, R., Hancock, J. T., Coffey, M. J., and Neill, S. J. (1996). Generation of active oxygen in elicited cells of *Arabidopsis thaliana* is mediated by a NADPH oxidase-like enzyme, *FEBS Letters,* 382:213-217.

Desjardins, A. E., Spencer, G. F., Plattner, R. D., and Beremand, M. N. (1989). Furanocoumarin phytoalexins, trichothecene toxins, and infection of *Pastinaca sativa* by *Fusarium sporotrichoides, Phytopathology,* 79:170-175.

Destefano-Beltran, L., Nagpala, P. G., Cetiner, M. S., Denny, T., and Jaynes, J. M. (1993). Using genes encoding novel peptides and proteins to enhance disease resistance in plants. In I. Chet (Ed.), *Biotechnology in Plant Disease Control,* Wiley-Liss, Inc., New York, pp. 175-189.

Devlin, W. S., and Gustine, D. L. (1992). Involvement of the oxidative burst in phytoalexin accumulation and the hypersensitive reaction, *Plant Physiol.,* 100:1189-1195.

Devys, M., Barbier, M., Kollmann, A., Rouxel, T., and Bousquet, I. F. (1990). Cyclobrassinin sulphoxide, a sulphur-containing phytoalexin from *Brassica juncea, Phytochemistry,* 90:1087-1088.

Devys, M., Barbier, M., Loiselet, I., Rouxel, T., Sarniguet, A., Kollmann, A., and Bousquet, J. F. (1988). Brassilexin, a novel sulphur-containing phytoalexin from *Brassica juncea* L. (Cruciferae), *Tetrahedron Lett.,* 29:6447-6448.

Dewick, P. M. (1975). Pterocarpan biosynthesis: Chalcone and isoflavone precursors of demethylhomopterocarpin and maackiain in *Trifolium pratense, Phytochemistry,* 14:979-982.

DeWit, P. J. G. M., and Kodde, E. (1981). Induction of polyacetylenic phytoalexins in *Lycopersicon esculentum* after inoculation with *Cladosporium fulvum, Physiol. Plant Pathol.,* 18:143-148.

Dhawale, S., Souciet, G., and Kuhn, D. N. (1989). Increase of chalcone synthase mRNA in pathogen-inoculated soybeans with race-specific resistance is different in leaves and roots, *Plant Physiol.,* 91:911-916.

Dhindsa, R. S., Plumb-Dhindsa, P., and Thorpe, T. A. (1981). Leaf senescence: Correlated with increased levels of membrane permeability and lipid peroxidation, and decreased levels of superoxide dismutase and catalase, *J. Expt. Bot.,* 32:93-101.

Didierjean, K., Frendo, P., and Burkard, G. (1992). Stress responses in maize: Sequence analysis of cDNAs encoding glycine-rich proteins, *Plant Mol. Biol.,* 18:847-849.

Dietrich, R. A., Delaney, T. P., Uknes, S. J., Ward, E. R., Ryals, J. A., and Dangl, J. L. (1994). *Arabidopsis* mutants simulating disease resistance response, *Cell,* 77:565-577.

Dimarcq, J. L., Zachary, D., Hoffmann, J. A., Hoffmann, D., and Reichhart, J. M. (1990). Insect immunity: Expression of the two major inducible antibacterial peptides, defensin and diptericin, in *Phormia terranovae, EMBO J.,* 9:2507-2515.

Dingwall, C., and Laskey, R. A. (1991). Nuclear targeting sequences—A consensus?, *TIBS,* 16:478-481.

Divecha, N., and Irvine, R. F. (1995). Phospholipid signaling, *Cell,* 80:269-278.

Dixelius, C. (1994). Presence of the pathogenesis-related proteins 2, Q and S in stressed *Brassica napus* and *B. nigra* plantlets, *Physiol. Mol. Plant Pathol.,* 44:1-8.

Dixon, M., Jones, D., Keddie, J., Thomas, C., Harrison, K., and Jones, J. (1996). The tomato Cf-2 disease resistance locus comprises two functional genes encoding leucine-rich repeat proteins, *Cell,* 84:451-459.

Dixon, R. A., Gerrish, C., Lamb, C. J., and Robbins, M. P. (1983). Elicitor-mediated induction of chalcone isomerase in *Phaseolus vulgaris, Planta,* 159:561-569.

Dixon, R. A., and Lamb, C. J. (1990). Molecular communications in interactions between plants and microbial pathogens, *Annu. Rev. Plant Physiol. Plant Mol. Biol.,* 41:339-367.

Dmitriev, A. P., Tverskoy, L. A., Kozlovsky, A. G., and Grodzinsky, D. M. (1990). Phytoalexins from onion and their role in disease resistance, *Physiol. Mol. Plant Pathol.,* 37:235-244.

Doares, S. H., Syrovets, F., Weiler, E. W., and Ryan, C. A. (1995). Oligogalacturonides and chitosan activate plant defensive genes through the octadecanoid pathway, *Proc. Natl. Acad. Sci. U.S.A.,* 92:4095-4098.

Doke, N. (1985). NADPH-dependent O_2^- generation in membrane fractions isolated from wounded potato tubers inoculated with *Phytophthora infestans, Physiol. Plant Pathol.,* 27:311-322.

Doke, N., Miura, Y., Sanchez, L. M., and Kawakita, K. (1994). Involvement of superoxide in signal transduction: Responses to attack by pathogens, physical and chemical shocks, and UV radiation. In C. H. Foyer and P. M. Mullineaux (Eds.), *Causes of Photooxidative Stress and Amelioration of Defense Systems in Plants,* CRC Press, Boca Raton, FL, pp. 177-197.

Dominov, J. A., Stenzler, L., Lee, S., Schwartz, J. J., Leisner, S., and Howell, S. H. (1992). Cytokinins and auxins control the expression of a gene in *Nicotiana plumbaginifolia* cells by feed-back regulation, *Plant Cell,* 4:451-461.

Dong, X., Mindrinos, M., Davis, K. R., and Ausubel, F. M. (1991). Induction of *Arabidopsis* defense genes by virulent and avirulent *Pseudomonas syringae* strains and by a cloned avirulence gene, *Plant Cell,* 3:61-72.

Dore, I., Legrand, M., and Cornelissen, B. J. C., and Bol, J. F. (1991). Subcellular localization of acidic and basic proteins in tobacco mosaic virus-infected tobacco, *Arch. Virol.,* 120:97-107.

Dorel, C., Hugouvieux-Cotte-Pattat, N., Robert-Baudouy, J., and Lojkowska, E. (1996). Production of *Erwinia chrysanthemi* pectinases in potato tubers showing high or low level of resistance to soft-rot, *Eur. J. Plant Pathol.,* 102:511-517.

Doty, S. L., Yu, M. C., Lundin, J. I., Heath, J. D., and Nester, E. W. (1996). Mutational analysis of the input domain of the VirA protein of *Agrobacterium tumefaciens, J. Bacteriol.,* 178:961-970.

Doubrava, N., Dean, R., and Kuc, J. (1988). Induction of systemic resistance to anthracnose caused by *Colletotrichum lagenarium* in cucumber by oxalates and extracts from spinach and rhubarb leaves, *Physiol. Mol. Plant Pathol.,* 33:69-79.

Douglas, C. J., Halperin, W., and Nester, E. W. (1982). *Agrobacterium tumefaciens* mutants affected in attachment to plant cells, *J. Bacteriol.,* 152:1265-1275.

Douglas, C., Hoffmann, H., Schluz, W., and Hahlbrock, K. (1987). Structure and elicitor or U.V.-light-stimulated expression of two 4-coumarate:CoA ligase genes in parsley, *EMBO J.,* 6:1189-1195.

Douglas, C. J., Staneloni, R. J., Rubin, R. A., and Nester, E. W. (1985). Identification and genetic analysis of an *Agrobacterium tumefaciens* chromosomal virulence region, *J. Bacteriol.,* 161:850-860.

Dow, J. M., Clarke, J. M., Milligan, D. E., Tang, J. L., and Daniels, M. (1990). Extracellular proteases from *Xanthomonas campestris* pv. *campestris, Appl. Environ. Microbiol.,* 56:2994-2998.

Dow, J. M., Fan, M. J., Newman, M. A., and Daniel, M. J. (1993). Differential expression of conserved protease genes in crucifer-attacking pathovars of *Xanthomonas campestris, Appl. Environ. Microbiol.,* 59:3996-4003.

Dow, J. M., Milligan, D. E., Jamieson, L., Barber, C. E., and Daniels, M. J. (1989). Molecular cloning of a polygalacturonate lyase gene from *Xanthomonas campestris* pv. *campestris* and role of the gene product in pathogenicity, *Physiol. Mol. Plant Pathol.,* 35:113-120.

Dow, J. M., Osbourn, A. E., Wilson, T. J. G., and Daniels, M. J. (1995). A locus determining pathogenicity of *Xanthomonas campestris* is involved in lipopolysaccharide biosynthesis, *Mol. Plant-Microbe Interact.,* 8:768-777.

Dreier, J., Meletzus, D., and Eichenlaub, R. (1997). Characterization of the plasmid encoded virulence region *pat-1* of phytopathogenic *Clavibacter michiganensis* subsp. *michiganensis, Mol. Plant-Microbe Interact.,* 10:195-206.

Droback, B. K., and Ferguson, I. B. (1985). Release of Ca^{2+} from plant hypocotyl microsomes by inositol-1,4,5-trisphosphorothioate, a stable analogue of inositol trisphosphate which mobilizes intracellular calcium, *Biochem. J.,* 259:645-650.

Drolet, G., Dumbroff, E. B., Legge, R. L., and Thompson, J. E. (1986). Radical scavenging properties of polyamines, *Phytochemistry,* 25:367-371.

Dron, M., Clouse, S. D., Lawton, M. A., Dixon, R. A., and Lamb, C. J. (1988). Glutathione and fungal elicitor regulation of a plant defense gene promoter in electroporated protoplasts, *Proc. Natl. Acad. Sci. U.S.A,* 85:6738-6742.

D'Souza-Ault, M. R., Cooley, M. B., and Kado, C. I. (1993). Analysis of the Ros repressor of *Agrobacterium virC* and *virD* operons: Molecular intercommunication between plasmid and chromosomal genes, *J. Bacteriol.,* 178:3486-3490.

Du, H., and Klessig, D. F. (1997). Role for salicylic acid in the activation of defense responses in catalase-deficient transgenic tobacco, *Mol. Plant-Microbe Interact.,* 10:922-925.

Dudley, M. W., Dueber, M. T., and West, C. A. (1986). Biosynthesis of the macrocyclic diterpene casbene in castor bean (*Ricinus communis* L.) seedlings. Changes in enzyme levels induced by fungal infection and intracellular localisation of the pathway, *Plant Physiol.,* 81:335-342.

Dugger, W. M., Palmer, R. L., and Black, C. C. (1991). Changes in β-1,3-glucan synthase activity in developing lima bean plants, *Plant Physiol.,* 97:569-573.

Duijff, B. J., Gianinazzi-Pearson, V., and Lemanceau, P. (1997). Involvement of the outer membrane lipopolysaccharides in the endophytic colonization of tomato root by biocontrol *Pseudomonas fluorescens* strain WC417r, *New Phytol.,* 135:325-334.

Dumas, E., Lherminier, J., Gianinazzi, S., White, R. F., and Antoniw, J. F. (1988). Immunocytochemical location of pathogenesis-related b1 protein induced in tobacco mosaic virus-infected or polyacrylic acid-treated tobacco plants, *J. Gen. Virol.,* 69:2687-2694.

During, K., Porsch, P., Fladung, M., and Lorz, H. (1993). Transgenic potato plants resistant to the phytopathogenic bacterium *Erwinia carotovora, Plant J.,* 3:587-598.

Durner, J., and Klessig, D. F. (1995). Inhibition of ascorbate peroxidase by salicylic acid and 2,6-dichloroisonicotinic acid, two inducers of plant defense responses, *Proc. Natl. Acad. Sci. U.S.A.,* 92:11312-11316.

Durner, J., Shah, J., and Klessig, D. F. (1997). Salicylic acid and disease resistance in plants, *Trends in Plant Science,* 2:266-274.

Dwyer, S. C., Legendre, L., Low, P. S., and Leto, T. L. (1995). Plant and human neutrophil oxidative burst complexes contain immunologically related proteins, *Biochim. Biophys. Acta,* 1289:231-237.

Dyer, J. H., Ryu, S. B., and Wang, X. (1994). Multiple forms of phospholipase D following germination and during leaf development of castor bean, *Plant Physiol.,* 105:715-724.

Ealing, P. M., and Casey, R. (1988). Complete amino acid sequence of a pea *(Pisum sativum)* seed lipoxygenase predicted from a near full-length cDNA, *Biochem. J.,* 253:915-918.

Ebel, J., Schmidt, W. E., and Loyal, R. (1984). Phytoalexin synthesis in soybean cells: Elicitor induction of phenylalanine ammonia-lyase and chalcone synthase mRNAs and correlation with phytoalexin accumulation, *Arch. Biochem. Biophys.,* 232:240-248.

Ecker, J. R. (1995). The ethylene signal transduction pathway in plants, *Science,* 268:667-675.

Ecker, J. R., and Davis, R. W. (1987). Plant defense genes are regulated by ethylene, *Proc. Natl. Acad. Sci. U.S.A.,* 84:5202-5206.

Edelbaum, O., Sher, N., Rubinstein, M., Novick, D., Tal, N., Moyer, M., Ward, E., Ryals, J., and Sela, I. (1991). Two antiviral proteins, gp 35 and gp 22, correspond to β-1,3-glucanase and an isoform of PR-5, *Plant Mol. Biol.,* 17:171-173.

Edwards, K., Cramer, C. L., Bolwell, G. P., Dixon, R. A., Schuch, W., and Lamb, C. J. (1985). Rapid transient induction of phenylalanine ammonia-lyase mRNA in elicitor-treated bean cells, *Proc. Natl. Acad. Sci. U.S.A,* 82:6731-6735.

Edwards, R., Blount, J. W., and Dixon, R. A. (1991). Glutathione and elicitation of the phytoalexin response in legume cell cultures, *Planta,* 184:403-409.

Elkind, Y., Edwards, R., Mavandad, M., Hedrick, S., Ribak, O., Dixon, R. A., and Lamb, C. J. (1990). Abnormal plant development and down regulation of phenyl-propanoid biosynthesis in transgenic tobacco containing a heterologous phenylalanine ammonia-lyase gene, *Proc. Natl. Acad. Sci. U.S.A.,* 87:9057-9061.

Ellis, J. S., Jennings, A. C., Edwards, L. A., Lamb, C. J., and Dixon, R. A. (1989). Defense gene expression in elicitor-treated cell suspension cultures of French bean cv. Imuna, *Plant Cell Rep.,* 8:504-507.

Elstner, E.F. (1982). Oxygen activation and oxygen toxicity, *Annu. Rev. Plant Physiol.,* 33:73-96.

Elstner, E. F., and Osswald, W. (1994). Mechanisms of oxygen activation during plant stress, *Proc. Royal Soc. Edinburgh,* 102B:131-154.

El-Turk, J., Asemota, O., Leymarie, J., Sallaud, C., Mesnage, S., Breda, C., Buffard, D., Kondorost, A., and Esnault, R. (1996). Nucleotide sequences of four pathogen-induced alfalfa peroxidase-encoding cDNAs, *Genes,* 170:213-216.

Enard, C., Diolez, A., and Expert, D. (1988). Systemic virulence of *Erwinia chrysanthemi* 3937 requires a functional iron assimilation system, *J. Bacteriol.,* 170:2419-2426.

Enkerli, J., Gisi, V. and Mosinger, E. (1993). Systemic acquired resistance to *Phytophthora infestans* in tomato and the role of pathogenesis related proteins, *Physiol. Mol. Plant Pathol.,* 43:161-171.

Epperlein, M. N., Noronha-Dutra, A. A., and Strange, R. N. (1986). Involvement of the hydroxyl radical in the abiotic elicitation of phytoalexins in legumes, *Physiol. Mol. Plant Pathol.,* 28:67-77.

Epple, P., Apel, K., and Bohlmann, H. (1995). An *Arabidopsis thaliana* thionin gene is inducible via a signal transduction pathway different from that for pathogenesis-related proteins, *Plant Physiol.,* 109:813-820.

Epple, P., Apel, K., and Bohlmann, H. (1997). ESTs reveal a multigene family for plant defensins in *Arabidopsis thaliana, FEBS Lett.,* 400:168-170.

Ercolani, G. L., and Crosse, J. E. (1966). The growth of *Pseudomonas phaseolicola* and related plant pathogens *in vivo, J. Gen. Microbiol.,* 45:429-439.

Ernst, D., Schraudner, M., Langebartels, C., and Sandermann, H. Jr. (1992). Ozone-induced changes of mRNA levels of β-1,3-glucanase, chitinase and "pathogenesis-related" protein 1b in tobacco plants, *Plant Mol. Biol.,* 20:673-682.

Eskin, N. A. M., Grossman, S., and Pinsky, A. (1977). Biochemistry of lipoxygenase in relation to food quality, *Crit. Rev. Food Sci. Natur.,* 9:1-41.

Esnault, R., Buffard, D., Breda, C., Sallaud, C., El-Turk, J., and Kondorosi, A. (1993). Pathological and molecular characterization of alfalfa interactions with compatible and incompatible bacteria, *Xanthomonas campestris* pv. *alfalfae* and *Pseudomonas syringae* pv. *pisi, Mol. Plant-Microbe Interact.,* 6:655-664.

Espelie, K. E., Franceshi, V. R., and Kolattukudy, P. E. (1986). Immunocytochemical localization and time course of appearance of an anionic peroxidase associated with suberization in wound healing potato tuber tissue, *Plant Physiol.,* 81:487-492.

Esquerre-Tugaye, M. T., Campargue, C., and Mazau, D. (1999). The response of plant cell wall hydroxyproline-rich glycoproteins to microbial pathogens and their elicitors. In S. K. Datta and S. Muthukrishnan (Eds.), *Pathogenesis-Related Proteins in Plants,* CRC Press, Boca Raton, FL, pp. 157-170.

Esquerre-Tugaye, M. T., Lafitte, C., Mazau, D., Toppan, A., and Touze, A. (1979). Cell surfaces in plant-microorganism interactions. II. Evidence for the accumulation of hydroxyproline-rich glycoproteins in the cell wall of diseased plants as a defense mechanism, *Plant Physiol.,* 64:320-326.

Essenberg, M., Grover, P. B. Jr., and Cover, E. C. (1990). Accumulation of antibacterial sesquiterpenoids in bacterially inoculated *Gossypium* leaves and cotyledons, *Phytochemistry,* 29:3107-3113.

Essenberg, M., Pierce, M. L., Hamilton, B., Cover, E. C., Scholes, V. E., and Richardson, P. E. (1992). Development of fluorescent hypersensitively necrotic cells containing phytoalexins adjacent to colonies of *Xanthomonas* pv. *malvacearum* in cotton leaves, *Physiol. Mol. Plant Pathol.,* 41:85-99.

Everdeen, D. S., Kiefer, S., Willard, J. J., Muldoon, E. P., Dey, P. M., Li, X. B., and Lamport, D. T. A. (1988). Enzymatic cross-linkage of monomeric extensin precursors *in vitro, Plant Physiol.,* 87:616-621.

Evtushenkov, A. N., and Fomichev, Y. K. (1996). Secretion of pectate lyases by cells of *Erwinia chrysanthemi* and *Erwinia carotovora* var. *atroseptica, Microbiology (New York),* 65:290-295

Expert, D., Enard, C., and Masclaux, C. (1996). The role of iron in host-pathogen interactions, *Trends in Microbiology,* 4:232-237.

Expert, D., Sauvage, C., and Neilands, J. B. (1992). Negative transcriptional control of iron transport in *Erwinia chrysanthemi* involves an iron-responsive two factor system, *Mol. Microbiol.,* 6:2009-2017.

Eyal, Y., and Fluhr, R. (1991). Cellular and molecular biology of pathogenesis-related proteins, *Oxford Surveys of Plant Molecular and Cellular Biology,* 7:223-254.

Eyal, Y., Meller, Y., Lev Yadun, S., and Fluhr, R. (1993). A basic-type *PR-1* promoter directs ethylene responsiveness, vascular and abscission zone-specific expression, *Plant J.,* 4:225-234.

Eyal, Y., Sagee, O., and Fluhr, R. (1992). Dark induced accumulation of a basic pathogenesis-related (PR-1) transcript and a light requirement for its induction by ethylene, *Plant Mol. Biol.,* 19:589-599.

Falkenstein, E., Groth, B., Mithofer, A., and Weiler, E. W. (1991). Methyl jasmonate and α-linolenic acid are potent inducers of tendril coiling, *Planta,* 185:316-322.

Fantl, W. J., Johnson, D. E., and Williams, L. T. (1993). Signalling by receptor tyrosine kinases, *Annu. Rev. Biochem.,* 62:453.

Farmer, E. (1994). Fatty acid signalling in plants and their associated microorganisms, *Plant Mol. Biol.,* 26:1423-1437.

Farmer, E. E., Caldelari, D., Pearce, G., Walker-Simmons, M. K., and Ryan, C. A. (1994). Diethyldithiocarbamic acid inhibits the octadecanoid signaling pathway for the wound induction of proteinase inhibitors in tomato leaves, *Plant Physiol.,* 106:337-342.

Farmer, E. E., Moloshok, T. D., Saxton, M. J., and Ryan, C. A. (1991). Oligosaccharide signaling in plants. Specificity of oligouronide-enhanced plasma membrane protein phosphorylation, *J. Biol. Chem.,* 266:3140-3415.

Farmer, E. E., Pearce, G., and Ryan, C. A. (1989). In vitro phosphorylation of plant plasma membrane proteins in response to the proteinase inhibitor inducing factor, *Proc. Natl. Acad. Sci. U.S.A.,* 86:1539-1542.

Farmer, E. E., and Ryan, C. A. (1990). Interplant communication: Airborne methyl jasmonate induces synthesis of proteinase inhibitors in plant leaves, *Proc. Natl. Acad. Sci. U.S.A.,* 87:7713-7716.

Farmer, E. E., and Ryan, C. A. (1992). Octadecanoid precursors of jasmonic acid activate the synthesis of wound-inducible proteinase inhibitors, *Plant Cell,* 4:129-134.

Fauth, M., Merten, A., Hahn, M. G., Jeblick, W., and Kauss, H. (1996). Competence for elicitation of H_2O_2 in hypocotyls of cucumber is induced by breaching the cuticle and is enhanced by salicylic acid, *Plant Physiol.,* 110:347-354.

Favaron, F., Castiglioni, C., and Di Lenna, P. (1993). Inhibition of some rot fungi polygalacturonases by *Allium cepa* L. and *Allium porum* L. extracts, *J. Phytopathol.,* 139:201-206.

Favaron, F., Castiglioni, C., D'Ovido, R., and Alghisi, P. (1997). Polygalacturonase inhibiting proteins from *Allium porrum* L., and their role in plant tissue against fungal endo-polygalacturonases, *Physiol. Mol. Plant Pathol.,* 50:403-417.

Favaron, F., D'Ovidio, R., Porceddu, E., and Alghisi, P. (1994). Purification and molecular characterization of a soybean polygalacturonase-inhibiting protein, *Planta,* 195:80-87.

Felix, G., Grosskoff, D. G., Regenass, M., and Boller, T. (1991). Rapid changes of protein phosphorylation are involved in transduction of the elicitor signal in plant cells, *Proc. Natl. Acad. Sci. U. S. A.,* 88:8831-8834.

Felix, G., and Meins, F. (1986). Developmental and hormonal regulation of β-1,3-glucanase in tobacco, *Planta,* 167:206-211.

Felix, G., and Meins, F. (1987). Ethylene regulation of β-1,3-glucanase in tobacco, *Planta,* 172:386-392.

Fellay, R., Rahme, L. G., Mindrinos, M. N., Frederick, R. D., Pisi, A., and Panopoulos, N. J. (1991). Genes and signals controlling the *Pseudomonas syringae* pv. *phaseolicola*-plant interaction. In H. Hennecke and D. P. S. Verma (Eds.), *Advances in Molecular Genetics of Plant-Microbe Interactions. Vol. 1.* Kluwer Academic Publishers, Dordrecht, The Netherlands, pp. 45-52.

Fenselau, S., Balbo, I., and Bonas, U. (1992). Determinants of pathogenicity in *Xanthomonas campestris* pv. *vesicatoria* are related to proteins involved in secretion in bacterial pathogens of animals, *Mol. Plant-Microbe Interact.,* 5:390-396.

Fenselau, S., and Bonas, U. (1995). Sequence and expression analysis of the *hrpB* pathogenicity operon of *Xanthomonas campestris* pv. *vesicatoria* which encodes eight proteins with similarity to components of the Hrp, Ysc, Spa, and Fli secretion systems, *Mol. Plant-Microbe Interact.,* 8:845-854.

Fernandez de Caleya, R., Gonzalez-Pascual. B., Gonzalez-Pascal, B., Garcia-Olmedo, F., and Carbonero, P. (1972). Susceptibility of phytopathogenic bacteria to wheat purothionins in vitro, *Appl. Microbiol.,* 23:998-1000.

Fett, W. F., and Osman, S. F. (1982). Inhibition of bacteria by the isoflavanoids glyceollin and coumestrol, *Phytopathology,* 72:755-760.

Feuillet, C., Schachacher, F., and Keller, B. (1997). Molecular cloning of a new receptor like kinase gene encoded at the Lr10 disease resistance locus of wheat, *Plant J.,* 11:45-52.

Feys, B. J. F., Benedetti, C. E., Penfold, C. N., and Turner, J. G. (1994). *Arabidopsis* mutants selected for resistance to the phytotoxin coronatine are male sterile, insensitive to methyl jasmonate, and resistant to a bacterial pathogen, *Plant Cell,* 6:751-759.

Fillingham, A. J., Wood, J., Bevan, J. R., Crute, I. R., Mansfield, J.W., Taylor, J. D., and Vivian, A. (1992). Avirulence genes from *Pseudomonas syringae* pathovars *phaseolicola* and *pisi* confer specificity towards both host and non-host species, *Physiol. Mol. Plant Pathol.,* 40:1-15.

Finberg, K. E., Muth, T. R., Young, S. P., Maken, J. B., Heitritter, S. M., Binns, A. N., and Banta, L. M. (1995). Interactions of VirB9, –10, and –11 with the membrane fraction of *Agrobacterium tumefaciens:* Solubility studies provide evidence for tight associations, *J. Bacteriol.,* 177:4881-4889.

Fink, J., Boman, A., Boman, H. G., and Merrifield, R. B. (1989). Design, synthesis, and antibacterial activity of cecropin-like model peptides, *Int. J. Peptide Protein Res.,* 33:412-421.

Finlay, B. B. (1994). Molecular and cellular mechanisms of *Salmonella* pathogenesis. In J. L. Dangl (Ed.), *Bacterial Pathogenesis of Plants and Animals: Molecular and Cellular Mechanisms,* Vol.192, Springer-Verlag, Berlin, pp.163-185.

Fischer, W., Christ, U., Baumgartner, M., Erismann, K. H. and Mosinger, E. (1989). Pathogenesis-related proteins of tomato. II. Biochemical and immunological characterization, *Physiol. Mol. Plant Path.,* 35:67-83.

Fladung, M., and Ballvora, A. (1992). Further characterization of *rolC* transgenic tetraploid potato clones, and influence of daylength and level of *rolC* expression on yield parameters, *Plant Breeding,* 109:18-27.

Fladung, M., and Gieffers, W. (1993). Resistance reactions of leaves and tubers of *rolC* transgenic tetraploid potato to bacterial and fungal pathogens. Correlation with sugar, starch and chlorophyll content, *Physiol. Mol. Plant Pathol.,* 42:123-132.

Flavell, R. B. (1994). Inactivation of gene expression in plants as a consequence of specific sequence duplication, *Proc. Natl. Acad. Sci. U.S.A.,* 91:3490-3496.

Flego, D., Pirhonen, M., Saarilahti, H., Palva, T. K., and Palva, E. T. (1997). Control of virulence gene expression by plant calcium in the phytopathogen *Erwinia carotovora, Mol. Microbiol.,* 25:831-838.

Fleming, A. W., Mandel, T., Hofmann, S., Sterk, P., deVries, S. C., and Kuhlemeier, C. (1992). Expression pattern of a tobacco lipid transfer protein gene within the shoot apex, *Plant J.,* 2:855-862.

Fleming, T. M., McCarthy, D. A., White, R. F., Antoniw, J. F., and Mikkelsen, J. D. (1991). Induction and characterization of some of the pathogenesis-related proteins in sugar beet, *Physiol. Mol. Plant Pathol.,* 39:147-160.

Florack, D. E. A., Allefs, S., Bollen, R., Bosch, D., Visser, B., and Stiekema, W. (1995). Expression of giant silk moth cecropin B genes in tobacco, *Transgenic Res.,* 4:132-141.

Florack, D. E. A., Dirkse, W. G., Visser, B., Heidekamp, F., and Stiekema, W. J. (1994). Expression of biologically active hordothionins in tobacco. Effects of pre- and pro-sequences at the amino and carboxyl termini of the hordothionin precursor on mature protein expression and sorting, *Plant Mol. Biol.,* 24:83-96.

Flott, B. E., Moerschbacher, B. M., and Reisener, H. J. (1989). Peroxidase isozyme patterns of resistant and susceptible wheat leaves following stem rust infection, *New Phytol.,* 111:413-421.

Forrest, R. S., and Lyon, G. D. (1990). Substrate degradation patterns of polygalacturonic acid lyase from *Erwinia carotovora* and *Bacillus polymyxa* and release of phytoalexin-eliciting oligosaccharides from potato cell walls, *J. Expt. Bot.,* 41:481-488.

Franza, T., and Expert, D. (1991). The virulence-associated chrysobactin iron uptake system of *Erwinia chrysanthemi* 3937 involves an operon encoding transport and biosynthetic functions, *J. Bacteriol.,* 173:6874-6881.

Fraser, R. S. S. (1981). Evidence for the occurrence of the "pathogenesis-related" proteins in leaves of healthy tobacco plants during flowering, *Physiol. Plant Pathol.,* 19:69-76.

Frederick, R. D., Chiu, J., Bennetzen, J. L., and Handa, A. K. (1997). Identification of a pathogenicity locus, rpfA, in *Erwinia carotovora* subsp. *carotovora* that encodes a two-component sensor-regulator protein, *Mol. Plant-Microbe Interact.,* 10:407-415.

Frederick, R. D., Majerczak, D. R., and Coplin, D. L. (1993). *Erwinia stewartii* WtsA, a positive regulator of pathogenicity gene expression, is homologous to *Pseudomonas syringae* pv. *phaseolicola* HrpS, *Mol. Microbiol.,* 9:477-485.

Frediani, M., Cremonini, R., Salvi, G., Caprari, C., Desiderio, A., D'Ovidio, R., Cervone, F., and De Lorenzo, G. (1993). Cytological localization of the *pgip* genes in the embryo suspensor cells of *Phaseolus vulgaris* L., *Theor. Appl. Genet.,* 87:369-373.

Fredrikson, K., and Larsson, C. (1989). Activation of 1,3-β-glucan synthase by Ca^{2+}, spermine and cellobiose. Localization of activator sites using inside-oat plasma membrane vesicles, *Physiol. Plant.,* 77:196-201.

Freeman, T. L. F., and San Francisco, M. J. D. (1994). Cloning of a galacturonic acid uptake gene from *Erwinia chrysanthemi* EC16, *FEMS Microbiol. Lett.,* 118: 101-106.

Fressmuth, M., and Gilman, A. G. (1991). G-proteins and the regulation of second messenger systems. In J. D. Wilson, E. Braunwald, K. J. Isselbacher, R. G. Petersdorf, J. B. Martin, A. S. Fauci, and R. F. Root (Eds.), *Principles of Internal Medicine,* New York, McGraw Hill, pp. 393-397.

Fridovich, I. (1986). Superoxide dismutases, *Adv. Enzymol.,* 58:62-97.

Friedrich, L., Lawton, K., Reuss, W., Masner, P., Specker, N., Gut Rella, M., Meier, B., Dincher, S., Staub, T., Uknes, S., Metraux, J. P., Kessmann, H., and Ryals, J. (1996). A benzothiadiazole derivative induces systemic acquired resistance in tobacco, *Plant J.,* 10:61-70.

Friedrich, L., Vernooij, B., Gaffney, T., Morse, A., and Ryals, J. (1995). Molecular and biochemical characterization of tobacco plants expressing a bacterial salicylate hydroxylase gene, *Plant Mol. Biol.,* 29:959-968.

Fristensky, B., Horovitz, D., and Hadwiger, L.A. (1988). cDNA sequences for pea disease resistance response genes, *Plant Mol. Biol.,* 11:713-715.

Fritzemeier, K. H., and Kindl, H. (1981). Coordinate induction by UV light of stilbene synthase, phenylalanine ammonia-lyase and cinnamate 4-hydroxylase in leaves of vitaceae, *Planta,* 151:48-52.

Froissard, D., Gough, C., Czernic, P., Schneider, M., Toppan, A., Roby, D., and Marco, Y. (1994). Structural organization of *str246C* and *str246N,* plant defense-related genes from *Nicotiana tabacum, Plant Mol. Biol.,* 26:515-521.

Frommel, M. I., Nowak, J., and Lazarovits, G. (1991). Growth enhancement and developmental modifications of *in vitro* grown potato (*Solanum tuberosum* ssp.

tuberosum) affected by a nonfluorescent *Pseudomonas* sp., *Plant Physiol.,* 96:928-936.

Fry, S. C. (1982). Isodityrosine, a new cross-linking amino acid from plant cell wall glycoprotein, *Biochem. J.,* 204:449-455.

Fry, S. C. (1986). Cross-linking of matrix polymers in the growing cell walls of angiosperms, *Annu. Rev. Plant Physiol.,* 37:165-186.

Fukuda, Y. (1997). Interaction of tobacco nuclear protein with an elicitor-responsive element in the promoter of a basic class I chitinase gene, *Plant Mol. Biol.,* 34:81-87.

Furuse, K., Takemoto, D., Doke, N., and Kawakita, K. (1999). Invovement of actin filament association in hypersensitive reactions in potato cells, *Physiol. Mol. Plant Pathol.,* 54:51-61.

Gabriel, D. W., Burges, A., and Lazo, G. (1986). Gene-for-gene interactions of five cloned avirulence genes from *Xanthomonas campestris* pv. *malvacearum* with specific resistance genes in cotton, *Proc. Natl. Acad. Sci. U. S. A.,* 83:6415-6419.

Gabriel, D. W., and Rolfe, B. G. (1990). Working models of specific recognition in plant-microbe interactions, *Annu. Rev. Plant Physiol.,* 28:365-391.

Gaffney, T., Friedrich, L., Vernooij, B., Negrotto, D., Nye, G., Uknes, S., Ward, E., Negrotto, D., Nye, G., Uknes, S., Ward, E., Kessmann, H., and Ryals, J. (1993). Requirement of salicylic acid for the induction of systemic acquired resistance, *Science,* 261:754-756.

Galan, J., Ginocchio, C., and Costeas, P. (1992). Molecular and functional characterization of the *Salmonella* invasion gene *invA:* Homology to members of a new protein family, *J. Bacteriol.,* 174:4338-4349.

Galvez, A., Masqueda, M., Martinez-Bueno, M., and Valdiva, E. (1991). Permeation of bacterial cells, permeation of cytoplasmic and artificial membrane vesicles, and channel formation on lipid bilayers by peptide antibiotic AS-48, *J. Bacteriol.,* 173:886-892.

Garcia-Bustos, J., Heitman, J., and Hall, M. N. (1991). Nuclear protein localization, *Biophys. Acta,* 1071:83-101.

Garcia-Garcia, F., Schmelzer, E., and Hahlbrock, K. (1994). Differential expression of chitinase and β-1,3-glucanase genes in various tissues of potato plants, *Z. fur Naturforschung,* 49C:195-203.

Garcia-Olmedo, F., Molina, A., Segura, A., and Moreno, M. (1995). The defensive role of nonspecific lipid-transfer proteins in plants, *Trends in Microbiology,* 3:72-74.

Garfinkel, D. J., and Nester, E. W. (1980). *Agrobacterium tumefaciens* mutants affected in crown gall tumorigenesis and octopine catabolism, *J. Bacteriol.,* 144:732-743.

Garg, V. K., and Douglas, T. J. (1983). Hydroxymethyl-glutaryl CoA reductase in plants. In J. R. Sabine (Ed.), *3-Hydroxy-3-Methylglutaryl CoenzymeA Reductase,* CRC Press, Boca Raton, FL, pp. 29-37.

Garibaldi, A., and Bateman, D. (1971). Pectic enzymes produced by *Erwinia chrysanthemi* and their effects on plant tissue, *Physiol. Plant Pathol.,* 1:25-40.

Ghigo, J. M., and Wandersman, C. (1994). A carboxy-terminal four-amino acid motif is required for secretion of the metalloprotease PrtG through the *Erwinia chrysanthemi* protease secretion pathway, *J. Biol. Chem.,* 269:8979-8985.

Gianinazzi, S., and Kassanis, B. (1974). Virus resistance in plants by polyacrylic acid, *J. Gen. Virol.,* 23:1-9.

Gibbon, M. J., Jenner, C., Mur, L. A. J., Puri, N., Mansfield, J. W., Taylor, J. D., and Vivian, A. (1997). Avirulence gene *avrPpiA* from *Pseudomonas syringae* pv. *pisi* is not required for full virulence on pea, *Physiol. Mol. Plant Pathol.,* 50:219-236.

Gilkes, N. J., Henrissat, B., Kilburn, D. G., Miller, R. C. Jr., and Warren, R. A. J. (1991). Domains in microbial β-1,4-endoglucanases: Sequence conservation, function and enzyme families, *Microbiol. Rev.,* 55:303-315.

Gillikin, J. W., Burkhart, W., and Graham, J. S. (1991). Complete amino acid sequence of a polypetide from *Zea mays* similar to the pathogenesis-related-1 family, *Plant Physiol.,* 96:1372-1375.

Glazebrook, J., and Ausubel, F. M. (1994). Isolation of phytoalexin-deficient mutants of *Arabidopsis thaliana* and characterization of their interactions with bacterial pathogens, *Proc. Natl. Acad. Sci. U.S.A.,* 91:8955-8959.

Glazebrook, J., Rogers, E. E., and Ausubel, F. M. (1996). Isolation of *Arabidopsis* mutants with enhanced disease susceptibility by direct screening, *Genetics,* 143:973-982.

Gnanamanickam, S. S., and Patil, S. S. (1977). Phaseotoxin suppresses bacterially induced hypersensitive reaction and phytoalexin synthesis in bean cultivars, *Physiol. Plant Pathol.,* 10:169-179.

Godiard, I., Froissard, D., Fournier, J., Axelos, M., and Marco, Y. J. (1991). Differential regulation in tobacco cell suspensions of genes involved in plant-bacteria interactions by pathogen-related signals, *Plant Mol. Biol.,* 17:409-413.

Godiard, L., Ragueh, F., Froissard, D., Leguay, J.J., Grosset, J., Chartier, Y., Meyer, Y. and Marco, Y. (1990). Analysis of the synthesis of several pathogenesis-related proteins in tobacco leaves infiltrated with water and with compatible and incompatible isolates of *Pseudomonas solanacearum, Mol. Plant-Microbe Interact.,* 3:207-213.

Godshall, M. A., and Lonergan, T. A. (1987). The effect of sugarcane extracts on the growth of the pathogenic fungus *Colletotrichum falcatum, Physiol. Mol. Plant Pathol.,* 30:299-308.

Goldberg, R., Liberman, M., Mathieu, C., Pierron., and Catesson, A. M. (1987). Development of epidermal cell wall peroxidase along the mungbean hypocotyl: Possible involvement in the cell wall stiffening process, *J. Exp. Bot.,* 38:1378-1390.

Goldsbrough, A. P., Albrecht, H., and Stratford, R. (1993). Salicylic acid-inducible binding of a tobacco nuclear protein to a 10 bp sequence which is highly conserved amongst stress-inducible genes, *Plant J.,* 3:563-571.

Gonzalez, C. F., Pettit, E. A., Valadez, V. A., and Provin, E. M. (1997). Mobilization, cloning, and sequence determination of a plasmid-encoded polygalacturonase from a phytopathogenic *Burkholderia (Pseudomonas) cepacia, Mol. Plant-Microbe Interact.,* 10:840-851.

Goodman, R. N. (1968). The hypersensitive reaction in tobacco: A reflection of changes in host cell permeability, *Phytopathology,* 58:872-875.

Goodman, R. N., Huang, P. Y., and White, J. A. (1976). Ultrastructural evidence for immobilization of an incompatible bacterium, *Pseudomonas pisi* in tobacco leaf tissue, *Phytopathology,* 66:754-764.

Gopalan, S., Bauer, D. W., Alfano, J. R., Loniello, A. O., He, S. Y., and Collmer, A. (1996). Expression of the *Pseudomonas syringae* avirulence protein AvrB in plant cells alleviates its dependence on the hypersensitive response and pathogenicity (Hrp) secretion system in eliciting genotype-specific hypersensitive cell death, *Plant Cell,* 8:1095-1105.

Gordon-Weeks, R., Sugars, J. M., Antoniw, J. F., and White, R. F. (1997). Accumulation of a novel PR1 protein in *Nicotiana langsdorfii* leaves in response to virus infection or treatment with salicylic acid, *Physiol. Mol. Plant Pathol.,* 50:263-273.

Gorlach, J., Volrath, S., Knauf-Beiter, G., Hengry, G., Beckhove, U., Kogel, K.-H., Oostendorp, M., Staub, T., Ward, E., Kessmann, H., and Ryals, J. (1996). Benzothiadiazole, a novel class of inducers of systemic acquired resistance, activates gene expression and disease resistance in wheat, *Plant Cell,* 8:629-643.

Gottstein, H. D., and Kuc, J. (1989). Induction of systemic resistance to anthracnose in cucumber by phosphates, *Phytopathology,* 79:176-179.

Gough, C. L., Dow, J. M., Barber, C. E., and Daniels, M. J. (1988). Cloning of two endoglucanases of *Xanthomonas campestris* pv. *campestris:* Analysis of the role of the major endoglucanase in pathogenesis, *Mol. Plant-Microbe Interact.,* 1:275-281.

Gough, C. L., Genin, S., Zischek, C., and Boucher, C. A. (1992). *hrp* genes of *Pseudomonas solanacearum* are homologous to pathogenicity determinants of animal pathogenic bacteria and are conserved among plant pathogenic bacteria, *Mol. Plant-Microbe Interact.,* 5:384-389.

Gowri, G., Paiva, N. L., and Dixon, R. A. (1991). Stress responses in alfalfa (*Medicago sativa* L.): Sequence analysis of phenylalanine ammonia-lyase (PAL) cDNA clones and appearance of PAL transcripts in elicitor-treated cell cultures and developing plants, *Plant Mol. Biol.,* 17:415-429.

Graham, J. S., Burkhart, W., Xiong, J., and Gillikin, J. (1992). Complete amino acid sequence of soybean leaf P21, *Plant Physiol.,* 98:163-165.

Graham, M. A., Marek, L. F., and Shoemaker, R. C. (1998). Analysis of resistance gene analog cDNA clones from soybean, *Sixth Intl. Conf. Plant and Animal Genome Research,* San Diego, California, p. 83.

Graham, M. Y., and Graham, T. L. (1991). Rapid accumulation of anionic peroxidases and phenolic polymers in soybean cotyledon tissues following treatment with *Phytophthora megasperma* f.sp. *glycinea* wall glucan, *Plant Physiol.,* 97:1445-1455.

Graham, T. L., and Graham, M. Y. (1991). Glyceollin elicitors induce major but distinctly different shifts in isoflavonoid metabolism in proximal and distal soybean cell populations, *Mol. Plant-Microbe Interact.,* 4:60-68.

Graham, T. L., and Graham, M. Y. (1996). Signaling in soybean phenylpropanoid responses. Dissection of primary, secondary, and conditioning effects of light, wounding, and elicitor treatments, *Plant Physiol.,* 110:1123-1133.

Graham, T. L., Sequeira, L., and Huang, T. R. (1977). Bacterial lipopolysaccharides as inducers of disease resistance in tobacco, *Appl. Environ. Microbiol.,* 34:424-432.

Granell, A., Belles, J. M., and Conejero, V. (1987). Induction of pathogenesis-related proteins in tomato by citrus exocortis viroid, silver ion, and ethephon, *Physiol. Mol. Plant Pathol.,* 31:83-90.

Granell, A., Pereto, J. G., Schindler, U., and Cashmore, A. R. (1992). Nuclear factors binding to the extensin promoter exhibit differential activity in carrot protoplasts and cells, *Plant Mol. Biol.,* 18:739-748.

Grant, M. R., Godiard, L., Straube, E., Ashfield, T., Lewald, J., Sattlier, A., Innes, R. W., and Dangl, J. L. (1995). Structure of the *Arabidopsis* RPM1 gene enabling dual specificity disease resistance, *Science,* 269:843-846.

Gray, J., Wang, J., and Gelvin, S. B. (1992). Mutation of the *miaA* gene of *Agrobacterium tumefaciens* results in reduced *vir* gene expression. *J. Bacteriol.,* 174:1086-1098.

Grayer, R. J., and Harborne, J. B. (1994). A survey of antifungal compounds from higher plants, 1982-1993, *Phytochemistry,* 37:19-42.

Green, R., and Fluhr, R. (1995). UV-B-induced PR-1 accumulation is mediated by active oxygen species, *Plant Cell,* 7:203-212.

Green, T. R., and Ryan, C. A. (1972). Wound-induced proteinase inhibitor in plant leaves: a possible defense mechanism against insects, *Science,* 175:776-777.

Greenberg, J. T., and Demple, B. (1989). A global response induced in *Escherichia coli* by redox cycling agents overlaps with that induced by peroxide stress, *J. Bacteriol.,* 171:3933-3939.

Greenberg, J. T., Guo, A., Klessig, D. F., and Ausubel, F. M. (1994). Programmed cell death in plants: A pathogen-triggered response activated coordinately with multiple defense functions, *Cell,* 77:551-563.

Gregory, P., Tingey, W. M., Ave' D. A., and Bouthyette, P. Y. (1986). Potato glandular trichomes: A physiochemical defense mechanism against insects. In M. B. Green and P. A. Heldin (Eds.), *Natural Resistance of Plants to Pests: Roles of Allelochemicals,* American Chemical Society, Washington, DC, pp. 160-167.

Grenier, J., and Asselin, A. (1990). Some pathogenesis-related proteins are chitosanases with lytic activity against fungal spores, *Mol. Plant-Microbe Interact.,* 3:401-407.

Grillo, S., Leone, A., Xu, Y., Tucci, M., Francione, R., Hasegawa, P. M., Monti, L., and Bressan, R. A. (1995). Control of osmotin gene expression by ABA and osmotic stress in vegetative tissues of wild-type and ABA-deficient mutants of tomato, *Physiol. Plantarum,* 93:498-504.

Grimault, V., Gelie, B., Lemattre, M., Prior, P., and Schmit, J. (1994). Comparative histology of resistant and susceptible tomato cultivars infected by *Pseudomonas solanacearum, Physiol. Mol. Plant Pathol.,* 44:105-123.

Grimault, V., Vian, B., Perino, C., Reis, D., and Bertheau, Y. (1997). Degradation patterns of pectic substrates related to the localization of bacterial pectate-lyases in the model *Erwinia chrysanthemi/Saintpaulia ionantha, Physiol. Mol. Plant Pathol.,* 51:45-62.

Grimm, C., Aufsatz, W., and Panopoulos, N. J. (1995). The *hrpRS* locus of *Pseudomonas syringae* pv. *phaseolicola* constitutes a complex regulatory unit, *Mol. Microbiol.,* 15:155-165.

Grimm, C., and Panopoulos, N. J. (1989). The predicted protein product of a pathogenicity locus from *Pseudomonas syringae* pv. *phaseolicola* is homologous to a highly conserved domain of several procaryotic regulatory proteins, *J. Bact.*, 171:5031-5038.

Grisebach, H. (1981). Lignins. In E. E. Conn (Ed.), *The Biochemistry of Plants*, Academic Press, New York, pp. 457-478.

Groom, Q. J., Torres, M. A., Fordham-Skelton, A. P., Hammond-Kosack, K. E., Robinson, N. J., and Jones, J. D. G. (1996). *RbohA*, a rice homologue of the mammalian *gp91phox* respiratory burst oxidase gene, *Plant J.*, 10:515-522.

Gross, G. G. (1978). Recent advances in the chemistry and biochemistry of lignin, *Recent Adv. Phytochem.*, 12:177-200.

Gross, G. G. (1980). The biochemistry of lignification, *Adv. Bot. Res.*, 8:25-63.

Gross, G. G., Janse, C., and Elstner, E. F. (1977). Involvement of malate, monophenols and the superoxide radical in hydrogen peroxide formation by isolated cell walls from horseradish (*Armoracia lapathifolia* Gilib.), *Planta*, 136:271-276.

Gross, R., Arico, B., and Rappuoli, R. (1989). Families of bacterial signal-transduction proteins, *Mol. Microbiol.*, 3:1661-1667.

Grosskopf, D. G., Felix, G., and Boller, T. (1990). K-252a inhibits the response of tomato cells to fungal elicitors in vivo and their microsomal protein kinase in vitro, *FEBS Lett.*, 275:177-180.

Gu, Q., Kawata, C. E., Morse, M. J., Wu, H. M., and Cheung, A.Y. (1992). A flower-specific cDNA encoding a novel thionin in tobacco, *Mol. Gen. Genet.*, 234:89-96.

Gubba, S., Xie, Y.-H., and Das, A. (1995). Regulation of *Agrobacterium tumefaciens* virulence gene expression: Isolation of a mutation that restores *virGD52E* function, *Mol. Plant-Microbe Interact.*, 8:788-791.

Gundlach, H., Muller, M. J., Kutchan, T., and Zenk, M. H. (1992). Jasmonic acid is a signal inducer in elicitor-induced plant cell cultures, *Proc. Natl. Acad. Sci. U.S.A.*, 89:2389-2393.

Guo, A., Reimers, P. J., and Leach, J. E. (1993). Effect of light on compatible interaction between *Xanthomonas oryzae* pv. *oryzae* and rice, *Physiol. Mol. Plant Pathol.*, 42:413-425.

Guo, Y., Delseny, M., and Puigdomenech, P. (1994). mRNA accumulation and promoter activity of the gene coding for a hydroxyproline-rich glycoprotein in *Oryza sativa*, *Plant Mol. Biol.*, 25:159-165.

Gurlitz, R. H. G., Lamb, P. W., and Matthysse, A. G. (1987). Involvement of carrot cell surface proteins in attachment of *Agrobacterium tumefaciens*, *Plant Physiol.*, 83:564-568.

Gururaj Rao, A. (1995). Antimicrobial peptides, *Mol. Plant-Microbe Interact.*, 8:6-13.

Gustine, D. L., Sherwood, R. T., Moyer, B. G., and Lukezic, F. I. (1990). Metabolites from *Pseudomonas corrugata* elicit phytoalexin biosynthesis in white clover, *Phytopathology*, 80:1427-1432.

Haahtela, K., Tarkka, E., and Korhonen, T. K. (1985). Type-1 fimbria-mediated adhesion of enteric bacteria to grass roots, *Appl. Environ. Microbiol.*, 49:1182-1185.

Habereder, H., Schroder, G., and Ebel, J. (1989). Rapid induction of phenylalanine ammonia-lyase and chalcone synthase mRNAs during fungus infection of soybean *(Glycine max)* roots or elicitor treatment of soybean cell cultures at the onset of phytoalexin synthesis, *Planta,* 177:58-65.

Hagiwara, H., Matsuoka, M., Ohshima, M., Watanabe, M., Hosokawa, D., and Ohashi, Y. (1993). Sequence-specific binding of protein factors to 2 independent promoter regions of the acidic tobacco pathogenesis-related-1 protein gene (PR-1), *Mol. Gen. Genet.,* 240:197-205.

Hahlbrock, K., Boudet, A. M., Chappell, J., Kreuzaler, F., Kuhn, D. N., and Ragg, H. (1983). Differential induction of mRNAs by light and elicitor in cultured plant cells. In O. Cifferi and L. Dure (Eds.), *Structure and Function of Plant Genomes,* NATO ASI Series, Chapman and Hall, London, pp. 15-23.

Hahlbrock, K., and Grisebach, H. (1979). Enzymic controls in the biosynthesis of lignin and flavonoids, *Annu. Rev. Plant Physiol.,* 30:105-130.

Hahlbrock, K., Lamb, C. J., Purwin, C., Ebel, J., Fautz, E., and Schafer, E. (1981). Rapid response of suspension-cultured parsley cells to the elicitor from *Phytophthora megasperma* var. *sojae.* Induction of the enzymes of general phenylpropanoid metabolism, *Plant Physiol.,* 67:768-773.

Hahlbrock, K., and Scheel, D. (1989). Physiology and molecular biology of phenylpropanoid metabolism, *Annu. Rev. Plant Physiol. Plant Mol. Biol.,* 40:347-369.

Hahn, M. G., Bucheli, P., Cervone, F., Doares, S. H., O'Neill, R. A., Darvill, A., and Albersheim, P. (1989). The roles of cell constituents in plant-pathogen interactions. In E. Nester and T. Kosuge (Eds.), *Plant-Microbe Interactions,* McGraw Hill, New York, pp. 131-181.

Hain, R., Bieseler, B., Kindl, H., Schroder, G., and Stocker, R. (1990). Expression of a stilbene synthase gene in *Nicotiana tabacum* results in synthesis of the phytoalexin resveratrol, *Plant Mol. Biol.,* 15:325-335.

Hain, R., Reif, H. J., Krause, E., Langebartels, R., Kindl, H., Vornau, B., Wiese, W., Schmetzer, E., and Schreier, P. H. (1993). Disease resistance results from foreign phytoalexin expression in a novel plant, *Nature,* 361:153-156.

Halliwell, B. (1978). Lignin synthesis: the generation of hydrogen peroxide and superoxide by horseradish peroxidase and its stimulation by manganese (II) and phenols, *Planta,* 140:81-88.

Halliwell, B. (1982). The toxic effects of oxygen on plant tissues. In Oberley (Ed.), *Superoxide Dismutase,* Vol. I, CRC Press, Boca Raton, pp. 89-123.

Halliwell, B., and Gutteridge, J. M. C. (1986). Iron and free radical reactions: two aspects of antioxidant protection, *Trends Biochem. Sci.,* 11:372-375.

Halliwell, B., and Gutteridge, J. M. C. (1989). *Free Radicals in Biology and Medicine,* Clarendon Press, Oxford, pp. 305.

Hammond-Kosack, K. E., and Jones, J. D. G. (1996). Resistance gene-dependent plant defense responses, *Plant Cell,* 8:1773-1781.

Hammond-Kosack, K. E., and Jones, J. D. G. (1997). Plant disease resistance genes, *Annu. Rev. Plant Mol. Biol.,* 48:573-607.

Hammond-Kosack, K. E., Silverman, P., Raskin, I., and Jones, J. D. G. (1996). *Avr* elicitors of *Cladosporium fulvum* induce changes in cell morphology, and ethylene and salicylic acid synthesis, in tomato plants carrying the corresponding *Cf*-disease resistance gene, *Plant Physiol.,* 110:1381-1394.

Han, D. C., Chen, C., Chen, Y., and Winans, S. C. (1992). Altered-function mutations of the transcriptional regulatory gene *virG* of *Agrobacterium tumefaciens, J. Bacteriol.,* 174:7040-7043.

Hargreaves, J. A. (1979). Investigations into the mechanism of mercuric chloride stimulated phytoalexin accumulation in *Phaseolus vulgaris* and *Pisum sativum, Physiol. Plant Pathol.,* 15:279-287.

Hargreaves, J. A. (1981). Accumulation of phytoalexins in cotyledons of French bean (*Phaseolus vulgaris* L.) following treatment with triton (I-octylphenol polyethoxyethanol) surfactants, *New Phytol.,* 87:733-741.

Hargreaves, J. A., and Bailey, J. A. (1978). Phytoalexin production by hypocotyls of *Phaseolus vulgaris* in response to constitutive metabolites released by damaged cells, *Physiol. Plant Pathol.,* 13:89-100.

Hargreaves, J. A., Mansfield, J. W., and Rossall, S. (1977). Changes in phytoalexin concentrations in tissues of the broad bean plant (*Vicia faba* L.), following inoculation with species of *Botrytis, Physiol. Plant Pathol.,* 11:227-242.

Harrison, M. J., Choudhary, A. D., Kooter, J., Lamb, C. J., and Dixon, R. A. (1991a). Stress-responses in alfalfa (*Medicago sativa* L.). 8. *cis*-elements and *trans*-acting factors for the quantitative expression of a bean chalcone synthase gene promoter in electroporated alfalfa protoplasts, *Plant Mol. Biol.,* 16:877-890.

Harrison, M. J., Lawton, M. A., Lamb, C. J., and Dixon, R. A. (1991b). Characterization of a nuclear protein that binds to three elements within the silencer region of a bean chalcone synthase gene, *Proc. Natl. Acad. Sci. U.S.A.,* 88:2515-2519.

Harrison, S. J., Curtis, M. D., McIntyre, C. L., Maclean, D. J., and Manners, J. M. (1995). Differential expression of peroxidase isogenes during the early stages of infection of the tropical forage legume *Stylosanthes humulis* by *Colletotrichum gloeosporioides, Mol. Plant-Microbe Interact.,* 8:398-406.

Hart, C. M., Fischer, B., Neuhaus, J. M., and Meins, F. Jr. (1992). Regulated inactivation of homologous gene expression in transgenic *Nicotiana sylvestris* plants containing a defense-related tobacco chitinase gene, *Mol. Gen. Genet.,* 235:179-188.

Hart, C. M., Nagy, F., and Meins, F. Jr. (1993). A 61 bp enhancer element of tobacco β-1,3-glucanase B gene interacts with one or more regulated nuclear proteins, *Plant Mol. Biol.,* 21:121-131.

He, S. Y., Bauer, D. W., Collmer, A., and Beer, S. V. (1994). Hypersensitive response elicited by *Erwinia amylovora* harpin requires active plant metabolism, *Mol. Plant-Microbe Interact.,* 7:289-292.

He, S. Y., Huang, H.-C., and Collmer, A. (1993). *Pseudomonas syringae* pv. *syringae* $harpin_{Pss}$: A protein that is secreted via the Hrp pathway and elicits the hypersensitive response in plants, *Cell,* 73:1255-1266.

Heath, J. D., Boulton, M. I., Raineri, D. M., Doty, S. L., Mushegian, A. R., Charles, T. C., Davies, J. W., and Nester, E. W. (1997). Discrete regions of the sensor protein VirA determine the strain-specific ability of *Agrobacterium* to agroinfect maize, *Mol. Plant-Microbe Interact.,* 10:221-227.

Heikinheimo, R., Flego, D., Pirhonen, M., Karlsson, M.-B., Eriksson, A., Mae, A., Koiv, V., and Palva, E. T. (1995). Characterization of a novel pectate lyase from *Erwinia carotovora* subsp. *carotovora, Mol. Plant-Microbe Interact.,* 8:207-217.

Heilbronn, J., Johnston, D. J., Dunbar, B., and Lyon, G. D. (1995). Purification of a metalloprotease produced by *Erwinia carotovora* ssp. *carotovora* and the degradation of potato lectin in vitro, *Physiol. Mol. Plant Pathol.,* 47:285-292.

Heitz, T., Bergey, D. R., and Ryan, C. A. (1997). A gene encoding a chloroplast-targeted lipoxygenase in tomato leaves is transiently induced by wounding, systemin, and methyl jasmonate, *Plant Physiol.,* 114:1085-1093.

Heitz, T., Geoffroy, P., Fritig, B., and Legrand, M. (1999). The PR-6 family: Proteinase inhibitors in plant-microbe and plant-insect interactions. In S. K. Datta and S. Muthukrishnan (Eds.), *Pathogenesis-Related Proteins in Plants,* CRC Press, Boca Raton, FL, pp. 247-260.

Heitz, T., Segond, S., Kauffmann, S., Geoffroy, P., Prasad, V., Brunner, F., Fritig, B., and Legrand, M. (1994). Molecular characterization of a novel tobacco pathogenesis-related (PR) protein: A new plant chitinase/lysozyme, *Mol. Gen. Genet.,* 245:246-254.

Hejgaard, J., Jacobsen, S., Bjorn, S. E., and Kragh, K. M. (1992). Antifungal activity of chitin-binding PR-4 type proteins from barley grain and stressed leaf, *FEBS Lett.,* 307:389-392.

Hejgaard, J., Jacobsen, S., and Svendsen, I. (1991). Two antifungal thaumatin-like proteins from barley grain, *FEBS Lett.,* 291:127-131.

Hendson, M., Hildebrand, D. C., and Schroth, M. N. (1992). Relatedness of *Pseudomonas syringae* pv. *tomato, Pseudomonas syringae* pv. *maculicola,* and *Pseudomonas syringae* pv. *antirrhini, J. Bacteriol.,* 73:455-464.

Henning, J., Dewey, R. E., Cutt, J. R., and Klessig, D. F. (1993). Pathogen, salicylic acid and developmental dependent expression of β-1,3-glucanase/GUS gene fusion in transgenic tobacco plants, *Plant J.,* 4:481-493.

Henning, J., Malamy, J., Grynkiewicz, G., and Klessig, D. F. (1993). Interconversion of the salicylic acid signal and its glucoside in tobacco, *Plant J.,* 4:593-600.

Henrissat, B., Heffron, S. E., Yoder, M. D., Lietzke, S. E., and Jurnak, F. (1995). Functional implications of structure-based sequence alignment of proteins in the extracellular pectate lyase super family, *Plant Physiol.,* 107:963-976.

Herbers, K., Conrads-Strauch, J., and Bonas, U. (1992). Race-specificity of plant resistance to bacterial spot disease determined by repetitive motifs in a bacterial avirulence protein, *Nature,* 356:172-174.

Herde, O., Atzorn, R., Fisahn, J., Wasternack, C., Willmitzer, L., and Pena-Cortes, H. (1996). Localized wounding by heat initiates the accumulation of proteinase inhibitor II in abscisic acid-deficient plants by triggering jasmonic acid biosynthesis, *Plant Physiol.,* 112:853-860.

Herlache, T. C., Hotchkiss, A. T., Burr, T. J., and Collmer, A. (1997). Characterization of the *Agrobacterium vitis pehA* gene and comparison of the encoded polygalacturonase with the homologous enzymes from *Erwinia carotovora* and *Ralstonia solanacearum, Appl. Environ. Microbiol.,* 63:338-346.

Herrera-Estrella, L., and Simpson, J. (1995). Genetically engineered resistance to bacterial and fungal pathogens, *World J. Microbiol. Biotechnol.,* 11:383-392.

Hetherington, A., and Trewavas, A. (1984). Activation of pea membrane protein kinase by calcium ions, *Planta,* 161:409-417.

Heu, S., and Hutcheson, S. W. (1991a). Molecular characterization of the *Pseudomonas syringae* pv. *syringae* 61 *hrmA* locus, *Phytopathology,* 81:1245.

Heu, S., and Hutcheson, S. W. (1991b). *Pseudomonas syringae* pv. *syringae hrp/hrm* genes encode avirulence functions in *P. syringae* pv. *glycinea* race4, *Phytopathology,* 81:702-703.

Heu, S., and Hutcheson, S. W. (1993). Nucleotide sequence and properties of the *hrmA* locus associated with the *Pseudomonas syringae* pv. *syringae hrp* gene cluster, *Mol. Plant-Microbe Interact.,* 6:553-564.

Higgins, C. F., Hiles, I. D., Salmond. G. P., Gill, D. R., Downie, J. A., Evans, I. J., Holland, I. B., Gray, L., Buckel, S. D., Bell, A. W., and Hermodson, M. A. (1986). A family of related ATP-binding subunits coupled to many distinct biological processes in bacteria, *Nature,* 323:448-450.

Higgins, V. J., and Smith, D. G. (1972). Separation and identification of two pterocarpanoid phytoalexins produced by red clover leaves, *Phytopathology,* 62:235-238.

Hightower, R., Baden, C., Penzes, E., and Dunsmuir, P. (1994). The expression of cecropin peptide in transgenic tobacco does not confer resistance to *Pseudomonas syringae* pv. *tabaci*, *Plant Cell Rep.,* 13:295-299.

Hignett, R. C., and Quirk, A. V. (1979). Properties of phytotoxic cell-wall components of plant pathogenic pseudomonads, *J. Gen. Microbiol.,* 110:77-81.

Hijwegen, T., Verhaar, M. A., and Zadoks, J. C. (1996). Resistance to *Sphaerotheca pannosa* in roses induced by 2,6-dichloroisonicotinic acid, *Plant Pathol.,* 45: 631-635.

Hildebrand, D. F. (1989). Lipoxygenases, *Physiol. Plant.,* 76:249-253.

Hildebrand, D. F., Hamilton-Kemp, T. R., Leggs, C. S., and Bookjans, G. (1988). Plant lipoxygenases: Occurrence, properties and possible functions, *Curr. Top. Plant Biochem. Physiol.,* 7:201-219.

Hildenbrand, S., and Ninnemann, H. (1994). Kinetics of phytoalexin accumulation in potato tubers of different genotypes infected with *Erwinia carotovora* ssp. *atroseptica, Physiol. Mol. Plant Pathol.,* 44:335-347.

Hille, A., Purwin, C., and Ebel, J. (1982). Induction of enzymes of phytoalexin synthesis in cultured soybean cells by an elicitor from *Phytophthora megasperma* f.sp. *glycinea, Plant Cell Rep.,* 1:123-127.

Hinsch, M., and Staskawicz, B. (1996). Identification of a new *Arabidopsis* disease resistance locus, RPS4, and cloning of the corresponding avirulence gene, *avrRps4,* from *Pseudomonas syringae* pv. *pisi, Mol. Plant-Microbe Interact.,* 9:55-61.

Hitchin, F. F., Jenner, C. E., Harper, S., Mansfield, J. W., Barber, C. E., and Daniels, M. J. (1989). Determinant of cultivar specific avirulence cloned from *Pseudomonas syringae* pv. *phaseolicola* race 3, *Physiol. Mol. Plant Pathol.,* 34:309-322.

Hodson, N., El-Masry, M. H., and Sigee, D. C. (1995a). Electron microscope studies on *Pseudomonas syringae* pv. *phaseolicola* isolated from inoculated tobacco leaves: Changes in flagellation, biomass, and elemental composition during the hypersensitive reaction, *J. Phytopathol.,* 143:361-367.

Hodson, A., Smith, A. R. W., and Hignett, R. C. (1995b). Characterization of an outer membrane preparation of *Pseudomonas syringae* pv. *mors-prunorum* and its biological activities in planta, *Physiol. Mol. Plant Pathol.,* 47:159-172.

Hodson, A., Smith, A. R. W., and Hignett, R. C. (1996). Outer membrane and rough lipopolysaccharide preparations of mutants of *Pseudomonas syringae* pv. *morsprunorum* C28 prevent hypersensitive necrosis and cause silvering in planta, *Physiol. Mol. Plant Pathol.,* 48:11-20.

Hoffland, E., Hakulinen, J., and Pelt, J. A. V. (1996). Comparison of systemic resistance induced by avirulent and nonpathogenic *Pseudomonas* species, *Phytopathology,* 86:757-762.

Hoffland, E., Pieterse, C. M. J., Bik, L., and Van Pelt, J.A. (1995). Induced systemic resistance is not associated with accumulation of pathogenesis-related proteins, *Physiol. Mol, Plant Pathol.,* 46:309-320.

Hoffman, R. M., and Turner, J. G. (1984).Occurrence and specificity of an endopoygalacturonase inhibitor in *Pisum sativum, Physiol. Plant Pathol.,* 24:49-59.

Hohn, B., Koukolikova-Nicola, Z., Bakkeren, G., and Grimsley, N. (1989). *Agrobacterium*-mediated gene transfer to monocots and dicots, *Genome,* 34:987-991.

Hoj, P. B., Hartman, D. J., Morrice, N. A., Doan, D. N. P., and Fincher, G. B. (1989). Purification of (1-3)-β-glucan endohydrolase isoenzyme II from germinated barley and determination of its primary structure from cDNA clone, *Plant Mol. Biol.,* 13:31-42.

Holliday, M. J., and Keen, N. T. (1982). The role of phytoalexins in the resistance of soybean leaves to bacteria: Effect of glyphosate on glyceollin accumulation, *Phytopathology,* 72:1470-1474.

Holton, T. A., and Cornish, E. C. (1995). Genetics and biochemistry of anthocyanin biosynthesis, *Plant Cell,* 7:1071-1083.

Hooft van Huijsduijnen, R. A. M., Kauffmann, S., Brederode, F. T., Cornelissen, B. J. C., Legrand, M., Fritig, B., and Bol, J. F. (1987). Homology between chitinases that are induced by TMV infection of tobacco, *Plant Mol. Biol.,* 9:411-420.

Hooft van Huijsduijnen, R. A. M., Van Loon, L. C., and Bol, J. F. (1986). Complementary DNA cloning of six messenger RNAs induced by TMV infection of tobacco and characterization of their translation products, *EMBO J.,* 5:2057-2061.

Hopkins, C. M., White, F. F., Choi, S. H., Guo, A., and Leach, J. E. (1992). Identification of a family of avirulense genes from *Xanthomonas oryzae* pv. *oryzae, Mol. Plant Microbe Interact.,* 5:451-459.

Horino, O., and Kaku, H. (1989). Defense mechanisms of rice against bacterial blight caused by *Xanthomonas campestris* pv. *oryzae.* In published proceedings, *Bacterial Blight in Rice,* International Rice Research Institute, Los Banos, Philippines, pp. 135-152.

Horn, M. A., Heinstein, P. F., and Low, P. S. (1989). Receptor-mediated endocytosis in plant cells, *Plant Cell,* 1:1003-1009.

Horns, T., and Bonas, U. (1996). The *rpoN* gene of *Xanthomonas campestris* pv. *vesicatoria* is not required for pathogenicity. *Mol. Plant-Microbe Interact.,* 9:856-859.

Hosokawa, D., and Ohashi, Y. (1988). Immunochemical localization of pathogenesis-related proteins secreted into the intercellular spaces of salicylate-treated tobacco leaves, *Plant Cell Physiol.,* 29:1035-1040.

Hotchkiss, A. T. Jr., Revear, L. G., and Hicks, K. B. (1996). Substrate polymerization pattern of *Pseudomonas viridiflava* SF-312 pectate lyase, *Physiol. Mol. Plant Pathol.,* 48:1-9.

Howe, G. A., Lightner, J., Browse, J., and Ryan, C. A. (1996). An octadecanoid pathway mutant (JL5) of tomato is compromised in signalling for defense against insect attack, *Plant Cell,* 8:2067-2077.

Hoyos, M. E., Stanley, C. M., He, S. Y., Pike, S., Pu, X.-A., and Novacky, A. (1996). The interaction of harpin$_{Pss}$ with plant cell walls, *Mol. Plant-Microbe Interact.,* 9:608-616.

Hrabak, E. M., and Willis, D. K. (1990). Molecular analysis of a locus from *Pseudomonas syringae* pv. *syringae* required for lesion formation, *Phytopathology,* 80:984.

Hrabak, E. M., and Willis, D. K. (1992). The *lemA* gene required for pathogenicity of *Pseudomonas syringae* pv. *syringae* on bean is a member of a family of two-component regulators, *J. Bacteriol.,* 174:3011-3020.

Hrabak, E. M., and Willis, D. K. (1993). Involvement of the *lemA* gene in production of syringomycin and protease by *Pseudomonas syringae* pv. *syringae, Mol. Plant-Microbe Interact.,* 6:368-375.

Hrubcova, M., Cvikrova, M., and Eder, J. (1994). Peroxidase activities and contents of phenolic acids in embryogenic and nonembryogenic alfalfa cell suspension cultures, *Biol. Plantarum,* 36:175-182.

Hu, C., Smith, R., and van Huystee, R. (1989). Biosynthesis and localization of peanut peroxidases. A comparison of the cationic and anionic isozymes, *J. Plant Physiol.,* 135:391-397.

Hu, N., Hung, M., Liao, C. T., and Lin, M. (1995). Subcellular location of XpsD, a protein required for extracellular protein secretion by *Xanthomonas campestris* pv. *campestris, Microbiology (Reading),* 141:1395-1406.

Hu, X., and Reddy, A. S. N. (1995). Nucleotide sequence of a cDNA clone encoding a thaumatin-like protein from *Arabidopsis, Plant Physiol.,* 107:305-306.

Huang, H.-C., He, S. Y., Bauer, D. W., and Collmer, A. (1992). The *Pseudomonas syringae* pv. *syringae* 61 *hrpH* product: An envelope protein required for elicitation of the hypersensitive response in plants, *J. Bacteriol.,* 174:6878-6885.

Huang, H.-C., Hutcheson, S. W., and Collmer, A. (1990). *TnphoA* tagging of *Pseudomonas syringae* pv. *syringae hrp* genes encoding potentially exported proteins, *Phytopathology,* 80:984 (Abstr.).

Huang, H.-C., Hutcheson, S. W., and Collmer, A. (1991). Characterization of the *hrp* cluster from *Pseudomonas syringae* pv. *syringae* 61 and Tn*phoA* tagging of genes encoding exported or membrane-spanning Hrp proteins, *Mol. Plant-Microbe Interact.,* 4:469-476.

Huang, H.-C., Lin, R. H., Chang, C. J., Collmer, A., and Deng, W. L. (1995). The complete *hrp* gene cluster of *Pseudomonas syringae* 61 includes two blocks of genes required for harpin$_{Pss}$ secretion that are arranged colinearly with *Yersinia ysc* homologs, *Mol. Plant-Microbe Interact.,* 5:733-746.

Huang, H.-C., Shuurink, R., Denny, T. P., Atkinson, M. M., Baker, C. J., Yucel, I., Hutcheson, S. W., and Collmer, A. (1988). Molecular cloning of a *Pseudomonas syringae* pv. *syringae* gene cluster that enables *Pseudomonas fluorescens* to elicit hypersensitive response in tobacco plants, *J. Bacteriol.,* 170:4748-4756.

Huang, H.-C., Xiao, Y., Lin, R.-H., Lu, Y., Hutcheson, S. W., and Collmer, A. (1993). Characterization of the *Pseudomonas syringae* pv. *syringae hrpJ* and *hrpI* genes: Homology of HrpI to a superfamily of proteins associated with protein translocation, *Mol. Plant-Microbe Interact.,* 6:515-520.

Huang, J. S. (1986). Ultrastructure of bacterial penetration in plants, *Annu. Rev. Phytopathol.,* 24:141-157.

Huang, J. S., and Goodman, R. N. (1970). The relationship of phosphatidase activity to the hypersensitive reaction in tobacco induced by bacteria, *Phytopathology,* 60:1020-1021.

Huang, Y., Helgeson, J. P., and Sequeira, L. (1989). Isolation and purification of a factor from *Pseudomonas solanacearum* that induces a hypersensitive-like response in potato cells, *Mol. Plant-Microbe Interact.,* 2:132-138.

Huang, Y., Morel, P., Powell, B., and Kado, C. (1990a). VirA, a coregulator of Ti-specified virulence genes, is phosphorylated in vitro, *J. Bacteriol.,* 172:1142-1144.

Huang, Y., Nordeen, R. O., Di, M., Owens, L. D., and McBeath, J. H. (1997). Expression of an engineered cecropin gene cassette in transgenic tobacco plants confers disease resistance to *Pseudomonas syringae* pv. *tabaci, Phytopathology,* 87:494-499.

Huang, Y., Xu, P., and Sequeira, L., (1990b). A second cluster of genes that specify pathogenicity and host response in *Pseudomonas solanacearum, Mol. Plant-Microbe Interact.,* 3:48-53.

Hugouvieux-Cotte-Pattat, N., Condemine, G., Nasser, W., and Reverchon, S. (1996). Regulation of pectinolysis in *Erwinia chrysanthemi, Annu. Rev. Microbiol.,* 50:213-257.

Hugouvieux-Cotte-Pattat, N., Dominguez, H., and Robert-Baudouy, J. (1992). Environmental conditions affect transcription of the pectinase genes of *Erwinia chrysanthemi* 3937, *J. Bacteriol.,* 174:7807-7818.

Hugouvieux-Cotte-Pattat, N., Quesneau, Y., and Robert-Baudouy, J. (1983). Aldohexuronate transport system in *Erwinia carotovora, J. Bacteriol.,* 154:663-668.

Huguet, E., and Bonas, U. (1997). *hrpF* of *Xanthomonas campestris* pv. *vesicatoria* encodes an 87-kDa protein with homology to NolX of *Rhizobium fredii, Mol. Plant-Microbe Interact.,* 10:488-498.

Hunt, M. D., Delaney, T. P., Dietrich, R. A., Weymann, K. B., Dangl, J. L., and Ryals, J. A. (1997). Salicylate–independent lesion formation in *Arabidopsis lsd* mutants, *Mol. Plant-Microbe Interact.,* 10:531-536.

Hunt, M. D., and Ryals, J. A. (1996). Systemic acquired resistance signal transduction, *Crit. Rev. Plant Sci.,* 15:583-606.

Hutcheson, S. W., Collmer, A., and Baker, C. J. (1989). Elicitation of the hypersensitive response by *Pseudomonas syringae, Physiol. Plant.,* 76:155-163.

Huynh, Q. K., Hironaka, C. M., Levine, E. B., Smith, C. E., Borgmeyer, J. R., and Shah, D. M. (1992). Antifungal proteins from plants. Purification, molecular cloning and antifungal properties of chitinases from maize seed, *J. Biol. Chem.*, 267:6635-6640.

Huynh, T. V., Dahlbeck, D., and Staskawicz, B. J. (1989). Bacterial blight of soybean: Regulation of a pathogen gene determining host cultivar specificity, *Science,* 245:1374-1377.

Hwang, B. K., Yoon, J.Y., Ibenthal, W. D., and Heitefuss, R. (1991). Soluble proteins, esterases and superoxide dismutase in stem tissue of pepper plants in relation to age-related resistance to *Phytophthora capsici, J. Phytopathol.,* 132:129-138.

Hwang, I., Lim, S.M., and Shaw, P. D. (1992). Cloning and characterization of pathogenicity genes from *Xanthomonas campestris* pv. *glycines, J. Bacteriol.,* 174: 1923-1931.

Ingham, J. L. (1973). Disease resistance in higher plants: The concept of pre-infectional and post-infectional resistance, *Phytopathol. Z.,* 78:314-335.

Innes, R. W., Bent, A. F., Kunkel, B. N., Bisgrove, S. R., and Staskawicz, B. J. (1993a). Molecular analysis of avirulence gene *avrRpt2* and identification of a putative regulatory sequence common to all known *Pseudomonas syringae* avirulence genes, *J. Bacteriol.,* 175:4859-4869.

Innes, R. W., Bisgrove, S. R., Smith, N. M., Bent, A. F., Staskawicz , B. J., and Liu, Y. C. (1993b). Identification of a disease resistance locus in *Arabidopsis* that is functionally homologous to the RPG1 locus of soybean, *Plant J.,* 4:813-820.

Irving, H. R., and Kuc, J. (1990). Local and systemic induction of peroxidase, chitinase and resistance in cucumber plants by K_2HPO_4, *Physiol. Mol. Plant Pathol.,* 37:355-366.

Ishige, F., Mori, H., Yamazaki, K., and Imaseki, H. (1993). Identification of a basic glycoprotein induced by ethylene in primary leaves of azuki bean as a cationic peroxidase, *Plant Physiol.,* 101:193-199.

Ishikawa, A., Tsubouchi, H., Iwasaki, Y., and Asahi, T. (1995). Molecular cloning and characterization of cDNA for the α subunit of a G protein from rice, *Plant Cell Physiol.,* 36:353-359.

Iturriaga, E. A., Leech, M. J., Paul Barratt, D. H., and Wang, T. L. (1994). Two ABA-responsive proteins from pea (*Pisum sativum* L.) are closely related to intracellular pathogenesis-related proteins, *Plant Mol. Biol.,* 24:235-240.

Jach, G., Gornhardt, B., Mundy, J., Logemann, J., Pinsdorf, E., Leah, R., Schell, J., and Maas, C. (1995). Enhanced quantiative resistance against fungal disease by combinatorial expression of different barley antifungal proteins in transgenic tobacco, *Plant J.,* 8:97-109.

Jacobsen, S., Mikkelsen, J. D., and Hejgaard, J. (1990). Characterization of two antifungal endochitinases from barley grain, *Physiol. Plant.,* 79:554-562.

Jahnen, W., and Hahlbrock, K. (1988). Cellular localization of nonhost resistance reactions of parsley *(Petroselinum crispum)* to fungal infection, *Planta,* 173:197-204.

Jakobek, J. L., and Lindgren, P. B. (1990). Hrp mutants of *Pseudomonas syringae* pv. *tabaci* activate the transcription of genes associated with disease resistance in bean, *Phytopathology,* 80:1010 (Abstr.).

Jakobek, J. L., and Lindgren, P. B. (1993). Generalized induction of defense responses in bean is not correlated with the induction of the hypersensitive response, *Plant Cell,* 5:49-56.

Jakobek, J. L., Smith, J. A., and Lindgren, P. B. (1993). Suppression of bean defense responses by *Pseudomonas syringae, Plant Cell,* 5:57-63.

Jalali, B. L., Singh, G., and Grover, R. K. (1976). Role of phenolics in bacterial blight resistance in cotton, *Acta Phytopath. Acad. Sinica Hungaria,* 11:81-83.

James, V., and Hugouvieux-Cotte-Pattat, N. (1996). Regulatory systems modulating the transcription of the pectinase genes of *Erwinia chrysanthemi* are conserved in *Escherichia coli, Microbiology (Reading),* 142:2613-2619.

Jann, K., and Jann, B. (1990). Bacterial adhesins, *Curr. Topics Microbiol. Immunol.* 151, Berlin/Heidelberg, Springer Verlag, pp. 209.

Jauneau, A., Cabin-Flaman, A., Verdus, M.-C., Ripoll, C., and Thellier, M. (1994). Involvement of calcium in the inhibition of endopoygalacturonase activity in epidermis cell wall of *Linum usitatissimum, Plant Physiol. Biochem.,* 32:839-846.

Jauris, S., Rucknagel, K. P., Schwartz, W. H., Kratzsch, P., Bromenmeier, K., and Staudenbauer, W. L. (1990). Sequence analysis of the *Clostridium stercorarium celZ* gene encoding a thermoactive cellulase (Avicellulase I): Identification of catalytic and cellulose binding domains, *Mol. Gen. Genet.,* 223:258-267.

Jayaswal, R. K., Veluthambi, K., Gelvin, S. B., and Slightom, J. L. (1987). Double-stranded cleavage of T-DNA and generation of single-stranded T-DNA molecules in *Escherichia coli* by a *virD*-encoded border-specific endonuclease from *Agrobacterium tumefaciens, J. Bacteriol.,* 169:5035-5045.

Jaynes, J. M. (1990). Lytic peptides portend an innovative age in the management and treatment of human disease, *Drug News Perspect.,* 3:69-78.

Jaynes, J. M., Nagpala, P., Destefano-Beltran, L., Huang, J. H., Kim, J. H., Denny, T., and Cetiner, S. (1993). Expression of a cecropin-B lytic peptide analog in transgenic tobacco confers enhanced resistance to bacterial wilt caused by *Pseudomonas solanacearum, Plant Sci.,* 89:43-53.

Jenner, C., Hitchin, E., Mansfield, J., Walters, K., Betteridge, P., Teverson, D., and Taylor, J. (1991). Gene-for-gene interactions between *Pseudomonas syringae* pv. *phaseolicola* and *Phaseolus, Mol. Plant-Microbe Interact.,* 4:553-562.

Jensen, L. G., Olsen, O., Kops, O., Wolf, N., Thomsen, K. K., and Wettstein, D. (1996). Transgenic barley expressing a protein-engineered, thermostable β-1,3-β-1,4-glucanase during germination, *Proc. Natl. Acad. Sci. U.S.A.,* 93:3487-3491.

Ji, C., and Kuc, J. (1995). Purification and characterization of an acidic β-1,3-glucanase from cucumber and its relationship to systemic disease resistance induced by *Colletotrichum lagenarium* and tobacco necrosis virus, *Mol. Plant-Microbe Interact.,* 8:899-905.

Ji, C., and Kuc, J. (1996). Antifungal activity of cucumber β-1,3-glucanase and chitinase, *Physiol. Mol. Plant Pathol.,* 49:257-265.

Jia, Y., Loh, Y. T., Zhou, J., and Martin, G. B. (1997). Alleles of *Pto* and *Fen* occur in bacterial speck-susceptible and fenthion-insensitive tomato cultivars and encode active protein kinases, *Plant Cell,* 9:61-73.

Jin, S.-G., Komari, T., Gordon, M. P., and Nester, E. W. (1987). Genes responsible for the supervirulence phenotype of *Agrobacterium tumefaciens* A281, *J. Bacteriol.,* 169:4417-4425.

Jin, S., Prusti, R. W., Roitsch, T., Ankenbauer, R. G., and Nester, E. W. (1990a). Phosphorylation of the VirG protein of *Agrobacterium tumefaciens* by the autophosphorylated VirA protein: Essential role in the biological activity of VirG, *J. Bacteriol.,* 172:4945-4950.

Jin, S., Roitsch, T., Ankenbauer, R., Gordon, M., and Nester, E. (1990b). The VirA protein of *Agrobacterium tumefaciens* is autophosphorylated and is essential for *vir* gene regulation, *J. Bacteriol.,* 172:525-530.

Jin, S., Roitsch, T., Christie, P. J., and Nester, E. W. (1990c). The regulatory VirG protein specifically binds to a *cis*-acting regulatory sequence involved in transcriptional activation of *Agrobacterium tumefaciens* virulence genes, *J. Bacteriol.,* 172:531-537.

Jin, S.-G., Song, Y. N., Pan, S. Q., and Nester, E. W. (1993). Characterization of a *virG* mutation that confers constitutive virulence gene expression in *Agrobacterium, Mol. Microbiol.,* 7:555-562.

Jing, Y., and Yang, H. S. (1996). The *Pseudomonas syringae hrp* regulation and secretion of multiple extracellular proteins, *J. Bacteriol.,* 178:6399-6402.

Johal, G. S., and Briggs, S. P. (1992). Reductase activity encoded by the HM1 disease resistance gene in maize, *Science,* 258:985-987.

Johannes, E., Brosnan, J. M., and Sanders, D. (1991). Calcium channels and signal transduction in cells, *Bioessay,* 13:331-336.

Johnson, P. F., and Mcknight, S. L. (1989). Eukaryotic transcriptional regulatory proteins, *Annu. Rev. Biochem.,* 58:799-839.

Johnston, D. J., Ramanathan, V., and Williamson, B. (1993). A protein from immature raspberry fruits which inhibits endopolygalacturonases from *Botrytis cinerea* and other microorganisms, *J. Expt. Bot.,* 44:971-976.

Jolles, P., and Jolles, J. (1984). What's new in lysozyme research?, *Mol. Cell Biochem.,* 63:165-189.

Jones, D. A., Thomas, C. A., Hammond-Kosack, K. E., Baliant-Kurti, P. A., and Jones, J. D. G. (1994). Isolation of the tomato Cf-9gene for resistance to *Cladosporium fulvum* by transposon tagging, *Science,* 266:789-793.

Jones, R. G., and Lunt, O. R. (1967). The function of calcium in plants, *Botanical Reviews,* 33:407-426.

Jones, S., Yu, B., Bainton, N. J., Birdsall, M., Bycroft, B. W., Chhabra, S. R., Cox, A. J. R., Golby, P., Reeves, P. J., Stephens, S., Winson, M. K., Salmond, G. P. C., Stewart, G. S. A. B., and Williams, P. (1993). The lux autoinducer regulates the production of exoenzyme virulence determinants in *Erwinia carotovora* and *Pseudomonas aeruginosa, EMBO J.,* 12:2477-2482.

Jongedijk, E., Tigelaar, H., Van Roekel, J. S. C., Bres-Vloemans, S. A., Dekker, L., Vanden Elzen, P. J. M., Cornelissen, B. J. C., and Melchers, L. S. (1995). Synergistic activity of chitinases and β-1,3-glucanases enhances fungal resistance in transgenic tomato plants, *Euphytica,* 85:173-175.

Joos, H. J., and Hahlbrock, K. (1992). Phenylalanine ammonia-lyase in potato (*Solanum tuberosum* L.), *Eur. J. Biochem.,* 204:621-629.

Joosten, M. H. A. J., Bergmans, C. J. B., Meulenhoff, E. J. S., Cornelissen, B. J. C., and de Wit, P. J. G. M. (1990). Purification and serological characterization of three basic 15 kD pathogenesis-related proteins from tomato, *Plant Physiol.,* 94:585-591.

Jorgensen, R. (1990). Altered gene expression in plants due to *trans* interactions between homologous genes, *Trends Biotechnol.,* 8:340-344.

Jung, J. L., Fritig, B., and Hahne, G. (1993). Sunflower (*Helianthus annuus* L.) pathogenesis-related proteins: Induction by aspirin (acetyl salicylic acid) and characterization, *Plant Physiol.,* 101:873-880.

Junghans, H., Dalkin, K., and Dixon, R. A. (1993). Stress responses in alfalfa (*Medicago sativa* L.). 15. Characterization and expression patterns of members of a subset of the chalcone synthase multigene family, *Plant Mol. Biol.,* 20:167-170.

Kamada, Y., and Muto, S. (1994a). Proteinase inhibitors inhibit stimulation of inositol phospholipid turnover and induction of phenylalanine ammonia-lyase in fungal elicitor-treated tobacco suspension culture cells, *Plant Cell Physiol.,* 35:405-409.

Kamada, Y., and Muto, S. (1994b). Stimulation by fungal elicitor of inositol phospholipid turnover in tobacco suspension culture cells, *Plant Cell Physiol.,* 35:397-404.

Kamoun, S., and Kado, C. I. (1990). A plant inducible gene of *Xanthomonas campestris* pv. *campestris* encodes an exocellular component required for growth in the host and hypersensitivity on nonhosts, *J. Bacteriol.,* 172:5165-5172.

Kamoun, S., Kamdar, H. V., Tola, E., and Kado, C. I. (1992). Incompatible interactions between crucifers and *Xanthomonas campestris* involve a vascular hypersensitive response: Role of the *hrpX* locus, *Mol. Plant-Microbe Interact.,* 5:22-33.

Karlempi, S. O., Airaksinen, K., Miettinen, A. T. E., Kokko, H. I., Holopainen, J. K., Karlelampi, L. V., and Karjalainen, R. O. (1994). Pathogenesis-related proteins in ozone-exposed Norway spruce (*Picea abies* (Karst) L.), *New Phytologist,* 126:81-99.

Karunanandaa, B., Singh, A., and Kao, T. H. (1994). Characterization of a predominantly pistil-expressed gene encoding a γ-thionin-like protein of *Petunia inflata, Plant Mol. Biol.,* 26:459-464.

Kauffmann, S., Legrand, M., Geoffroy, P., and Fritig, B. (1987). Biological function of "pathogenesis-related" proteins: Four PR proteins of tobacco have β-1,3-glucanase activity, *EMBO J.,* 6:3209-3212.

Kauss, H. (1987). Some aspects of calcium-dependent regulation in plant metabolism, *Annu. Rev. Plant Physiol.,* 38:47-72.

Kauss, H., Jeblick, W., Ziegler, J., and Krabler, W. (1994). Pretreatment of parsley (*Petroselinum crispum* L.) suspension cultures with methyl jasmonate enhances elicitation of activated oxygen species, *Plant Physiol.,* 105:89-94.

Kauss, H., Waldmann, T., Jeblick, N., Euler, G., Ranjeva, D., and Domard, A. (1990). Ca^{2+} is an important but not the only signal in callose synthesis induced by chitosan, saponins, and polyene antibiotics. In B. J. J. Lugteberg (Ed.), *Signal molecules in plant-microbe interactions,* Springer-Verlag, Berlin, pp. 107-116.

Kauss, H., Waldmann, T., Jeblick, W., and Takemoto, J.Y. (1991). The phytotoxin syringomycin elicits Ca^{2+}-dependent callose synthesis in suspension-cultured cells of *Catharanthus roseus, Physiol. Plant.*, 81:134-138.

Kawamata, S., Yamada, T., Tanaka, Y., Sriprasertsak, P., Kato, H., Ichinose, Y., Shiraishi, T., and Oku, H. (1992). Molecular cloning of phenylalanine ammonia-lyase cDNA from *Pisum sativum, Plant Mol. Biol.*, 20:167-170.

Kearney, B., Ronald, P. C., Dahlbeck, D., and Staskawicz, B. J. (1988). Molecular basis for evasion of plant host defense in bacterial spot disease of pepper, *Nature*, 332:541-543.

Kearney, B., and Staskawicz, B. J. (1990). Widespread distribution and fitness contribution of the *Xanthomonas campestris* avirulence gene *avrBs2, Nature*, 346:385-386.

Keefe, D., Hinz, U., and Meins, F. Jr. (1990). The effect of ethylene on the cell-type-specific and intracellular localization of β-1,3-glucanase and chitinase in tobacco leaves, *Planta*, 182:43-51.

Keen, N. T. (1990). Gene-for-gene complementarity in plant-pathogen interactions, *Annu. Rev. Genet.*, 24:447-463.

Keen, N. T. (1992). The molecular biology of disease resistance, *Plant Mol. Biol.*, 19:109-122.

Keen, N. T., Boyd, C., and Henrissat, B. (1996). Cloning and characterization of a xylanase gene from corn strains of *Erwinia chrysanthemi, Mol. Plant-Microbe Interact.*, 9:651-657.

Keen, N. T., and Buzzell, R. I. (1991). New disease resistance genes in soybean against *Pseudomonas syringae* pv. *glycinea:* Evidence that one of them interacts with a bacterial elicitor, *Theor. and Appl. Genet.*, 81:133-138.

Keen, N. T., Dahlbeck, D., Staskawicz, B., and Belser, W. (1987). Molecular cloning of pectate lyase genes from *Erwinia chrysanthemi* and their expression in *Escherichia coli, J. Bacteriol.*, 159:825-831.

Keen, N. T., and Kennedy, B. W. (1974). Hydroxyphaseollin and related isoflavanoids in the hypersensitive resistance reaction of soybeans to *Pseudomonas glycinea, Physiol. Plant Pathol.*, 4:173-185.

Keen, N. T., and Legrand, M. (1980). Surface of glycoproteins: Evidence that may function as the race specific phytoalexin elicitors of *Phytophthora megasperma* f. sp. *glycinea, Physiol. Plant Pathol.*, 17:175-192.

Keen, N.T., and Staskawicz, B. (1988). Host range determinants in plant pathogens and symbionts, *Annu. Rev. Microbiol.*, 42:421-440.

Keen, N. T., Tamaki, S., Kobayashi, D., Gerhold, D., Stayton, M., Shen, H., Gold, S., Lorang, J., Thordal-Christensen, H., Dahlbeck, D., and Staskawicz, B. (1990). Bacteria expressing avirulence gene D produce a specific elicitor of the soybean hypersensitive reaction, *Mol. Plant-Microbe Interact.*, 3:112-121.

Keith, L. W., Boyd, C., Keen, N. T., and Partidge, J. E. (1997). Comparison of *avrD* alleles from *Pseudomonas syringae* pv. *glycinea, Mol. Plant-Microbe Interact.*, 10:416-422.

Kelemu, S., and Leach, J. E. (1990). Cloning and characterization of an avirulence gene from *Xanthomonas campestris* pv. *oryzae, Mol. Plant-Microbe Interact.*, 3:59-65.

Keller, H., Blein, J. P., Bonnet, P., and Ricci, P. (1996a). Physiological and molecular characteristics of elicitin-induced systemic acquired resistance in tobacco, *Plant Physiol.,* 110:365-376.

Keller, H., Bonnet, P., Galiana, E., Pruvot, L., Freidrich, L., Ryals, J., and Ricci, P. (1996b). Salicylic acid mediates elicitin-induced systemic acquired resistance, but not necrosis in tobacco, *Mol. Plant-Microbe Interact.,* 9:696-703.

Kellmann, J. W., Kleinow, T., Engelhardt, K., and Philipp, C. (1996). Characterization of two class II chitinase genes from peanut and expression studies in transgenic tobacco plants, *Plant Mol. Biol.,* 30:351-358.

Kemner, J. M., Liang, X. Y., and Nester, E. W. (1997). The *Agrobacterium tumefaciens* virulence gene chvE is part of a putative ABC-type sugar transport operon, *J. Bacteriol.,* 179:2452-2458.

Keppler, L. D., Atkinson, M. M., and Baker, C. J. (1989). Active oxygen production during a bacteria-induced hypersensitive reaction in tobacco suspension cells, *Phytopathology,* 79:974-978.

Keppler, L. D., and Baker, C. J. (1989). O_2^--initiated lipid peroxidation in a bacteria-induced hypersensitive reaction in tobacco cell suspensions, *Phytopathology,* 79:555-562.

Keppler, L. D., Baker, C. J., and Atkinson, M. M. (1989). Active oxygen production during a bacteria-induced hypersensitive reaction in tobacco suspension cells, *Phytopathology,* 79:974-978.

Keppler, L. D., and Novacky, A. (1986). Involvement of membrane lipid peroxidation in the development of a bacterially induced hypersensitive reaction, *Phytopathology,* 76:104-108.

Keppler, L. D., and Novacky, A. (1987). The initiation of membrane lipid peroxidation during bacteria-induced hypersensitive reaction, *Physiol. Mol. Plant Pathol.,* 30:233-245.

Keppler, L. D., and Novacky, A. (1989). Changes in cucumber cotyledon membrane lipid fatty acids during paraquat treatment and a bacteria-induced hypersensitive reaction, *Phytopathology,* 79:705-708.

Kernan, A., and Thornburg, R. W. (1989). Auxin levels regulate the expression of a wound-inducible proteinase inhibitor II-chloramphenicol acyl transferase gene fusion in vitro and in vivo, *Plant Physiol.,* 91:73-78.

Kerr, L. D., Inoue, J., and Verma, I. M. (1992). Signal transduction: The nuclear target, *Current Opinion Cell Biol.,* 4:496-501.

Kessmann, H., and Barz, W. (1986). Elicitation and suppression of phytoalexin and isoflavone accumulation in cotyledons of *Cicer arietinum* L. as caused by wounding and by polymeric components from the fungus, *Ascochyta rabiei, J. Phytopathol.,* 117:321-335.

Kessmann, H., Choudhary, A. D., and Dixon, R. A. (1990). Stress responses in alfalfa (*Medicago sativa* L.). III. Induction of medicarpin and cytochrome P 450 enzyme activities in elicitor-treated cell suspension cultures and protoplasts, *Plant Cell Rep.,* 9:38-41.

Kessmann, H., Staub, T., Hoffmann, C., Maetzke, T., Herzog, J., Ward, E., Uknes, S., and Ryals, J. (1994a). Induction of systemic acquired disease resistance in plants by chemicals, *Annu. Rev. Phytopathol.,* 32:439-459.

Kessmann, H., Staub, T., Ligon, J., Oostendorp, M., and Ryals, J. (1994b). Activation of systemic acquired resistance in plants, *Eur. J. Plant Pathol.,* 100:359-369.

Kessmann, H., Staub, T., and Ryals, J. (1996). Benzothiadiazole induces disease resistance in Arabidopsis by activation of the systemic acquired resistance signal transdyction pathway, *Plant J.,* 10:71-82.

Khush, G. S., Bacalangco, E., and Ogawa, T. (1990). A new gene for resistance to bacterial blight from *O. longistaminata, Rice Genetics Newslett.,* 7:121-122.

Kiedrowski, S., Kawalleck, P., Hahlbrock, K., Somssich, I. E., and Dangl, J. (1992). Rapid activation of a novel plant defense gene is strictly dependent on the *Arabidopsis RPM1* disease resistance locus, *EMBO J.,* 11:4677-4684.

Kim, J. D., and Hwang, B. K. (1995). Differential induction of pathogenesis-related proteins in the compatible and incompatible interactions of tomato leaves with *Xanthomonas campestris* pv. *vesicatoria, Korean J. Plant Pathology,* 11:53-60.

Kim, J. F., Wei, Z. M., and Beer, S. V. (1997). The *hrpA* and *hrpC* operons of *Erwinia amylovora* encode components of a type III pathway that secretes harpin, *J. Bacteriol.,* 179:1690-1697.

Kim, S. G., and Yoo, J. Y. (1992). Changes in protein patterns resulting from infection of rice leaves with *Xanthomonas oryzae* pv. *oryzae, Mol. Plant-Microbe Interact.,* 5:356-360.

Kim, S. H., Terry, M. E., Hoops, P., Dauwalder, M., and Roux, S. J. (1988). Production and characterization of monoclonal antibodies to wall-localized peroxidases from corn seedlings, *Plant Physiol.,* 88:1446-1453.

Kim, S. R., Choi, J. L., Costa, M. A., and An, G. (1992). Identification of G-box sequence as an essential element for methyl jasmonate response of potato proteinase inhibitor II promoter, *Plant Physiol.,* 99:627-631.

Kim, Y. J., and Hwang, B. K. (1996). Purification, N-terminal amino acid sequencing and antifungal activity of chitinases from pepper stems treated with mercuric chloride, *Physiol. Mol. Plant Pathol.,* 48:417-432.

Kim, Y. J., and Hwang, B. K. (1997). Isolation of a basic 34 kilodalton β-1,3-glucanase with inhibitory activity against *Phytophthroa capsici* from pepper stems, *Physiol. Mol. Plant Pathol.,* 50:103-115.

King, G. J., Turner, V. A., Hussey, C. E., Wurtele, E. S., and Lee, S. M. (1988). Isolation and characterization of a tomato cDNA clone which codes for a salt-induced protein, *Plant Mol. Biol.,* 10:401-412.

Kiraly, Z., El-Zahaby, H. M., and Klement, Z. (1997). Role of extracellular polysaccharide (EPS) slime of plant pathogenic bacteria in protecting cells to reactive oxygen species, *J. Phytopathol.,* 145:59-68.

Kiraly, Z., Ersek, T., Barna, B., Adam, A., and Gullner, G. (1991). Pathophysiological aspects of plant disease resistance, *Acta Phytopath. Hung.,* 26:233-250.

Kishore, G. M., and Somerville, C. R. (1993). Genetic engineering of commercially useful biosynthetic pathways in transgenic plants, *Current Opinion in Biotechnology,* 4:152-158.

Kita, N., Boyd, C. M., Garrett, M. R., Jurnak, F., and Keen, N. T. (1996). Differential effect of site-directed mutations in *pelC* on pectate lyase activity, plant tissue maceration, and elicitor activity, *J. Biol. Chem.,* 271:26529-26535.

Kitten, T., and Willis, D. K. (1996). Suppression of a sensor kinase-dependent phenotype in *Pseudomonas syringae* by ribosomal proteins L35 and L20, *J. Bacteriol.,* 178:1548-1555.

Klement, Z. (1982). Hypersensitivity. In M. S. Mount and G. H. Lucy, (Eds.), *Phytopathogenic Prokaryotes,* Vol. 2, Academic Press, New York, pp. 149-179.

Kloepper, J. W., Tuzun, S., and Kuc, J. A. (1992). Proposed definitions related to induced disease resistance, *Biocontrol Sci. Technol.,* 2:349-351.

Kloepper, J. W., Tuzun, S., Liu, L., and Wei, G. (1993). Plant-growth promoting rhizobacteria as inducers of systemic resistance. In R.D. Lumsden and J.L. Waughn (Eds.), *Pest Management: Biologically Based Technologies,* ASC Cong. Proc. Series, American Chemical Society Press, Washington, DC, pp. 156-165.

Klotz, M. G., and Hutcheson, S. W. (1992). Multiple periplasmic catalases in phytopathogenic strains of *Pseudomonas syringae, Appl. Environ. Microbiol.,* 58:2468-2473.

Klotz, M. G., Kim, Y. C., Katsuwon, J., and Anderson, A. J. (1995). Cloning, characterization and phenotypic expression in *Escherichia coli* of *catF,* which encodes the catalytic subunit of catalase isozyme CatF of *Pseudomonas syringae, Appl. Microbiol. Biotechnol.,* 43:656-666.

Kmecl, A., Mauch, F., Winzeler, M., and Dudler, R. (1995). Quantitative field resistance of wheat to powdery mildew and defense reactions at the seedling stage: Identification of potential markers, *Physiol. Mol. Plant Pathol.,* 47:185-199.

Knoester, M., Van Loon, L. C., Den Heuvel, J. V., Hennig, J., Bol, J. F., and Linthorst, H. J. M. (1998). Ethylene-insensitive tobacco lacks nonhost resistance against soil-borne fungi, *Proc. Natl. Acad. Sci. U.S.A.,* 95:1933-1937.

Knogge, W., Kombrink, E., Schmelzer, E., and Hahlbrock, K. (1987). Occurrence of phytoalexins and other putative defense-related substances in uninfected parsley plants, *Planta,* 174:279-287.

Knoop, V., Staskawicz, B., and Bonas, U. (1991). Expression of the avirulence gene *avrBs3* from *Xanthomonas campestris* pv. *vesicatoria* is not under the control of *hrp* genes and is independent of plant factors, *J. Bacteriol.,* 173:7142-7150.

Kobayashi, D. Y., Tamaki, S. J., and Keen, N. T. (1989). Cloned avirulence genes from the tomato pathogen *Pseudomonas syringae* pv. *tomato* confer cultivar specificity on soybean, *Proc. Natl. Acad. Sci. U. S. A.,* 86:157-161.

Kobayashi, D. Y., Tamaki, S. J., Trollinger, D. J., Gold, S., and Keen, N.T. (1990a). A gene from *Pseudomonas syringae* pv. *glycinea* with homology to avirulence gene D from *P. s.* pv. *tomato* but devoid of the avirulence phenotype, *Mol. Plant-Microbe Interact.,* 3:103-111.

Kobayashi, D. Y., Tamaki, S. J., and Keen, N. T. (1990b). Molecular characterization of avirulence gene D from *Pseudomonas syringae* pv. *tomato, Mol. Plant-Microbe Interact.,* 3:94-102.

Kobe, B., and Deisenhofer, J. (1993). Crystal structure of porcine ribonuclease inhibitor, a protein with leucine-rich repeats, *Nature,* 366:751-756.

Koch, E., Meier, B. M., Eiben, H. G., and Slusarenko, A. (1992). A lipoxygenase from leaves of tomato (*Lycopersicon esculentum* Mill.) is induced in response to plant pathogenic pseudomonads, *Plant Physiol.,* 99:571-575.

Kodama, O., Miyakawa, J., Akatsuka, T., and Kiyosawa, S. (1992). Sakuranetin, a flavanone phytoalexin from ultraviolet-irradiated rice leaves, *Phytochemistry,* 31:3807-3809.

Kodama, O., Suzuki, T., Miyakawa, J., and Akatsuka, T. (1988). Ultraviolet-induced accumulation of phytoalexins in rice leaves, *Agric. Biol. Chem.,* 52: 2469-2473.

Kogel, K-H., Beckhove, U., Dreschers, J., Munch, S., and Romme, Y. (1994). Acquired resistance in barley. The resistance mechanism induced by 2,6-dichloroisonicotinic acid is a phenocopy of a genetically based mechanism governing race-specific powdery mildew resistance, *Plant Physiol.,* 106:1269-1277.

Kohle, H., Jeblick, W., Poten, F., Blaschek, W., and Kauss, H. (1985). Chitosan-elicited callose synthesis in soybean cells as a Ca^{2+}-dependent process, *Plant Physiol.,* 77:544-551.

Koiwa, H., Bressan, R. A., and Hasegawa, P. M. (1997). Regulation of protease inhibitors and plant defense, *Trends in Plant Science,* 2:379-384.

Koiwa, H., Sato, F., and Yamada, Y. (1994). Characterization of accumulation of tobacco PR-5 proteins by IEF-immunoblot analysis, *Plant Cell Physiol.,* 35:821-827.

Kolattukudy, P. E. (1981). Structure, biosynthesis and biodegradation of cutin and suberin, *Annu. Rev. Plant Physiol.,* 32:539-567.

Kolattukudy, P. E., Crawford, M. S., Woloshuk, C. P., Ettinger, W. F., and Soliday, C. I. (1987). The role of cutin, the plant cuticular hydroxy fatty acid polymer, in the fungal interaction with plants. In G. Fuller and W. D. Nes (Eds.), *Ecology and Metabolism of Plant Lipids, American Chemical Society,* Washington DC, pp. 152-175.

Kolattukudy, P. E., Ettinger, W. F., and Sebastian, J. (1987). Cuticular lipids in plant-microbe interactions. In P.K.Stumpf, J. B. Mudd, and W. D. Nes (Eds.), *The Metabolism, Structure, and Function of Plant Lipids,* Plenum Press, New York, pp. 473-480.

Komari, T., Halperin, W., and Nester, E. W. (1986). Physical and functional map of supervirulent *Agrobacterium tumefaciens* tumor-inducing plasmid pTiBo542. *J. Bacteriol.,*166:88-94.

Kombrink, E., and Hahlbrock, K. (1986). Responses of cultured parsley cells to elicitors from phytopathogenic fungi: Timing and dose dependency of elicitor-induced reactions, *Plant Physiol.,* 81:216-221.

Korhonen, T. K., Haahtela, K., Pirkola, A., and Parkkinen, J. (1988). A N-acetyllactosamine-specific cell-binding activity in a plant pathogen, *Erwinia rhapontici, FEBS Lett.,* 236:163-166.

Kpëmoua, K., Boher, B., Nicole, M., Calatayud, P., and Geiger, J. P. (1996). Cytochemistry of defense responses in cassava infected by *Xanthomonas campestris* pv. *manihotis, Can. J. Microbiol.,* 42:1131-1143.

Kragh, K. M., Jacobsen, S., and Mikkelsen, J. D. (1990). Induction, purification and characterization of barley leaf chitinase, *Plant Sci.,* 71:55-68.

Kragh, K. M., Jacobsen, S., Mikkelsen, J. D., and Nielson, K. A. (1991). Purification and characterization of three chitinases and one β-1,3-glucanase accumulating in medium of cell suspension cultures of barley (*Hordeum vulgare* L.), *Plant Sci.,* 76:65-77.

Kratka, J. (1987). Activity of cell wall destroying enzymes and growth regulators in *Clavibacter michiganense* pv. *insidiosum, Zentralbl. Mikrobiol.,* 142:527-534.

Kristensen, B. K., Brandt, J., Bojsen, K., Thordal-Christensen, H., Kerby, K. B., Collinge, D. B., Mikkelsen, J. D., and Rasmussen, S. K. (1997). Expression of a defense-related intercellular barley peroxidase in transgenic tobacco, *Plant Sci.,* 122:173-176.

Kuc, J. (1982). Induced immunity to plant disease, *BioScience,* 32:854-860.

Kuc, J. (1983). Induced systemic resistance in plants to diseases caused by fungi and bacteria. In J. Bailey and B. Deverall (Eds.), *The Dynamics of Host Defense,* Academic Press, Sydney, pp. 191-221.

Kuc, J. (1987). Plant immunization and its applicability for disease control. In I. Chet (Ed.), *Innovative Approaches to Plant Disease Control,* John Wiley and Sons, New York, pp. 255-274.

Kuc, J. (1995). Systemic induced resistance, *Aspects of Applied Biology,* 42:235-242.

Kuhn, D., Chappell, J., Boudet, A., and Hahlbrock, K. (1984). Induction of phenylalanine ammonia-lyase and 4-coumarate: CoA ligase mRNAs in cultured plant cells by UV light or fungal elicitor, *Proc. Natl. Acad. Sci. U.S.A.,* 81:1102-1106.

Kunkel, B. N., Bent, A. F., Dahlbeck, D., Innes, R. W., and Staskawicz, B. J. (1993). RPS2, an *Arabidopsis* disease resistance locus specifying recognition of *Pseudomonas syringae* strains expressing the avirulence gene *avrRpt2, Cell,* 5:865-875.

Kurosaki, F., Tsurusawa, Y., and Nishi, A. (1987a). Breakdown of phosphtidylinositol during the elicitation of phytoalexin production in cultured carrot cells, *Plant Physiol.,* 85:601-604.

Kurosaki, F., Tsurusawa, Y., and Nishi, A. (1987b). The elicitation of phytoalexins by Ca^{2+} and cyclic AMP in carrot cells, *Phytochemistry,* 26:1919-1923.

Kyostio, S. R. M., Cramer, C. L., and Lacy, G. H. (1991). *Erwinia carotovora* subsp. *carotovora* extracellular protease: Characterization and nucleotide sequence of the gene, *J. Bacteriol.,* 173:6537-6546.

La Rosa, P. C., Chen, Z., Nelson, D. E., Singh, N. K., Hasegawa, P. M., and Bressan, R. A. (1992). Osmotin gene expression is posttranscriptionally regulated, *Plant Physiol.,* 100:409-415.

Laby, R. J., and Beer, S. V. (1990). The *hrp* gene cluster of *Erwinia amylovora* shares DNA homology with other bacteria, *Phytopathology,* 80:1038-1039.

Laby, R. J., and Beer, S. V. (1992). Hybridization and functional complementation of the *hrp* gene cluster from *Erwinia amylovora* strain Ea321 with DNA of other bacteria, *Mol. Plant-Microbe Interact.,* 5:412-419.

Laby, R. J., Zumoff, C. H., Sneath, B. J., Bauer, D. W., and Beer, S. V. (1989). Cloning and preliminary characterization of an *hrp* gene cluster from *Erwinia amylovora, Phytopathology*, 79:1211 (Abstr.).

Lacomme, C., and Roby, D. (1996). Molecular cloning of a sulfotransferase in *Arabidopsis thaliana* and regulation during development and in response to infection with pathogenic bacteria, *Plant Mol. Biol.,* 30:995-1008.

Lagrimini, L. M. (1991). Wound-induced deposition of polyphenols in transgenic plants overexpressing peroxidase, *Plant Physiol.,* 96:577-583.

Lagrimini, L. M., Bradfords, S., and Rothstein, S. (1990). Peroxidase-induced wilting in transgenic tobacco plants, *Plant Cell,* 2:7-18.

Lagrimini, L. M., Burkhart, W., Moyer, M., and Rothstein, S. (1987). Molecular cloning of complementary DNA encoding the lignin-forming peroxidase from tobacco: molecular analysis and tissue-specific expression, *Proc. Natl. Acad. Sci. U.S.A.,* 84:7542-7546.

Lagrimini, L. M., Vaughan, J., Erb, W. A., and Miller, S. A. (1993). Peroxidase overproduction in tomato: Wound-induced polyphenol deposition and disease resistance, *Hort. Sci.,* 28:218-221.

Laine, M. J., Nakhei, H., Dreier, J., Lehtila, K., Meletzus, D., Eichenlaub, R., and Metzler, M. C. (1996). Stable transformation of the Gram-positive phytopathogenic bacterium *Clavibacter michiganensis* subsp. *sepedonicus* with several cloning vectors, *Appl. Environ. Microbiol.,* 62:1500-1506.

Lamb, C. J. (1994). Plant disease resistance genes in signal perception and transduction, *Cell,* 76:419-422.

Lamb, C. (1996). A ligand-receptor mechanism in plant-pathogen recognition, *Science,* 274:2038-2039.

Langcake, P., and Pryce, R. J. (1977). A new class of phytoalexins from grapevines, *Experientia,* 33:151-152.

Langford, A. N. (1948). Autogenous necrosis in tomatoes immune from *Cladosporium fulvum* Cooke, *Can. J. Res.,* 26:35-64.

Larson, R. A. (1988). The antioxidants of higher plants, *Phytochemistry,* 27:969-978.

Law, M. Y., Charles, S. A., and Halliwell, B. (1983). Glutathione and ascorbic acid in spinach (*Spinacia oleracea*) chloroplasts, *Biochem. J.,* 210:899-903.

Lawrence, C. B., Joosten, M. H. A. J., and Tuzun, S. (1996). Differential induction of pathogenesis-related proteins in tomato by *Alternaria solani* and the association of a basic chitinase isozyme with resistance, *Physiol. Mol. Plant Pathol.,* 48:361-377.

Lawrence, G. L., Finnegan, J. E., Ayliffe, M. A., and Ellis, J. G. (1995). The L6 gene for flax rust resistance is related to the *Arabidopsis* R gene *RPS2* and the tobacco viral resistance gene *N*, *Plant Cell,* 7:1195-1206.

Lawson, S. G., Mason, T. L., Sabin, R. D., Sloan, M. E., Drake, R. R., Haley, D. E., and Wasserman, B. P. (1989). UDP-glucose:(1-3)-β-glucan synthase from *Daucus carota* L., *Plant Physiol.,* 90:101-109.

Lawton, K., Friedrich, L., Hunt, M., Weymann, K., Delaney, T., Kessmann, H., Staub, T., and Ryals, J. (1996). Benzothiadiazole induces disease resistance in *Arabidopsis* by activation of the systemic acquired resistance signal transduction pathway, *Plant J.,* 10:71-82.

Lawton, K., Weymann, K., Friedrich, L., Vernooij, B., Uknes, S., and Ryals, J. (1995). Systemic acquired resistance in *Arabidopsis* requires salicylic acid but not ethylene, *Mol. Plant-Microbe Interact.,* 8:863-870.

Lawton, K. A., Beck, J., Potter, S., Ward, E., and Ryals, J. (1994a). Regulation of cucumber class III chitinase gene expression, *Mol. Plant-Microbe Interact.,* 7:48-57.

Lawton, K. A., Potter, S. L., Uknes, S., and Ryals, J. (1994b). Acquired resistance signal transduction in *Arabidopsis* is ethylene independent, *Plant Cell,* 6:581-588.

Lawton, M. A., Dixon, R. A., Hahlbrock, K., and Lamb, C. J. (1983a). Elicitor induction of mRNA activity: Rapid effects of elicitor on phenylalanine ammonia-lyase and chalcone synthase mRNA activities in bean cells, *Eur. J. Biochem.,* 130:131-139.

Lawton, M. A., Dixon, R. A., Hahlbrock, K., and Lamb, C. J. (1983b). Rapid induction of phenylalanine ammonia-lyase and of chalcone synthase synthesis in elicitor-treated plant cells, *Eur. J. Biochem.,* 129:593-601.

Lawton, M. A., Dixon, R. A., and Lamb, C. J. (1980). Elicitor nodulation of the turnover of L-phenylalanine ammonia-lyase in French bean cell suspension cultures, *Biochemica et. Biophysica Acta,* 633:162-175.

Lawton, M. A., and Lamb, C. J. (1983). Transcriptional activation of plant defense genes by fungal elicitor, wounding and infection, *Mol. Cell. Biol.,* 7:335-341.

Leach, J. E., Centrell, M. A., and Sequeira, L. (1982). A hydroxyproline-rich bacterial agglutinin from potato: Extraction, purification, and characterization, *Plant Physiol.,* 70:1353-1358.

Leach, J. E., Guo, A., Reimers, P., Choi, S. H., Hopkins, C. M., and White, F. F. (1994). Physiology of resistant interactions between *Xanthomonas oryzae* pv. *oryzae* and rice. In C. I. Kado and J. H. Crosa (Eds.), *Molecular Mechanisms of Bacterial Virulence,* Kluwer Academic Publishers, Dordrecht, The Netherlands, pp. 551-560.

Leach, J. E., Sherwood, J., Fulton, R. W., and Sequeira, L. (1983). Comparison of soluble proteins associated with disease resistance induced by bacterial lipopolysaccharide and by viral necrosis, *Physiol. Plant Pathol.,* 23:377-385.

Leah, R., Tommerup, H., Svendsen, I., and Mundy, J. (1991). Biochemical and molecular characterization of three antifungal proteins from barley seed, *J. Biol. Chem.,* 266:1564-1573.

Lee, K. L., Hess, K. M., Dudley, M.W., Lynn, D. G., Joerger, R. D., and Binns, A. N. (1992). Mechanism of phenol activation of *Agrobacterium* virulence genes: Identification of phenol binding proteins, *Proc. Natl. Acad. Sci. U.S.A.,* 89:8666-8670.

Lee, S. W., Heinz, R., Robb, J., and Nazar, R. N. (1994). Differential utilization of alternate initiation sites in a plant defense gene responding to environmental stimuli, *Eur. J. Biochem.,* 226:109-114.

Lee, Y. K., and Hwang, B. K. (1996). Differential induction and accumulation of β-1,3-glucanase and chitinase isoforms in the intercellular space and leaf tissues of pepper by *Xanthomonas campestris* pv. *vesicatoria* infection, *J. Phytopathol.,* 144:79-87.

Lee, Y., Jin, S., Sim, W., and Nester, E. W. (1995). Genetic evidence for direct sensing of phenolic compounds by the VirA protein of *Agrobacterium tumefaciens, Proc. Natl. Acad. Sci. U.S.A.,* 92:12245-12249.

Leeman, M., Den Ouden, F. M., Van Pelt, J. A., Dirkx, F. P. M., Steijl, H., Bakker, P. A. H. M., and Schippers, B. (1996). Iron availability affects induction of systemic resistance to *Fusarium* wilt of radish by *Pseudomonas fluorescens, Phytopathology,* 86:149-155.

Leeman, M., Van Pelt, J. A., Den Ouden, F. M., Heinsbroek, M., Bakker, P. A. H. M., and Schippers, B. (1995a). Induction of systemic resistance by *Pseudomonas fluorescens* in radish cultivars differing in susceptibility to *Fusarium* wilt, using a novel bioassay, *Eur. J. Plant Pathol.,* 101:655-664.

Leeman, M., Van Pelt, J. A., Den Ouden, F. M., Heinsbroek, M., Bakker, P. A. H. M., and Schippers, B. (1995b). Induction of systemic resistance in cucumber against fusarium wilt of radish by llipopolysaccharides of *Pseudomonas fluorescens, Phytopathology,* 85:1021-1027.

Legendre, L., Heinstein, P. F., and Low, P. S. (1992). Evidence for participation of GTP-binding proteins in elicitation of the rapid oxidative burst in cultured soybean cells, *J. Biol. Chem.,* 267:20140-20147.

Legendre, L., Reuter, S., Heinstein, P. F., and Low, P. S. (1993). Characterization of the oligogalacturonide-induced oxidative burst in cultured soybean *(Glycine max)* cells, *Plant Physiol.,* 102:233-240.

Legrand, M., Kauffman, S., Geoffroy, P., and Fritig, B. (1987). Biological function of pathogenesis-related proteins: Four tobacco pathogenesis-related proteins are chitinases, *Proc. Natl. Acad. Sci. U.S.A.,* 84:6750-6754.

Lehmann, J., Atzorn, R., Bruckner, C., Reinbothe, S., Leopold, J., Wasternack, C., and Parthier, B. (1995). Accumulation of jasmonate, abscisic acid, specific transcripts and proteins in osmotically stressed barley leaf segments, *Planta,* 197:156-162.

Leigh, J. A., and Coplin, D. (1992). Exopolysaccharides in plant-bacterial interactions, *Annu. Rev. Microbiol.,* 46:307-346.

Leister, R. T., Ausubel, F. M., and Katagiri, F. (1996). Molecular recognition of pathogen occurs inside of plant cells in plant disease resistance specified by the *Arabidopsis* genes *RPS2* and *RPM1, Proc. Natl. Acad. Sci. U.S.A.,* 93:15497-15502.

Leon, J., Lawton, M. A., and Raskin, I. (1995). Hydrogen peroxide stimulates salicylic acid biosynthesis in tobacco, *Plant Physiol.,* 108:1673-1678.

Lerner, D. R., and Raikhel, N. V. (1992). The gene for stinging nettle lectin (*Urtica dioica* agglutinin) encodes both a lectin and a chitinase, *J. Biol. Chem.,* 267: 11085-11091.

Leroux, B., Yanofsky, M. F., Winans, S. C., Ward, S. C., Ward, J. E., Ziegler, S. F., and Nester, E. W. (1987). Characterization of the *virA* locus of *Agrobacterium tumefaciens:* A transcriptional regulator and host range determinant, *EMBO J.,* 6:849-856.

Leubner-Metzger, G., and Meins, F. Jr. (1999). Functions and regulation of plant β-1,3-glucanases (PR2). In S. K. Datta and S. Muthukrishnan (Eds.), *Pathogenesis-Related Proteins in Plants,* CRC Press, Boca Raton, FL, pp. 49-76.

Levine, A., Pennell, R. I., Alvarez, M. E., Palmer, R., and Lamb, C. (1996). Calcium-mediated apoptosis in a plant hypersensitive disease resistance response, *Current Biology,* 6:427-437.

Levine, A., Tenhaken, R., Dixon, R., and Lamb, C. (1994). H_2O_2 from the oxidative burst orchestrates the plant hypersensitive disease resistance response, *Cell,* 79:583-593.

Liang, X., Dron, M., Cramer, C. L., Dixon, R. A., and Lamb, C. J. (1989a). Differential regulation of phenylalanine ammonia-lyase genes during plant development and by environmental cues, *J. Biol. Chem.,* 264:14486-14492.

Liang, X., Dron, M., Schmid, J., Dixon, R. A., and Lamb, C. J. (1989b). Developmental and environmental regulation of a phenylalanine ammonia-lyase-β-glucuronidase gene fusion in transgenic tobacco plants, *Proc. Natl. Acad. Sci. U.S.A.,* 86:9284-9288.

Liao, C.-H. (1989). Analysis of pectate lyase produced by soft rot bacteria associated with spoilage of vegetables, *Appl. Environ. Microbiol.,* 55:1677-1683.

Liao, C.-H., Gaffney, T. D., Bradley, S.-P., and Wong, L.-J. C. (1996). Cloning of a pectate lyase gene from *Xanthomonas campestris* pv. *malvacearum* and comparison of its sequence relationship with *pel* genes of soft-rot *Erwinia* and *Pseudomonas, Mol. Plant-Microbe Interact.,* 9:14-21.

Liao, C.-H., Huang, H.-Y., and Chatterjee, A. K. (1988). An extracellular pectate lyase is the pathogenicity factor of the soft-rotting bacterium *Pseudomonas viridiflava, Mol. Plant-Microbe Interact.,* 1:199-206.

Liao, C.-H., McCallus, D. E., and Fett, W. F. (1994). Molecular characterization of two gene loci required for production of the key pathogenicity factor pectate lyase by *Pseudomonas viridiflava, Mol. Plant-Microbe Interact.,* 7:391-400.

Liao, C.-H., Sasaki, K., Nagahashi, G., and Hicks, K. B. (1992). Cloning and characterization of a pectate lyase gene from the soft-rotting bacterium *Pseudomonas viridiflava, Mol. Plant-Microbe Interact.,* 5:301-308.

Lidell, M. C., and Hutcheson, S. W. (1994). Characterization of the *hrpJ* and *hrpU* operons of *Pseudomonas syringae* pv. *syringae* Pss 61: Similarity with components of enteric bacteria involved in flagellar biogenesis and demonstration of their role in $\text{Harpin}_{\text{Pss}}$ secretion, *Mol. Plant-Microbe Interact.,* 7:488-497.

Lielder, C. D., Kling, S. D., and Reed, D. J. (1986). Antioxidant protection of phospholipid bilayers by tocopherol. Control of α-tocopherol status and lipid peroxidation by ascorbic acid and glutathione, *J. Biol.Chem.,* 261:12114-12119.

Ligterink, W., Kroj, T., Nieden, U. Z., Hirt, H., and Scheel, D. (1997). Receptor-mediated activation of a MAP kinase in pathogen defense of plants, *Science,* 276:2054-2057.

Lin, K. C., Bushnell, W. R., Szabo, L. J., and Smith, A. G. (1996). Isolation and expression of a host response gene family encoding thaumatin-like proteins in incompatible oat-stem rust fungus interactions, *Mol. Plant-Microbe Interact.,* 9:511-522.

Lin, W., Anuratha, C. S., Datta, K., Potrykus, I., Muthukrishnan, S., and Datta, S. K. (1995). Genetic engineering of rice for resistance to sheath blight, *Biotechnology,* 13:686-691.

Lindeberg, M., and Collmer, A. (1992). Analysis of eight *out* genes in a cluster required for pectic enzyme secretion by *Erwinia:* Sequence comparison with secretion genes from other Gram-negative bacteria, *J. Bacteriol.,* 174:7385-7397.

Lindeberg, M., Salmond, G. P. C., and Collmer, A. (1996). Complementation of deletion mutations in a cloned functional cluster of *Erwinia chrysanthemi out* genes with *Erwinia carotovora out* homologues reveals *OutC* and *OutD* as candidate gate keepers of species-specific secretion of proteins via the type II pathway, *Mol. Microbiol.,* 20:175-190.

Lindgren, P. B., Fredrick, R., Govindarajan, A. G., Panopoulos, N. J., Staskawicz, B. J., and Lindow, S. E. (1989). An ice nucleation reporter gene system: Identification of inducible pathogenicity genes in *Pseudomonas syringae* pv. *phaseolicola, EMBO J.,* 8:1291-1301.

Lindgren, P. B., and Jakobek, J. L. (1990). Inoculation of bean with Hrp mutants of *Pseudomonas syringae* pv. *tabaci* alters susceptibility to *P. s. phaseolicola, Phytopathology,* 80:1010 (Abstr.).

Lindgren, P. B., Panopoulos, N. J., Staskawicz, B. J., and Dahlbeck, D. (1988). Genes required for pathogenicity and hypersensitivity are conserved and interchangeable among pathovars of *Pseudomonas syringae, Mol. Gen. Genet.,* 211:499-506.

Lindgren, P. B., Panopoulos, N. J., Willis, D. K., and Peet, R. C. (1984). Analysis of Vir$^-$HR$^-$Tn5 insertion mutants of *Pseudomonas syringae* pv. *syringae, Phytopathology,* 74:837 (Abstr.).

Lindgren, P. B., Peet, R. C., and Panapoulos, N. J. (1986). Gene cluster of *Pseudomonas syringae* pv. *phaseolicola* controls pathogenicity on bean plants and hypersensitivity on nonhost plants, *J. Bacteriol.,* 168:515-522.

Lindsay, J. A., Zhang, H., Kaseki, H., Morisaki, N., Sato, T., and Cornwell, D. G. (1985). Fatty acid metabolism and cell proliferation. VII. Antioxidant effects of tocopherols and their quinones, *Lipids,* 20:151-157.

Liners, F., and Van Cutsem, P. (1991). Immunocytochemical localization of homogalacturonic acid in plant cell walls, *Micron Microscopy Acta,* 22:265-266.

Linthorst, H. J. M., Danhash, N., Brederode, F. T., Van Kan, J. A. L., De Wit, P. J. G. M., and Bol, J. F. (1991). Tobacco and tomato PR proteins homologous to Win and Pro-Hevein lack the "Hevein" domain, *Mol. Plant-Microbe Interact.* 4:586-592.

Linthorst, H. J. M., Melchers, L. S., Mayer, A., Van Rockel, J. S. C., Cornelissen, B. J. C. and Bol, J. F. (1990a). Analysis of gene families encoding acidic and basic β-1,3-glucanases of tobacco, *Proc. Natl. Acad. Sci. U.S.A.,* 87:8756-8760.

Linthorst, H. J. M., Van Loon, L. C., Van Rossum, C. M. A., Mayer, A., Bol. J. F., Van Roekel, C., Meulenhoff, E. J. S., and Cornelissen, B. J. C. (1990b). Analysis of acidic and basic chitinases from tobacco and *Petunia* and their constitutive expression in transgenic tobacco, *Mol. Plant-Microbe Interact.,* 3:252-258.

Lippincott, B. B., and Lippincott, J. A. (1969). Bacterial attachment to a specific wound site as an essential stage in tumor initiation by *Agrobacterium tumefaciens, J. Bacteriol.,* 97:620-628.

Liu, C.-N., Steck, T. R., Habeck, L. L., Meyer, J. A., and Gelvin, S. B. (1993). Multiple copies of *virG* allow induction of *Agrobacterium tumefaciens vir* genes and T-DNA processing at alkaline pH, *Mol. Plant-Microbe Interact.,* 6:144-156.

Liu, D., Narasimhan, M. L., Xu, Y., Raghothama, K. G., Hasegawa, P. M., and Bressan, R. A. (1995). Fine structure and function of the osmotin gene promoter, *Plant Mol. Biol.,* 29:1015-1026.

Liu, D., Raghothama, K. M., Hasegawa, P. M., and Bressan, R. A. (1994). Osmotin overexpression in potato delays development of disease symptoms, *Proc. Natl. Acad. Sci. U.S.A.,* 91:1888-1892.

Liu, D., Rhodes, D., D'Urzo, M. P., Xu, Y., Narasimhan, M. L., Hasegawa, P. M., Bressan, R. A., and Abad, L. (1996). In vivo and in vitro activity of truncated osmotin that is secreted into the extracellular matrix, *Plant Science,* 121:123-131.

Liu, L., Kloepper, J. W., and Tuzun, S. (1995a). Induction of systemic resistance in cucumber against fusarium wilt by plant growth promoting rhizobacteria, *Phytopathology,* 85:695-698.

Liu, L., Kloepper, J. W., and Tuzun, S. (1995b). Induction of systemic resistance in cucumber against bacterial angular leaf spot by plant growth-promoting rhizobacteria, *Phytopathology,* 85:843-847.

Liu, L., Kloepper, J. W., and Tuzun, S. (1995c). Induction of systemic resistance in cucumber by plant growth promoting rhizobacteria. Duration of protection and effect of host resistance on protection and root colonization, *Phytopathology,* 85:1064-1068.

Liu, Y., Chatterjee, A., and Chatterjee, A. K. (1994). Nucleotide sequence and expression of a novel pectate lyase gene *(pel-3)* and a closely linked endopolygalacturonase gene *(peh-1)* of *Erwinia carotovora* subsp. *carotovora* 71, *Appl. Environ. Microbiol.,* 60:2545-2552.

Liu, Y., Cui, Y., Mukherjee, A., and Chatterjee, A. K. (1997). Activation of the *Erwinia carotovora* subsp. *carotovora* pectin lyase structural gene *pnlA:* A role for RdgB, *Microbiology (Reading),* 143:705-712.

Liu, Y., Murata, H., Chatterjee, A., and Chatterjee, A. K. (1993). Characterization of a novel regulatory gene *aepA* that controls extracellular enzyme production in the phytopathogenic bacterium *Erwinia carotovora* subsp. *carotovora, Mol. Plant-Microbe Interact.,* 6:299-308.

Livne, B., Faktor, O., Zeitoune, S., Edelbaum, O., and Sela, I. (1997). TMV-induced expression of tobacco β-glucanase promoter activity is mediated by a single, inverted, GCC motif, *Plant Sci.,* 130:159-169.

Loewen, P. C., and Triggs-Raine, B. L. (1984). Genetic mapping of *katF,* a locus that with *katE* affects the synthesis of a second catalase species in *Escherichia coli, J. Bacteriol.,* 160:668-675.

Logemann, J., Jach, G., Tommerup, H., Mundy, J., and Schell, J. (1992). Expression of a barley ribosome-inactivating protein leads to increased fungal protection in transgenic tobacco plants, *Biotechnology,* 10:305-308.

Logemann, J., Melchers, L. S., Trigelaar, H., Sela-Buurlage, M. B., Ponstein, A. S., van Roekel, J. S. C., Bres-Vloemans, S. A., Dekker, I., Cornelissen, B. J. C., van den Elzen, P. J. M., and Jongedijk, E. (1994). Synergistic activity of chitinase and β-1,3-glucanase enhances *Fusarium* resistance in transgenic tomato plants, *J. Cell Biochem.,* 18A:88.

Loh, Y. T., and Martin, G. B. (1995). The *Pto* bacterial resistance gene and the *Fen* insecticide sensitivity gene encode functional protein kinase with serine/ threonine specificity, *Plant Physiol.,* 108:1735-1739.

Lois, R., Dietrich, A., and Hahlbrock, K. (1989a). A phenylalanine ammonia-lyase gene from parsley: Structure, regulation and identification of elicitor and abiotic phytoalexin inducers, *Physiol. Plant Pathol.,* 23:163-173.

Lois, R., Dietrich, A., Hahlbrock, K., and Schulz, W. (1989b). A phenylalanine ammonia-lyase gene from parsley: Structure, regulation, and identification of elicitor and light-responsive *cis*-acting elements, *EMBO J.,* 8:1641-1648.

Lojkowska, E., Masclaux, C., Boccara, M., Robert-Baudouy, J., and Hugouvieux-Cotte-Pattat, N. (1995). Characterization of the *pelL* gene encoding a novel pectate lyase of *Erwinia chrysanthemi* 3937, *Mol. Microbiol.,* 16:1183-1195.

Long, M., Barton-Willis, P., Staskawicz, B. J., Dahlbeck, D., and Keen, N. T. (1985). Further studies on the relationship between glyceollin accumulation and the resistance of soybean leaves to *Pseudomonas syringae* pv. *glycinea, Phytopathology,* 75:235-239.

Long, S. R. (1989). *Rhizobium*-legume nodulation: Life together in the underground, *Cell,* 56:203-214.

Long, S. R., and Staskawicz, B. J. (1993). Procaryotic plant parasites, *Cell,* 73:921-935.

Longland, A. C., Slusarenko, A. J., and Friend, J. (1992). Pectolytic enzymes from interactions between *Pseudomonas syringae* pv. *phaseolicola* and French bean *(Phaseolus vulgaris), J. Phytopathol.,* 134:75-86.

Lopez-Lopez, M. J., Lielana, E., Marcilia, P., and Beltra, R. (1995). Resistance induced in potato tubers by treatment with acetylsalicylic acid to soft rot produced by *Erwinia carotovora* subsp. *carotovora, J. Phytopathol.,* 143:719-724.

Lorang, J. L., Shen, H., Kobayashi, D., Cooksey, D., and Keen, N. T. (1994). *avrA* and *avrE* in *Pseudomonas syringae* pv. *tomato* PT23 play a role in virulence on tomato plants, *Mol. Plant-Microbe Interact.,* 7:508-515.

Lorang, J. M., and Keen, N. T. (1995). Characterization of *avrE* from *Pseudomonas syringae* pv. *tomato*: A *hrp*-linked avirulence locus consisting of at least two transcriptional units, *Mol. Plant-Microbe Interact.,* 8:49-57.

Lord, J. M. (1985). Synthesis and intracellular transport of lectin and storage protein precursors in endosperm from castor bean, *Eur. J. Biochem.,* 146:403-409.

Lotan, T., and Fluhr, R. (1990a). Function and regulated accumulation of plant pathogenesis-related proteins, *Symbiosis,* 8:33-46.

Lotan, T., and Fluhr, R. (1990b). Xylanase, a novel elicitor of pathogenesis-related proteins in tobacco uses a non-ethylene pathway for induction, *Plant Physiol.,* 93:811-817.

Lotan, T., Ori, N., and Fluhr, R. (1989). Pathogenesis-related proteins are developmentally regulated in tobacco flowers, *Plant Cell,* 1:881-887.

Loughrin, J. H., Hamilton-Kemp, T. R., Burton, H. R., and Andersen, R. A. (1993). Effect of diurnal sampling on the headspace composition of detached *Nicotiana suaveolens* flowers, *Phytochemistry,* 30:1417-1419.

Louhelain, J., Haahtela, K., Lindroos, O., Nurmiaho-Lassila, E. L., Korhonen, E. L., and Korhonen, T. K. (1990). Adhesion of *Erwinia rhapontici* to plant surfaces, *Int. Symp. Mol. Genet. Plant-Microbe Interact. 5th. Int. Symp.,* Interlaken, Switzerland, (Abstr.).

Lozoya, E., Hoffmann, H., Douglas, C., Schulz, W., Scheel, D., and Hahlbrock, K. (1988). Primary structures and catalytic properties of isoenzymes encoded by the two 4-coumarate: CoA ligase genes in parsley, *Eur. J. Biochem.,* 176:661-667.

Lummerzheim, M., de Oliveira, D., Castresana, C., Miguens, F. C., Louzada, E., Roby, D., Van Montagu, M., and Timmerman, B. (1993). Identification of compatible and incompatible interactions between *Arabidopsis thaliana* and *Xanthomonas campestris* pv. *campestris* and characterization of the hypersensitive response, *Mol. Plant-Microbe Interact.,* 6:532-544.

Lund, P., and Dunsmuir, P. (1992). A plant signal sequence enhances the secretion of bacterial ChiA in transgenic tobacco, *Plant Mol. Biol.,* 18:47-53.

Lund, P., Lee, R. Y., and Dunsmuir, P. (1989). Bacterial chitinase is modified and secreted in transgenic tobacco, *Plant Physiol.,* 91:130-135.

Lynn, D. G., and Chang, M. (1990). Phenolic signals in cohabitation: Implications for plant development, *Annu. Rev. Plant Physiol. Plant Mol. Biol.,* 41:497-526.

Lyon, B. R., Hill, M. K., Kota, R., and Lyon, K. J. (1998). Isolation and characterization of genes associated with enhanced tolerance to phytopathogenic fungi in cotton, *Sixth Intl. Conf. Plant and Animal Genome Research,* San Diego, California, p. 86.

Lyon, C. E., Lyon, G. D., and Robertson, W. M. (1989). Observations on the structural modification of *Erwinia carotovora* subsp. *atroseptica* in rotted potato tuber tissue, *Physiol. Mol. Plant Pathol.,* 34:181-187.

Lyon, F. M., and Wood, R. K. S. (1975). Production of phaseollin, coumestrol and related compounds in bean leaves inoculated with *Pseudomonas* spp., *Physiol. Plant Pathol.,* 6:117-124.

Lyon, F. M., and Wood, R. K. S. (1976). The hypersensitive reaction and other responses of bean leaves to bacteria, *Ann. Bot.,* 40:479-491.

Lyon, G. D. (1972). Occurrence of rishitin and phytuberin in potato tubers inoculated with *Erwinia carotovora* var. *atroseptica, Physiol. Plant Pathol.,* 2:411-416.

Lyon, G. D. (1978). Attenuation by divalent cations of the effect of the phytoalexin rishitin on *Erwinia carotovora* var. *atroseptica, J. Gen. Microbiol.,* 109:5-10.

Lyon, G. D. (1989). The biochemical basis of resistance of potatoes to soft rot *Erwinia* spp.—a review, *Plant Pathology,* 38:313-339.

Ma, H., Yanofsky, M. F., and Meyerowitz, E. M. (1991). Isolation and sequence analysis of TGA1 cDNAs encoding a tomato G protein α subunit, *Gene,* 107:719-724.

Ma, Q.-S., Chang, M.-F., Tang, J.-L., Feng, J.-X., Fan, M.-J., Han, B., and Liu,T. (1988). Identification of DNA sequences involved in host specificity in the pathogenesis of *Pseudomonas solanacearum* strain T2005, *Mol. Plant-Microbe Interact.,* 1:169-174.

Madamanchi, N. R., and Kuc, J. (1991). Induced systemic resistance in plants. In G. T. Cole and H. C. Hoch (Eds.), *Induced Systemic Resistance in Plants,* Plenum Press, New York, pp. 347-362.

Maddaloni, M., Forlani, F., Balmas, V., Donini, G., Stasse, L., Corazza, L., and Motto, M. (1997). Tolerance to the fungal pathogen *Rhizoctonia solani* AG4 of transgenic tobacco expressing the maize ribosome inactivating protein B-32, *Transgenic Res.,* 6:393-402.

Mader, M., and Amberg-Fisher, V. (1982). Role of peroxidase in the lignification of tobacco cells. I. Oxidation of nicotinamide adenine nucleotide and formation of hydrogen peroxide by cell wall peroxidases, *Plant Physiol.,* 70:1128-1131.

Magasanik, B. (1982). Genetic control of nitrogen assimilation in bacteria, *Annu. Rev. Genet.,* 16:135-168.

Maggio, A., D'Urzo, M. P., Abad, L. R., Takeda, S., Hasegawa, P. M., and Bressan, R. A. (1996). Large quantities of recombinant PR-5 proteins from the extracellular matrix of tobacco: Rapid production of microbial-recalcitrant proteins, *Plant Mol. Biol. Reporter,* 14:249-260.

Magwa, M. L., Lindner, W. A., and Brand, J. M. (1993). Guttation fluid peroxidases from *Helianthus annuus, Phytochemistry,* 32:251-253.

Mahe, B., Masclaux, C., Rauscher, L., Enard, C., and Expert, D. (1995). Differential expression of two siderophore-dependent iron acquisition pathways in *Erwinia chrysanthemi* 3937: Characterization of a novel ferrisiderophore permease of the ABC transporter family, *Mol. Microbiol.,* 18:33-43.

Maher, E. A., Bate, N. J., Ni, W., Elkind, Y., Dixon, R. A., and Lamb, C. J. (1994). Increased disease susceptibility of transgenic tobacco plants with suppressed levels of preformed phenylpropanoid products, *Proc. Natl. Acad. Sci. U.S.A.,* 91:7802-7806.

Malamy, J., Carr, J. P., Klessig, D. F., and Raskin, I. (1990). Salicylic acid: A likely endogenous signal in the resistance response of tobacco to viral infection, *Science,* 250:1001-1004.

Malamy, J., Sanchez-Casas, P., Hennig, J., Guo, A., and Klessig, D. F. (1996). Dissection of the salicylic acid signaling pathway in tobacco, *Mol. Plant-Microbe Interact.,* 9:474-482.

Malehorn, D. E., Borgmeyer, J. R., Smith, C. E., and Shah, D. M. (1994). Characterization and expression of an antifungal zeamatin-like protein *(Zlp)* gene from *Zea mays, Plant Physiol.,* 106:1471-1481.

Mansfield, J. W. (1982). Role of phytoalexins in disease resistance. In J. A. Bailey and J. W. Mansfield (Eds.), *Phytoalexins,* Blackie, Glascow, pp. 253-288.

Mansfield, J. W., and Brown, I. R. (1986). The biology of interactions between plant and bacteria. In J. A. Bailey, (Ed.), *Biology and Molecular Biology of Plant-Pathogen Interactions,* Springer Verlag, Berlin, pp. 71-98.

Mansfield, J., Jenner, C., Hockenhull, R., Bennett, M. A., and Stewart, R. (1994). Characterization of *avrPphE,* a gene for cultivar-specific avirulence from *Pseudomonas syringae* pv. *phaseolicola* which is physically linked to *hrpY,* a new *hrp* gene identified in the halo-blight bacterium, *Mol. Plant-Microbe Interact.,* 7:726-739.

Marchant, R., Davey, M. R., Lucas, J. A., Lamb, C. J., Dixon, R. A., and Power, J. B. (1998). Expression of a chitinase transgene in rose (*Rosa hybrida* L.) reduces development of blackspot disease (*Diplocarpon rosae* Wolf), *Mol. Breed.,* 4:187-194.

Marco, Y. J., Ragueh, F., Godiard, L., and Froissard, D. (1990). Transcriptional activation of 2 classes of genes during the hypersensitive reaction of tobacco leaves infiltrated with an incompatible isolate of the phytopathogenic bacterium *Pseudomonas solanacearum, Plant Mol. Biol.,* 15:145-154.

Martin, C., and Paz-Ares, J. (1997). Myb transcription factors in plants, *Trends in Genet.,* 13:67-73.

Martin, G. B., Brommonschenkel, S. H., Chunwongse, J., Frary, A., Ganal, M. W., Spivey, R., Wu, T., Earle, E. D., and Tanksley, S. D. (1993a). Map-based cloning of a protein kinase gene conferring disease resistance in tomato, *Science,* 262:1432-1436.

Martin, G. B., Frary, A., Wu, T., Brommonschenkel, S., Chunwongse, J., Earle, E. D., and Tanksley, S. D. (1994). A member of the tomato *Pto* gene family confers sensitivity to fenthion resulting in rapid cell death, *Plant Cell,* 6:1543-1552.

Martin, G. B., de Vicente, M. C., and Tanksley, S. D. (1993b). High-resolution linkage analysis and physical characterization of the *Pto* bacterial resistance locus in tomato, *Mol. Plant-Microbe Interact.,* 6:26-34.

Martin, M., and Dewick, P. M. (1980). Biosynthesis of pterocarpan, isoflavan and coumestan metabolites of *Medicago sativa:* The role of an isoflav-3-ene, *Phytochemistry,* 19:2341-2346.

Marty, P., Jouan, B., Bertheau, Y., Vian, B., and Goldberg, R. (1997). Charge density in stem cell walls of *Solanum tuberosum* genotypes and susceptibility to blackleg, *Phytochemistry,* 44:1435-1441.

Masclaux, C., Hugouvieux-Cotte-Pattat, N., and Expert, D. (1996). Iron is a triggering factor for differential expression of *Erwinia chrysanthemi* strain 3937 pectate lyases in pathogenesis of African violets, *Mol. Plant-Microbe Interact.,* 9:108-205.

Masoud, S. A., Zhu, Q., Lamb, C., and Dixon, R. A. (1996). Constitutive expression of an inducible β-1,3-glucanase in alfalfa reduced disease severity caused by the oomycete pathogen *Phytophthora megasperma* f. sp. *medicaginis,* but does not reduce severity of chitin-containing fungi, *Transgenic Res.,* 5:313-323.

Matheson, I. B., Etheridge, R. D., Kratowich, N. R., and Lee, J. (1975). The quenching of singlet oxygen by amino acids and proteins, *Photochem. Photobiol.,* 21:165-171.

Mathieu, Y., Kurkdjian, A., Xia, H., Guern, J., Koller, A., Spiro, M. D., O'Neill, M., Albersheim, P., and Darvill, A. (1991). Membrane responses induced by oligogalacturonides in suspension-cultured tobacco cells, *Plant J.,* 1:333-343.

Matsuoka, M., Yamamoto, N., Kano-Murakami, Y., Tanaka, Y., Ozeki, Y., Hirano, H., Kagawa, H., Oshima, M., and Ohashi, Y. (1987). Classification and structural comparison of full-length cDNAs for pathogenesis-related proteins, *Plant Physiol.,* 85:942-946.

Matthysse, A. G. (1987a). Characterization of nonattaching mutants of *Agrobacterium tumefaciens, J. Bacteriol.,* 169:313-323.

Matthysse, A. G. (1987b). Effect of plasmid pSa and the auxin on attachment of *Agrobacterium tumefaciens* to carrot cells, *Appl. Environ. Microbiol.,* 53: 2574-2582.

Matthysse, A. G., Yarnall, H. A., and Young, N. (1996). Requirement for genes with homology to ABC transport systems for attachment and virulence of *Agrobacterium tumefaciens, J. Bacteriol.,* 178:5302-5308.

Matton, D. P., and Brisson, N. (1989). Cloning, expression, and sequence conservation of pathogenesis-related gene transcripts of potato, *Mol. Plant-Microbe Interact.,* 2:325-331.

Matton, D. P., Prescott, G., Bertrand, C., Camirand, A., and Brisson, N. (1993). Identification of *cis*-acting elements involved in the regulation of the pathogenesis-related gene STH-2 in potato, *Plant Mol. Biol.,* 22:279-291.

Mauch, F., Hadwiger, L. A., and Boller, T. (1988a). Antifungal hydrolases in pea tissue. I. Purification and characterization of two chitinases and two β-1,3-glucanases differentially regulated during development and in response to fungal infection, *Plant Physiol.,* 87:325-333.

Mauch, F., Mauch-Mani, B., and Boller, T. (1988b). Antifungal hydrolases in pea tissue. II. Inhibition of fungal growth by combinations of chitinase and β-1,3-glucanase, *Plant Physiol.,* 88:936-942.

Mauch, F., Meehl, J. B., and Staehelin, L. A. (1992). Ethylene-induced chitinase and β-1,3-glucanase accumulate specifically in the lower epidermis and along the vascular strands of bean leaves, *Planta,* 186:367-375.

Mauch-Mani, B., and Slusarenko, A. J. (1994). Systemic acquired resistance in *Arabidopsis thaliana* induced by a predisposing infection with a pathogenic isolate of *Fusarium oxysporum, Mol. Plant-Microbe Interact.,* 7:378-383.

Maurhofer, M., Hase, C., Meuwly, P., Metraux, J. P., and Defago, G. (1994). Induction of systemic resistance of tobacco to tobacco necrosis virus by the root colonizing *Pseudomonas fluorescens* strain CHAO: Influence of the *gacA* gene and of pyoverdine production, *Phytopathology,* 84:139-146.

May, M. J., Hammond-Kosack, K. E., and Jones, J. D. G. (1996). Involvement of reactive oxygen species, glutathione metabolism and lipid peroxidation of the *Cf*-gene-dependent defense response of tomato cotyledons induced by race-specific elicitors of *Cladosporium fulvum, Plant Physiol.,* 110:1367-1379.

May, M. J., Parker, J. E., Daniels, M. J., Leaver, C. J., and Christopher, S. (1996). An *Arabidopsis* mutant depleted in glutathione shows unaltered responses to fungal and bacterial pathogens, *Mol. Plant-Microbe Interact.,* 9:349-356.

Mazau, D., and Esquerre-Tugaye, M. T. (1986). Hydroxyproline-rich glycoprotein accumulation in the cell walls of plants infected by various pathogens, *Physiol. Mol. Plant Pathol.,* 29:147-157.

Mazzucchi, U. (1983). Recognition of bacteria by plants. In J. A. Callow (Ed.), *Biochemical Plant Pathology,* John Wiley, Chichester, pp. 299-324.

Mazzucchi, U., Bazzi, C., and Pupilo, P. (1979). The inhibition of susceptible and hypersensitive reactions by protein-lipopolysaccharide complexes from phytopathogenic pseudomonads: Relationship to polysaccharide antigenic determinants, *Physiol. Plant Pathol.,* 14:19-30.

Mazzuchi, U., and Pupillo, P. (1976). Prevention of confluent hypersensitive necrosis in tobacco leaves by a bacterial protein-lipopolysaccharide complex, *Physiol. Plant Pathol.,* 9:101-102.

McConkey, G. A., Waters, A. P., and McCutchan, T. F. (1990). The generation of genetic diversity in malaria parasites, *Annu. Rev. Microbiol.,* 44:479-498.

McGuire, R. G., and Kelman, A. (1986). Calcium in potato cell walls in relation to tissue maceration by *Erwinia carotovora* pv. *atroseptica, Phytopathology,* 78:401-406.

McGurl, B., Orozco-Cardenas, M., Pearce, G., and Ryan, C. A. (1994). Overexpression of the prosystemin gene in transgenic tomato plants generates a systemic signal that constitutively induces proteinase inhibitor synthesis, *Proc. Natl. Acad. Sci. U.S.A.,* 91:9799-9802.

McIntyre, C. L., Bettenay, H. M., and Manners, J. M. (1996). Strategies for the suppression of peroxidase gene expression in tobacco. II. In vivo suppression of peroxidase activity in transgenic tobacco using ribozyme and antisense constructs, *Transgenic Res.,* 5:263-268.

McLean, B. G., Greene, E. A., and Zambryski, P. C. (1994). Mutants of *Agrobacterium* VirA that activate *vir* gene expression in the absence of the inducer acetosyringone, *J. Biol. Chem.,* 269:2645-2651.

McMillan, G. P., Headley, D., Fyffe, L., and Perombelon, M. C. M. (1993a). Potato resistance to soft-rot erwinias is related to cell wall pectin esterification, *Physiol. Mol. Plant Pathol.,* 42:279-289.

McMillan, G. P., Headley, D., and Perombelon, M. C. M. (1993b). Purification to homogeneity of extracellular polygalacturonase and pectate lyase isoenzymes of *Erwinia carotovora* subsp. *atroseptica* by column chromatography, *J. Appl. Bacteriol.,* 73:83-86.

McMurchy, R. A., and Higgins, V. J. (1984). Trifolirhizin and maackiain in red clover: Changes in *Fusarium roseum* 'Avenacearum' infected roots and in vitro effects on the pathogen, *Physiol. Plant Pathol.,* 25:229-238.

McNeil, M., Darvill, A. G., Fry, S. C., and Albersheim, P. (1984). Structure and function of the primary cell walls of higher plants, *Annu. Rev. Biochem.,* 53:625-663.

Meeley, R. B., and Walton, J. D. (1991). Enzymatic detoxification of HC-toxin, the host selective cyclic peptide from *Cochliobolus carbonum, Plant Physiol.,* 97:1080-1086.

Meena, R., Radhajeyalakshmi, R., Marimuthu, T., Vidhyasekaran, P., Sabitha, D., and Velazhahan, R. (2000). Induction of pathogenesis-related proteins, phenolics, and phenylalanine ammonia-lyase in groundnut by *Pseudomonas fluorescens, J. Plant Diseases and Protection,* 107:514-527.

Mehdy, M. C. (1994). Active oxygen species in plant defense against pathogens, *Plant Physiol.,* 105:467-472.

Mehdy, M., and Lamb, C. J. (1987). Chalcone isomerase cDNA cloning and mRNA induction by fungal elicitor, wounding and infection, *EMBO J.,* 6:1527-1533.

Meier, B., Shaw, N., and Slusarenko, A. J. (1993). Spatial and temporal accumulation of defense gene transcripts in bean *(Phaseolus vulgaris)* leaves in relation to bacteria-induced hypersensitive cell death, *Mol. Plant-Microbe Interact.,* 6:453-466.

Meins, F., and Ahl, P. (1989). Induction of chitinase and β-1,3-glucanase in tobacco plants infected with *Pseudomonas tabaci* and *Phytophthora parasitica* var. *nicotianae, Plant Sci.,* 61:155-161.

Meins, F. Jr., Neuhaus, J. M., Sperisen, C., and Ryals, J. (1992). The primary structure of plant pathogenesis-related glucanohydrolases and their genes. In T. Boller and F. Meins Jr. (Eds.), *Genes Involved in Plant Defense,* Springer-Verlag, Vienna/New York, pp. 245-282.

Melchers, L. S., Apotheker-deGroot, M., Van der Knaap, J. A., Ponstein, A. S., Sela-Buurlage, M. B., Bol, J. F., Cornelissen, B. J. C., Van den Elzen, P. J. M., and Linthorst, H. J. M. (1994). A new class of tobacco chitinases homologous to bacterial exo-chitinases displays antifungal activity, *Plant J.,* 5:469-480.

Melchers, L. S., Regensburg-Tuink, T. J. G., Bourret, R., Sedee, N. J. A., Schilperoot, R. A., and Hooykaas, P. J. J. (1989a). Membrane topology and functional analysis of the sensory protein VirA of *Agrobacterium tumefaciens, EMBO J.,* 8:1919-1925.

Melchers, L. S., Regensburg-Tuink, T. J. G., Schilperoot, R. A., and Hooykaas, P. J. J. (1989b). Specificity of signal molecules in the activation of *Agrobacterium* virulence gene expression, *Mol. Microbiol.,* 3:969-977.

Melchers, L. S., Sela-Buurlage, M. B., Vloemans, S. A., Woloshuk, C. P., Van Roekel, J. S. C., Pen, J., van den Elzen, P. J. M., and Cornelissen, B. J. C. (1993). Extracellular targeting of the vacuolar tobacco proteins AP 24, chitinase and β-1,3-glucanase in transgenic plants, *Plant Mol. Biol.,* 21:583-593.

Melchers, L. S., Thompson, D. V., Idler, K. B., Saskia, T. C., Neuteboom, S. T. C., De Maagd, R. A., Schilperoort, R. A., and Hooykaas, P. J. J. (1987). Molecular characterization of the virulence gene *virA* of the *Agrobacterium tumefaciens* octopine Ti plasmid, *Plant Mol. Biol.,* 11:227-237.

Melchers, L. S., Thompson, D. V., Idler, K. B., Schilperoort, R. R., and Hooykaas, P. J. J. (1986). Nucleotide sequence of the virulenae gene *virG* of the *Agrobacterium tumefaciens* octopine Ti plasmid: Significant homology between *virG* and the regulatory genes *ompR, phoB,* and *dye* of *E. coli, Nucleic Acids Res.,* 14:9933-9942.

Meletzus, D., Bermpohl, A., Dreier, J., and Eichenlaub, R. (1993). Evidence for plasmid-encoded virulence factors in the phytopathogenic bacterium *Clavibacter michiganense* subsp. *michiganensis* NCPPB382, *J. Bacteriol.,* 175:2131-2136.

Meller, Y., Sessa, G., Eyal, Y., and Fluhr, R. (1993). DNA-protein interactions on a *cis*-DNA element essential for ethylene regulation, *Plant Mol. Biol.,* 23:453-463.

Mellon, J. E., and Helgeson, J. P. C. (1982). Interaction of a hydroxyproline-rich glycoprotein from tobacco callus with potential pathogens, *Plant Physiol.,* 70:401-405.

Memelink, J., Hoge, J. H. C., and Schilperoort, R. A. (1987). Cytokinin stress changes the developmental regulation of several defence-related genes in tobacco, *EMBO J.,* 6:3579-3583.

Memelink, J., Linthorst, H. J. M., Schilperoot, R. A., and Hoge, J. H. C. (1990). Tobacco genes encoding acidic and basic isoforms of pathogenesis-related proteins display different expression patterns, *Plant Mol. Biol.,* 14:119-126.

Mendez, E., Moreno, A., Colilla, F., Pelaez, F., Limas, G. G., Mendez, R., Soriano, F., Salinas, M., and de Haro, C. (1990). Primary structure and inhibition of protein synthesis in eukaryotic cell-free system of a novel thionin, γ-hordothionin, from barley endosperm, *Eur. J. Biochem.,* 194:533-539.

Messiaen, J., Read, N. D., Van Cutsem, P., and Trewavas, A. J. (1993). Cell wall oligogalacturonides increase cytosolic free calcium in carrot protoplasts, *J. Cell Sci.,* 104:365-371.

Metraux, J. P., Ahl-Goy, P., Staub, T., Speich, J., Steinemann, A., Ryals, J., and Ward, E. (1991). Induced systemic resistance in cucumber in response to 2,6-dichloroisonicotinic acid and pathogens. In I. H. Hennecke and D. P. S. Verma (Eds.), *Advances in Molecular Genetics of Plant-Microbe Interactions,* Vol. I, Kluwer Academic Publishers, Dordrecht, The Netherlands, pp. 432-439.

Metraux, J. P., and Boller, T. (1986). Local and systemic induction of chitinase in cucumber plants in response to viral, bacterial and fungal infections, *Physiol. Mol. Plant Pathol.,* 28:161-169.

Metraux, J. P., Burkhart, W., Moyer, M., Dincher, S., Middlesteadt, W., Williams, S., Payne, G., Carnes, M., and Ryals, J. (1989). Isolation of a complementary DNA encoding a chitinase with structural homology to a bifunctional lysozyme/chitinase, *Proc. Natl. Acad. Sci. U.S.A.,* 86:896-900.

Metraux, J. P., Signer, H., Ryals, J., Ward, E., Wyss-Benz, M., Gaudin, J., Raschdorf, K., Schmid, E., Blum, W., and Inverardi, B. (1990). Increase in salycylic acid at the onset of systemic acquired resistance in cucumber, *Science,* 250:1004-1006.

Metts, J., West, J., Doares, S. H., and Matthysse, A. G. (1991). Characterization of three *Agrobacterium tumefaciens* avirulent mutants with chromosomal mutations that affect induction of *vir* genes *J. Bacteriol.,* 173:1080-1087.

Meuwly, P., Molders, W., Buchala, A., and Metraux, J. P. (1995). Local and systemic biosynthesis of salicylic acid in infected cucumber plants, *Plant Physiol.,* 109:1107-1114.

Meyer, A., Miersch, O., Buttner, C., Dathe, W., and Sembdner, G. (1984). Occurrence of the plant growth regulator jasmonic acid in plants, *J. Plant Growth Regul.,* 3:1-8.

Meyer, J. M., Azelvandre, P., and Georges, C. (1992). Iron metabolism in *Pseudomonas:* Salicylic acid, a siderophore of *Pseudomonas fluorescens* CHAO, *Biofactors,* 4:23-27.

Michelmore, R. (1996). Flood warning-resistance genes unleashed, *Nature Genet.,* 14:376-378.

Michiels, T., and Cornelis, G. R. (1991). Secretion of hybrid proteins by the *Yersinia* Yop export system, *J. Bacteriol.,* 173:1677-1685.

Michiels, T., Vanooteghem, J. C., Lambert de Rouvroit, C., China, B., Gustin, A., Boudry, P., and Cornelis, G. R. (1991). Analysis of *virC,* an operon involved in the secretion of Yop proteins by *Yersinia enterocolitica, J. Bacteriol.,* 173:4994-5009.

Midland, S. L., Keen, N. T., Sims, J. J., Midland, M. M., Stayton, M. M., Burton, V., Smith, M. J., Mazzola, E. P., Graham, K. J., and Clardy, J. (1993). The structure of syringolides 1 and 2, novel C-glycosidic elicitors from *Pseudomonas syringae* pv. *tomato, J. Org. Chem.,* 58:2940-2945.

Midoh, N., and Iwata, M. (1996). Cloning and characterization of a probenazole-inducible gene for an intracellular pathogenesis-related protein in rice, *Plant Cell Physiol.,* 37:9-18.

Mila, I., Scalbert, A., and Expert, D. (1996). Iron withholding by plant phenols and resistance to pathogens and rots, *Phytochemistry,* 42:1551-1555.

Milat, M. L., Ducruet, J. M., Ricci, P., Marty, F., and Blein, J. P. (1991). Physiological and structural changes in tobacco leaves treated with cryptogein, a proteinaceous elicitor from *Phytophthora cryptogea, Phytopathology,* 81:1364-1368.

Miller, W., Mindrinos, M. N., Rahme, L. G., Frederick, R. D., Grimm, C., Gressman, R., Kyriakides, X., Kokkinidis, M., and Panapoulos, N. J. (1993). *Pseudomonas syringae* pv. *phaseolicola*-plant interactions: Host-pathogen signalling through cascade control of *hrp* gene expression. In E. W. Nester and D. P. S. Verma

(Eds.), *Advances in Molecular Genetics of Plant-Microbe Interactions,* Kluwer Academic Press, Dordrecht, pp. 267-274.

Mills, D., and Hammerschlag, F. A. (1993). Effect of cecropin B on peach pathogens, protoplasts and cells, *Plant Sci.,* 93:143-150.

Mills, D., and Mukhopadhyay, P. (1990). Organization of the *hrpM* locus of *Pseudomonas syringae* pv. *syringae.* In S. Silver, A. M. Chakrabarty, B. Iglewski, and S. Kaplan (Eds.), *Pseudomonas: Biotransformations, Pathogenesis, and Evolving Biotechnology,* American Society for Microbiology, Washington, DC, pp. 74-81.

Mills, D., and Niepold, F. (1987). Molecular analysis of *Pseudomonas syringae* pv. *syringae.* In S. Nishimura, C. P. Vance, and N. Doke (Eds.), *Molecular Determinants of Plant Diseases,* Japan Scientific Societies Press, Tokyo, pp. 185-200.

Mills, D., Niepold, F., and Zuber, M. (1985). Cloned sequence controlling colony morphology and pathogenesis of *Pseudomonas syringae.* In I. Sussex, A. Ellingboe, M. Crouch, and R. Malmberg (Eds.), *The Genetics of Plant Cell/Cell Interactions,* Cold Spring Harbor Laboratory, Cold Spring Harbor, NY, pp. 97-102.

Mills, P. R., and Wood, R. K. S. (1984). The effects of polyacrylic acid, acetylsalicylic acid, and salicylic acid on resistance of cucumber to *Colletotrichum lagenarium, Phytopathol. Z.,* 11:209-216.

Milosevic, N., and Slusarenko, A. J. (1996). Active oxygen metabolism and lignification in the hypersensitive response in bean, *Physiol. Mol. Plant Pathol.,* 49:143-158.

Minami, E., Kuchitsu, K., He, D. Y., Kouchi, H., Midoh, N., Ohtsuki, Y., and Shibuya, N. (1996). Two novel genes rapidly and transiently activated in suspension-cultured rice cells by treatment with *N*-acetylchitoheptaose, a biotic elicitor for phytoalexin production, *Plant Cell Physiol.,* 37:563-567.

Minardi, P., Fede, A., Mazzucchi, U. (1989). Protection induced by protein lipopolysaccharide complexes in tobacco leaves: Role of protected tissue free-space solutes, *J. Phytopathol.,* 121:211-220.

Mindrinos, M., Katagiri, F., Yu, G.-L., and Ausubel, F. M. (1994). The *A. thaliana* disease resistance gene RPS2 encodes a protein containing a nucleotide-binding site and leucine-rich repeats, *Cell,* 78:1089-1099.

Minsavage, G. V., Dahlbeck, D., Whalen, M. C., Kearney, B., Bonas, U., Staskawicz, B. J., and Stall, R. E. (1990). Gene-for-gene relationships specifying disease resistance in *Xanthomonas campestris* pv. *vesicatoria*-pepper interactions, *Mol. Plant-Microbe Interact.,* 3:41-47.

Mitra, A., and Zhang, Z. (1994). Expression of a human lactoferrin cDNA in tobacco cells produces antibacterial protein(s), *Plant Physiol.,* 106:977-981.

Mittler, R., Shulaev, V., and Lam, E. (1995). Co-ordinated activation of programmed cell death and defense mechanisms in transgenic tobacco plants expressing a bacterial proton pump, *Plant Cell,* 7:29-42.

Moerschbacher, B. M., Noll, U. M., Flott, B. E., and Reisener, H. J. (1988). Lignin biosynthetic enzymes in stem rust infected, resistant and susceptible near isogenic wheat lines, *Physiol. Mol. Plant Pathol.,* 33:33-46.

Moerschbacher, B. M., Noll, U., Gorrichon, L., and Reisener, H. J. (1990). Specific inhibition of lignification breaks hypersensitive resistance to stem rust, *Plant Physiol.,* 93:465-470.

Moesta, P., and Grisebach, H. (1982). L-2-Aminooxy-3-phenyl-propionic acid inhibits phytoalexin accumulation in soybean with concomitant loss of resistance against *Phytophthora megasperma* f.sp. *glycinea, Physiol.Plant Pathol.,* 21:65-70.

Moesta, P., and West, C. A. (1985). Casbene synthetase: Regulation of phytoalexin biosynthesis in *Ricinus communis* L. seedlings. Purification of casbene synthetase and regulation of its biosynthesis during elicitation, *Arch. Biochem. Biophys.,* 238:325-333.

Mohammed, F., and Sehgal, O. P. (1987). Alterations in leaf proteins accompanying virus-induced hypersensitive reaction in *Phaseolus vulgaris* L. cv. Pinto, *Phytopathology,* 77:1765.

Mohan, R., Bajar, A. M., and Kollattukudy, P. E. (1993). Induction of a tomato anionic peroxidase gene *(tap1)* by wounding in transgenic tobacco and activation of *tap1* GUS and *tap2*/GUS chimeric gene fusions in transgenic tobacco by wounding and pathogen attack, *Plant Mol. Biol.,* 21:339-354.

Moiseyev, G., Beintema, J. J., Fedoreyeva, L. I., and Yakovlev, G. I. (1994). High sequence similarity between a ribonuclease from ginseng calluses and fungus-elicited proteins from parsley indicates that intracellular pathogenesis-related (IPR) proteins are ribonucleases, *Planta,* 193:470-472.

Molders, W., Buchala, A., and Metraux, J. P. (1996). Transport of salicylic acid in tobacco necrosis virus-infected cucumber plants, *Plant Physiol.,* 112:787-792.

Molina, A., Ahl-Goy, P., Fraile, A., Sanchez-Monge, R., and Garcia-Olmedo, F. (1993a). Inhibition of bacterial and fungal plant pathogens by thionins of types I and II, *Plant Sci.,* 92:169-177.

Molina, A., and Garcia-Olmedo, F. (1993). Developmental and pathogen-induced expression of three barley genes encoding lipid transfer proteins, *Plant J.,* 4:983-991.

Molina, A., and Garcia-Olmedo, F. (1994). Expression of genes encoding thionins and lipid-transfer proteins: A combinatorial model for the responses of defense genes to pathogens. In G. Coruzzi and P. Puigdomenech (Eds.), *Plant Molecular Biology,* Springer Verlag, Heidelberg, pp. 235-244.

Molina, A., and Garcia-Olmedo, F. (1997). Enhanced tolerance to bacterial pathogens caused by the transgenic expression of barley lipid transfer protein LTP2, *Plant J.,* 12:669-675.

Molina, A., Segura, A., and Garcia-Olmedo, F. (1993b). Lipid transfer proteins (nsLTPs) from barley and maize leaves are potent inhibitors of bacterial and fungal plant pathogens, *FEBS Lett.,* 316:119-122.

Moloshok, T., Pears, G., and Ryan, C. A. (1992). Oligouronide signaling of proteinase inhibitor genes in plants: Structure-activity relationships of di- and trigalacturonide acids and their derivatives, *Arch. Biochem. Biophys.,* 294:731-734.

Monde, K., Kishimoto, M., and Takasugi, M. (1992). Yurinelide, a novel 3-benzylidene-1,4-benzodioxin-2(3H)-one phytoalexin from *Lilium maximowiczii, Tetrahedron Letters,* 33:5395-5398.

Monde, K., Saski, K., Shirata, A., and Takasugi, M. (1990). 4-Methoxy-brassinin, a sulphur-containing phytoalexin from *Brassica oleracea, Phytochemistry,* 29:1499-1500.

Montalbini, P. (1992). Changes in xanthine oxidase activity in bean leaves induced by *Uromyces phaseoli* infection, *J. Phytopathol.,* 134:218-228.

Montanelli, C., and Nascari, G. (1991). Introduction of an antibacterial gene in potato (*Solanum tuberosum* L.) using a binary vector in *Agrobacterium rhizogenes, J. Genet. Breed.,* 45:307-316.

Montillet, J. L., and Degousee, N. (1991). Hydroperoxydes induce glyceollin accumulation in soybean, *Plant Physiol. Biochem.,* 29:689-694.

Morales, M., Pedren, M. A., Barcelo, A. R., and Calderon, A. A. (1993). Oxidation of flavonol and flavonol glycosides by a hypodermal peroxidase isozyme from Gamay Rouge grape *Vitis vinifera* berries, *J. Sci. Food Agri.,* 62:385-391.

Moran, F., Nasuno, S., and Starr, M. P. (1968). Extracellular and intracellular polygalacturonic acid trans-eliminase of *Erwinia carotovora, Arch. Biochem. Biophys.,* 123:298-306.

Morel, F., Doussiere, J., and Vignais, P. V. (1991). The superoxide-generating oxidase of phagocytic cells: Physiological, molecular and pathological aspects, *Eur. J. Biochem.,* 201:523-546.

Moreno, M., Segura, A., and Garcia-Olmedo, F. (1994). Pseudothionin-St1, a potato peptide active against potato pathogens, *Eur. J. Biochem.,* 223:135-139.

Morgham, A. T., Richardson, P. E., Essenberg, M., and Covers, E. C. (1988). Effects of continuous dark upon ultrastructure, bacterial population and accumulation of phytoalexins during interactions between *Xanthomonas campestris* pv. *malvacearum* and bacterial blight susceptible and resistant cotton, *Physiol. Mol. Plant Pathol.,* 32:141-162.

Mouradov, A., Mouradova, E., and Scott, K. J. (1994). Gene family encoding basic pathogenesis-related 1 proteins in barley, *Plant Mol. Biol.,* 26:503-507.

Mouradov, A., Petrasovits, L., Davidson, A., and Scott, K. J. (1993). A cDNA clone for a pathogenesis-related protein 1 from barley, *Plant Mol. Biol.,* 23:439-442.

Mourgues, F., Brisset, M. N., and Chevreau, E. (1998). Strategies to improve plant resistance to bacterial diseases through genetic engineering, *TIBTECH,* 16:203-210.

M'Piga, P., Belanger, R. R., Paulitz, T. C., and Benhamou, N. (1997). Increased resistance to *Fusarium oxysporum* f. sp. *radicis-lycopersici* in tomato plants treated with the endophytic bacterium *Pseudomonas fluorescens* strain 6328, *Physiol. Mol. Plant Pathol.,* 50:301-320.

Mukherjee, A., Cui, Y., Liu, Y., and Chatterjee, A. K. (1997). Molecular characterization and expression of the *Erwinia carotovora* $hrpN_{Ecc}$ gene, which encodes an elicitor of the hypersensitive reaction, *Mol. Plant-Microbe Interact.,* 10:462-471.

Mukopadhyay, P., Williams, J., and Mills, D. (1988). Molecular analysis of a pathogenicity locus in *Pseudomonas syringae* pv. *syringae, J. Bacteriol.,* 170:5479-5488.

Muller, K. O., and Borger, H. (1940). Experimentelle untersuchungen uber die *Phytophthora*-resistenz der kartoffel, *Arb. Biol. Reichsasnstalt. Landw. Forstw. Berlin,* 23:189-231.

Mulya, K., Takikawa, Y., and Tsuyumu, S. (1996). The presence of regions homologous to *hrp* cluster in *Pseudomonas fluorescens* PBG32R, *Ann. Phytopath. Soc. Japan,* 62:355-359.

Mundy, J., Leah, R., Boston, R., Endo, Y., and Stirpe, F. (1994). Genes encoding ribosome-inactivating proteins, *Plant Mol. Biol. Reporter,* 12:S60-S62.

Mundy, J., and Rogers, J. C. (1986). Selective expression of a probable amylase/protease inhibitor in barley aleurone cells: Comparison to the barley amylase/subtilisin inhibitor, *Planta,* 169:51-63.

Munnik, T., Arisz, S. A., de Vrije, T., and Musgrave, A. (1995). G protein activation stimulates phospholipase D signaling in plants, *Plant Cell,* 7:2197-2210.

Mur, L. A. J., Naylor, G., Warner, S. A. J., Sugars, J. M., White, R. F., and Draper, J. (1996). Salicylic acid potentiates defense gene expression in tissue exhibiting acquired resistance to pathogen attack, *Plant J.,* 9:559-571.

Murata, H., Chatterjee, A., Liu, Y., and Chatterjee, A. K. (1994). Regulation of the production of extracellular pectinase, cellulase, and protease in the soft rot bacterium *Erwinia carotovora* subsp. *carotovora:* Evidence that *aepH* of *E. carotovora* subsp. *carotovora 71* activates gene expression in *E. carotovora* subsp. *atroseptica,* and *Escherichia coli, Appl. Environ. Microbiol.,* 60:3150-3159.

Murata, H., McEvoy, J. L., Chatterjee, A., Collmer, A., and Chatterjee, A. K. (1991). Molecular cloning of an *aepA* gene that activates production of extracellular pectolytic, cellulolytic, and proteolytic enzymes in *Erwinia carotovora* subsp. *carotovora, Mol. Plant-Microbe Interact.,* 4:239-246.

Nakajima, H., Muranaka, T., Ishige, F., Akutsu, K., and Oeda, K. (1997). Fungal and bacterial disease resistance in transgenic plants expressing human lysozyme, *Plant Cell Rep.,* 16:674-679.

Nakayachi, O. (1995). Transposon mutagenesis of *Xanthomonas oryzae* pv. *oryzae* and partial characterization of isolated mutants expressing reduced virulence, *Bulletin of Research Institute of Agricultural Sources,* Ishikawa Agricultural College, No. 4:75-85.

Napoli, C., and Staskawicz, B. J. (1987). Molecular characterization and nucleic acid sequence of an avirulence gene from race 6 of *Pseudomonas syringae* pv. *glycinea, J. Bacteriol.,* 169:572-578.

Narvaez-Vasquez, J., Orozco-Cardenas, M. L., and Ryan, C. A. (1994). Sulfhydryl reagent modulates systemic signaling for wound-induced and systemin-induced proteinase inhibitor synthesis, *Plant Physiol.,* 105:725-730.

Narvaez-Vasquez, J., Pearce, G., Orozco-Cardenas, M. L., Franceschi, V. R., and Ryan, C. A. (1995). Autoradiographic and biochemical evidence for the systemic translocation of systemin in tomato plants, *Planta,* 195:593-600.

Nasser, W., Condemine, G., Plantier, R., Anker, D., and Robert-Baudouy, J. (1991). Inducing properties of analogs of 2-keto-3-deoxygluconate on the expression of pectinase genes of *Erwinia chrysanthemi, FEMS Microbiol. Lett.,* 81:73-78.

Nasser, W., De Tapia, M., and Burkard, G. (1990). Maize pathogenesis-related proteins: characterization and cellular distribution of 1,3-β-glucanases and chitinases induced by brome mosaic virus infection or mercuric chloride treatment, *Physiol. Mol. Plant Pathol.,* 36:1-14.

Nasser, W., Reverchon, S., Condemine, G., and Robert-Baudouy, J. (1994). Specific interactions of *Erwinia chrysanthemi kdgR* repressor with different operators of genes involved in pectinolysis, *J. Mol. Biol.,* 236:427-440.

Nasser, W., Reverchon, S., and Robert-Baudouy, J. (1992). Purification and characterization of the kdgR protein, a major repressor of pectinolysis genes of *Erwinia chrysanthemi, Mol. Microbiol.,* 6:257-265.

Nassuth, A., and Sanger, H. L. (1986). Immunological relationship between "pathogenesis-related" leaf proteins from tomato, tobacco and cowpea, *Virus Res.,* 4:229-242.

Neale, A. D., Wahleithner, J. A., Lund, B., Bonnett, H. T., Kelly, A., Meeks-Wagner, D. R., Peacock, W. J., and Dennis, E. S. (1990). Chitinase, β-1,3-glucanase, osmotin, and extensin are expressed in tobacco explants during flower formation, *Plant Cell,* 2:673-684.

Neema, C., Laulhere, J. P., and Expert, D. (1993). Iron deficiency induced by chrysobactin in *Saintpaulia ionantha* leaves inoculated with *Erwinia chrysanthemi, Plant Physiol.,* 102:967-973.

Neff, N. T., Binns, A. N., and Brandt, C. (1987). Inhibitory effects of a pectin-enriched tomato cell wall fraction on *Agrobacterium tumefaciens* binding and tumor formation, *Plant Physiol.,* 83:525-528.

Nelson, A. J., Doerner, P. W., Zhu, Q., and Lamb, C. J. (1994). Isolation of a monocot 3-hydroxy-3-methylglutaryl coenzymeA reductase gene that is elicitor-inducible, *Plant Mol. Biol.,* 25:401-412.

Nelson, P. E., and Dickey, R. S. (1970). Histopathology of plants infected with vascular bacterial pathogens, *Annu. Rev. Phytopathol.,* 8:259-280.

Nemestothy, G. S., and Guest, D. I. (1990). Phytoalexin accumulation, phenylalanine ammonia-lyase activity and ethylene biosynthesis in fosetyl-Al treated resistant and susceptible tobacco cultivars infected with *Phytophthora nicotinae* var. *nicotinae, Physiol. Mol. Plant Pathol.,* 37:207-219.

Neuenschwander, U., Vernooij, B., Friedrich, L., Uknes, S., Kessmann, H., and Ryals, J. (1995). Is hydrogen peroxide a second messenger of salicylic acid in systemic acquired resistance?, *Plant J.,* 8:227-233.

Neuhaus, G., Neuhaus-Uri, G., Katagiri, F., Seipel, K., and Chua, N. H. (1994a). Tissue-specific expression of *as-1* in transgenic tobacco, *Plant Cell,* 6:827-834.

Neuhaus, J. M. (1999). Plant chitinases. In S. K. Datta and S. Muthukrishnan (Eds.), *Pathogenesis-Related Proteins in Plants,* CRC Press, Boca Raton, FL, pp. 77-105.

Neuhaus, J. M., Ahl-Goy, P., Hinz, U., Flores, S., and Meins, F. (1991). High-level expression of a tobacco chitinase gene in *Nicotiana sylvestris.* Susceptibility of transgenic plants to *Cercospora nicotianae, Plant Mol. Biol.,* 16:141-151.

Neuhaus, J. M., Pietrzak, M., and Boller, T. (1994b). Mutation analysis of the C-terminal vacuolar targeting peptide of tobacco chitinase: low specificity of the sorting system and graduate transition between intracellular retention and secretion into the extacellular space, *Plant J.,* 5:45-54.

Neuhaus, J. M., Sticher, L., Meins, F. Jr., and Boller, T. (1991). A short C-terminal sequence is necessary and sufficient for the targeting of chitinases to the plant vacuole, *Proc. Natl. Acad. Sci. U.S.A,* 88:10362-10366.

Newman, M.-A., Conrads-Strauch, J., Scofield, G., Daniels, M. J., and Dow, J. M. (1994). Defense-related gene induction in *Brassica campestris* in response to defined mutants of *Xanthomonas campestris* with altered pathogenicity, *Mol. Plant-Microbe Interact,* 7:553-563.

Newman, M.-A., Daniels, J. M., and Dow, J. M. (1995). Lipopolysaccharide from *Xanthomonas campestris* induces defense-related gene expression in *Brassica campestris, Mol. Plant-Microbe Interact.,* 8:778-780.

Newman, M.-A., Daniels, M. J., and Dow, J. M. (1997). The activity of lipidA and core components of bacterial lipopolysaccharides in the prevention of the hypersensitive response in pepper, *Mol. Plant-Microbe Interact.,* 10:926-928.

Ni, W., and Trelease, R. N. (1991). Post-transcriptional regulation of catalase isozyme expression in cotton seeds, *Plant Cell,* 3:737-744.

Nicholson, R. L., Kollipara, S. S., Vincent, J. R., Lyon, P. C., and Cadena-Gomez, G. (1987). Phytoalexin synthesis in the sorghum mesocotyl in response to infection by pathogenic and nonpathogenic fungi, *Proc. Natl. Acad. Sci. U.S.A,* 84:5520-5524.

Niderman, T., Genetet, I., Bruyere, T., Gees, R., Stintzi, A., Legrand, M., Fritig, B., and Mosinger, E. (1995). Pathogenesis-related PR-1 proteins are antifungal. Isolation and characterization of three 14-kilodalton proteins of tomato and of a basic PR-1 of tobacco with inhibitory activity against *Phytophthora infestans, Plant Physiol.,* 108:17-27.

Niebel, A., de Almeida Engler, J., Tire, C., Engler, G., Van Montagu, M., and Gheysen, G. (1993). Induction patterns of an extensin gene in tobacco upon nematode infection, *Cell,* 5:1697-1710.

Niebel, A., Heungens, K., Barthels, N., Van Montagu, M. V., and Gheysen, G. (1995). Characterization of a pathogen-induced potato catalase and its systemic expression upon nematode and bacterial infection, *Mol. Plant-Microbe Interact.,* 8:371-378.

Nielsen, K. K., Bojsen, K., Collinge, D. B., and Mikkelsen, J. D. (1994a). Induced resistance in sugar beet against *Cercospora beticola*: Induction by dichloroisonicotinic acid is independent of chitinase and β-1,3-glucanase transcript accumulation, *Physiol. Mol. Plant Pathol.,* 45:89-99.

Nielsen, K. K., Bojsen, K., Roepstorff, P., and Mikkelsen, J. D. (1994b). A hydroxyproline-containing class IV chitinase of sugar beet is glycosylated with xylose, *Plant Mol. Biol.,* 25:241-257.

Nielsen, K. K., Mikkelsen, J. D., Kragh, K. M., and Bojsen, K. (1993). An acidic class III chitinase in sugar beet induction by transgenic tobacco plants, *Mol. Plant-Microbe Interact.,* 6:495-506.

Niemann, G. J. (1993). The anthranilamide phytoalexins of the caryophyllaceae and related compounds, *Phytochemistry,* 34:319-328.

Niepold, F., Anderson, D., and Mills, D. (1985). Cloning determinants of pathogenesis from *Pseudomonas syringae* pathovar *syringae, Proc. Natl. Acad. Sci. U.S.A,* 82:406-410.

Nigg, E. A., Baeuerle, P. A., and Luhrmann, R. (1991). Nuclear import-export: In search of signals and mechanisms, *Cell,* 66:15-22.

Nikaidou, N., Kamio, Y., and Izaki, K. (1993). Molecular cloning and nucleotide sequence of the pectate lyase gene from *Pseudomonas marginalis* N6301, *BioSci. Biotechnol. Biochem.,* 57:957-960.

Niki, T., Mitsuhara, I., Seo, S., Ohtsubo, N., and Ohasshi, Y. (1998). Antagonistic effect of salicylic acid and jasmonic acid on the expression of pathogenesis-related (PR) protein genes in wounded mature tobacco leaves, *Plant Cell Physiol.,* 39:500-507.

Nissinen, R., Lai, F.-M., Laine, M. J., Bauer, P. J., Reilley, A. A., Li, X., De Boer, S. H., Ishimaru, C. A., and Metzler, M. C. (1997). *Clavibacter michiganensis* subsp. *sepedonicus* elicits hypersensitive response in tobacco and secretes hypersensitive response-inducing protein(s), *Phytopathology,* 87:678-684.

Nitti, G., Orru, S., Bloch, C. Jr., Morhy, L., Marino, G., and Pucci, P. (1995). Amino acid sequence and disulphide-bridge pattern of three γ-thionins from *Sorghum bicolor, Eur. J. Biochem.,* 228:250-256.

Nizan, R., Barash, I., Valinsky, L., Lichter, A., and Manulis, S. (1997). The presence of *hrp* genes on the pathogenicity-associated plasmid of the tumorigenic bacterium *Erwinia herbicola* pv. *gypsophilae, Mol. Plant-Microbe Interact.,* 10:677-682.

Nojiri, H., Sugimori, M., Yamane, H., Nishimura, Y., Yamada, A., Shibuya, N., Kodama, O., Murofushi, N., and Omori, T. (1996). Involvement of jasmonic acid in elicitor-induced phytoalexin production in suspension-cultured rice cells, *Plant Physiol.,* 110:387-392.

Norelli, J. L., Aldwinckle, H. S., Destefano-Beltran, L., and Jaynes, J. M. (1994a). Transgenic 'Malling 26' apple expressing the attacin E gene has increased resistance to *Erwinia amylovora, Euphytica,* 77:123-128.

Norelli, J. L., Aldwinckle, H. S., Destefano-Beltran, L., and Jaynes, J. M. (1994b). Transgenic apple plants containing lytic proteins have increased resistance to *Erwinia amylovora, J. Cell. Biochem.* Suppl., 18A:89.

Nothnagel, E. A., McNeil, M., Albersheim, P., and Dell, A. (1983). Host-pathogen interactions. XXII. A galacturonic acid oligosaccharide from plant cell walls elicits phytoalexins, *Plant Physiol.,* 71:916-926.

Nurmiaho-Lassila, E. L., Rantala, E., and Romantschuk, M. (1991). Pilus-mediated adsorption of *Pseudomonas syringae* to the surface of bean leaves, *Micron. Microsc. Acta,* 22:71-72.

Nurnberger, T., Nennsteil, D., Jabs, T., Sacks, W. R., Hahlbrock, K., and Scheel, D. (1994). High affinity binding of a fungal oligopeptide elicitor to parsley plasma membranes triggers multiple defense responses, *Cell,* 78:449-460.

Nuss, L., Mahe, A., Clark, A. J., Grisvard, J., Dron, M., Cervone, F., and De Lorenzo, G. (1996). Differential accumulation of PGIP-inhibiting protein mRNA in two near-isogenic lines of *Phaseolus vulgaris* L. upon infection with *Colletotrichum lindemuthianum, Physiol. Mol. Plant Pathol.,* 48:83-89.

Oba, K., Tatematsu, H., Yamashita, G., and Uritani, I. (1976). Induction of furanoterpene production and formation of the enzyme system from mevalonate to isopentenyl pyrophosphate in sweet potato root tissue injured by *Ceratocystis fimbriata* and by toxic chemicals, *Plant Physiol.,* 58:51-56.

O'Connell, K. P., and Handelsman, J. (1989). *chvA* locus may be involved in export of neutral cyclic β-1,2-linked D-glucan from *Agrobacterium tumefaciens, Mol. Plant-Microbe Interact.,* 2:11-16.

O'Connell, R. J., Brown, I. R., Mansfield, J. W., Bailey, J. A., Mazau, D., Rumeau, D., and Esquerre-Tugaye, M. T. (1990). Immunocytochemical localization of hydroxyproline-rich glycoproteins accumulating in melon and bean at sites of resistance to bacteria and bean at sites of resistance to bacteria and fungi, *Mol. Plant-Microbe Interact.,* 2:33-40.

O'Donnell, P. J., Calvert, C., Atzorn, R., Wasternack, C., and Willmitzer, L. (1996). Ethylene as a signal mediating the wound response of tomato plants, *Science,* 274:1914-1917.

O'Garro, L. W., and Charlemange, E. (1994). Comparison of bacterial growth and activity of glucanase and chitinase in pepper leaf and flower tissue infected with *Xanthomonas campestris* pv. *vesicatoria, Physiol. Mol. Plant Pathol.,* 45:181-188.

Ohana, P., Benziman, M., and Delmer, D. P. (1993). Stimulation of callose synthesis in vivo correlates with changes in intracellular distribution of the callose synthase activator β-furfuryl-β-glucoside, *Plant Physiol.,* 101:187-191.

Ohana, P., Delmer, D. P., Volman, G., Steffens, J. C., Matthews, D. E., and Benziman, M. (1992). β-Furfuryl-β-glucoside: An endogenous activator of higher plant UDP-glucose: (1-3)-β-glucan synthase. Biological activity, distribution, and in vitro synthesis, *Plant Physiol.,* 98:708-715.

Ohashi, Y., and Matsuoka, M. (1987). Localization of pathogenesis-related proteins in the epidermis and intercellular spaces of tobacco leaves after their induction by potassium salicylate or tobacco mosaic virus infection, *Plant Cell Physiol.,* 28:1227-1235.

Ohashi, Y., and Ohshima, M. (1992). Stress-induced expression of genes for pathogenesis-related proteins in plants, *Plant Cell Physiol.,* 33:819-826.

Ohashi, Y., and Shimomura, T. (1982). Modification of cell membranes of leaves systemically infected with tobacco mosaic virus, *Physiol. Plant Pathol.,* 20:125-128.

Ohl, S., Apotheker-de Groot, M., van der Knaap, J. A., Ponstein, A. S., Sela-Buurlage, M. B., Bol, J. F., Cornelissen, B. J. C., Linthorst, H. J. M., and Melchers, L. S. (1994). A new-class of tobacco chitinases homologous to bacterial exo-chitinases is active against fungi in vitro, *J. Cellular Biochemistry Suppl.,* 18A:90.

Ohme-Takagi, M., and Shinshi, H. (1990). Structure and expression of a tobacco ß-1,3-glucanase gene, *Plant Mol. Biol.,* 15:941-946.

Ohme-Takagi, M., and Shinshi, H. (1995). Ethylene-inducible DNA binding proteins that interact with an ethylene-responsive element, *Plant Cell,* 7:173-182.

Ohno, Y., Okuda, S., Natsuaki, T., and Teranaka, M. (1992). Control of bacterial seedling blight of rice by fluorescent *Pseudomonas* spp, *Proc. Kanto-Tosan Plant Prot. Soc.,* 39:9-11.

Ohta, H., Shida, K., Peng, Y.-L., Furusawa, I., Shishiyama, J., Aibara, S., and Morita, Y. (1991). A lipoxygenase pathway is activated in rice after infection with the rice blast fungus *Magnaporthe grisea, Plant Physiol.,* 97:94-98.

Oi, X., and Mort, A. J. (1990). Co-solubilization of hydroxyproline and pectin. Is there a link between the two?, *Plant Physiol.,* 93:S-92.

Oku, T., Alvarez, A. M., and Kado, C. (1995). Conservation of the hypersensitivity-pathogenicity regulatory gene *hrpX* of *Xanthomonas campestris* and *X. oryzae, DNA Sequence,* 5:245-249.

Omidiji, O., and Ehimidu, J. (1990). Changes in the content of antibacterial isorhamnetin 3-glucoside and quercetin 3'-glucoside following inoculation of onion (*Allium cepa* L. cv. Red Creole) with *Pseudomonas cepacia, Physiol. Mol. Plant Pathol.,* 37:287-292.

Ori, N., Eshed, Y., Paran, Y., Presting, G., Aviv, D., Tanksley, S., Zamir, D., and Fluhr, R. (1997). The I2C family from the wilt disease resistance locus I2 belongs to the nucleotide binding, leucine-rich repeat superfamily of plant resistance genes, *Plant Cell,* 9:521-532.

Ori, N., Sessa, G., Lotan, T., Himmelhoch, S., and Fluhr, R. (1990). A major stylar matrix polypetide (sp 41) is a member of the pathogenesis-related proteins superclass, *EMBO J.,* 9:3429-3436.

Orlandi, W. E., Hutcheson, S. W., and Baker, C. J. (1992). Early physiological responses associated with race-specific recognition in soybean leaf tissue and cell suspensions treated with *Pseudomonas syringae* pv. *glycinea, Physiol. Mol. Plant Pathol.,* 40:173-180.

Orr, J. D., Edwards, R., and Dixon, R. A. (1993). Stress responses in alfalfa (*Medicago sativa* L.). XIV. Changes in the level of phenylpropanoid pathway intermediates in relation to regulation of L-phenylalnine ammonia-lyase in elicitor-treated cell-suspension cultures, *Plant Physiol.,* 101:847-856.

Oshima, M., Itoh, H., Matsuoka, M., Murakami, T., and Ohashi, Y. (1990). Analysis of stress-induced or salicylic-induced expression of the pathogenesis-related 1a protein gene in transgenic tobacco, *Plant Cell,* 2:95-106.

Oshima, M., Matsuoka, M., Yamamoto, N., Tanaka, Y., KanoMurakami, Y., Ozeki, Y., Kato, A., Harad, N., and Ohashi, Y. (1987). Nucleotide sequence of the PR-1 gene of *Nicotiana tabacum, FEBS Lett.,* 225:243-246.

Ottow, J. C. G. (1975). Ecology, physiology and genetics of fimbriae and pili, *Annu. Rev. Microbiol.,* 29:79-108.

Pagel, W., and Heitfuss, R. (1989). Calcium content and cell wall polygalacturonans in potato tubers of cultivars with different susceptibilities to *Erwinia carotovora* subsp. *atroseptica, Physiol. Mol. Plant Pathol.,* 18:361-387.

Palme, K. (1992). Molecular analysis of plant signalling elements: Relevance of eukaryotic signal transduction models, *International Review of Cytology,* 132:223-283.

Palomaki, T., and Saarilahti, H. T. (1995). The extreme C-terminal is required for secretion of both the native polygalacturonase (PehA) and PehA-Bla hybrid proteins in *Erwinia carotovora* subsp. *carotovora, Mol. Microbiol.,* 17:449-459.

Palva, T. K., Holmstrom, K.-O., Hemo, P., and Palvio, E. T. (1993). Induction of plant defense response by exoenzymes of *Erwinia carotovora* subsp. *carotovora, Mol. Plant-Microbe Interact.,* 6:190-196.

Palva, T. K., Hurtig, M., Saindrenan, P., and Palva, E. T. (1994). Salicylic acid induced resistance to *Erwinia carotovora* in tobacco, *Mol. Plant-Microbe Interact.,* 7:356-363.

Pan, S. Q., Ye, X. S., and Kuc, J. (1989). Direct detection of β-1,3-glucanase isozymes on polyacrylamide electrophoresis and isoelectrofocusing gels, *Anal. Biochem.,* 182:136-140.

Pan, S. Q., Ye, X. S., and Kuc, J. (1991). Association of β-1,3-glucanase activity and isoform pattern with systemic resistance to blue mold in tobacco induced by stem injection with *Peronospora tabacina* or leaf inoculation with tobacco mosaic virus, *Physiol. Mol. Plant Pathol.,* 39:25-39.

Panopoulos, N. J., Lindgren, P. B., Willis, D. K., and Peet, R. C. (1985). Clustering and conservation of genes controlling the interactions of *Pseudomonas syringae* pathovars with plants. In I. Sussex, A. Ellingboe, M. Crouch, and R. Malmberg (Eds.), *Current Communications in Molecular Biology: Plant Cell/Cell Interactions,* Cold Spring Harbor Laboratory, Cold Spring Harbor, NY, pp. 69-75.

Park, H., Denbow, C. J., and Cramer, C. L. (1992). Structure and nucleotide sequence of tomato HMG2 encoding 3-hydroxy-3-methyl-glutaryl coenzyme A reductase, *Plant Mol. Biol.,* 20:327-331.

Park, W. S., Denbow, C. J., and Cramer, C. L. (1989). Molecular cloning of a tomato HMG CoA reductase gene and its defense-related expression, *Phytopathology,* 79:1198.

Parker, J. E., Szabo, V., Staskawicz, B. J., Lister, C., Dean, C., Daniels, M. J., and Jones, J. D. G. (1993). Phenotypic characterization and molecular mapping of the *Arabidopsis thaliana* locus RPP5, determining disease resistance to *Peronospora parasitica, Plant J.,* 4:821-831.

Parkinson, J. S., and Kofoid, E. C. (1992). Communication modules in bacterial signaling proteins, *Annu. Rev. Genet.,* 26:71-112.

Parkos, C. A., Allen, R. A., Cochrane, C. G., and Jesaitis, A. J. (1987). Purified cytochrome-b from human granulocyte plasma membrane is comprised of 2 polypeptides with relative molecular weights of 91,000 and 22,000, *J. Clin. Invest.,* 80:732-742.

Parsot, C., Menard, R., Gounon, P., and Sansonetti, P. (1995). Enhanced secretion through the *Shigella flexneri* Mxi-Spa translocon leads to assembly of extracellular proteins into macromolecular structures, *Mol. Microbiol.,* 16:291-300.

Patzlaff, M., and Barz, W. (1978). Peroxidatic degradation of flavanones, *Z. fur Naturforschung,* 33C:675-684.

Pautot, V., Holzer, F. M., and Walling, L. (1991). Differential expression of tomato proteinase inhibitor I and II genes during bacterial pathogen invasion and wounding, *Mol. Plant-Microbe Interact.,* 4:284-292.

Pavlovkin, J., Novacky, A., and Ullrich-Eberius, C. I. (1986). Membrane potential changes during bacteria-induced hypersensitive reaction, *Physiol. Mol. Plant Pathol.,* 28:125-135.

Paxton, J. D. (1980). A new working definition of the term "phytoalexin," *Plant Dis.,* 64:734.

Paxton, J. D. (1981). Phytoalexins—a working redefinition, *Phytopathol. Z.,* 101: 106-109.

Payne, G., Ahl, P., Moyer, M., Harper, A., Beck, J., Meins, F. Jr., and Ryals, J. (1990a). Isolation of complementary DNA clones encoding pathogenesis-related proteins P and Q, two acidic chitinases from tobacco, *Proc. Natl. Acad. Sci. U.S.A.,* 87:98-102.

Payne, G., Middlesteadt, W., Desai, N., Williams, S., Dincher, S., Carnes, M., and Ryals, J. (1989). Isolation and sequence of a genomic clone encoding the basic form of pathogenesis-related protein 1 from *Nicotiana tabacum, Plant Mol. Biol.,* 12:595-596.

Payne, G., Middlesteadt, W., Williams, S., Desai, N., Parks, D., Dincher, S., Carnes, M., and Ryals, J. (1988). Isolation and nucleotide sequence of a novel cDNA clone encoding the major form of pathogenesis-related protein R, *Plant Mol. Biol.,* 11:223-234.

Payne, G., Ward, E., Gaffney, T., Ahl Goy, P., Moyer, M., Harper, A., Meins, F., and Ryals, J. (1990b). Evidence for a third structural class of β-1,3-glucanase in tobacco, *Plant Mol. Biol.,* 15:797-808.

Payne, J. H., Schoedel, C., Keen, N. T., and Collmer, A. (1987). Multiplication and virulence in plant tissues of *Escherichia coli* clones producing pectate lyase isozymes PLb and Ple at high levels and of an *Erwinia chrysanthemi* mutant deficient in Ple, *Appl. Environ. Microbiol.,* 53:2315-2320.

Pazour, G. J., and Das, A. (1990). *VirG,* an *Agrobacterium tumefaciens* transcriptional activator, initiates translation at a UUG codon and is a sequence-specific DNA-binding protein, *J. Bacteriol.,* 172:1241-1249.

Pazour, G. J., Ta, C. N., and Das, A. (1991). Mutants of *Agrobacterium tumefaciens* with elevated *vir* gene expression, *Proc. Natl. Acad Sci. U.S.A.,* 88:6941-6945.

Pearce, G., Johnson, S., and Ryan, C. A. (1993). Structure-activity of deleted and substituted systemin, an 18-amino acid polypeptide inducer of plant defensive genes, *J. Bio. Chem.,* 268:212-216.

Pearce, R. B., Edwards, P. P., Green, T. L., Anderson, P. A., Fisher, B. J., Carpenter, T. A., and Hall, L. D. (1997). Immobilized long-lived free radicals at the host-pathogen interface in sycamore (*Acer pseudoplatanus* L.), *Physiol. Mol. Plant Pathol.,* 50:371-390.

Pellegrini, L., Rohfritsch, O., Fritig, B., and Legrand, M. (1994). Phenylalanine ammonia-lyase in tobacco. Molecular cloning and gene expression during the hypersensitive reaction to tobacco mosaic virus and the response to a fungal elicitor, *Plant Physiol.,* 106:877-886.

Pena-Cortes, H., Fisahn, J., and Willmitzer, L. (1995). Signals involved in wound-induced proteinase inhibitor II gene expression in tomato and potato plants, *Proc. Natl. Acad. Sci. U.S.A.,* 92:4106-4113.

Pena-Cortes, H., Prat, S., Atzorn, R., Wasternack, C., and Willmitzer, L. (1996). Abscisic acid-deficient plants do not accumulate proteinase inhibitor II following systemin treatment, *Planta,* 198:447-451.

Peng, M., and Kuc, J. (1992). Peroxidase-generated hydrogen peroxide as source of antifungal activity in vitro and on tobacco leaf disks, *Phytopathology,* 82:696-699.

Pennazio, S., Colaracio, D., Roggero, P., and Lenzi, R. (1987). Effect of salicylate stress on the hypersensitive reaction of asparagus bean to tobacco necrosis virus, *Physiol. Mol. Plant Pathol.,* 30:347-357.

Pennazio, S., and Roggero, P. (1990). Induction of pathogenesis-related proteins by spermidine exogenously supplied to detached tobacco leaves, *Biologia Plantarum (Praha),* 32:241-246.

Penninckx, I. A. M. A., Eggermont, K., Terras, F. R. G., Thomma, B. P. H. J., De Samblanx, G. W., Buchala, A., Metraux, J. P., Manners, J. M., and Broekaert, W. F. (1996). Pathogen-induced systemic activation of a plant defensin gene in *Arabidopsis* follows a salicylic acid independent pathway, *Plant Cell,* 8:2309-2323.

Penninckx, I. A. M. A., Thomma, B. P. H. J., Buchala, A., Metraux, J. P., and Broekaert, W. F. (1998). Concomitant activation of jasmonate and ethylene responses is required for induction of a plant defensin gene in *Arabidopsis, Plant Cell,* 10:2103-2113.

Perombelon, M. C. M., and Lowe, R. (1980). Studies on the initiation of bacterial soft rot in potato tubers, *Potato Research,* 18:64-82.

Peumans, W. J., and Van Damme, E. J. M. (1995). Lectins as plant defense proteins, *Plant Physiol.,* 109:347-352.

Pfaffmann, H., Hartmann, E., Brightman, A. O., and Morre, D. J. (1987). Phosphotidylinositol specific phospholipase C of plant stems. Membrane associated activity concentrated in plasma membranes, *Plant Physiol.,* 85:1151-1155.

Pfitzner, U. M., and Goodman, H. M. (1987). Isolation and characterization of cDNA clones encoding pathogenesis-related proteins from tobacco mosaic virus-infected plants, *Nucleic Acids Res.,* 15:4449-4465.

Pfitzner, U. M., Pfitzner, A. J. P., and Goodman, H. M. (1988). DNA sequence analysis of a PR-1a gene from tobacco. Molecular relationship of heat shock and pathogen responses in plants, *Mol. Gen. Genet.,* 211:290-295.

Pierce, M., and Essenberg, M. (1987). Localization of phytoalexins in fluorescent mesophyll cells isolated from bacterial blight-infected cotton cotyledons and separated from other cells by fluorescence-activated cell sorting, *Physiol. Mol. Plant Pathol.,* 31:273-290.

Pierpoint, W. S., Gordon-Weeks, R., and Jackson, P. J. (1992). The occurrence of the thaumatin-like, pathogenesis related protein, PR-5, in intercellular fluids from *Nicotiana* species and from an interspecific *Nicotiana* hybrid, *Physiol. Mol. Plant Pathol.,* 41:1-10.

Pierpoint, W. S., Jackson, P. J., and Evans, R. M. (1990). The presence of a thaumatin-like protein, a chitinase and a glucanase among the pathogenesis-related proteins of potato *(Solanum tuberosum), Physiol. Mol. Plant Pathol.,* 36:325-338.

Pierpoint, W. S., Robinson, N. P., and Leason, M. P. (1981). The pathogenesis-related proteins of tobacco: Their induction by viruses in intact plants and their induction by chemicals in detached leaves, *Physiol. Plant Pathol,* 19:85-95.

Pieterse, C. M. J., Van Wees, S. C. M., Hoffland, E., Van Pelt, J. A., and Van Loon, L. C. (1996). Systemic resistance in *Arabidopsis* induced by biocontrol bacteria is independent of salicylic acid accumulation and pathogenesis-related gene expression, *Plant Cell,* 8:1225-1237.

Pineiro, M., Diaz, I., Rodriquez-Palenzuela, P., Titarenko, E., and Garcia-Olmedo, F. (1995). Selective disulphide linkage of plant thionins with other proteins, *FEBS Lett.,* 369:239-242.

Pirhonen, M., Flego, D., Heikinheimo, R., and Palva, E. T. (1993). A small diffusible signal molecule is responsible for the global control of virulence and exoenzyme production in the plant pathogen *Erwinia carotovora, EMBO J.,* 12:2467-2476.

Pirhonen, M. U., Lidell, M. C., Rowley, D. L., Lee, S. W., Jin, S., Liang, Y., Silverstone, S., Keen, N. T., and Hutcheson, S. W. (1996). Phenotypic expression of *Pseudomonas syringae avr* genes in *E. coli* is linked to the activities of the *hrp*-encoded secretion system, *Mol. Plant-Microbe Interact.,* 9:252-260.

Pirhonen, M., and Palva, E. T. (1988). Occurrence of bacteriophage T4 receptor in *Erwinia carotovora, Mol. Gen. Genet.,* 214:170-172.

Pirhonen, M., Saarilahti, H., Karlsson, M. B., and Palva, E. T. (1991). Identification of pathogenicity determinants of *Erwinia carotovora* subsp. *carotovora* by transposon mutagenesis, *Mol. Plant-Microbe Interact.,* 4:276-283.

Pishchik, V. N., Chernyaeva, I. I., Vorobev, N. I., and Lazarev, A. M. (1996). Characteristic features of virulent and avirulent strains of *Erwinia carotovora, Microbiology (New York),* 65:232-237.

Pissavin, C., Robert-Baudouy, J., and Hugouvieux-Cotte-Pattat, N. (1996). Regulation of *pelZ,* a gene of the *pelB- pelC* cluster encoding a new pectate lyase of *Erwinia chrysanthemi* 3937, *J. Bacteriol.,* 178:7187-7196.

Plano, G. V., Barve, S. S., and Straley, S. C. (1991). LcrD, a membrane-bound regulator of the *Yersinia pestitis* low-calcium response, *J. Bacteriol.,* 173:7293-7303.

Politis, D. J., and Goodman, R. N. (1978). Localized cell wall appositions: Incompatibility response of tobacco leaf cells to *Pseudomonas pisi, Phytopathology,* 68:309-316.

Ponstein, A. S., Bres-Vloemans, A. A., Sella-Buurlage, M. B., Cornelissen, B. J. C., and Melchers, L. S. (1994). The "Missing" class I PR-4 protein from tobacco exhibits antifungal activity, *J. Cellular Biochemistry,* Suppl., 18A:90.

Pontier, D., Godiard, L., Marco, Y., and Roby, D. (1994). *Hsr203J,* a tobacco gene whose activation is rapid, highly localized and specific for incompatible plant/pathogen interactions, *Plant J.,* 5:507-521.

Popham, P., Pike, S., and Novacky, A. (1993). Membrane potential alterations induced by the bacterial HR-elicitor harpin, *Plant Physiol. Suppl.,* 102:111.

Popham, P., Pike, S., and Novacky, A. (1995). The effect of harpin from *Erwinia amylovora* on the plasmalemma of suspension-cultured tobacco vells, *Physiol. Mol. Plant Pathol.,* 47:39-50.

Powell, B., and Kado, C. (1990). Specific binding of VirG to the *vir* box requires a C-terminal domain and exhibits a minimum concentration threshold, *Mol. Microbiol.,* 4:1-8.

Powell, R. G., Te Packe, M. R., Plattner, R. D., White, J. F., and Clement, S. L. (1994). Isolation of resveratrol from *Festuca versuta* and evidence for the widespread occurrence of this stilbene in the Poaceae, *Phytochemistry,* 35:335-338.

Pozsar-Hajnal, K., and Polcasek-Racz, M. (1975). Determination of pectin methyl esterase, polygalacturonase and pectic substances in some fruits and vegetables, Part I—study into the pectolytic enzyme content in tomatoes, *Acta Alimentaria,* 4:271-289.

Praillet, T., Nasser, W., Robert-Baudouy, J., and Reverchon, S. (1996). Purification and functional characterization of PecS: A regulator of virulence factor synthesis in *Erwinia chrysanthemi, Mol. Microbiol.,* 20:391-402.

Prasad, T. K., Anderson, M. D., Martin, M. A., and Stewart, C. R. (1994). Evidence for chilling-induced oxidative stress in maize seedlings and a regulatory role for hydrogen peroxide, *Plant Cell,* 6:65-74.

Press, C. M., Wilson, M., and Kloepper, J. W. (1996). Salicylate and plant growth-promoting rhizobacteria-mediated induced systemic disease resistance. In G. Stacey, B. Mullin, and P. M. Gresshoff (Eds.), *Proc. 8th Int. Congr. Mol. Plant-Microbe Interact.,* The University of Tennesse, Knoxville, Abstract A-17.

Press, C. M., Wilson, M., Tuzun, S., and Kloepper, J. W. (1997). Salicylic acid produced by *Serratia marcescens* 90-166 is not the primary determinant of induced systemic resistance in cucumber or tobacco, *Mol. Plant-Microbe Interact.,* 10:761-768.

Pressey, R. (1996). Polygalacturonase inhibitors in bean pods, *Phytochemistry,* 42:1267-1270.

Preston, G., Huang, H.-C., He, S. Y., and Collmer, A. (1995). The HrpZ proteins of *Pseudomonas syringae* pvs. *syringae, glycinea,* and *tomato* are encoded by an operon containing *Yersinia ysc* homologs and elicit the hypersensitive response in tomato but not soybean, *Mol. Plant-Microbe Interact.,* 8:717-732.

Preston, T. F., Rice, J. D., Ingram, L. O., and Keen, N. T. (1992). Differential polymerization mechanisms of pectate lyases secreted by *Erwinia chrysanthemi* EC16, *J. Bacteriol.,* 174:2039-2042.

Pu, X. A., and Goodman, R. N. (1993). Attachment of agrobacteria to grape cells, *Appl. Environ. Microbiol.,* 59:2572-2577.

Pugsley, A. P. (1993). The complete general protein secretory pathway in Gram-negative bacteria, *Microbiol. Revs.,* 57:50-108.

Pugsley, A. P., d'Enfert, C., Reyss, T., and Kornacker, M. G. (1990). Genetics of extracellular protein secretion by gram-negative bacteria, *Annu. Rev. Genet.,* 24:67-90.

Puri, N., Jenner, C., Bennett, M., Stewart, R., Mansfield, J., Lyons, N., and Taylor, J. (1997). Expression of *avrPphB,* an avirulence gene from *Pseudomonas syringae* pv. *phaseolicola,* and the delivery of signals causing the hypersensitive reaction in bean, *Mol. Plant-Microbe Interact.,* 10:247-256.

Puvanesarajah, V., Schell, F. M., Stacey, G., Douglas, C. J., and Nester, E. W. (1985). Role for 2-linked β-D-glucan in the virulence of *Agrobacterium tumefaciens, J. Bacteriol.,* 164:102-106.

Py, B., Salmond, G. P. C., Chippaux, M., and Barras, F. (1991). Secretion of cellulases in *Erwinia chrysanthemi* and *Erwinia carotovora* is species specific, *FEMS Microbiol. Lett.,* 79:315-322.

Pyee, J., Yu, H., and Kolattukudy, P. E. (1994). Identification of a lipid transfer protein as the major protein in the surface wax of broccoli *(Brassica oleracea)* leaves, *Arch. Biochem. Biophys.,* 311:460-468.

Qin, X. F., Holuigue, L., Horvath, D. M., and Chua, N. H. (1994). Immediate early transcription activation by salicylic acid via the cauliflower mosaic virus *as-1* element, *Plant Cell,* 6:863-874.

Raaijmakers, J. M., Leeman, M., Van Oorschot, M. M. P., Van der Sluis, I., Schippers, B., and Bakker, P. A. H. M. (1995). Dose-response relationship in biological control of fusarium wilt of radish by *Pseudomonas* sp., *Phytopathology,* 85:1075-1081.

Rabindran, R., and Vidhyasekaran, P. (1996). Development of a formulation of *Pseudomonas fluorescens* PfALR 2 of management of rice sheath blight, *Crop Protection,* 15:715-721.

Rahimi, S., Perry, R. N., and Wright, D. J. (1996). Identification of pathogenesis-related proteins induced in leaves of potato plants infected with potato cyst nematodes, *Globodera* species, *Physiol. Mol. Plant Pathol.,* 49:49-59.

Rahme, L. G., Mindrinos, M. N., and Panopoulos, N. J. (1991). The genetic and transcriptional organization of the *hrp* cluster of *Pseudomonas syringae* pathovar *phaseolicola, J. Bacteriol.,* 173:575-586.

Rahme, L. G., Mindrinos, M. N., and Panopoulos, N. J. (1992). Plant and environmental sensory signals control the expression of *hrp* genes in *Pseudomonas syringae* pv. *phaseolicola, J. Bacteriol.,* 174:3499-3507.

Raikhel, N. V., and Wilkins, T. A. (1987). Isolation and characterization of a cDNA clone encoding wheat germ agglutinin, *Proc. Natl. Acad. Sci. U.S.A.,* 84:6745-6749.

Raina, S., Raina, R., Venkatesh, T. V., and Das, H. K. (1995). Isolation and characterization of a locus from *Azospirillum brasilense* Sp7 that complements the tumorigenic defect of *Agrobacterium tumefaciens chvB* mutant, *Mol. Plant-Microbe Interact.,* 8:322-326.

Ranjeva, R., and Boudet, A. M. (1987). Phosphorylation of proteins in plants: Regulatory effects and potential involvement in stimulus/response coupling, *Annu. Rev. Plant Physiol.,* 38:73-93.

Rasi-Caldogno, F., Pugliarello, M. C., Olivari, C., and De Michelis, M. I. (1993). Controlled proteolysis mimics the effect of fusicoccin on the plasma membrane H^+-ATPase, *Plant Physiol.,* 103:391-398.

Raskin, I., Skubatz, Z., Tang, W., and Meeuse, B. J. D. (1990). Salicylic acid levels in thermogenic and non-thermogenic plants, *Ann. Bot.,* 66:376-373.

Rasmussen, J. B., Hammerschmidt, R., and Zook, M. N. (1991). Systemic induction of salicylic acid accumulation in cucumber after inoculation with *Pseudomonas syringae* pv. *syringae, Plant Physiol.,* 97:1342-1347.

Raupach, G. S., Liu, L., Murphy, J. F., Tuzun, S., and Kloepper, J. W. (1996). Induced systemic resistance in cucumber and tomato against cucumber mosaic cucumovirus using plant growth-promoting rhizobacteria (PGPR), *Plant Dis.,* 80:1107-1108.

Raz, R., Cretin, C., Puigdomenech, P., and Martinez-Izquierdo, J. A. (1991). The sequence of a hydroxyproline-rich glycoprotein gene from *Sorghum vulgare, Plant Mol. Biol.,* 16:365-367.

Raz, V., and Fluhr, R. (1993). Ethylene signal is transduced via protein phosphorylation events in plants, *Plant Cell,* 5:523-530.

Ream, W. (1989). *Agrobacterium tumefaciens* and interkingdom genetic exchange, *Annu. Rev. Phytopathol.,* 27:227-234.

Rebmann, G., Hertig, C., Bull, J., Mauch, F., and Dudler, R. (1991a). Cloning and sequencing of cDNAs encoding a pathogen-induced putative peroxidase of wheat (*Triticum aestivum* L.), *Plant Mol. Biol.,* 16:329-331.

Rebmann, G., Mauch, F., and Dudler, R. (1991b). Sequence of a wheat cDNA encoding a pathogen-induced thaumatin-like protein, *Plant Mol. Biol.,* 17:283-285.

Reeves, P. J., Whitcombe, D., Wharam, S., Gibson, M., Allison, G., Bunce, N., Barallon, R., Douglas, P., Mulholland, V., Stevens, S., Walker, D., and Salmond, G. P. C. (1993). Molecular cloning and characterization of 13 *out* genes from *Erwinia carotovora* ssp. *carotovora:* Genes encoding members of a general secretion pathway (GSP) wide-spread in gram negative bacteria, *Mol. Microbiol.,* 8:443-456.

Reimers, P. J., Guo, A., and Leach, J. E. (1992). Increased activity of a cationic peroxidase associated with an incompatible interaction between *Xanthomonas oryzae* pv. *oryzae* and rice *(Oryza sativa), Plant Physiol.,* 99:1044-1050.

Reimers, P. J., and Leach, P. E. (1991). Race-specific resistance to *Xanthomonas oryzae* pv. *oryzae* conferred by bacterial blight resistance gene *Xa-10* in rice *(Oryza sativa)* involves accumulation of a lignin-like substance in host tissues, *Physiol. Mol. Plant Pathol.,* 38:39-55.

Reimers, P. J., Ringl, C., and Dudler, R. (1992). Complementary DNA cloning and sequence analysis of a pathogen-induced peroxidase from rice, *Plant Physiol.,* 100:1611-1612.

Reimmann, C., and Dudler, R. (1993). cDNA cloning and sequence analysis of a pathogen-induced thaumatin-like protein from rice *(Oryza sativa), Plant Physiol.,* 101:1113-1114.

Reinecke, T., and Kindl, H. (1994). Inducible enzymes of the 9,10-dihydrophenanthrene pathway. Sterile orchid plants responding to fungal infection, *Mol. Plant-Microbe Interact.,* 7:449-454.

Ren, Y. Y., and West, C. A. (1992). Elicitation of diterpene biosynthesis in rice (*Oryza sativa* L.) by chitin, *Plant Physiol.,* 99:1169-1178.

Renelt, A., Colling, C., Hahlbrock, K., Nurnberger, T., Parker, J. E., Sacks, W. R., and Scheel, D. (1993). Studies on elicitor recognition and signal transduction in plant defense, *J. Exp. Bot.,* Suppl. 44:257-268.

Reverchon, S., Huang, Y., Bourson, C., and Robert-Baudouy, J. (1989). Nucleotide sequences of the *Erwinia chrysanthemi ogl* and *pelE* genes negatively regulated by the *kdgR* gene product, *Gene,* 85:125-134.

Reverchon, S., Nasser, W., and Robert-Baudouy, J. (1991). Characterization of *kdgR,* a gene of *Erwinia chrysanthemi* that regulates pectin degradation, *Mol. Microbiol.,* 5:2203-2216.

Reverchon, S., Nasser, W., and Robert-Baudouy, J. (1994). *pecS:* A locus controlling pectinase, cellulase and blue pigment production in *Erwinia chrysanthemi, Mol. Microbiol.,* 11:1127-1139.

Reymond, P., Grunberger, S., Paul, K., Mueller, M., and Farmer, E. E. (1995). Oligogalacturonide defense signals in plants: Large fragments interact with the plasma membrane in vitro, *Proc. Natl. Acad. Sci. U.S.A.,* 92:4145-4149.

Rich, J. J., Kinscherf, T. G., Kitten, T., and Willis, D. K. (1994). Genetic evidence that the *gacA* gene encodes the cognate response regulator for the *lemA* sensor in *Pseudomonas syringae, J. Bacteriol.,* 176:7468-7475.

Rich, P. R., and Bonner, W. D. Jr. (1978). The sites of superoxide anion generation in higher plant mitochondria, *Arch. Biochem. Biophys.,* 188:206-213.

Richard, K., Montserrat, A., Hoebeke, J., Meeks-Wagner, R., and Tran Thanh Van, K. (1992). Immunological evidence of thaumatin-like proteins during tobacco floral differentiation, *Plant Physiol.,* 98:337-342.

Rickauer, M., Brodschelm, W., Bottin, A., Veronesi, C., Grimal, H., and Esquerre-Tugaye, M. T. (1997). The jasmonate pathway is involved differentially in the regulation of different defence responses in tobacco cells, *Planta,* 202:155-162.

Rickauer, M., Fournier, J., Pouenat, M. L., Berthalon, E., Bottin, A., and Esquerre-Tugaye, M. T. (1990). Early changes in ethylene synthesis and lipoxygenase activity during defense induction in tobacco cells, *Plant Physiol. Biochem.,* 28: 647-653.

Ried, J. L., and Collmer, A. (1988). Construction and characterization of an *Erwinia chrysanthemi* mutant with directed deletions in all of the pectate lyase structural genes, *Mol. Plant-Microbe Interact.,* 1:32-38.

Ritter, C., and Dangl, J. L. (1995). The *avrRpm1* gene of *Pseudomonas syringae* pv. *maculicola* is required for virulence on *Arabidopsis, Mol. Plant-Microbe Interact.,* 8:444-453.

Robb, J., Lee, S. W., Mohan, R. and Kolattukudy, P. E. (1991). Chemical characterization of stress-induced vascular coating, *Plant Physiol.,* 97:528-536.

Robbins, M. P., Bolwell, G. P., and Dixon, R. A. (1985). Metabolic changes in elicitor-treated bean cells. Selectivity of enzyme induction in relation to phytoalexin accumulation, *Eur. J. Biochem.,* 148:563-569.

Rober, K. C. (1989). Untersuchungen zur Dynamik der Polyphenol-und Phytoalexin synthese faulein-fizierter kartoffelknollen, *Biochemie und Physiologie der Pflanzen,* 184:277-284.

Roberts, D. P., Berman, P. M., Allen, C., Stromberg, V. K., Lacy, G. H., and Mount, M. S. (1986a). *Erwinia carotovora:* Molecular cloning of a 3.4 kilobase DNA fragment mediating production of pectate lyases, *Can. J. Plant Pathol.,* 8:17-27.

Roberts, D. P., Berman, P. M., Allen, C., Stromberg, V. K., Lacy, G. H., and Mount, M. S. (1986b). Requirement for two or more *Erwinia carotovora* subsp. *carotovora* pectolytic gene products for maceration of potato tuber tissue by *Escherichia coli, J. Bacteriol.,* 167:279-284.

Roberts, D. P., Denny, T. P., and Schell, M. A. (1988). Cloning of the *egl* gene of *Pseudomonas solanacearum* and analysis of its role in phytopathogenicity, *J. Bacteriol.,* 170:1445-1451.

Roberts, K. (1994). The plant extracellular matrix: In a new expansive mood, *Curr. Opinions Cell Biology,* 6:688-694.

Roberts, W. K., and Selitrennikoff, C. P. (1988). Plant and bacterial chitinases differ in antifungal activity, *J. Gen. Microbiol.,* 134:169-176.

Roberts, W. K., and Selitrennikoff, C. P. (1990). Zeamatin, an antifungal protein from maize with membrane-permeabilizing activity, *J. Gen. Microbiol.,* 136:1771-1778.

Robertson, B. (1986). Elicitors of the production of lignin-like compounds in cucumber hypocotyls, *Physiol. Mol. Plant Pathol.,* 28:137-148.

Robertson, W. M., Lyon, G. D., and Henry, C. E. (1985). Modification of the outer surface of *Erwinia carotovora* ssp. *atroseptica* by the phytoalexin rishitin, *Can. J. Microbiol.,* 31:1108-1112.

Robertson Crews, J. L., Colby, S., and Matthysse, A. G. (1990). *Agrobacterium rhizogenes* mutants that fail to bind to plant cells, *J. Bacteriol.,* 172:6182-6188.

Roby, D., Broglie, K., Cressman, R., Biddle, P., Chet, I., and Broglie, R. (1990). Activation of bean chitinase promoter in transgenic tobacco plants by phytopathogenic fungi, *Plant Cell,* 2:999-1007.

Roby, D., Toppan, A., and Esquerre-Tugaye, M. T. (1985). Cell surfaces in plant-microorganism interactions. V. Elicitors of fungal and of plant origin trigger the synthesis of ethylene and of cell wall hydroxyproline-rich glycoprotein in plants, *Plant Physiol.,* 77:700-704.

Rogers, E. E., Glazebrook, J., and Ausubel, F. M. (1996). Mode of action of the *Arabidopsis thaliana* phytoalexin camalexin and its role in *Arabidopsis*-pathogen interactions, *Mol. Plant-Microbe Interact.,* 9:748-757.

Rogers, K. R., Albert, F., and Anderson, A. J. (1988). Lipid peroxidation is a consequence of elicitor activity, *Plant Physiol.,* 86:547-553.

Roggen, H. P., and Stanley, R. G. (1969). Cell-wall hydrolysing enzymes in wall formation as measured by pollen-tube extension, *Planta,* 84:295-303.

Rogowsky, P. M., Close, T. J., Chimera, J. A., Shaw, J. J., and Kado, C. I. (1987). Regulation of the *vir* genes of *Agrobacterium tumefaciens* plasmid pTiC58, *J. Bacteriol.,* 169:5101-5112.

Rogowsky, P. M., Powell, B. S., Shirasu, K., Lin, T.-S., Morel, P., Zyprian, E. M., Steck, T. R., and Kado, C. I. (1990). Molecular characterization of the *vir* regulon of *Agrobacterium tumefaciens:* Complete nucleotide sequence and gene organization of the 28.63 kbp region cloned as a single unit, *Plasmid,* 23:85-106.

Roine, E., Wei, W. S., Yuan, J., Nurmiaho-Lassila, E. L., Kalkkinen, N., Romantschuk, M., and He, S. Y. (1997). Hrp pilus: An hrp-dependent bacterial surface appendage produced by *Pseudomonas syringae* pv. *tomato* DC 3000, *Proc. Natl. Acad. Sci. U.S.A.,* 94:3459-3464.

Romantschuk, M. (1992). Attachment of plant pathogenic bacteria to plant surfaces, *Annu. Rev. Phytopathol.,* 30:225-243.

Romantschuk, M., and Bamford, D. H. (1986). The causal agent of halo blight in bean, *Pseudomonas syringae* pv. *phaseolicola,* attaches to stomata via its pili, *Microb. Pathol.,* 1:139-148.

Romantschuk, M., Nurmiaho-Lassila, E. L., and Rantala, E. (1991). Pilus-mediated adsorption of *Pseudomonas syringae* to bean leaves, *Phytopathology,* 81:1245.

Rommens, C. M., Salmeron, J. M., Oldroyd, G. E., and Staskawicz, B. J. (1995). Inter generic transfer and functional expression of the tomato disease resistance gene *Pto, Plant Cell,* 7:1537-1544.

Ronald, P. C. (1997). The molecular basis of disease resistance in rice, *Plant Mol. Biol.,* 35:179-186.

Ronald, P., Albano, B., Tabien, R., Abenes, L., Wu, K., McCouch, S. R., and Tanksley, S. D. (1992a). Genetic and physical analysis of the rice bacterial blight resistance locus, *Xa-21, Mol. Gen. Genet.,* 236:113-120.

Ronald, P.C., Salmeron, J., Carland, F. M., and Staskawicz, B. J. (1992b). Cloned avirulence gene *avrPto* induces disease resistance in tomato cultivars containing the *Pto* resistance, *J. Bacteriol.,* 174:1604-1611.

Rong, L., Carpita, N. C., Mort, A., and Gelvin, S. B. (1994). Soluble cell wall compounds from carrot roots induce the *picA* and *pgl* loci of *Agrobacterium tumefaciens, Mol. Plant-Microbe Interact.,* 7:6-14.

Rong, L., Karcher, S. J., and Gelvin, S. B. (1991). Genetic and molecular analyses of *picA,* a plant-inducible locus on the *Agrobacterium tumefaciens* chromosome, *J. Bacteriol.,* 173:5110-5120.

Rong, L., Karcher, S. J., O'Neal, K., Hawes, M. C., Yerkes, C. D., Jayaswal, R. K., Hallberg, C. A., and Gelvin, S. B. (1990). *PicA,* a novel plant-inducible locus on the *Agrobacterium tumefaciens* chromosome, *J. Bacteriol.,* 172:5828-5836.

Rosqvist, R., Magnusson, K. E., and Wolf-Watz, H. (1994). Target cell contact triggers expression and polarized transfer of *Yersinia* YopE cytotoxin into mammalian cells, *EMBO J.,* 13:964-972.

Ross, A. F. (1961). Systemic acquired resistance induced by localized virus infections in plants, *Virology,* 14:340-358.

Rothfield, L., and Pearlman-Kothencz, M. (1969). Synthesis and assembly of bacterial membrane components. A lipopolysaccharide phospholipid-protein complex excreted by living bacteria, *J. Mol. Biol.,* 44:477-492.

Rousseau-Limouzin, M., and Fritig, B. (1991). Induction of chitinases, 1,3-β-glucanases and other pathogenesis-related proteins in sugar beet leaves upon infection with *Cercospora beticola, Plant Physiol. Biochem.,* 29:105-118.

Rouster, J., Leah, R., Mundy, J., Cameron-Mills, V. (1997). Identification of a methyl jasmonate-responsive region in the promoter of a lipoxygenase I gene expressed in barley grain, *Plant J.,* 11:513-523.

Rudolph, K., El-Banoby, M., and Gross, M. (1987). Effect of extracellular polysaccharides on bacterial multiplication '*in planta*'. In E. L. Civerolo, A. Collmer, R. E. Davis, and A. G. Gillaspie (Eds.), *Plant Pathogenic Bacteria,* M. Nijhoff Publishers, Dordrecht, pp. 597-598.

Ruiz-Medrano, R., Jimenez-Moraila, B., Herrera-Estrella, L. and Rivera-Bustamante, R. F. (1992). Nucleotide sequence of an osmotin-like cDNA induced in tomato during viroid infection, *Plant Mol. Biol.,* 20:1199-1202.

Rumeau, D., Maher, E. A., Kelman, A., and Showalter, A. M. (1990). Extensin and phenylalanine ammonia-lyase gene expression altered in potato tubers in response to wounding, hypoxia and *Erwinia carotovora* infection, *Plant Physiol.,* 93:1134-1139.

Rumeau, D., Mazau, D., Panabieres, F., Delseny, M., and Esquerre-Tugaye, M. T. (1988). Accumulation of hydroxyproline-rich glycoprotein mRNAs in infected or ethylene treated melon plants, *Physiol. Mol. Plant Pathol.,* 33:419-428.

Ryals, J., Neuenschwander, U. H., Willits, M. G., Molina, A., Steiner, H. Y., and Hunt, M. D. (1996). Systemic acquired resistance, *Plant Cell,* 8:1809-1819.

Ryals, J., Uknes, S., and Ward, E. (1994). Systemic acquired resistance, *Plant Physiol.,* 104:1109-1112.

Ryan, C. A. (1987). Oligosaccharide signalling in plants, *Annu. Rev. Cell Biol.,* 3:295-317.

Ryan, C. A. (1988). Oligosaccharides as recognition signals for the expression of defensive genes in plants, *Biochemistry,* 27:8879-8898.

Ryan, C. A. (1990). Protease inhibitors in plants: Genes for improving defenses against insects and pathogens, *Annu. Rev. Phytopathol.,* 28:425-449.

Ryan, C. A. (1992). The search for the proteinase inhibitor-inducing factor, PIIF, *Plant Mol. Biol.,* 19:123-133.

Ryan, C. A., and Farmer, E. E. (1991). Oligosaccharide signals in plants: A current assessment, *Annu. Rev. Plant Physiol. Mol. Biol.,* 42:651-674.

Ryder, T. B., Cramer, C. L., Bell, J. N., Robbins, M. P., Dixon, R. A., and Lamb, C. J. (1984). Elicitor rapidly induces chalcone synthase mRNA in *Phaseolus vulgaris* cells at the onset of the phytoalexin response, *Proc. Natl. Acad. Sci. U.S.A,* 81:5724-5728.

Ryder, T. B., Hedrick, S. A., Bell, J. N., Liang, X., Clouse, S. D., and Lamb, C. J. (1987). Organization and differential activation of a gene family encoding the plant defense enzyme chalcone synthase in *Phaseolus vulgaris, Mol. Gen. Genet.,* 210:219-233.

Saarilahti, H. T., Pirhonen, M., Karlsson, M.-B., Flego, D., and Palva, E. T. (1992). Expression of *pehA-bla* gene fusions in *Erwinia carotovora* subsp. *carotovora* and isolation of regulatory mutants affecting polygalacturonase production, *Mol. Gen. Genet.,* 234:81-88.

Sakamoto, K., Tada, Y., Yokozeki, Y., Akagi, H., Hayashi, N., Fujimura, T., and Ichikawa, N. (1999). Chemical induction of disease resistance in rice is correlated with the expression of a gene encoding a nucleotide binding site and leucine-rich repeats, *Plant Mol. Biol.,* 40:847-855.

Salch, Y. P., and Shaw, P. D. (1988). Isolation and characterization of pathogenicity genes of *Pseudomonas syringae* pv. *tabaci, J. Bacteriol.,* 170:2584-2591.

Sallaud, C., Zuanazzi, J., El-Turk, J., Leymarie, J., Breda, C., Buffard, D., de Kozak, I., Rayet, P., Husson, P., Kondorosi, A., and Esnault, R. (1997). Gene expression is not systematically linked to phytoalexin production during alfalfa leaf interaction with pathogenic bacteria, *Mol. Plant-Microbe Interact.,* 10:257-267.

Salmeron, J. M., Oldroyd, G. E., Rommens, C. M., Scofield, S. R., Kim, H. S., Lavelle, D. T., Dahlbeck, D., and Staskawicz, B. J. (1996). Tomato *Prf* is a member of the leucine-rich repeat class of plant disease resistance genes and lies embedded within the *Pto* kinase gene cluster, *Cell,* 86:123-133.

Salmeron, J. M., and Staskawicz, B. J. (1993). Molecular characterization and *hrp*-dependence of the avirulence gene *avrPto* from *Pseudomonas syringae* pv. *tomato, Mol. Gen. Genet.,* 239:6-16.

Salmond, G. P. C. (1994). Secretion of extracellular virulence factors by plant pathogenic bacteria, *Annu. Rev. Phytopathol.,* 32:181-200.

Salmond, G. P. C., Golby, P., and Jones, S. (1994). Global regulation of *Erwinia carotovora* virulence factor production, *Adv. Mol. Gene. Plant-Microbe Interactions,* 3:13-20.

Salmond, G. P. C., and Reeves, P. J. (1993). Membrane traffic wardens and protein secretion in Gram-negative bacteria, *Trends Biochem. Sci.,* 18:7-12.

Samac, D. A., Hironaka, C. M., Yallaly, P. E., and Shah, D. M. (1990). Isolation and characterization of the genes encoding basic and acidic endochitinase in *Arabidopsis thaliana, Plant Physiol.,* 93:907-914.

Samborski, D. J., and Forsyth, F. R. (1960). Inhibition of rust development on detached wheat leaves by metabolites, antimetabolites, and enzyme poisons, *Can. J. Bot.,* 38:467-476.

San Francisco, M. J. D., and Keenan, R. W. (1993). Uptake of galacturonic acid in *Erwinia chrysanthemi* EC16, *J. Bacteriol.,* 175:4263-4265.

San Francisco, M. J. D., Xiang, Z.-X., and Keenan, R. W. (1996). Digalacturonic acid uptake in *Erwinia chrysanthemi, Mol. Plant-Microbe Interact.,* 9:144-147.

Sano, H., and Ohashi, Y. (1995). Involvement of small GTP-binding proteins in defence signal-transduction pathways of higher plants, *Proc. Natl. Acad. Sci. U.S.A.,* 92:4138-4144.

Sano, H., Seo, S., Koizumi, N., Kini, T., Iwamura, H., and Ohashi, Y. (1996). Regulation by cytokinins of endogenous levels of jasmonic acid and salicylic acid in mechanically wounded tobacco plants, *Plant and Cell Physiol.,* 37:762-769.

Sano, H., Seo, S., Orudgev, E., Youssefian, S., Ishizuka, K., and Ohashi, Y. (1994). Expression of small GTP binding protein in transgenic tobacco elevates endogenous cytokinin levels, abnormally induces salicylic acid in response to wounding, and increases resistance to tobacco mosaic virus infection, *Proc. Natl. Acad. Sci. U.S.A.,* 91:10556-10560.

Sasser, M. (1982). Inhibition by antibacterial compounds of the hypersensitive reaction induced by *Pseudomonas pisi* in tobacco, *Phytopathology,* 72:1513-1517.

Sato, F., Kitajima, S., Koyama, T., and Yamada, Y. (1996). Ethylene-induced gene expression of osmotin-like protein, a neutral isoform of tobacco PR-5, is mediated by the AGCCGCC *cis*-sequence, *Plant Cell Physiol.,* 37:249-255.

Sato, N., Yoshizawa, Y., Miyazaki, H., and Murai, A. (1985). Antifungal activity to *Phytophthora infestans* and toxicity to tuber tissue of several potato phytoalexins, *Ann. Phytopathol. Soc. Japan,* 51:494-497.

Sauvage, C., and Expert, D. (1994). Differential regulation by iron of *Erwinia chrysanthemi* pectate lyases: Pathogenicity of iron transport regulatory *(cbr)* mutants, *Mol. Plant-Microbe Interact.,* 7:71-77.

Sawczyc, M. K., Barber, C. C., and Daniels, M. J. (1989). The role in pathogenicity of some related genes in *Xanthomonas campestris* pathovars *campestris* and *translucens:* A shuttle strategy for cloning genes required for pathogenicity, *Mol. Plant-Microbe Interact.,* 2:249-255.

Sayler, R. J., Wei, G., Kloepper, J. W., and Tuzun, S. (1994). Induction of β-1,3-glucanases and chitinases in tobacco by seed treatment with select strains of plant growth promoting rhizobacteria, *Phytopathology,* 84:1107-1108.

Scandalios, J .G. (1993). Oxygen stress and superoxide dismutases, *Plant Physiol.,* 101:7-12.

Scandalios, J. G. (1994). Regulation and properties of plant catalases. In C. H. Foyer and P. M. Mullineaux, (Eds.), *Causes of Photooxidative Stress and Amelioration of Defense Systems in Plants,* CRC Press, Boca Raton, FL, pp. 275-316.

Schaffrath, P., Scheinpflug, H., and Reisener, H. J. (1995). An elicitor from *Pyricularia oryzae* induces resistance responses in rice: Isolation, characterization and physiological properties, *Physiol. Mol. Plant Pathol.*, 46:293-307.

Schaffrath, U., Freydl, E., and Dudler, R. (1997). Evidence for different signaling pathways activated by inducers of acquired resistance in wheat, *Mol. Plant-Microbe Interact.*, 10:779-783.

Schaller, A., and Ryan, C. A. (1995). Systemin—a polypeptide defense signal in plants, *BioEssays*, 18:27-33.

Scheeren-Groot, E. P., Rodenburg, K. W., den Dulk-Ras, A., Turk, S. C. H. J., and Hooykaas, P. J. J. (1994). Mutational analysis of the transcriptional activator VirG of *Agrobacterium tumefaciens*, *J. Bacteriol.*, 176:6418-6426.

Schell, M. A. (1993). Molecular biology of the LysR family of transcriptional regulators, *Annu. Rev. Microbiol.*, 47:597-626.

Schell, M. A., Roberts, D. P., and Denny, T. P. (1988). Analysis of the *Pseudomonas solanacearum* polygalacturonase encoded by *pglA* and its involvement in phytopathogenicity, *J. Bacteriol.*, 170:4501-4508.

Schell, S. A. (1987). Purification and characterization of an endoglucanase from *Pseudomonas solanacearum*, *Appl. Environ. Microbiol.*, 53:2237-2241.

Schmele, I., and Kauss, H. (1990). Enhanced activity of the plasma membrane localized callose synthase in cucumber leaves with induced resistance, *Physiol. Mol. Plant Pathol.*, 37:221-228.

Schmelzer, E., Borner, H., Grisebach, H., Ebel, J., and Hahlbrock, K. (1984). Phytoalexin synthesis in soybean *(Glycine max)*. Similar time course of mRNA induction in hypocotyls infected with a fungal pathogen and in cell cultures treated with fungal elicitor, *FEBS Lett.*, 172:59-63.

Schmid, J., Doerner, P. W., Clouse, S. D., Dixon, R. A., and Lamb, C. J. (1990). Developmental and environmental regulation of a bean chalcone synthase promoter in transgenic tobacco, *Plant Cell*, 2:619-631.

Schmid, P. S., and Feucht, W. (1980). Tissue-specific oxidative browning of polyphenols by peroxidase in cherry shoots, *Gartenbauwissenschaft*, 45:68-73.

Schnabelrauch, L. S., Kieliszewski, M., Upham, B. L., Alizedeh, H., and Lamport, D. T. A. (1996). Isolation of pI4.6 extensin peroxidase from tomato cell suspension cultures and identification of Val-Tyr-Lys as putative intermolecular crosslink site, *Plant J.*, 9:477-489.

Schneider, M., Schweizer, P., Meuwly, P., and Metraux, J. P. (1996). Systemic acquired resistance in plants, *Int. Rev. Cytol.*,168:303-340.

Schneider, S., and Ullrich, W. R. (1994). Differential induction of resistance and enhanced enzyme activities in cucumber and tobacco caused by treatment with various abiotic and biotic inducers, *Physiol. Mol. Plant Pathol.*, 45:715-721.

Schroder, G., Brown, J. W. S., and Schroder, J. (1988). Molecular analysis of resveratrol synthase, cDNA, genomic clones and relationship with chalcone synthase, *Eur. J. Biochem.*, 172:161-169.

Schroeder, J. I., and Hagiwara, S. (1989). Cytosolic calcium regulates ion channels in the plasma membrane of *Vicia faba* guard cells, *Nature*, 338:427-430.

Schufflebottom, D., Edwards, K., Schuch, W., and Bevan, M. (1993). Transcription of two members of a gene family encoding phenylalanine ammonia-lyase leads to remarkably different cell specificities and induction patterns, *Plant J.,* 3:835-845.

Schulte, R., and Bonas, U. (1992a). A *Xanthomonas* pathogenicity locus is induced by sucrose and sulfur-containing amino acids, *Plant Cell,* 4:79-86.

Schulte, R., and Bonas, U. (1992b). Expression of the *Xanthomonas campestris* pv. *vesicatoria hrp* gene cluster, which determines pathogenicity and hypersensitivity on pepper and tomato, is plant-inducible, *J. Bacteriol.,* 174:815-823.

Schwake, R., and Hager, A. (1992). Fungal elicitors induce a transient release of active oxygen species from cultured spruce cells that is dependent on Ca^{2+} and protein-kinase activity, *Planta,* 187:136-141.

Schweizer, P., Buchala, A., and Metraux, J. P. (1997a). Gene-expression patterns and levels of jasmonic acid in rice treated with the resistance inducer 2,6-dichloroisonicotinic acid, *Plant Physiol.,* 115:61-70.

Schweizer, P., Buchala, A., Silverman, P., Seskar, M., Raskin, I., and Metraux, J. P. (1997b). Jasmonate-inducible genes are activated in rice by pathogen attack without a concomitant increase in endogenous jasmonic acid levels, *Plant Physiol.,* 114:79-88.

Schweizer, P., Hunziker, W., and Mosinger, E. (1989). cDNA cloning, in vitro transcription and partial sequence analysis of mRNAs from winter wheat (*Triticum aestivum* L.) with induced resistance to *Erysiphe graminis* f. sp. *tritici, Plant Mol. Biol.,* 12:643-654.

Scofield, S. R., Tobias, C. M., Rathjen, J. P., Chang, J. H., Lavelle, D. T., Michelmore, R. W., and Staskawicz, B. J. (1996). Molecular basis of gene-for-gene specificity in bacterial speck disease of tomato, *Science,* 274:2063-2065.

Segal, A. W., and Abo, A. (1993). The biochemical basis of the NADPH oxidase of the phagocytes, *Trends Biochem. Sci.,* 18:43-47.

Seguchi, K., Kurotaki, M., Sekido, S., and Yamaguchi, I. (1992). Action mechanism of N-cyanomethyl-2-chloroisonicotinamide in controlling rice blast disease, *J. Pesticide Sci.,* 17:107-113.

Segura, A., Moreno, M., and Garcia-Olmedo, F. (1993). Purification and antipathogenic activity of lipid transfer proteins (LTPs) from the leaves of *Arabidopsis* and spinach, *FEBS Lett.,* 332:243-246.

Sehgal, O. P., and Mohamed, F. (1990). Pathogenesis-related proteins. In C. L. Mandahar (Ed.), *Plant Viruses-Pathology,* Vol. II, CRC Press, Boca Raton, FL, pp. 65-83.

Sehgal, O. P., Rieger, R., and Mohamed, F. (1991). Induction of bean PR-4d-type protein in divergent plant species after infection with tobacco ringspot virus and its relationship with tobacco PR-5, *Phytopathology,* 81:215-219.

Seibertz, B., Logemann, J., Willmitzer, L., and Schell, J. (1989). Cis-analysis of the wound-inducible promoter wun1 in transgenic tobacco plants and histochemical localization of its expression, *Plant Cell,* 1:961-968.

Sela-Buurlage, M. B., Ponstein, A. S., Bres-Vloemans, S. A., Melchers, L. S., Van Den Elzen, P. J. M., and Cornelissen, B. J. C. (1993). Only specific tobacco *(Nicotiana tabacum)* chitinases and β-1,3-glucanases exhibit antifungal activity, *Plant Physiol.,* 101:857-863.

Sembdner, G., and Partier, B. (1993). The biochemistry and the physiological and molecular actions of jasmonates, *Annu. Rev. Plant Physiol. Plant Mol. Biol.,* 44:569-589.

Seo, H. S., Kim, H. Y., Jeong, J. Y., Lee, S. Y., Cho, J. J., and Bahk, J. D. (1995). Molecular cloning and characterization of *RGA1* encoding a G-protein α subunit from rice (*Oryza sativa* L. IR-36), *Plant Mol. Biol.,* 27:1119-1131.

Sequeira, L. (1976). Induction and suppression of the hypersensitive reaction caused by phytopathogenic bacteria: Specific and non-specific components. In R. K. S. Wood and A. Graniti (Eds.), *Specificity in Plant Diseases,* Plenum Press, New York, pp. 289-309.

Sequeira, L. (1983). Mechanisms of induced resistance in plants, *Annu. Rev. Microbiol.,* 37:51-79.

Sequeira, L., and Graham, T. L. (1977). Agglutination of avirulent strains of *Pseudomonas solanacearum* by potato lectin, *Physiol. Plant Pathol.,* 11:43-54.

Serhan, C., Anderson, P., Goodman, E., Dunham, P., and Weissmann, G. (1981). Phosphatidate and oxidized fatty acids are calcium ionophores, *J. Biol. Chem.,* 256:2736-2741.

Seskar, M., Shulaev, V., and Raskin, I. (1998). Endogenous methyl salicylate in pathogen-inoculated tobacco plants, *Plant Physiol.,* 116:387-392.

Sessa, G., Meller, Y., and Fluhr, R. (1995). A GCC element and a G-box motif participate in ethylene-induced expression of the PRB-1b gene, *Plant Mol. Biol.,* 28:145-153.

Shah, J., and Klessig, D. F. (1996). Identification of a salicylic acid-responsive element in the promoter of the tobacco pathogenesis-related β-1,3-glucanase gene, PR-2d, *Plant J.,* 10:1089-1101.

Shah, J., Tsui, F., and Klessig, D. F. (1997). Characterization of a *s*alicylic *a*cid-*i*nsensitive mutant *(sai1)* of *Arabidopsis thaliana,* identified in a selective screen utilizing the SA-inducible expression of the *tms2* gene, *Mol. Plant-Microbe Interact.,* 10:69-78.

Sharma, P., Borja, D., Stougaard, P., and Lonneborg, A. (1993). PR-proteins accumulating in spruce roots infected with a pathogenic *Pythium* sp. isolate include chitinases, chitosanases, and β-1,3-glucanases, *Physiol. Mol. Plant Pathol.,* 43:57-67.

Sharma, Y. K., Hinojos, C. M., and Mehdy, M. C. (1992). cDNA cloning, structure and expression of a novel pathogenesis-related protein in bean, *Mol. Plant-Microbe Interact.,* 5:89-95.

Sharma, Y. K., Leon, J., Raskin, I., and Davis, K. R. (1996). Ozone-induced responses in *Arabidopsis thaliana:* The role of salicylic acid in the accumulation of defense-related transcripts and induced resistance, *Proc. Natl. Acad. Sci. U.S.A.,* 93:5099-5104.

Sharma, Y. K., and Mehdy, M. E. (1992). Early cellular redox changes mediate the induction of phytoalexin synthesis pathway mRNAs by fungal elicitor, *Plant Physiol.,* 99:S-24.

Sharon, A., Ghirlando, R., and Gressel, J. (1992). Isolation, purification, and identification of 2-(*p*-hydroxyphenoxy)-5,7-dihydroxychromone: A fungal induced phytoalexin from *Cassia obtusifolia, Plant Physiol.,* 98:303-308.

Sharon, A., and Gressel, J. (1991). Elicitation of a flavanoid phytoalexin accumulation in *Cassia obtusifolia* by a mycoherbicide: Estimation by $AlCl_3$ spectrofluorimetry, *Pestic. Biochem. Physiol.,* 41:142-149.

Sharp, J. A., Albersheim, P., Ossowski, P., Pilotti, A., Garegg, P., and Lindberg, B. (1984). Comparison of the structures and elicitor activities of a synthetic and a mycelial-wall derived hexa-(β-D-glucopyranosyl)-D-glucitol, *J. Biol. Chem.,* 9:11341-11345.

Sharrock, K. R., and Labavitch, J. M. (1994). Polygalacturonase inhibitors of Bartlett pear fruits: Differential effects on *Botrytis cinerea* polygalacturonase isozymes, and influence on products of fungal hydrolysis of pear cell walls and on ethylene induction in cell culture, *Physiol. Mol. Plant Pathol.,* 45:305-319.

Shen, H., and Keen, N. T. (1993). Characterization of the promoter of avirulence gene D from *Pseudomonas syringae* pv. *tomato, J. Bacteriol.,* 175:5916-5924.

Shevchik,V. E., Bortoli-German, I., Robert-Baudouy, J., Robinet, S., Barras, F., and Condemine, G. (1995). Differential effect of *dsbA* and *dsbC* mutations on extracellular enzyme secretion in *Erwinia chrysanthemi, Mol. Microbiol.,* 16:745-753.

Shevchik, V. E., Condemine, G., Hugouvieux-Cotte-Pattat, N., and Robert-Baudouy, J. (1996). Characterization of pectin methylesterase B, an outer membrane lipoprotein of *Erwinia chrysanthemi* 3937, *Mol. Microbiol.,* 19:455-466.

Shevchik, V. E., Robert-Baudouy, J., and Condemine, G. (1997). Specific interaction between *OutD,* an *Erwinia chrysanthemi* outer membrane protein of the general secretory pathway, and secreted proteins, *EMBO J.,* 16:3007-3016.

Shibata, D., Steczko, J., Dixon, J. E., Andrews, P. C., Hermodson, M., and Axelrod, B. (1988). Primary structure of soybean lipoxygenase-2, *J. Biol. Chem.,* 263: 6816-6821.

Shibata, D., Steczko, J., Dixon, J. E., Hermodson, M., Yazdanparast, R., and Axelrod, B. (1987). Primary structure of soybean lipoxygenase-1, *J. Biol. Chem.,* 262: 10008-10085.

Shimoda, N., Toyoda-Yamamoto, A., Aoki, S., and Machida, Y. (1993). Genetic evidence for an interaction between the VirA sensor protein and the ChvE sugar-binding protein of *Agrobacterium, J. Biol. Chem.,* 268:26552-26558.

Shimoda, N., Toyoda-Yamamoto, A., Nagamine, J., Usami, S., Katayama, M., Sakagami, Y., and Machida, Y. (1990). Control of expression of *Agrobacterium vir* genes by synergistic actions of phenolic signal molecules and monosaccharides, *Proc. Natl. Acad. Sci. U.S.A.,* 87:6684-6688.

Shinshi, H., Neuhaus, J. M., Ryals, J., and Meins, F. Jr. (1990). Structure of a tobacco endochitinase gene: Evidence that different chitinase genes can arise by transposition of sequences encoding a cysteine-rich domain, *Plant Mol. Biol.,* 14:357-368.

Shinshi, H., Wenzler, H., Neuhaus, J. M., Felix, G., and Hofsteenge, J. (1988). Evidence for N- and C-terminal processing of a plant-defence related enzyme. Primary structure of tobacco prepro-β-1,3-glucanase, *Proc. Natl. Acad. Sci. U.S.A.,* 85:5541-5545.

Shintaku, M. H., Kluepfel, D. A., Yacoub, A., and Patil, S. S. (1989). Cloning and partial characterization of an avirulence determinant from race 1 of *Pseudomonas syringae* pv. *phaseolicola, Physiol. Mol. Plant Pathol.,* 35:313-322.

Shiraishi, T., Araki, M., Yoshioka, H., Kobayashi, I., Yamada, T., Ichinose, Y., Kunoh, H., and Oku, H. (1991). Inhibition of ATPase activity in pea plasma membranes in situ by a suppressor from a pea pathogen, *Mycospharella pinodes, Plant Cell Physiol.,* 32:1067-1075.

Shirasu, K., Dixon, R. A., and Lamb, C. (1996). Signal transduction in plant immunity, *Curr. Opin. Immunol.,* 8:3-7.

Shirasu, K., Nakajima, H., Rajasekhar, V. K., Dixon, R. A., and Lamb, C. (1997). Salicylic acid potentiates an agonist-dependent gain control that amplifies pathogen signals in the activation of defense mechanisms, *Plant Cell,* 9:261-270.

Showalter, A. M. (1993). Structure and function of plant cell wall proteins, *Plant Cell,* 5:9-23.

Showalter, A. M., Bell, J. N., Cramer, C. L., Bailey, J. A., Varner, J. E., and Lamb, C. J. (1985). Accumulation of hydroxyproline-rich glycoprotein mRNAs in response to fungal elicitor and infection, *Proc. Natl. Acad. Sci. U.S.A.,* 82:6551-6555.

Showalter, A. M., and Varner, J. E. (1989). Plant hydroxyproline-rich glycoproteins. In A. Marcus (Ed.), *The Biochemistry of Plants,* Vol.15, Academic Press, New York, pp. 485-520.

Showalter, A. M., Zhou, J., Rumeau, D., Worst, S. G., and Varner, J. E. (1991). Tomato extensin and extensin-like cDNAs: Structure and expression in response to wounding, *Plant Mol. Biol.,* 16:547-565.

Shulaev, V., Leon, J., and Raskin, I. (1995). Is salicylic acid a translocated signal of systemic acquired resistance in plants?, *Plant Cell,* 7:1691-1701.

Shulaev, V., Silverman, P., and Raskin, I. (1997). Airborne signaling by methyl salicylate in plant pathogen resistance, *Nature,* 385:718-727.

Siedow, J. N. (1991). Plant lipoxygenase: Structure and function, *Annu. Rev. Plant Physiol. Plant Mol. Biol.,* 42:145-188.

Siegel, S. M., and Halpern, L. A. (1965). Effects of peroxide on permeability and their modification by indoles, vitamin E, and other substances, *Plant Physiol.,* 40:792-796.

Siegrist, J., Jeblick, W., and Kauss, H. (1994). Defense responses in infected and elicited cucumber (*Cucumis sativus* L.) hypocotyl segments exhibiting acquired resistance, *Plant Physiol.,* 105:1365-1374.

Silverman, P., Nuckles, E., Ye, X. S., Kuc, J., and Raskin, I. (1993). Salicylic acid, ethylene, and pathogen resistance in tobacco, *Mol. Plant-Microbe Interact.,* 6:775-781.

Silverman, P., Seskar, M., Kanter, D., Schweizer, P., Metraux, J. P., and Raskin, I. (1995). Salicylic acid in rice: Biosynthesis, conjugation, and possible role, *Plant Physiol.,* 108:633-639.

Simmons, C. R., Litts, J. C., Huang, N., and Rodriguez, R. L. (1992). Structure of a rice β-glucanase gene regulated by ethylene, cytokinin, wounding, salicylic acid and fungal elicitors, *Plant Mol. Biol.,* 18:33-45.

Simonich, M. T., and Innes, R. (1995). A disease resistance gene in *Arabidopsis* with specificity for the *avrPph3* gene of *Pseudomonas syringae* pv. *phaseolicola, Mol. Plant-Microbe Interact.,* 8:637-640.

Singh, N. K., Bracker, C. A., Hasegawa, P. M., Handa, A. K., Buckel, S., Hermodson, M. A., Pfankoch, E., Regnier, F. E., and Bressan, R. A. (1987). Characterization of osmotin: A thaumatin-like protein associated with osmotic adaptation in plant cells, *Plant Physiol.,* 85:529-536.

Singh, N. K., Nelson, D. E., Kuhn, D., Hasegawa, P. M., and Bressan, R. A. (1989). Molecular cloning of osmotin and regulation of its expression by ABA and adaptation to low water potential, *Plant Physiol.,* 90:1096-1101.

Slusarenko, A. J., Croft, K. P. C., and Voisey, C. R. (1991). Biochemical and molecular events in the hypersensitive response of bean to *Pseudomonas syringae* pv. *phaseolicola.* In C. J. Smith (Ed.), *Biochemistry and Molecular Biology of Host-Pathogen Interactions,* Clarendon Press, Oxford, U.K., pp. 126-143.

Slusarenko, A. J., and Longland, A. C. (1986). Changes in gene activity during the hypersensitive response in *Phaseolus vulgaris* cv. Red Mexican to an avirulent race1 isolate of *Pseudomonas syringae* pv. *phaseolicola, Physiol. Mol. Plant Pathol.,* 29:79-94.

Smidt, J., and Kosuge, T. (1978). The role of indole-3-acetic acid accumulation by alpha methyl tryptophan-resistant mutants of *Pseudomonas savastanoi* in gall formation on oleanders, *Physiol. Plant Pathol.,* 13:203-214.

Smith, D. A. (1982). Toxicity of phytoalexins. In J. A. Bailey and J. W. Mansfield (Eds.), *Phytoalexins,* Blackie, Glasgow, pp. 218-252.

Smith, J. A., Hammerschmidt, R., and Fulbright, D. W. (1991). Rapid induction of systemic resistance in cucumber by *Pseudomonas syringae* pv. *syringae, Physiol. Mol. Plant Pathol.,* 33:255-261.

Smith, J. A., and Metraux, J. P. (1991). *Pseudomonas syringae* pv. *syringae* induces systemic resistance to *Pyricularia oryzae* in rice, *Physiol. Mol. Plant Pathol.,* 39:451-461.

Smith, M. J., Mazzola, E. P., Sims, J. J., Midland, S. L., Keen, N. T., Burton, V., and Stayton, M. M. (1993). The syringolides: Bacterial C-glycosyl lipids that trigger plant disease resistance, *Tetrahedron Lett.,* 34:223-226.

Smith, S. G., Wilson, T. J. G., Dow, J. M., and Daniels, M. J. (1996). A gene for superoxide dismutase from *Xanthomonas campestris* pv. *campestris* and its expression during bacterial-plant interactions, *Mol. Plant-Microbe Interact.,* 9:584-593.

Sneath, B. J., Howson, J. M., and Beer, S. V. (1990). A pathogenicity gene from *Erwinia amylovora* encodes a predicted protein product homologous to a family of procaryotic response regulators, *Phytopathology,* 80:1038 (Abstr.).

Sock, J., Rohringer, R., and Kang, Z. (1990). Extracellular β-1,3-glucanases in stem rust-affected and abiotically stressed wheat leaves. Immunocytochemical localization of the enzyme and detection of isozymes in gels by activity staining with dye-labeled laminarin, *Plant Physiol.,* 94:1376-1389.

Somssich, I. E., Schmelzer, E., Bollmann, J., and Hahlbrock, K. (1986). Rapid activation by fungal elicitor of genes encoding "pathogenesis-related" proteins in cultured parsely cells, *Proc. Natl. Acad Sci. U.S.A.,* 83:2427-2430.

Somssich, I. E., Schmelzer, E., Kawalleck, P., and Hahlbrock, K. (1988). Gene structure and in situ transcript localization of pathogenesis-related protein 1 in parsley, *Mol. Gen. Genet.*, 213:93-98.

Song, W.-Y., Pi, L.-Y., Wang, G.-L., Gardner, J., Holsten, T., and Ronald, P. C. (1997). Evolution of the rice *Xa21* disease resistance gene family, *Plant Cell*, 9:1279-1287.

Song, W.-Y., Wang, G.-L., Chen, L.-L., Kim, H.-S., Pi, L.-Y., Holsten, T., Gardner, J., Wang, B., Zhai, W.-X., Zhu, Li-H., Fauquet, C., and Ronald, P. (1995). A receptor kinase-like protein encoded by the rice disease resistance gene, *Xa 21*, *Science*, 257:1804-1806.

Spanswick, R. M. (1981). Electrogenic ion pumps, *Annu. Rev. Plant Physiol.*, 32:267-289.

Spencer, P. A., and Towers, G. H. N. (1988). Specificity of signal compounds detected by *Agrobacterium tumefaciens*, *Phytochemistry*, 27:2781-2785.

Stab, M. R., and Ebel, J. (1987). Effects of Ca^{2+} on phytoalexin induction by fungal elicitor in soybean cells, *Arch. Biochem. Biophys.*, 257:416-423.

Stachel, S. E., Messens, E., Van Montagu, M., and Zambryski, P. (1985). Identification of the signal molecules produced by wounded plant cells that activate T-DNA transfer in *Agrobacterium tumefaciens*, *Nature*, 318:624-629.

Stachel, S. E., and Nester, E. W. (1986). The genetic and transcriptional organization of the *vir* region of Ti plasmid of *Agrobacterium tumefaciens*, *EMBO J.*, 5:1445-1454.

Stachel, S. E., Nester, E. W., and Zambryski, P. C. (1986). A plant cell factor induces *Agrobacterium tumefaciens vir* gene expression, *Proc. Natl. Acad. Sci. U.S.A.*, 83:379-383.

Stachel, S. E., Timmerman, B., and Zambryski, P. (1987). Activation of *Agrobacterium tumefaciens vir* gene expression generates multiple T-strand molecules from the pTiA6T-region: Requirement for 5' *virD* products, *EMBO J.*, 6:857-863.

Stachel, S., and Zambryski, P. (1986). *virA* and *virG* control the plant-induced activation of the T-DNA transfer process of *A. tumefaciens*, *Cell*, 46:325-333.

Stack, J. P., Mount, M. S., Berman, P. M., and Hubbard, J. P. (1980). Pectic enzymes complex from *Erwinia carotovora:* A model for degradation and assimilation of host pectic fractions, *Phytopathology*, 70:262-272.

Stall, R. E., and Cook, A. A. (1979). Evidence that bacterial contact with the plant cell is necessary for the hypersensitive reaction but not the susceptible reaction, *Physiol. Plant Pathol.*, 14:77-84.

Stanford, A., Bevan, M., and Northcote, D. (1989). Differential expression within a family of novel wound-induced genes in potato, *Mol. Gen. Genet.*, 215:200-208.

Staskawicz, B. J., Ausubel, F. M., Baker, B. J., Ellis, J. G., and Jones, J. D. G. (1995). Molecular genetics of plant disease resistance, *Science*, 268:661-667.

Staskawicz, B. J., Dahlbeck, D., and Keen, N. T. (1984). Cloned avirulence gene of *Pseudomonas syringae* pv. *glycinea* determines race-specific incompatibility on *Glycine max* (L.) Merr., *Proc. Natl. Acad. Sci. U.S.A.*, 81:6024-6028.

Staskawicz, B. J., Dahlbeck, D., Keen, N., and Napoli, C. (1987). Molecular characterization of cloned avirulence genes from race O and race 1 of *Pseudomonas syringae* pv. *glycinea*, *J. Bacteriol.*, 169:5789-5794.

Staswick, P. E. (1992). Jasmonate, genes, and fragrant signals, *Plant Physiol.,* 99:804-807.

Steffani, E., Bazzi, C., and Mazzucchi, U. (1994). Modification of the *Pseudomonas syringae* pv. *tabaci*-tobacco leaf interaction by bacterial oligosaccharides, *Physiol. Mol. Plant Pathol.,* 45:397-406.

Steffens, M., Ettl, F., Kranz, D., and Kindl, H. (1989). Vanadate mimics effects of fungal cell wall in eliciting gene activation in plant cell cultures, *Planta,* 177:160-168.

Steinberger, E. M., and Beer, S. V. (1988). Creation and complementation of pathogenicity mutants of *Erwinia amylovora, Mol. Plant-Microbe Interact.,* 1:135-144.

Steiner, H., Hultmark, D., Engstrom, A., Bennich, H., and Boman, H. G. (1981). Sequence and specificity of two antibacterial proteins involved in insect immunity, *Nature,* 292:246-248.

Stemmer, W. P. C., and Sequeira, L. (1987). Fimbriae of phytopathogenic and symbiotic bacteria, *Phytopathology,* 77:1633-1639.

Sterk, P., Booij, H., Schellekens, G. A., Van Kammen, A., and De Vries, S. C. (1991). Cell-specific expression of the carrot EP2 lipid transfer protein gene, *Plant Cell,* 3:907-921.

Stermer, B. A., and Bostock, R. M. (1987). Involvement of 3-hydroxy-3-methylglutaryl coenzyme A reductase in the regulation of sesquiterpenoid phytoalexin synthesis in potato, *Plant Physiol.,* 84:404-408.

Stermer, B. A., Edwards, L. A., Edington, B. V., and Dixon, R. A. (1991). Analysis of elicitor-inducible transcripts encoding 3-hydroxy-3-methylglutaryl coenzyme A reductase in potato, *Physiol. Mol. Plant Pathol.,* 39:135-145.

Stermer, B. A., Schmid, J., Lamb, C. J., and Dixon, R. A. (1990). Infection and stress activation of bean chalcone synthase promoters in transgenic tobacco, *Mol. Plant-Microbe Interact.,* 3:381-388.

Stiefel, V., Perez-Grau, L. I., Alberico, F., Girlat, E., Ruiz-Avila, L., Ludevid, M. D., and Puigdomenech, P. (1989). Molecular cloning of cDNAs encoding a putative cell wall protein from *Zea mays* and immunological identification of related polypeptides, *Plant Mol. Biol.,* 11:483-493.

Stintzi, A., Heitz, T., Kauffmann, S., Legrand, M., and Fritig, B. (1991). Identification of a basic pathogenesis-related thaumatin-like protein of virus-infected tobacco as osmotin, *Physiol. Mol. Plant Pathol.,* 38:137-146.

Stintzi, A., Heitz, T., Prasad, V., Weidemann-Merdinoglu, S., Kauffmann, S., Geoffrey, P., Legrand, M., and Fritig, B. (1993). Plant "pathogenesis-related" proteins and their role in defense against pathogens, *Biochimie,* 75:687-706.

Stock, J. B., Hinfa, A. J., and Stock, A. M. (1989). Protein phosphorylation and regulation of adaptive responses in bacteria, *Microbiol. Rev.,* 50:450-490.

Stock, J. B., Surette, M. G., Levit, M., and Park, P. (1995). Two-component signal transduction systems: Structure-function relationships and mechanisms of catalysts. In J. H. Hoch and T. J. Silhavy (Eds.), *Two-Component Signal Transduction,* American Society for Microbiology, Washington, D C, pp. 25-51.

Stoddardt, R. W. (1984). *The Biosynthesis of Polysaccharides,* Croom Helm, London and Sydney, p. 354.

Stratilova, E., Dzurova, M., Markovic, O., and Jornvall, H. (1996). An essential tyrosine residue of *Aspergillus* polygalacturonase, *FEBS Lett.,* 382:164-166.

Strobel, N. E., Ji, C., Gopalan, S., Kuc, J. A., and He, S. Y. (1996). Induction of systemic acquired resistance in cucumber by *Pseudomonas syringae* pv. *syringae* 61 $HrpZ_{Pss}$ protein, *Plant J.,* 9:431-439.

Stromberg, A. (1996). Expression of genes related to systemic acquired resistance in potato, *Nordisk Jordbruksforskning,* 78:5.

Stuart, L. S., and Harris, T. H. (1942). Bactericidal and fungicidal properties of a crystalline protein isolated from unbleached wheat flour, *Cereal Chem.,* 19:288-300.

Subramaniam, R., Despres, C., and Brisson, N. (1997). A functional homolog of mammalian protein kinase C participates in the elicitor-induced defense response in potato, *Plant Cell,* 9:653-664.

Subramaniam, R., Reinold, S., Molitor, E., and Douglas, C. J. (1993). Structure, inheritance, and expression of hybrid poplar *(Populus trichocapta* × *Populus deltoides)* phenylalanine ammonia-lyase genes, *Plant Physiol.,* 102:71-83.

Summermatter, K., Meuwly, P., Molders, W., and Metraux, J. P. (1994). Salicylic acid levels in *Arabidopsis thaliana* after treatments with *Pseudomonas syringae* or synthetic inducers, *Acta Hortic.,* 381:367-370.

Summermatter, K., Sticher, L., and Metraux, J. P. (1995). Systemic responses in *Arabidopsis thaliana* infected and challenged with *Pseudomonas syringae* pv. *syringae, Plant Physiol.,* 108:1379-1385.

Sun, T. J., Essenberg, M., and Melcher, U. (1989). Photoactivated DNA nicking, enzyme inactivation, and bacterial inhibition by sesquiterpenoid phytoalexins from cotton, *Mol. Plant-Microbe Interact.,* 2:139-147.

Surgery, N., Robert-Baudouy, J., and Condemine, G. (1996). The *Erwinia chrysanthemi pecT* gene regulates pectinase gene expression, *J. Bacteriol.,* 178:1593-1599.

Sutherland, M. W. (1991). The generation of oxygen radicals during host plant responses to infection, *Physiol. Mol. Plant Pathol.,* 39:79-93.

Suzuki, H., Oba, K., and Uritani, I. (1975). The occurrence and some properties of 3-hydroxy-3-methylglutaryl coenzyme A reductase in sweet potato roots infected by *Ceratocystis fimbriata, Physiol. Plant Pathol.,* 7:265-276.

Swanson, J., Kearney, B., Dahlbeck, D., and Staskawicz, B. (1988). Cloned avirulence gene of *Xanthomonas campestris* pv. *vesicatoria* complements spontaneous race-change mutants, *Mol. Plant-Microbe Interact.,* 1:5-9.

Swart, S., Lugtenberg, B. J. J., Smit, G., and Kune, J. W. (1994). Rhicadhesin-mediated attachment and virulence of an *A. tumefaciens chvB* mutant can be restored by growth in a highly osmotic medium, *J. Bacteriol.,* 176:3816-3819.

Swarup, S., DeFeyter, R., Brlansky, R. H., and Gabriel, D. W. (1991). A pathogenicity locus from *Xanthomonas citri* enables strains from several pathovars of *X. campestris* to elicit cankerlike lesions on citrus, *Phytopathology,* 81:802-809.

Swarup, S., Yang, Y., Kingsley, M. T., and Gabriel, D. W. (1992). A *Xanthomonas citri* pathogenicity gene, pthA, pleiotropically encodes gratuitous avirulence on nonhosts, *Mol. Plant-Microbe Interact.,* 5:204-213.

Swegle, M., Hwang, J. K., Lee, G., and Muthukrishnan, S. (1989). Identification of an endochitinase cDNA clone from barley aleurone cells, *Plant Mol. Biol.,* 12:403-412.

Swoboda, I., Scheiner, O., Heberle-Bors, E., and Vicente, O. (1995). cDNA cloning and characterization of three genes in the *BetvI* gene family that encode pathogenesis-related proteins, *Plant Cell and Environ.,* 18:865-874.

Swords, K. M. M., Dahlbeck, D., Kearney, B., Roy, M., and Staskawicz, B. J. (1996). Spontaneous and induced mutations in a single open reading frame alter both virulence and avirulence in *Xanthomonas campestris* pv. *vesicatoria avrBs2, J. Bacteriol.,* 178:4661-4669.

Swords, K. M. M., and Staehelin, L. A. (1993). Complementary immunolocalization patterns of cell wall hydroxyproline-rich glycoproteins studied with the use of antibodies directed against different carbohydrate epitopes, *Plant Physiol.,* 102: 891-901.

Sykes, L., and Matthysse, A. G. (1988). Differences in attachment of the biotypes of *Agrobacterium tumefaciens* and *A. rhizogenes* to carrot suspension culture cells, *Phytopathology,* 78:1322-1326.

Tabei, Y., Kitade, S., Nishizawa, Y., Kikuchi, N., Kayano, T., Hibi, T., and Akutsu, K. (1998). Transgenic cucumber plants harboring a rice chitinase gene exhibit enhanced resistance to gray mold *(Botrytis cinerea), Plant Cell Reports,* 17:159-164.

Takahama, U. (1988). Oxidation of flavonols by hydrogen peroxide in epidermal and guard cells of *Vicia faba* L., *Plant Cell Physiol.,* 29:433-438.

Takahashi, H., Chen, Z., Du, H., Liu, Y., and Klessig, D. (1997). Development of necrosis and activation of disease resistance in transgenic tobacco plants with severely reduced catalase levels, *Plant J.,* 11:993-1005.

Takahashi, T., and Doke, N. (1984). Aspects of infection and multiplication of an extracellular polysaccharide-deficient mutant of *Xanthomonas campestris* pv. *citri* and a possible role of EPS in host-parasite interaction, *Ann. Phytopath. Soc. Japan,* 50:417.

Takahashi, Y., Niwa, Y., Machida, Y., and Nagata, T. (1990). Location of the *cis*-acting auxin-responsive region in the promoter of the *par* gene from tobacco mesophyll protoplasts, *Proc. Natl. Acad. Sci. U.S.A.,* 87:8013-8016.

Takasugi, M., Monde, K., Katsui, N., and Shirata, A. (1987). Spirobrassinin, a novel sulfur-containing phytoalexin from the daikon *Raphanus sativus* L. var. *hortensis* (Cruciferae), *Chemical Letters,* 1631-1632.

Takasugi, M., Okinaka, S., Katsui, N., Masamune, T., Shirata, A., and Ohuchi, M. (1985). Isolation and structure of lettucenin A, a novel guaianolide phytoalexin from *Lactuca sativa* var. *capitata* (Compositae), *J. Chem. Soc., Chemical Communications,* 10:621-622.

Takeda, S., Sato, F., Ida, K., and Yamada, Y. (1991). Nucleotide sequence of a cDNA for osmotin-like protein from cultured tobacco cells, *Plant Physiol.,* 97:844-846.

Tal, B., and Robeson, D. J. (1986). The induction by fungal inoculation of ayapin and scopoletin biosynthesis in *Helianthus annuus, Phytochemistry,* 25:77-79.

Tamaki, S., Dahlbeck, D., Staskawicz, B., and Keen, N. T. (1988). Characterization and expression of two avirulence genes cloned from *Pseudomonas syringae* pv. *glycinea, J. Bacteriol.,* 170:4846-4854.

Tamara, S., Hanawa, F., Harada, Y., and Muzutani, J. (1988). A fungitoxin inducibly produced by dandelion leaves treated with cupric chloride, *Agric. Biol. Chem.,* 52:2947-2948.

Tamogami, S., Randeep, R., and Kodama, O. (1997). Phytoalexin production by amino acid conjugates of jasmonic acid through induction of narigenin-7-O-methyltransferase, a key enzyme on phytoalexin biosynthesis in rice (*Oryza sativa* L.), *FEBS Lett.,* 401:239-242.

Tang, J. L., Liu, Y.-N., Barber, C. E., Dow, J. M., Wootton, J. C., and Daniels, M. J. (1991). Genetic and molecular analysis of a cluster of *rpf* genes involved in positive regulation of synthesis of extracellular enzymes and polysaccharide in *Xanthomonas campestris* pathovar *campestris, Mol. Gen. Genet.,* 226:409-417.

Tang, X., Frederick, R. D., Zhou, J., Halterman, D. A., Jia, Y., and Martin, G. B. (1996). Initiation of plant disease resistance by physical interaction of AvrPto and Pto kinase, *Science,* 274:2060-2063.

Tardy, F., Nasser, W., Robert-Baudouy, J., and Hugouvieux-Cotte-Pattat, N. (1997). Comparative analysis of the five major *Erwinia chrysanthemi* pectate lyases: Enzyme characteristics and potential inhibitors, *J. Bacteriol.,* 179: 2503-2511.

Templeton, M. D., and Lamb, C. J. (1988). Elicitors and defense gene activation, *Plant Cell Environ.,* 11:395-401.

Temsah, M., Bertheau, Y., and Vian, B. (1991). Pectate-lyase fixation and pectate disorganization visualized by immunocytochemistry in *Saintpaulia ionantha* infected by *Erwinia chrysanthemi, Cell Biol. Int. Rep.,* 15:611-620.

Tenhaken, R., Levine, A., Brisson, L. F., Dixon, R. A., and Lamb, C. (1995). Function of the oxidative burst in hypersensitive disease resistance, *Proc. Natl. Acad. Sci. U.S.A.,* 92:4158-4163.

Terras, F. R. G., Eggermont, K., Kovaleva, V., Raikhel, N. V., Osborn, R. W., Kester, A., Rees, S. B., Torrekens, S., Leuven, F. V., Vanderleyden, J., Commue, B. P. A., and Broekaert, W. F. (1995). Small cysteine-rich antifungal proteins from radish: Their role in host defense, *Plant Cell,* 7:573-578.

Terras, F. R. G., Goderis, I. J., Van Leuven, F., Vanderleyden, J., Cammue, B. P. A., and Broekaert, W. F. (1992a). In vitro antifungal activity of a radish (*Raphanus sativus* L.) seed protein homologous to nonspecific lipid transfer protein, *Plant Physiol.,* 100:1055-1058.

Terras, F. R. G., Schoofs, H. M. E., De Bolle, M. F. C., Van Leuven, F., Rees, S. B., Vanderleyden, J., Cammue, B. P. A., and Brockaert, W. F. (1992b). Analysis of two novel classes of plant antifungal proteins from radish (*Raphanus sativus* L.) seeds, *J. Biol. Chem.,* 267:15301-15309.

Terras, F. R. G., Schoofs, H. M. E., Thevissen, K., Osborn, R. W., Vanderlyden, J., Cammue, B. P. A., and Broekaert, W. F. (1993). Synergistic enhancement of the antifungal activity of wheat and barley thionins by radish and oilseed rape 2S albumins and by barley trypsin inhibitors, *Plant Physiol.,* 103:1311-1319.

Thain, J. F., Doherty, H. M., Bowles, D. J., and Wildon, D. C. (1990). Oligosaccharides that induce proteinase inhibitor activity in tomato plants cause depolarisation of tomato leaf cells, *Plant Cell Environ.,* 13:569-574.

Thilmony, L., Chen, Z., Bressan, R. A., and Martin, G. B. (1995). Expression of the tomato *Pto* gene in tobacco enhances resistance to *Pseudomonas syringae* pv. *tabaci* expressing *avrPto, Plant Cell,* 7:1529-1536.

Thoma, S., Kaneko, Y., and Somerville, C. (1993). A non-specific lipid transfer protein from *Arabidopsis* is a cell wall protein, *Plant J.,* 3:427-436.

Thomas, J. D., Reeves, P. J., and Salmond, G. P. C. (1997). The general secretion pathway of *Erwinia carotovora*: Analysis of the membrane topology of OutC and OutF, *Microbiology (Reading),* 143:713-720.

Thomashow, M. F., Karlinsey, J. E., Marks, J. R., and Hurlbert, R. E. (1987). Identification of a new virulence locus in *Agrobacterium tumefaciens* that affects polysaccharide composition and cell attachment, *J. Bacteriol.,* 169:3209-3216.

Thomma, B. P. H. J., and Broekaert, W. F. (1998). Tissue-specific expression of plant defensin genes PDF2.1 and PDF2.2 in *Arabidopsis thaliana, Plant Physiol. Biochem.,* 36:533.

Thomma, B. P. H. J., Eggermont, K., Penninckx, I. A. M. A., Mauch-Mani, B., Vogelsang, R., Cammue, B. P. A., and Broekaert, W. F. (1998). Separate jasmonate-dependent and salicylate-dependent defense response pathways in *Arabidopsis* are essential for resistance to distinct microbial pathogens, *Proc. Natl. Acad, Sci. U.S.A.,* 95:15107-15111.

Thompson, J. E., Legge, R. L., and Barber, R. F. (1987). The role of free radicals in senescence and wounding, *New Phytol.,* 105:317-344.

Thordal-Christensen, H., Brandt, J., Cho, B. H., Rasmussen, S. K., Gregersen, P. L., Smedegaard-Petersen, V., and Collinge, D. B. (1992). cDNA cloning and characterization of two barley peroxidase transcripts induced differentially by the powdery mildew fungus *Erysiphe graminis, Physiol. Mol. Plant Pathol.,* 40:395-409.

Tiburzy, R., and Reisener, H. J. (1990). Resistance of wheat to *Puccinia graminis* f. sp. *tritici:* Association of the hypersensitive reaction with the cellular accumulation of lignin-like material and callose, *Physiol. Mol. Plant Pathol.,* 36:109-120.

Tietjen, K. G., and Matern, U. (1984). Induction and suppression of phytoalexin synthesis in cultured cells of safflower *Carthamus tinctorius* L. by metabolites of *Alternaria carthami* Chowdhury, *Arch. Biochem. Biophys.,* 229:136-144.

Tobias, C. M. (1996). A kinase suicide squad in tomato, *Trend Plant Sci.,* 1:133-134.

Tolbert, N. E. (1982). Leaf peroxisomes, *Ann. NY. Acad. Sci.,* 386:254-268.

Tomijama, K., Sakai, R., and Otani, Y. (1967). Phenol metabolism in relation to disease resistance of potato tubers, *Plant Cell Physiol.,* 8:1-13.

Toppan, A., and Esquerre-Tugaye, M. T. (1984). Cell surfaces in plant-microorganism interactions. IV. Fungal glycopeptides which elicit the synthesis of ethylene in plants, *Plant Physiol.,* 75:1133-1138.

Torli, K. U., Mitsukawa, N., Osumi, T., Matsuura, Y., Yokoyama, R., Whittler, R. F., and Komeda, Y. (1996). The *Arabidopsis* ERECTA gene encodes a putative receptor protein kinase with extracellular leucine-rich repeats, *Plant Cell,* 8:735-746.

Tornero, P., Conejero, V., and Vera, P. (1997). Identification of a new pathogen-induced member of the subtilisin-like processing protease family from plants, *JBC Online* 272(22):14412-14419.

Toubart, P., Desiderio, A., Salvi, G., Cervone, F., Daroda, L., De Lorenzo, G., Darvill, A. G., and Albersheim, P. (1992). Cloning and characterization of the gene encoding the endopolygalacturonase-inhibiting protein (PGIP) of *Phaseolus vulgaris* L., *Plant J.,* 2:367-373.

Toyoda, K., Shiraishi, T., Yamada, T., Ichinose, Y., and Oku, H. (1993). Rapid changes in polyphosphoinositide metabolism in pea in response to fungal signals, *Plant Cell Physiol.,* 34:729-735.

Toyoda, K., Shiraishi, T., Yoshioka, H., Yamad, T., Ichinose, Y., and Oku, H. (1992). Regulation of polyphosphoinositide metabolism in pea plasma membranes by elicitor and suppressor from a pea pathogen, *Mycosphaerella pinodes, Plant Cell Physiol.,* 33:445-452.

Trewavas, A., and Gilroy, S. (1991). Signal induction in plant cells, *Trend. Genet.,* 7:356-361.

Trigalet, A., and Demery, D. (1986). Invasiveness in tomato plants of Tn5-induced avirulent mutants of *Pseudomonas solanacearum, Physiol. Mol. Plant Pathol.,* 28:423-430.

Tronsmo, A. M., Gregersen, P., Hjeljord, L., Sandal, T., Bryngelsson, T., and Collinge, D. B. (1993). Cold-induced disease resistance. In B. Fritig and M. Legrand (Eds.), *Mechanisms of Plant Defense Responses,* Kluwer Academic Publishers, Dordrecht, p. 369.

Truchet, G., Roche, P., Lerouge, P., Vasse, J., Camut, S., De Billy, F., Prome, J. C., and Denarie, J. (1991). Sulphated lipo-oligosaccharide signals of *Rhizobium meliloti* elicit root nodule organogenesis in alfalfa, *Nature,* 351:670-673.

Trudel, J., Porvin, C., and Asselin, A. (1995). Secreted hen lysozyme in transgenic tobacco: Recovery of bound enzyme and in vitro growth inhibition of plant pathogens, *Plant Sci.,* 106:55-62.

Tsugita, A. (1971). Phage lysozyme and other lytic enzymes. In P. D. Boyer (Ed.), *The Enzymes,* Academic Press, New York, pp. 344-411.

Tsuji, J., Jackson, E. P., Gage, D. A., Hammerschmidt, R., and Somerville, S. C. (1992). Phytoalexin accumulation in *Arabidopsis thaliana* during the hypersensitive reaction to *Pseudomonas syringae* pv. *syringae, Plant Physiol.,* 98:1304-1309.

Tsuji. J., and Somerville, S. C. (1992). First report of natural infection of *Arabidopsis thaliana* by *Xanthomonas campestris* pv. *campestris, Plant Dis.,* 76:539.

Tsuyumu, S., and Chatterjee, A. K. (1984). Pectin lyase production in *Erwinia chrysanthemi* and other soft rot *Erwinia* species, *Physiol. Plant Pathol.,* 24:291-302.

Tu, J., Ona, I., Zhang, Q., Mew, T. W., Khush, G. S., and Datta, S. K. (1998). Transgenic rice variety 'IR72' with *Xa21* is resistant to bacterial blight, *Theor. Appl. Genet.,* 97:31-36.

Turk, S. C. H. J., Melchers, L. S., den Dulk-Ras, H., Regensburg-Tuink, A. J. G., and Hooykaas, P. J. J. (1991). Environmental conditions differentially affect *vir* gene induction in different *Agrobacterium* strains. Role of the VirA sensor protein, *Plant Mol. Biol.,* 16:1051-1059.

Turk, S. C. H. J., van Lange, R. P., Sonneveld, E., and Hooykaas, P. J. J. (1993). The chimeric VirA-Tar receptor is locked into a highly responsive state, *J. Bacteriol.,* 175:5706-5709.

Turner, P., Barber, E., and Daniels, M. (1985). Evidence for clustered pathogenicity genes in *Xanthomonas campestris* pv. *campestris, Mol. Gen. Genet.,* 199:338-343.

Tuzun, S., Rao M. N., Vogeli, U., Schardl, C. I., and Kuc, J. (1989). Induced systemic resistance to blue mold: Early induction and accumulation of β-1,3-glucanases, chitinases and other pathogenesis-related proteins (b-proteins) in immunized tobacco, *Phytopathology,* 79:979-983.

Uknes, S., Dincher, S., Friedrich, L., Negrotto, D., Williams, S., Thompson-Taylor, H., Potter, S., Ward, E., and Ryals, J. (1993a). Regulation of pathogenesis-related protein-1a gene expression in tobacco, *Plant Cell,* 5:159-169.

Uknes, S., Mauch-Mani, B., Moyer, M., Potter, S., Williams, S., Dincher, S., Chandler, D., Slusarenko, A., Ward, E., and Ryals, J. (1992). Acquired resistance in *Arabidopsis, Plant Cell,* 4:645-656.

Uknes, S., Winter, A. M., Delaney, T., Vernooij, B., Morse, A., Friedrich, L., Nye, G., Potter, S., Ward, E., and Ryals, J. (1993b). Biological induction of systemic acquired resistance in *Arabidopsis, Mol. Plant-Microbe Interact.,* 6:692-698.

Ulmasov, T., Hagen, G., and Guilfoyle, T. (1994). The *ocs* element in the soybean GH2/4 promoter is activated by both active and inactive auxin and salicylic acid analogues, *Plant Mol. Biol.,* 26:1055-1064.

Vale, G. P., Torrigiani, E., Gatti, A., Delogu, G., Porta-Puglia, A., Vannacci, G., and Cattivelli, L. (1994). Activation of genes in barley roots in response to infection by two *Drechslera graminea* isolates, *Physiol. Mol. Plant Pathol.,* 44:207-215.

Vallad, G., Rivkin, M. I., Vallajos, E., and McClean, P. (1998). Cloning of *Pto*-like sequences in common bean, *Sixth Intl. Conf. Plant and Animal Genome Research,* San Diego, California, p. 137.

Valon, C., Smalle, J., Goodman, H. M., and Giraudat, J. (1993). Characterization of an *Arabidopsis thaliana* gene *(TMKL1)* encoding a putative transmembrane protein with an unusual kinase-like domain, *Plant Mol. Biol.,* 23:415-421.

Van de Locht, U., Meier, I., Hahlbrock, K., and Somssich, H. (1990). A 125 bp promoter fragment is sufficient for strong elicitor-mediated gene activation in parsley, *EMBO J.,* 9:2945-2950.

Van de Rhee, M. D., and Bol, J. F. (1993). Induction of the tobacco *PR-1a* gene by virus infection and salicylate treatment involves an interaction between multiple regulatory elements, *Plant J.,* 3:71-82.

Van de Rhee, M. D., Lemmers, R., and Bol, J. F. (1993). Analysis of regulatory elements involved in stress-induced and organ-specific expression of tobacco acidic and basic β-1,3-glucanase genes, *Plant Mol. Biol.,* 21:451-461.

Van de Rhee, M. D., Van Kan, J. A. L., Gonzalez-Jaen, M. T., and Bol, J. F. (1990). Analysis of regulatory elements involved in the induction of two tobacco genes by salicylate treatment and virus infection, *Plant Cell,* 2:357-366.

Van den Ackerveken, G., Marois, E., and Bonas, U. (1996). Recognition of the bacterial avirulence protein AvrBs3 occurs inside the host plant cell, *Cell,* 87: 1307-1316.

Van den Bulcke, M., Bauw, G., Castresana, C., Van Montague, M., and Vande Kerckhove, J. (1989). Characterization of vacuolar and extracellular β-(1,3)-glucanases of tobacco: Evidence for a strictly compartmentalized plant defense system, *Proc. Natl. Acad. Sci. U.S.A,* 86:2673-2677.

Van Doorn, J., Boonekamp, P. M., and Oudega, B. (1994). Partial characterization of fimbriae of *Xanthomonas campestris* pv. *hyacinthi, Mol. Plant-Microbe Interact.,* 7:334-344.

Van Etten, H. D., Mansfield, J. W., Bailey, J. F., and Farmer, E. E. (1994). Two classes of plant antibiotics: Phytoalexins versus "phytoanticipins," *Plant Cell,* 6:1191-1192.

Van Etten, H. D., Matthews, D. E., and Matthews, P. S. (1989). Phytoalexin detoxification: Importance for phytopathogenicity and practical application, *Annu. Rev. Phytopathol.,* 27:143-164.

Van Etten, H. D., Sandrock, R. W., Wasmam, C. C., Soby, S. D., Mellusky, K., and Wang, P. (1995). Detoxification of phytoanticipins and phytoalexins by phytopathogenic fungi, *Can. J. Bot.,* 73(Suppl.1):S518-S525.

Van Gijsegem, F. (1989). Relationship between the *pel* genes of the *pelADE* cluster in *Erwinia chrysanthemi* strain B374, *Mol. Microbiol.,* 3:1415-1424.

Van Gijsegem, F., Genin, S., and Boucher, C. (1994). Conservation of secretion pathways for pathogenicity determinants of plant and animal pathogens, *Trends Microbiol.,* 1:175-181.

Van Gijsegem, F., Gough, C., Zischek, C., Niqueux, E., Arlat, M., Genin, S., Barberis, P., German, S., Castello, P., and Boucher, C. (1995). The *hrp* gene locus of *Pseudomonas solanacearum,* which controls the production of a type III secretion system, encodes eight proteins related to components of the bacterial flagellar biogenesis complex, *Mol. Microbiol.,* 15:1095-1114.

Van Gijsegem, F., Somssich, I. E., and Scheel, D. (1995). Activation of defense-related genes in parsley leaves by infection with *Erwinia chrysanthemi, Eur. J. Plant Pathol.,* 101:549-559.

Van Huytsee, R. B. (1987). Some molecular aspects of plant peroxidase biosynthetic studies, *Annu. Rev. Plant Physiol.,* 38:205-219.

Van Kan, J. A. L., Cornelissen, B. J. C., and Bol, J. F. (1988). A virus-inducible gene encoding a glycine-rich protein shares putative regulatory elements with the ribulose bisphosphate carboxylase small subunit gene, *Mol. Plant-Microbe Interact.,* 1:107-112.

Van Kan, J. A. L., Joosten, M. H. A. J., Wagemakers, C. A. M., Van den Berg-Velthuis, G. C. M., and de Wit, P. J. G. M. (1992). Differential accumulation of mRNAs encoding intercellular and intracellular PR proteins in tomato induced by virulent and avirulent races of *Cladosporium fulvum, Plant Mol. Biol.,* 20:513-527.

Van Loon, L. C. (1976). Specific soluble leaf proteins in virus-infected tobacco plants are not normal constituents, *J. Gen. Virol.,* 30:375-379.

Van Loon, L. C. (1999). Occurrence and properties of plant pathogenesis-related proteins. In S. K. Datta and S. Muthukrishnan (Eds.), *Pathogenesis-Related Proteins in Plants,* CRC Press, Boca Raton, FL, pp. 1-19.

Van Loon, L. C., and Antoniw, J. F. (1982). Comparison of the effects of salicylic acid and ethephon with virus-induced hypersensitivity and acquired resistance in tobacco, *Neth. J. Plant Pathol.,* 88:237-256.

Van Loon, L. C., and Gerritsen, Y. A. M. (1989). Localization of pathogenesis-related proteins in infected and non-infected leaves of Samsun NN tobacco during the hypersensitive reaction to tobacco mosaic virus, *Plant Sci.,* 63:131-140.

Van Loon, L. C., Gerritsen, Y. A. M., and Ritter, C. E. (1987). Identification, purification and characterization of pathogenesis-related proteins from virus-infected Samsun NN tobacco leaves, *Plant Mol. Biol.,* 9:593-609.

Van Loon, L. C., Pierpoint, W. S., Boller, T., and Conejero, V. (1994). Recommendations for naming plant pathogenesis-related proteins, *Plant Mol. Biol. Reporter,* 12:245-264.

Van Loon, L. C., and Van Strien, E. A. (1999). The families of pathogenesis-related proteins, their activities, and comparative analysis of PR-1 type proteins, *Physiol. Mol. Plant Pathol.,* 55:85-97.

Van Montagu, M., Babiychuk, E., Herouart, D., Kushnir, S., Van Camp, W., Willekens, H., and Inze, D. (1994). Signal transduction in oxidative stress, *J. Cellular Biochem.* Suppl., 18A:76.

Van Parijs, J., Broekaert, W. F., Goldstein, I. J., and Peumans, W. J. (1991). Hevein-an antifungal protein from rubber-tree *(Hevea brasiliensis)* latex, *Planta,* 183: 258-264.

Van Peer, R., Niemann, G. J., and Schippers, B. (1991). Induced resistance and phytoalexin accumulation in biological control of fusarium wilt of carnation by *Pseudomonas* sp. strain WCS417r, *Phytopathology,* 81:728-734.

Van Peer, R., and Schippers, B. (1992). Lipopolysaccharides of plant growth-promoting *Pseudomonas* spp. strain WCS 417r induce resistance in carnation to *Fusarium* wilt, *Neth. J. Plant Pathol.,* 98:129-139.

Van Wees, S. C. M., Pieterse, C. M. J., Trijssenaar, A., Van't Westende, Y. A. M., Hartog, F., and Van Loon, L. C. (1997). Differential induction of systemic resistance in *Arabidopsis* by biocontrol bacteria, *Mol. Plant-Microbe Interact.,* 10:716-724.

Varin, L., De Luca, V., Ibrahim, R. K., and Brisson, N. (1992). Molecular characterization of two plant flavonol sulfotransferases, *Proc. Natl. Acad. Sci. U.S.A.,* 89:1286-1290.

Varin, L., and Ibrahim, R. K. (1992). Novel flavonol 3-sulfotransferase, *J. Biol. Chem.,* 267:1858-1863.

Vattanaviboon, P., Praituan, W., and Monkolsuk, S. (1995). Growth phase dependent resistance to oxidative stress in the phytopathogen *Xanthomonas oryzae* pv. *oryzae, Can. J. Microbiol.,* 41:1043-1047.

Velazhahan, R., Chen-Cole, K., Anuratha, C. S., and Muthukrishnan, S. (1998). Induction of thaumatin like proteins (TLPs) in *Rhizoctonia solani*-infected rice and characterization of two new cDNA clones, *Physiol. Plantarum,* 102:21-28.

Veluthambi, K., Jayaswal., R. K., and Gelvin, S. B. (1987). Virulence genes A, G, and D mediate the double stranded border cleavage of T-DNA from the *Agrobacterium tumefaciens* Ti plasmid, *Proc. Natl. Acad. Sci. U.S.A.,* 84:1881-1885.

Veluthambi, K., Ream, W., and Gelvin, S. (1988). Virulence genes, borders, and overdrive generate single stranded T-DNA molecules from the A6Ti plasmid of *Agrobacterium tumefaciens, J. Bacteriol.,* 170:1523-1532.

Venere, R. J. (1980). Role of peroxidase in cotton resistant to bacterial blight, *Plant Sci. Lett.,* 20:47-56.

Venere, R. J., Wang, X., Dyer, J. H., and Zheng, L. (1993). Role of peroxidase in cotton resistant to bacterial blight: purification and immunological analysis of phospholipaseD from castor bean endosperm, *Plant Sci. Lett.,* 306:486-488.

Vera, P., and Conejero, V. (1988). Pathogenesis-related proteins in tomato. P69 as an alkaline endoproteinase, *Plant Physiol.,* 87:58-63.

Vera, P., Hernandez-Yago, J., and Conejero V. (1989). Pathogenesis-related P1 (P14) Protein. Vacuolar and apoplastic localization in leaf tissue from tomato plants infected with citrus exocortis viroid: In vitro synthesis and processing, *J. Gen. Virol.,* 70:1933-1942.

Vera-Estrella, R., Barkla, B.J., Higgins, V. J., and Blumwald, E. (1994). Plant defense response to fungal pathogens. Activation of host-plasma membrane H^+-ATPase by elicitor-induced enzyme dephosphorylation, *Plant Physiol.,* 104:209-215.

Vera-Estrella, R., Blumwald, E., and Higgins, V. J. (1992). Effect of specific elicitors of *Cladosporium fulvum* on tomato cell suspension cells: Evidence for the involvement of active oxygen species, *Plant Physiol.,* 99:1208-1215.

Vernade, D., Herrera-Estrella, A., Wang, K., and Van Montagu, M. (1988). Glycine betaine allows enhanced induction of the *Agrobacterium tumefaciens vir* genes by acetosyringone at low pH, *J. Bacteriol.,* 170:5822-5829.

Vernooij, B., Friedrich, L., Ahl-Goy, P., Staub, T., Kessmann, H., and Ryals, J. (1995). 2,6-Dichloroisonicotinic acid-induced resistance to pathogens without the accumulation of salicylic acid, *Mol. Plant-Microbe Interact.,* 8:228-234.

Vernooij, B., Friedrich, L., Morse, A., Resist, R., Kolditz-Jawhar, R., Ward, E., Uknes, S., Kessmann, H., and Ryals, J. (1994). Salicylic acid is not the translocated signal responsible for inducing systemic acquired resistance but is required in signal transduction, *Plant Cell,* 6:959-965.

Vesper, S. J. (1987). Production of pili (fimbriae) by *Pseudomonas fluorescens* and correlation with attachment to corn roots, *Appl. Environ. Microbiol.,* 53:1397-1405.

Vian, B., Reis, D., Gea, L., and Grimault, V. (1996). The cell wall, first barrier or interface for microorganism: in situ approaches to understanding interactions. In M. Nicole and V. Gianinazzi-Pearson (Eds.), *Histology, Ultrastructure and Molecular Cytology of Plant-Microorganisms Interactions,* Kluwer Academic Publishers, Dordrecht, pp. 99-115.

Vian, B., Temsah, M., Reis, D., and Roland, J. C. (1992). Colocalization of the cellulose framework and cell wall matrix in the helicoidal constructions, *J. Microscopy,* 166:111-122.

Vianello, A., and Macri, F. (1991). Generation of superoxide anion and hydrogen peroxide at the surface of plant cells, *J. Bioenergetics Membranes,* 23:409-423.

Vick, B. A., and Zimmerman, D. C. (1984). Biosynthesis of jasmonic acid by several plant species, *Plant Physiol.,* 75:458-461.

Vick, B. A., and Zimmerman, D. C. (1987). Pathways of fatty acid hydroperoxide metabolism in spinach leaf chloroplasts, *Plant Physiol.,* 85:1073-1078.

Vidal, S., De Leon, I. P., Denecke, J., and Palva, E. T. (1997). Salicylic acid and the plant pathogen *Erwinia carotovora* induce defence genes via antagonistic pathways, *Plant J.,* 11:115-123.

Vidhyasekaran, P. (1988a). *Physiology of Disease Resistance in Plants,* Vol. I, CRC Press, Boca Raton, FL, p. 149.

Vidhyasekaran, P. (1988b). *Physiology of Disease Resistance in Plants,* Vol. II, CRC Press, Boca Raton, FL, p. 128.

Vidhyasekaran, P. (1990). *Physiology of Disease Resistance in Field Crops,* Today and Tomorrow's Publishers, New Delhi, p. 137.

Vidhyasekaran, P. (1993a). Chitinase gene and pathogenesis-related proteins for crop disease management. In P. Vidhyasekaran (Ed.), *Genetic Engineering, Tissue Culture, and Molecular Biology for Crop Pest and Disease Management,* Daya Publishing House, Delhi, pp. 41-63.

Vidhyasekaran, P. (1993b). Defense genes for crop disease management. In P. Vidhyasekaran (Ed.), *Genetic Engineering, Tissue Culture, and Molecular Biology for Crop Pest and Disease Management,* Daya Publishing House, Delhi, pp. 17-30.

Vidhyasekaran, P. (1993c). *Principles of Plant Pathology,* CBS Publishers, Delhi, p. 166.

Vidhyasekaran, P. (1997). *Fungal Pathogenesis in Plants and Crops,* Marcel Dekker, New York, p. 553.

Vidhyasekaran, P. (1998). Molecular biology of pathogenesis and induced systemic resistance, *Indian Phytopath.,* 51:111-120.

Vidhyasekaran, P., Alvenda, M. E., and Mew, T. W. (1989). Physiological changes in rice seedlings induced by *Xanthomonas campestris* pv. *oryzae, Physiol. Mol. Plant Pathol.,* 35:391-402.

Vidhyasekaran, P., Borromeo, E. S., and Mew, T. W. (1986). Host-specific toxin production by *Helminthosporium oryzae, Phytopathology,* 76:261-266.

Vidhyasekaran, P., Kamala, N., Ramanathan, A., Rajappan, K., Paranidharan, V., and Velazhahan, R. (2001). Induction of systemic resistance by *Pseudomonas fluorescens* Pf1 against *Xanthomonas oryzae* pv. *oryzae* in rice leaves, *Phyoparasitica,* 29:2, pp. 155-166.

Vidhyasekaran, P., Ling, D. H., Borromeo, E. S., and Mew, T. W. (1990). Selection of brown spot resistant rice plants from *Helminthosporium oryzae* toxin resistant calluses, *Ann. Appl. Biol.,* 117:515-523.

Vidhyasekaran, P., Parambaramani, C., and Durairaj, P. (1971). Pectolytic enzymes of *Xanthomonas malvacearum, Indian J. Microbiol.,* 11:93-96.

Vidhyasekaran, P., Rabindran, R., Muthamilan, M., Nayar, K., Rajappan, K., Subramanian, N., and Vasumathi, K. (1997). Development of a powder formulation of *Pseudomonas fluorescens* for control of rice blast, *Plant Pathology,* 46:291-297.

Vidhyasekaran, P., Velazhahan, R., and Balasubramanian, P. (2000). Biological control of crop diseases exploiting genes involved in systemic induced resistance. In R. K. Upadhyay, K. G. Mukherji, and B. P. Chamola (Eds.), *Biocontrol*

Potential and Its Exploitation in Sustainable Agriculture. Volume 1: Crop Diseases, Weeds, and Nematodes, Kluwer Academic/Plenum Publishers, New York, pp. 1-8.

Vierheilig, H., Alt, M., Neuhaus, J. M., Boller, T., and Wiemken, A. (1992). Colonization of transgenic *Nicotiana sylvestris* plants, expressing different forms of *Nicotiana tabacum* chitinase, by the root pathogen Rhizoctonia solani and by the mycorrhizal symbiont *Glomus mosseae, Mol. Plant-Microbe Interact.,* 6:261-264.

Vigers, A. J., Roberts, W. K., and Selitrennikoff, C. F. (1991). A new family of plant antifungal proteins, *Mol. Plant-Microbe Interact.,* 4:315-323.

Vigers, A. J., Wiedemann, S., Roberts, W. K., Legrand, M., Selitrennikoff, C. P., and Fritig, B. (1992). Thaumatin-like pathogenesis-related proteins are antifungal, *Plant Sci.,* 83:155-161.

Vivian, A., Atherton, G., Bevan, J., Crute, I., Muir, L., and Taylor, J. (1989). Isolation and characterization of cloned DNA conferring specific avirulence in *Pseudomonas syringae* pv. *pisi* to pea *(Pisum sativum)* cultivars carrying the resistance allele, R2, *Physiol. Mol. Plant Pathol.,* 34:335-344.

Vivian, A., and Gibbon, M. J. (1997). Avirulence genes in plant-pathogenic bacteria: Signals or weapons?, *Micobiology (Reading),* 143:693-704.

Vivian, A., and Mansfield, J. (1993). A proposal for a uniform genetic nomenclature for avirulence genes in phytopathogenic pseudomonads, *Mol. Plant-Microbe Intertact.,* 6:9-10.

Vogeli, U., Meins, F., and Boller, T. (1988). Co-ordinated regulation of chitinase and β-1,3-glucanase in bean leaves, *Planta,* 174:364-372.

Vogeli-Lange, R., Fruendt, C., Hart, C. M., Beffa, R., Nagy, F., and Meins, F. Jr. (1994a). Evidence for a role of β-1,3-glucanase in dicot seed germination, *Plant J.,* 5:273-278.

Vogeli-Lange, R., Fruendt, C., Hart, C. M., Nagy, F., and Merias, F. Jr. (1994b). Developmental, hormonal and pathogenesis-related regulation of the tobacco class I β-1,3-glucanase B promoter, *Plant Mol. Biol.,* 25:299-311.

Vogeli-Lange, R., Hansen-Gehri, A., Boller, T., and Meins, F. Jr. (1988). Induction of the defense-related glucanohydrolases, β-1,3-glucanase and chitinase, by tobacco mosaic virus infection of tobacco leaves, *Plant Sci.,* 54:171-176.

Vogelsang, R., and Barz, W. (1993). Purification, characterization and differential regulation of a β-1,3-glucanase and two chitinases from chickpea (*Cicer arietinum* L.), *Planta,* 189:60-69.

Voisey, C. R., and Slusarenko, A. J. (1989). Chitinase mRNA and enzyme activity in *Phaseolus vulgaris* (L.) increase more rapidly in response to avirulent than to virulent cells of *Pseudomonas syringae* pv. *phaseolicola, Physiol. Mol. Plant Pathol.,* 35:403-412.

Wagner, V. T., and Matthysse, A. G. (1992). Involvement of a vitronectin-like protein in attachment of *Agrobacterium tumefaciens* to carrot suspension culture cells, *J. Bacteriol.,* 174:5999-6003.

Walker, D. S., Reeves, P. J., and Salmond, G. P. C. (1994). The major secreted cellulase, CelV, of *Erwinia carotovora* subsp. *carotovora* is an important soft rot virulence factor, *Mol. Plant-Microbe Interact.,* 7:425-431.

Walker, J. C. (1994). Structure and function of the receptor-like protein kinases of higher plants, *Plant Mol. Biol.,* 26:1599-1609.

Wallis, F. M. (1977). Ultrastructural histopathology of tomato plants infected with *Corynebacterium michiganense, Physiol. Plant Pathol.,* 11:333-342.

Wallis, F. M., and Truter, S. J. (1978). Histopathology of tomato plants infected with *Pseudomonas solanacearum,* with emphasis on ultrastructure, *Physiol. Plant Pathol.,* 13:307-317.

Walter, M. H., Liu, J. W., Grand, C., Lamb, C. J., and Hess, D. (1990). Bean pathogenesis-related (PR) proteins deduced from elicitor-induced transcripts are members of a ubiquitous new class of conserved PR proteins including pollen allergens, *Mol. Gen. Genet.,* 222:353-360.

Walters, K., Maroofi, A., Hitchin, E., and Mansfield, J. (1990). Gene for pathogenicity and ability to cause the hypersensitive reaction cloned from *Erwinia amylovora, Physiol. Mol. Plant Pathol.,* 36:509-527.

Wandersman, C. (1992). Secretion across the bacterial outer membrane, *Trends Genet.,* 8:317-322.

Wandersman, C., Delepelaire, P., Letoffe, P., and Schwartz, M. (1987). Characterization of *Erwinia chrysanthemi* extracellular proteases: Cloning and expression of the protease genes in *Escherichia coli, J. Bacteriol.,* 169:5046-5053.

Waney, V. R., Kingsley, M. T., and Gabriel, D. W. (1991). *Xanthomonas campestris translucens* genes determining host-specific virulence and general virulence on cereals identified by Tn5-*gusA* insertion mutagenasis, *Mol. Plant-Microbe Interact.,* 4:623-627.

Wang, G.-L., Holsten, T. E., Song, W. Y., Wang, H. P., and Ronald, P. C. (1995). Construction of a rice bacterial artificial chromosome library and identification of clones linked to the *Xa21* disease resistance locus, *Plant J.,* 7:525-533.

Wang, G.-L., Song, W.-Y., Runan, D.-L., Sideris, S., and Ronald, P. C. (1996). The cloned *Xa* 21, confers resistance to multiple *Xanthomonas oryzae* pv. *oryzae* isolates in transgenic plants, *Mol. Plant-Microbe Interact.,* 9:850-855.

Wang, Zi-X., Yano, M., Yamanouchi, U., Iwamoto, M., Monna, L., Hayasaka, H., Katayose, Y., and Sasaki, T. (1999). The *Pib* gene for rice blast resistance belongs to the nucleotide binding and leucine-rich repeat class of plant disease resistance genes, *Plant J.,* 19:55-56.

Wanner, L., Mittal, S., and Davis, K. R. (1993). Recognition of the avirulence gene *avrB* from *Pseudomonas syringae* pv. *glycinea* by *Arabidopsis thaliana, Mol. Plant-Microbe Interact.,* 6:582-591.

Ward, E. R., Payne, G. P., Moyer, M. B., Williams, S. C., Dincher, S. S., Sharkey, K. C., Beck, J. J., Taylor, H. T., Ahl-Goy, P., Meins, F. Jr., and Ryals, J. A. (1991a). Differential regulation of β-1,3-glucanase messenger RNAs in response to pathogen infection, *Plant Physiol.,* 96:390-397.

Ward, E. R., Uknes, S. J., Williams, S. C., Dincher, S. S., Wiederhold, D. L., Alexander, D. C., Ahl-Goy, P., Metraux, J. P., and Ryals, J. (1991b). Co-ordinate gene activity in response to agents that induce systemic acquired resistance, *Plant Cell,* 3:1085-1094.

Warner, S. A. J., Gill, A., and Draper, J. (1994). The developmental expression of the asparagus intracellular PR protein (AoPR1) gene correlates with sites of phenylpropanoid biosynthesis, *Plant J.,* 6:31-43.

Warner, S. A. J., Scott, R., and Draper, J. (1992). Characterization of a wound induced transcript from the monocot asparagus that shares similarity with a class of intracellular pathogenesis-related (PR) proteins, *Plant Mol. Biol.,* 19:555-561.

Wasternack, C., and Parthier, B. (1997). Jasmonate-signalled plant gene expression, *Trends Plant Sci.,* 2:302-307.

Watanabe, T., Sekizawa, Y., Shimura, H., Suzuki, Y., Matsumoto, M., Iwata, M., and Mase, S. (1979). Effects of Probenazole (Oryzemate) on rice plants with reference to controlling rice blast, *J. Pestic. Sci.,* 4:53-58.

Waterkeyn, L. (1967). Sur l'existence d'un stade callosiqué presente par la au cours de la cytokinese, *C. R. Acad. Sci. Paris,* D 265:1792-1794.

Weber, J., Olsen, O., Wegener, C., and von Wettstein, D. (1996). Digalacturonates from pectin degradation induced tissue responses against potato soft rot, *Physiol. Mol. Plant Pathol.,* 48:389-401.

Wegener, C., Bartling, S., Olsen, O., Weber, J., and von Wettstein, D. (1996). Pectate lyase in transgenic potatoes confers preactivation of defence against *Erwinia carotovora, Physiol. Mol. Plant Pathol.,* 49:359-376.

Wei, G., Kloepper, J. W., and Tuzun, S. (1991). Induction of systemic resistance of cucumber to *Colletotrichum orbiculare* by select strains of plant growth-promoting rhizobacteria, *Phytopathology,* 81:1508-1512.

Wei, G., Kloepper, J. W., and Tuzun, S. (1996). Induced systemic resistance to cucumber diseases and increased plant growth by plant growth-promoting rhizobacteria under field conditions, *Phytopathology,* 86:221-224.

Wei, Z. M., and Beer, S. V. (1990). Functional homology between a locus of *Escherichia coli* and the *hrp* gene cluster of *Erwinia amylovora, Phytopathology,* 80:1039 (Abstr.).

Wei, Z. M., and Beer, S. V. (1993). HrpI of *Erwinia amylovora* functions in secretion of Harpin and is a member of a new protein family, *J. Bacteriol.,* 175:7958-7967.

Wei, Z. M., and Beer, S. V. (1995). *hrpL* activates *Erwinia amylovora hrp* gene transcription and is a member of the ECF subfamily of sigma factors, *J. Bacteriol.,* 177:6201-6210.

Wei, Z. M., and Beer, S. V. (1996). Harpin from *Erwinia amylovora* induces plant resistance. In W. G. Bonn (Ed.), *VII International Workshop on Fire Blight,* Ontario, Canada, Acta Horticulturae No. 411, pp. 427-431.

Wei, Z. M., Laby, R. J., Zumoff, C. H., Bauer, D. W., He, S. Y., Collmer, A., and Beer, S. V. (1992a). Harpin, elicitor of the hypersensitive response produced by the plant pathogen *Erwinia amylovora, Science,* 257:85-88.

Wei, Z. M., Sneath, B. J., and Beer, S. V. (1992b). Expression of *Erwinia amylovora hrp* genes in response to environmental stimuli, *J. Bacteriol.,* 174:1675-1882.

Wendehenne, D., Durner, J., Chen, Z., and Klessig, D. F. (1998). Benzothiadiazole, an inducer of plant defenses, inhibits catalase and ascorbate peroxidase, *Phytochemistry,* 47:651-657.

Wengelnik, K., and Bonas, U. (1996). HrpXv, an AraC-type regulator, activates expression of five out of the six loci in the *hrp* cluster of *Xanthomonas campestris* pv. *vesicatoria, J. Bacteriol.,* 178:3462-3469.

Wengelnik, K., Marie, C., Russel, M., and Bonas, U. (1996a). Expression and localization of HrpA1, a protein of *Xanthomonas campestris* pv. *vesicatoria* essential for pathogenicity and induction of the hypersensitive reaction, *J. Bacteriol.,* 178:1061-1069.

Wengelnik, K., Van den Ackerveken, G., and Bonas, U. (1996b). HrpG, a key *hrp* regulatory protein of *Xanthomonas campestris* pv. *vesicatoria* is homologous to two-component response regulators, *Mol. Plant-Microbe Interact.,* 9:704-712.

West, C. A., Moesta, P., Jin, D. F., Lois, A. F., and Wickham, K. A. (1985). The role of pectic fragments of the plant cell wall in the response to biological stress. In J. Key and T. Kosuge (Eds.), *Cellular and Molecular Biology of Plant Stress,* Alan R. Liss Press, New York, pp. 335-349.

Weymann, K., Hunt, M., Uknes, S., Neuenschwander, U., Lawton, K., Steiner, H. Y., and Ryals, J. (1995). Suppression and restoration of lesion formation in *Arabidopsis lsd* mutants, *Plant Cell,* 7:2013-2022.

Whalen, M., Innes, R., Bent, A., and Staskawicz, B. (1991). Identification of *Pseudomonas syringae* pathogens of *Arabidopsis thaliana* and a bacterial gene determining avirulence on both *Arabidopsis* and *soybean, Plant Cell,* 3:49-59.

Whalen, M. C., Stall, R. E., and Staskawicz, B. J. (1988). Characterization of a gene from a tomato pathogen determining hypersensitive resistance in a non-host species and genetic analysis of this resistance in bean, *Proc. Natl. Acad. Sci. U.S.A.,* 85:6743-6747.

Whalen, M. C., Wang, J. F., Carland, F. M., Heiskell, M. E., Dahlbeck, D., Minsavage, G. V., Jones, J. B., Scott, J. W., Stall, R. E., and Staskawicz, B. J. (1993). Avirulence gene *avrRxv* from *Xanthomonas campestris* pv. *vesicatoria* specifies resistance on tomato line Hawaii 7998, *Mol. Plant-Microbe Interact.,* 6:616-627.

White, R. F. (1979). Acetylsalicylic acid (aspirin) induces resistance to tobacco mosaic virus in tobacco, *Virology,* 99:410-412.

White, R. F., Antoniw, J. F., Carr, J. P., and Woods, R. D. (1983). The effects of aspirin and polyacrylic acid on the multiplication and spread of TMV in different cultivars of tobacco and without the N-gene, *Phytopath. Z.,* 107:224-232.

White, R. F., Reynolds, H. C., Sugars, J. M., and Antoniw, J. F. (1996). The suppression of mannitol induced pathogenesis-related protein accumulation in tobacco by antibiotics, *Physiol. Mol. Plant Pathol.,* 49:389-393.

Whitham, S., Dinesh-Kumar, S. P., Choi, D., Hehl, R., Corr, C., and Baker, B. (1994). The product of the tobacco mosaic virus resistance gene *N:* Similarity to Toll and the interleukin-I receptor, *Cell,* 78:1101-1115.

Whitham, S., McCormick, S., and Baker, B. (1996). The *N* of tobacco confers resistance to tobacco mosaic virus in transgenic tomato, *Proc. Natl. Acad. Sci. U.S.A.,* 93:8776-8781.

Whitmore, F. W. (1978). Lignin-protein complex catalyzed by peroxidase, *Plant Sci. Lett.,* 13:241-245.

Wijnsma, R., Go, J. T., van Weerden, I. N., Harkes, P. A. A., Verpoorte, R., and Svenden, A. B. (1985). Anthraquinones as phytoalexins in cell and tissue cultures of *Cinchona* spec., *Plant Cell Rep.*, 4:241-244.

Wilkins, T. A., Bednarek, S. Y., and Raikhel, N. V. (1990). Role of propeptide glycan in post-translational processing and transport of barley lectin to vacuoles in transgenic tobacco, *Plant Cell*, 2:301-313.

Willekens, H., Van Camp, W., Van Montagu, M., Inze, D., Sandermann, H. Jr., and Langebartels, C. (1994a). Ozone, sulfur dioxide, and ultraviolet B have similar effects on mRNA accumulation of antioxidant genes in *Nicotiana plumbaginifolia* (L.)., *Plant Physiol.*, 106:1007-1014.

Willekens, H., Villarroel, R., Van Camp, W., Van Montagu, M., and Inze, D. (1994b). Molecular identification of catalases from *Nicotiana plumbaginifolia* (L.), *FEBS Lett.*, 352:79-83.

Willis, D. K., Hrabak, E. M., Rich, J. J., Barta, T. M., Lindow, S. E., and Panopoulos, N. J. (1990). Isolation and characterization of a *Pseudomonas syringae* pv. *syringae* mutant deficient in lesion formation on bean, *Mol. Plant-Microbe Interact.*, 3:149-156.

Willis, D. K., Rich, J. J., and Hrabak, E. M. (1991). *hrp* genes of phytopathogenic bacteria, *Mol. Plant-Microbe Interact.*, 4:132-138.

Willis, J. W., Engwall, J. K., and Chatterjee, A. K. (1987). Cloning of genes for *Erwinia carotovora* pectolytic enzymes and further characterization of the polygalacturonases, *Phytopathology*, 77:1199-1205.

Wilson, C., Franklin, J. D., and Otto, B. E. (1987). Fruit volatiles inhibitory to *Monilinia fructicola* and *Botrytis cinerea*, *Plant Dis.*, 71:316-319.

Winans, S. (1990). Transcriptional induction of an *Agrobacterium* regulatory gene at tandem promoters by plant-released phenolic compounds, phosphate starvation, and acidic growth media, *J. Bacteriol.*, 172: 2433-2438.

Winans, S. C., Ebert, P. R., Stachel, S. E., Gordon, M. P., and Nester, E. W. (1986). A gene essential for *Agrobacterium* virulence is homologous to a family of positive regulatory loci, *Proc. Natl. Acad. Sci. U.S.A.*, 83:8278-8282.

Winans, S. C., Kerstetter, R. A., and Nester, E. W. (1988). Transcriptional regulation of the *virA* and *virG* genes of *Agrobacterium tumefaciens*, *J. Bacteriol.*, 170:4047-4054.

Winans, S. C., Kerstetter, R. A., Ward, J. E., and Nester, E. W. (1989). A protein required for transcriptional regulation of *Agrobacterium* virulence genes spans the cytoplasmic membrane, *J. Bacteriol.*, 171:1616-1622.

Wingate, V. P. M., Lawton, M. A., and Lamb, C. J. (1988). Glutathione causes a massive and selective induction of plant defense genes, *Plant Physiol.*, 87:206-210.

Wingate, V. P. M., Norman, P. M., and Lamb, C. J. (1990). Analysis of the cell surface of *Pseudomonas syringae* pv. *glycinea* with monoclonal antibodies, *Mol. Plant-Microbe Interact.*, 3:408-416.

Wingender, R., Rohrig, H., Horicke, C., Wing, D., and Schell, J. (1989). Differential regulation of soybean chalcone synthase genes in plant defence, symbiosis and upon environmental stimuli, *Mol. Gen. Genet.*, 218:315-322.

Wojtaszek, P., Trethowan, J., and Bolwell, J. P. (1995). Specificity in the immobilization of cell wall proteins in response to different elicitor molecules in suspension-cultured cells of French bean (*Phaseolus vulgaris* L.), *Plant Mol. Biol.*, 28:1075-1087.

Wolff, S. P., Garner, A., and Dean, R. T. (1986). Free radicals, lipids and protein degradation, *Trends Biochem. Sci.*, 11:27-31.

Wolfgang, B., and Wolfgang, H. (1975). Metabolism of flavonoids. In J. B. Harborne, T. J. Mabry and H. Mabry (Eds.), *The Flavonoids*, Chapman and Hall, London, pp. 916-969.

Woloshuk, C. P., Meulenhoff, J. S., Sela-Buurlage, M., Van Den Elzen, P. J. M., and Cornelissen, B. J. C. (1991). Pathogen-induced proteins with inhibitory activity towards *Phytophthora infestans*, *Plant Cell*, 3:619-628.

Wolter, M., Hollricher, K., Salamini, F., and Schulze-Lefert, P. (1993). The *mlo* resistance alleles to powdery mildew infection in barley trigger a developmentally controlled defence mimic phenotype, *Mol. Gen. Genet.*, 239:122-128.

Wood, J. R., Vivian, A., Jenner, C. J., Mansfield, J. W., and Taylor, J. D. (1994). Detection of a gene in pea controlling nonhost resistance to *Pseudomonas syringae* pv. *phaseolicola*, *Mol. Plant-Microbe Interact.*, 7:534-537.

Worrall, D., Hird, D. L., Hodge, R., Paul, W., Draper, J., and Scott, R. (1992). Premature dissolution of the microsporocyte callose wall causes male sterility in transgenic tobacco, *Plant Cell*, 4:759-771.

Wu, A., Harriman, R. W., Frost, D. J., Read, S. M., and Wasserman, B. P. (1991). Rapid enrichment of CHAPS-solubilized UDP-glucose: (1,3)-β-glucan (callose) synthase from *Beta vulgaris* L. by product entrapment, *Plant Physiol.*, 97:684-692.

Wu, G., Shortt, B. J., Lawrence, E. B., Levine, E. B., Fitzsimmons, K. C., and Shah, D. M. (1995). Disease resistance conferred by expression of a gene encoding H_2O_2-generating glucose oxidase in transgenic potato plants, *Plant Cell*, 7:1357-1368.

Wu, Y. X., and Zeng, F. H. (1995). Study on the correlation of resistance and susceptibility to bacterial blight with the active oxygen and the defensive enzymes in hybrid rice seedlings, *J. Tropical Subtropical Bot.*, 3:69-73.

Wubben, J. P., Eukelboom, C. A., and De Wit, P. J. G. M. (1993). Accumulation of pathogenesis-related proteins in the epidermis of tomato leaves infected by *Cladosporium fulvum*, *Neth. J. Pl. Path.*, 3(Suppl.):231-239.

Wubben, J. P., Joosten, J. A. L., and De Wit, P. J. G. M. (1992). Subcellular localization of plant chitinases and β-1,3-glucanases in *Cladosporium fulvum* (syn. *Fulvia fulva*)-infected tomato leaves, *Physiol. Mol. Plant Pathol.*, 41:23-32.

Wycoff, K. L., Powell, P. A., Gonzales, R. A., Corbin, D. R., Lamb, C., and Dixon, R. A. (1995). Stress activation of a bean hydroxyproline-rich glycoprotein promoter is superimposed on a pattern of tissue-specific developmental expression, *Plant Physiol.*, 109:41-52.

Wyman, J. G., and Van Etten, H. D. (1978). Antibacterial activity of selected isoflavonoids, *Phytopathology*, 68:583-589.

Xiao, Y., Heu, S., Yi, J., and Hutcheson, S. W. (1994). Identification of a putative alternate sigma factor and characterization of multicomponent regulatory cascade controlling the expression of *Pseudomonas syringae* pv. *syringae* Pss61 *hrp* and *hrmA* genes, *J. Bacteriol.*, 176:1025-1026.

Xiao, Y., and Hutcheson, S. W. (1991a). Environmental regulation of *Pseudomonas syringae* pv. *syringae* 61 *hrp* gene cluster, *Phytopathology,* 81:1245-1246.

Xiao, Y., and Hutcheson, S. W. (1991b). Transcriptional organization and expression of the *Pseudomonas syringae* pv. *syringae* 61 *hrp* gene cluster, *Phytopathology,* 81:1245-1246.

Xiao, Y., and Hutcheson, S. W. (1994). A single promoter sequence recognized by a newly identified alternate sigma factor directs expression of pathogenicity and host range determinants in *Pseudomonas syringae, J. Bacteriol.,* 176:3089-3091.

Xiao, Y., Lu, Y., Heu, S., and Hutcheson, S. W. (1992). Organization and environmental regulation of the *Pseudomonas syringae* pv. *syringae* 61 hrp cluster, *J. Bacteriol.,* 174:1734-1741.

Xu, D., McElroy, D., Thornburg, R. W., and Wu, R. (1993). Systemic induction of a potato *pin2* promoter by wounding, methyl jasmonate, and abscisic acid in transgenic rice plants, *Plant Mol. Biol.,* 22:573-588.

Xu, P., Leong, S., and Sequeira, L. (1988). Molecular cloning of genes that specify virulence in *Pseudomonas solanacearum, J. Bacteriol.,* 170:617-622.

Xu, Y., Chang, P. L., Liu, D., Narasimhan, L., Raghothama, K. G., Hasegawa, P. M., and Bressan, R. A. (1994). Plant defense genes are synergistically induced by ethylene and methyl jasmonate, *Plant Cell,* 6:1077-1085.

Yalpani, N., and Raskin, I. (1993). Salicylic acid: A systemic signal in plant disease resistance, *Trends Microbiol.,* 1:88-92.

Yalpani, N., Silverman, P., Wilson, M. A., Kleier, D. A., and Raskin, I. (1991). Salicylic acid is a systemic signal and an inducer of pathogenesis-related proteins in virus-infected tobacco, *Plant Cell,* 3:809-818.

Yamada, A., Shibuya, N., Kodama, O., and Akatsuka, T. (1993). Induction of phytoalexin formation in suspension-cultured rice cells by *N*-acetyl-chitooligosaccharides, *Biosci. Biotech. Biochem.,* 57:405-409.

Yamada, M. (1992). Lipid transfer proteins in plants and microorganisms, *Plant Cell Physiol.,* 33:1-6.

Yamada, T., Hashimoto, H., Shiraishi, T., and Oku, H. (1989). Suppression of pisatin, phenylalanine ammonia-lyase mRNA, and chalcone synthase mRNA accumulation by a putative pathogenicity factor from the fungus *Mycospherella pinodes, Mol. Plant-Microbe Interact.,* 2:256-261.

Yamane, H., Takagi, H., Abe, H., Yokota, T., and Takahashi, N. (1981). Identification of jasmonic acid in three species of higher plants and its biological activities, *Plant Cell Physiol.,* 22:689-697.

Yamazaki, H., Ishizuka, O., and Hoshina, T. (1996). Relationship between resistance to bacterial wilt and nutrient uptake in tomato seedlings, *Soil Sci. Plant Nutrition,* 42:203-208.

Yang, Y., DeFeyter, R., and Gabriel, D. W. (1994). Host-specific symptoms and increased release of *Xanthomonas citri* and *X. campestris* pv. *malvacearum* from leaves are determined by the 102-bp tandem repeats of *pthA* and *avrb6,* respectively, *Mol. Plant-Microbe Interact.,* 7:345-355.

Yang, Y., and Gabriel, D. W. (1995). *Xanthomonas* avirulence/pathogenicity gene family encodes functional plant nuclear targeting signals, *Mol. Plant-Microbe Interact.,* 8:627-631.

Yang, Y., and Klessig, D. F. (1996). Isolation and characterization of a tobacco mosaic virus-inducible *myb* oncogene homolog from tobacco, *Proc. Natl. Acad. Sci. U.S.A.,* 93:14972-14977.

Yang, Y., Yuan, Q., and Gabriel, D. W. (1996). Watersoaking function(s) of *XcmH1005* are redundantly encoded by members of the *Xanthomonas avr/pth* gene family, *Mol. Plant-Microbe Interact.,* 9:105-113.

Yang, Z., Cramer, C. L., and Lacy, G. H. (1989). System for simultaneous study of bacterial and plant genes in soft rot of potato, *Mol. Plant-Microbe Interact.,* 2:195-201.

Yang, Z., Cramer, C. L., and Lacy, G. H. (1992). *Erwinia carotovora* subsp. *carotovora* pectic enzymes: In planta gene activation and roles in soft-rot pathogenesis, *Mol. Plant-Microbe Interact.,* 5:104-112.

Yang, Z., Park, H., Lacy, G. H., and Cramer, C. L. (1991). Differential activation of potato 3-hydroxy-3-methylglutaryl coenzyme A reductase genes by wounding and pathogen challenge, *Plant Cell,* 3:397-405.

Yanofsky, M. F., Porter, S. G., Young, C., Albright, L. M., Gordon, M. P., and Nester, E. W. (1986). The *virD* operon of *Agrobacterium tumefaciens* encodes a site-specific endonuclease, *Cell,* 47:471-477.

Yao, C., Conway, W. S., and Sams, C. E. (1995). Purification and characterization of a polygalacturonase-inhibiting protein from apple fruit, *Phytopathology,* 85:1373-1377.

Yao, C., Wei, G., Zehnder, G. W., Shelby, R. A., and Kloepper, J. W. (1994). Induced systemic resistance against bacterial wilt of cucumber by select plant growth promoting rhizobacteria, *Phytopathology,* 84:1082.

Ye, X. S., Pan, S. Q., and Kuc, J. (1989). Pathogenesis-related proteins and systemic resistance to blue mold and tobacco mosaic virus induced by tobacco mosaic virus, *Peronospora tabacina* and aspirin, *Physiol. Mol. Plant Pathol.,* 35:161-175.

Ye, X. S., Pan, S. Q., and Kuc, J. (1990). Association of pathogenesis-related proteins and activities of peroxidase, β-1,3-glucanase and chitinase with systemic induced resistance to blue mould of tobacco but not to systemic tobacco mosaic virus, *Physiol. Mol. Plant Pathol.,* 36:523-536.

Yi, S.Y., and Hwang, B. K. (1996). Differential induction and accumulation of β-1,3-glucanase and chitinase isoforms in soybean hypocotyls and leaves after compatible and incompatible infection with *Phytophthora megasperma* f. sp. *glycinea, Physiol. Mol. Plant Pathol.,* 48:179-192.

Yoshida, H., Konishi, K., Nakagawa, T., Sekido, S., and Yamaguchi, I. (1990). Characteristics of N-phenylsulfonyl-2-chloroisonicotinamide as an anti-rice blast agent, *J. Pesticide Sci.,* 15:199-203.

Yoshikawa, M. (1978). Diverse modes of action of biotic and abiotic phytoalexin elicitors, *Nature,* 275:546-547.

Yoshikawa, M., Tsuda, M., and Takeuchi, Y. (1993). Resistance to fungal diseases in transgenic tobacco plants expressing the phytoalexin elicitor-releasing factor, β-1,3-endoglucanase, from soybean, *Naturwissenchaften,* 80:417-420.

Yoshioka, H., Shiraishi, T., Kawamata, S., Nasu, K., Yamada, T., Ichinose, Y., and Oku, H. (1992). Orthovanadate suppresses accumulation of phenylalanine ammonia-lyase mRNA and chalcone synthase mRNA in pea epicotyls induced by elicitor from *Mycosphaerella pinodes, Plant Cell Physiol.,* 33:201-204.

Youle, D., and Cooper, R. M. (1987). Possible determinants of pathogenicity of *Erwinia amylovora*: Evidence for an induced toxin, *Acta Horticulturae,* 217:161-166.

Young, D. H., and Sequeira, L. (1986). Binding of *Pseudomonas solanacearum* fimbriae to tobacco leaf cell walls and its inhibition by bacterial extracellular polysaccharides, *Physiol. Mol. Plant Pathol.,* 28:393-402.

Young, S. A., Guo, A., Guikema, J. A., White, F. F., and Leach, J. E. (1995). Rice cationic peroxidase accumulates in xylem vessels during incompatible interactions with *Xanthomonas oryzae* pv. *oryzae, Plant Physiol.,* 107:1333-1341.

Young, S. A., Wang, X., and Leach, J. E. (1996). Changes in the plasma membrane distribution of rice phospholipase D during resistant interactions with *Xanthomonas oryzae* pv. *oryzae, Plant Cell,* 8:1079-1090.

Yu, G. L., Katagiri, F., and Ausubel, F. M. (1993a). *Arabidopsis* mutations at the *RPS2* locus result in loss of resistance to *Pseudomonas syringae* strains expressing the avirulence gene *avrRpt2, Mol. Plant-Microbe Interact.,* 6:434-443.

Yu, L. M., Lamb, C. J., and Dixon, R. A. (1993b). Purification and biochemical characterization of proteins which bind to the H-box *cis*-element implicated in transcriptional activation of plant defense genes, *Plant J.,* 3:805-816.

Yu, Y. G., Buss, G. R., and Saghai Maroof, M. A. (1996). Isolation of a superfamily of candidate disease-resistance genes in soybean based on a conserved nucleotide-binding site, *Proc. Natl. Acad. Sci. U.S.A.,* 93:11751-11756.

Yucel, I., Boyd, C., Debnam, Q., and Keen, N. T. (1994a). Two different classes of *avrD* alleles occur in pathovars of *Pseudomonas syringae, Mol. Plant-Microbe Interact.,* 7:131-139.

Yucel, I., and Keen, N. T. (1994). Amino acid residues required for the activity of *avrD* alleles, *Mol. Plant-Microbe Interact.,* 7:140-147.

Yucel, I., Midland, S. C., Sims, J. J., and Keen, N. T. (1994b). Class I and class II *avrD* alleles direct the production of different products in gram-negative bacteria, *Mol. Plant-Microbe Interact.,* 7:148-150.

Yucel, I., Slaymaker, D., Boyd, C., Murillo, J., Buzzell, R. I., and Keen, N. T. (1994c). Avirulence gene *avrPphc* from *Pseudomonas syringae* pv. *phaseolicola* 3121: A plasmid-borne homologue of *avrC* closely linked to an *avrD* allele, *Mol. Plant-Microbe Interact.,* 7:677-679.

Zambryski, P. C. (1992). Chronicles from the *Agrobacterium*-plant cell DNA transfer story, *Annu. Rev. Plant Physiol. Plant Mol. Biol.,* 43:465-490.

Zdor, R. E., and Anderson, A. J. (1992). Influence of root colonizing bacteria on the defense responses of bean, *Plant Soil,* 140:99-107.

Zhang, L., and Birch, R. G. (1997). Mechanisms of biocontrol by *Pantoea dispersa* of sugarcane leaf scald disease caused by *Xanthomonas albilineans, J. Appl. Microbiol.,* 82:448-454.

Zhang, Z. G., Collinge, D. B., and Thordal-Christensen, H. (1995). Germin-like oxalate oxidase, a H_2O_2-producing enzyme, accumulates in barley attacked by the powdery mildew fungus, *Plant J.,* 8:139-145.

Zhang, Z. Y., Coyne, D. P., and Mitra, A. (1996). Gene transfer for enhancing plant disease resistance to bacterial pathogens, *Annu. Report Bean Improvement Cooperative,* 39:52-53.

Zhou, J., Loh, Y. T., Bressan, R. A., and Martin, G. B. (1995). The tomato gene *Pti1* encodes a serine/threonine kinase that is phosphorylated by *Pto* and is involved in the hypersensitive response, *Cell,* 83:925-935.

Zhou, J., Rumeau, D., and Showalter, A. M. (1992). Isolation and characterization of two wound-regulated tomato extensin genes, *Plant Mol. Biol.,* 20:5-17.

Zhou, J. M., Tang, X. Y., and Martin, G. B. (1997). The Pto kinase conferring resistance to tomato bacterial speck disease interacts with proteins that bind a *cis*-element of pathogenesis-related genes, *EMBO J.,* 16:3207-3218.

Zhou, T., and Paulitz, T. C. (1994). Induced resistance in the biocontrol of *Pythium aphanidermatum* by *Pseudomonas* spp. on cucumber, *J. Phytopathol.,* 142:51-63.

Zhu, B., Chen, T. H. H., and Li, A. H. (1993b). Expression of an ABA-responsive osmotin-like gene during the induction of freezing tolerance in *Solanum commersonii, Plant Mol. Biol.,* 21:729-735.

Zhu, B., Chen, T. H. H., and Li, P. H. (1995a). Activation of two osmotin-like protein genes by abiotic stimuli and fungal pathogen in transgenic potato plants, *Plant Physiol.,* 108:929-937.

Zhu, B., Chen, T. H. H., and Li, P. H. (1996). Analysis of late-blight disease resistance and freezing tolerance in transgenic potato plants expressing sense and antisense genes for an osmotin-like protein, *Planta,* 198:70-77.

Zhu, Q., Dabi, T., Beeche, A., Yamamoto, R., Lawton, M. A., and Lamb, C. (1995b). Cloning and properties of a rice gene encoding phenylalanine ammonia-lyase, *Plant Mol. Biol.,* 29:535-550.

Zhu, Q., Doerner, P. W., and Lamb, C. J. (1993b). Stress induction and developmental regulation of a rice chitinase promoter in transgenic tobacco, *Plant J.,* 3:203-212.

Zhu, Q., Maher, E. A., Masoud, S., Dixon, R. A., and Lamb, C. J. (1994). Enhanced protection against fungal attack by constitutive co-expression of chitinase and glucanase genes in transgenic tobacco, *Biotechnology,* 12:807-812.

Zocchi, G., and Rabotti, G. (1993). Calcium transport in membrane vesicles isolated from maize coleoptiles. Effect of indoleacetic acid and fusicoccin, *Plant Physiol.,* 101:135-139.

Zook, M. N., and Kuc, J. A. (1991). Induction of sesquiterpene cyclase and suppression of squalene synthetase activity in elicitor treated or fungal infected potato tuber tissue, *Physiol. Mol. Plant Pathol.,* 39:377-390.

Zorreguieta, A., Geremia, R. A., Cavaignac, S., Cangelosi, G. A., Nester, E. W., and Ugalde, R. A. (1988). Identification of the product of an *Agrobacterium tumefaciens* chromosomal virulence gene, *Mol. Plant-Microbe Interact.*, 1:121-127.

Zorreguieta, A., and Ugalde, R. (1986). Formation in *Rhizobium* and *Agrobacterium* spp. of a 235-kilodalton protein intermediate in β-D (1,2) glucan synthesis, *J. Bacteriol.*, 167:947-951.

Zu, P., Wang, J., and Fincher, G. B. (1992). Evolution and differential expression of the (1-3)-β-glucan endohydrolase-encoding gene family in barley, *Hordeum vulgare*, *Gene*, 120:157-165.

Index

Page numbers followed by the letter "f" indicate figures; those followed by the letter "t" indicate tables.